Electronics for the Modern Scientist

Paul B. Brown
Physiology Department
West Virginia University Medical School
Morgantown, West Virginia

Gunter N. Franz
Physiology Department
West Virginia University Medical School
Morgantown, West Virginia

Howard Moraff
New York State College of Veterinary Medicine
Cornell University
Ithaca, New York

Elsevier
New York • Amsterdam • Oxford

Elsevier Science Publishing Co., Inc.
52 Vanderbilt Avenue, New York, New York 10017

Sole distributors outside the United States and Canada:
Elsevier Science Publishers B. V.
P. O. Box 211, 1000 AE Amsterdam, The Netherlands

Library of Congress Cataloging in Publication Data

Brown, Paul Burton, 1942-
Electronics for the modern scientist.

Bibliography: p.
Includes index.
1. Electronics. I. Franz, Gunter N.
II. Moraff, Howard. III. Title.
TK7816.B763 621.381 81-15101
ISBN 0-444-00660-5 AACR2

Copy Editor: Richard P. Ross
Desk Editor: Louise Calabro Schreiber
Compositor: Science Typographers, Inc.
Printer: Halliday Lithograph

Manufactured in the United States of America

Contents

Preface

This textbook is predicated on the belief that modern scientists have an increasing need for a working knowledge of the principles of electronic design, signal analysis, and linear system theory. This belief has only begun to find acceptance among science educators since the inception of integrated circuits and the widespread research application of signal analysis and linear system theory. The introduction of these subjects to the scientific curriculum has been hampered by a scarcity of qualified teachers and by competition with more traditional subjects in an already crowded field of courses. Nevertheless, the more forward-looking universities are at least offering such material as elective subjects, and it is inevitable that eventually it will be required for graduate degrees in chemistry, physics, and some of the biological sciences.

Changes have occurred in scientific research that dictate the need for a knowledge of electronics and system analysis. Perhaps the most obvious is the increased electronic sophistication of research equipment itself. More and more electronic instrumentation is in use in all fields of science, and increasing numbers of instruments incorporate microcomputers. Of course not all scientists need to modify their instrumentation or do their own repairs. However, an understanding of the design and working principles of sophisticated instrumentation will enable any investigator to make more effective use of it. Moreover, the researcher who understands these principles will be better able to specify any needed modifications to a design engineer and to assure their proper implementation. The researcher who cannot do this will be limited not only in the types of experiments he or she can perform, but even in the types that can be conceived.

The need for a knowledge of system analysis is even more pressing. Theorizing in science involves the formulation of models. System analysis entails the formal development of such models, both at the descriptive level (when the behavior of a system is described, without any reference to underlying mechanisms) and at the level of hypothetical mechanism. In the field of physiology, for example, some researchers talk of physiological control systems with positive or negative feedback, and are often unaware of the simplest principles of control system theory, let alone its established applications to physiology. A biochemist, when taking periodic samples of blood in order to determine the time course of the appearance or disappearance of a biochemical substrate, must understand how to select the sampling frequency so as to avoid missing important frequency components in the deduced time-varying signal: too low a frequency may actually introduce spurious frequency components, resulting in a waveform quite different from the actual time course.

We believe that a scientific education must not only enable the student to understand past research but also equip him or her for a long scientific career. The areas of electronics and system analysis are as essential to the modern scientist's education as elementary computer science, calculus, physics, or statistics and probability. There is already a large body of research literature that is inaccessible to many graduates because they cannot understand the concepts or notation in system analytical approaches to research problems; their education is, in this sense, obsolete before they have their degree. Many medical students today are ill prepared to grasp the physical concepts underlying the theory of nerve membrane potentials because they do not have an adequate understanding of of electric principles; their education is obsolete before they even enter medical school!

The most important concepts in this textbook can be learned in one semester by a college junior or senior; all of the material can be mastered in two semesters. It is our hope that, after taking a course using this text, students will have sufficient understanding of system analysis and electronics design to be able to master more advanced material, as needed, either on their own or with minimal guidance. Since the text is intended for all scientific disciplines, we assume only a familiarity with elementary calculus, which all scientists need anyway. Even the student unfortunate enough not to have had calculus can understand most of the material if the teacher properly structures the course.

Every teacher will have his or her own preferred approach and will select different portions of this text and order them in different sequences. We have developed what we believe to be a logical sequence of material for didactic purposes, and we have deemphasized material that, in our view, will be considered optional in shorter courses by using smaller print. This textbook presents only theory: there is no material on construction techniques and no description of the many commercially available integrated circuits, and there are no laboratory exercises. To include these, although they are important, would make this text impractically large. The text covers a broad range of topics in electronics, but no attempt was made to be comprehensive. Instead, we hope that by presenting a "principles" approach to key concepts and topics in electronics, we shall make the reader comfortable with the subject and encourage him or her to go to other reference sources for in-depth treatments of particular areas of interest.

Paul B. Brown

Electronics for the Modern Scientist

1

Circuit Elements, Impedance, Network Analysis

Gunter N. Franz

Electronic instruments are designed for so many different applications that it would seem impossible to reduce the rules of design to a few basic principles. However, if we examine individual electronic instruments we notice that most of them are really constructed of similar components, the main difference being the way in which the components are interconnected to form the whole instrument. It is customary to call a purposefully connected set of electronic components an *electronic circuit*, or simply a *circuit*. The components are connected according to a *wiring diagram* or *schematic*.

The various types of components are represented by simple graphic symbols (rather than a replica of their actual appearance) in the wiring diagram. To simplify analysis and design of circuits, we often assume that the component symbols stand for *idealized* components. These idealized components are simple conceptual models of the real "hardware" and are called *circuit elements* or *model elements*. Circuits made up of idealized circuit elements are also referred to as *networks*.

In order to analyze a circuit or network we need to proceed from two sets of quantitative relations. The first is based on the quantitative description of the individual elements: the *element laws*. The second is derived from the way in which the circuit elements are interconnected to form the whole circuit: the description of *network topology*. Both sets are combined to give us the *network equations*. The solution of the network equations completes the analysis of the given network. This chapter deals with the basic tools of network analysis.

1.1 Circuit Elements: Basic Concepts and Definitions, Element Laws

We characterize circuit elements according to their manner of handling electric energy: *active* elements are sources of electric energy, whereas *passive* elements store, couple, dissipate or otherwise transform electric energy. Next we consider their complexity. Because current flow requires the current to enter the circuit element at one end (terminal) and leave it at another, the least complex elements must have two terminals, which together make up a port. Passive *two-terminal elements* or *one-ports* either store electromagnetic energy or dissipate it into heat. The coupling of electric energy, as in a transformer, requires *two-ports*, i.e., elements with three or four terminals. Active two-terminal elements are independent sources of electric energy. Controlled sources, such as amplifiers, generally require three or four terminals.

The quantitative characterization of an element requires that we know the relations between the currents entering (or leaving) its terminals and the associated port voltages. The resulting *element law* takes the form of an equation or graph describing the interdependence of a single current–voltage pair in the case of two-terminal elements.

In order to facilitate circuit analysis, the individual model elements are represented in the circuit diagram by special symbols and by an indication of the polarities of the voltages and currents. In Fig. 1.1a an arbitrary two-terminal element (represented by the rectangle) is shown with the measurement pair necessary to define the element law. If the current enters terminal 1, the measuring instrument in series (ammeter) shows a positive deflection. This direction of current flow implies that the electric potential at terminal 1 (e_1) is higher than that at 2 (e_2), hence the measuring instrument in parallel (voltmeter) shows a positive deflection, and we assign the associated positive polarity (+) to terminal 1 and a negative one (−) to terminal 2. It is important to use these conventions consistently in order to prevent

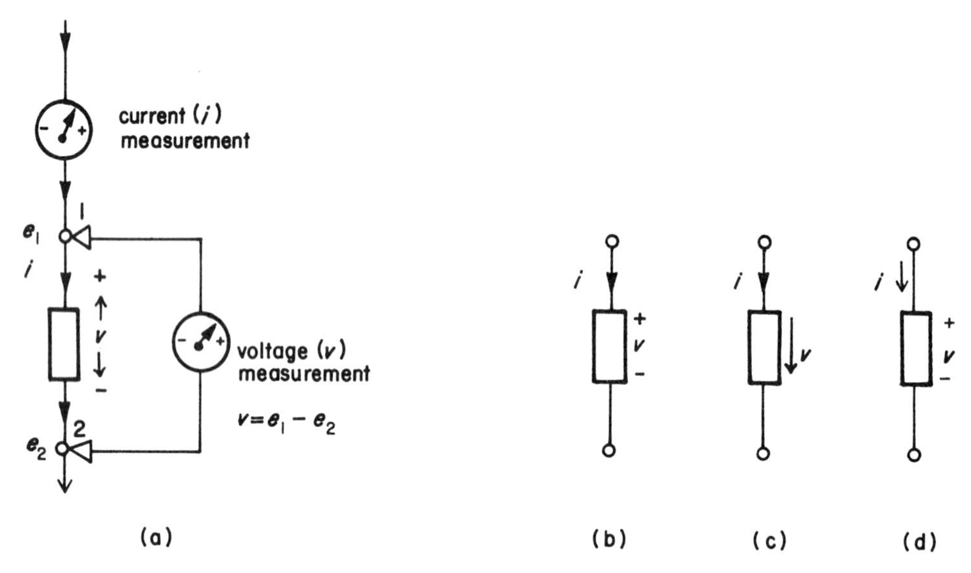

Figure 1.1 Polarities or associated reference directions for a general two-terminal element (one-port): **(a)** measurement of voltage–current pair; **(b)** preferred reference designation used in this chapter; **(c)** and **(d)** alternative reference designations.

errors. Figures 1.1b to 1.1d show three common ways of indicating the *reference directions* (*polarities*, *sign conventions*) for a two-terminal element.

REMARK: The convention to assign a positive sign or "direction" for the current when a net transfer of charge occurs from the terminal of higher electrical potential 1 to that of lower potential 2 is entirely arbitrary. The direction of flow of the actual charge carriers will depend on their polarity. Positive charges (such as positive ions in an electrolyte) will flow in the same direction as the current arrow in Fig. 1.1. Negative charges, such as electrons in metallic conductors, will flow in the *opposite* direction and accomplish the same net transfer of charges. For the purposes of circuit analysis it is immaterial how the charge transfer actually occurs. The *operational definitions* according to Fig. 1.1a are sufficient. Hence current is still indicated as flowing from 1 (+) to 2 (−) even if electrons flow in the opposite direction.

Table 1.1 contains the letter symbols and units of measurement of the electric variables of interest to us. The strict physical definitions of these quantities can be obtained from any good physics textbook.

It is customary to add prefixes to electric units for magnitude scaling by factors of 1000. Table 1.2 lists the factors and prefixes.

In order to facilitate future discussions of units of measurement we introduce the following short-hand notation:

> The expression $[x]$ means "unit of measurement or dimension for quantity x," where x can be a variable or a parameter.

Table 1.1 Units of Electrical Variables

Variable	Letter Symbol	Unit	Unit Symbol
Charge	Q, q	coulomb (ampere-seconds)[a]	C (A·s)[a]
Current	I, i	ampere	A
Voltage, electrical potential difference	V, v E, e U, u	volt	V
Power	N, n P, p	watt (volt-ampere)[a] (joule/sec)[a]	W (V·A)[a] ($J \cdot s^{-1}$)[a]
Magnetic flux	Φ, ϕ	weber (volt-seconds)[a]	Wb (V·s)[a]

[a]The units in parentheses are numerically identical to the units listed first (International System of Units or SI Units).

Example 1.1. Use of the dimension or unit symbol

$[v]$ = volt (V)

$[i]$ = ampere (A)

After this presentation of introductory concepts we proceed to the discussion of the major types of two-terminal elements: dissipators, stores, and sources of energy.

Table 1.2 Factors and Unit Prefixes[a]

Factor	Unit Prefix	Symbol
10^{12}	tera	T
10^{9}	giga	G
10^{6}	mega	M
10^{3}	kilo	k
10^{-3}	milli	m
10^{-6}	micro	μ
10^{-9}	nano	n
10^{-12}	pico	p
10^{-15}	femto	f
10^{-18}	atto	a

[a]Examples: 10^{-3} A = 1 mA; 10^{3} V = 1 kV; 0.63 A = 630 mA; 1200 W = 1.2 kW.

1.1.1 Dissipation: Resistors

Dissipation refers to the fact that practical conductors cannot carry electric currents without loss of energy. The energy drop per unit charge or voltage difference across the ends of the conductor is directly proportional to the current carried by the conductor according to Ohm's law. For the charge transfer within the conductor we may here assume a simple frictional resistance model. The friction encountered by the moving charges causes the degradation of electric energy into heat. In fact, the energy thus converted per unit time, or the *electric power dissipated*, equals the product of current (through the conductor) and voltage (across the conductor).

The appropriate circuit model element for dissipation is the *resistor*. In most instances, the idealization of this element rests on two assumptions: (1) Ohm's law is obeyed, and (2) temperature effects are negligible. (Aside from resistance changes, excessive heating may actually destroy a resistor. In order to avoid that, resistors are specified with a *power rating* or *wattage* (in watts) which should not be exceeded.)

REMARK: We assume that the wires connecting the various circuit elements in a circuit have negligible resistance. Hence the resistor symbol (Fig. 1.2a) is only used when "concentrated" resistance of appreciable magnitude has to be considered.

1.1.1.1 Ohm's Law and Linear Resistance

We assume that resistors are "ohmic," i.e., they obey Ohm's law. The symbol for such resistors with the associated reference directions is given in Fig. 1.2a. (The symbol in Fig. 1.2b is sometimes used. We reserve this symbol for unspecified or generalized two-terminal elements). The characteristic in Fig. 1.2c is a graphic description of the element law. From the straight line through the origin we deduce the following element law.

Element Law for Linear Resistors (Ohm's Law)

$$v = Ri \tag{1.1}$$

or

$$i = \frac{1}{R} v = Gv \tag{1.2}$$

where R = resistance and G = conductance

REMARK: With a resistor we may associate either the resistance parameter R or the conductance parameter G. These quantities give equivalent characterizations to the same physical process; accordingly, we use only one circuit symbol as in Fig. 1.2a. By adding the letters R or G we can indicate which parameter we are using.

Quantitatively, the proportionality constant or device parameter resistance R (or conductance G) is determined by an associated measurement pair for voltage and

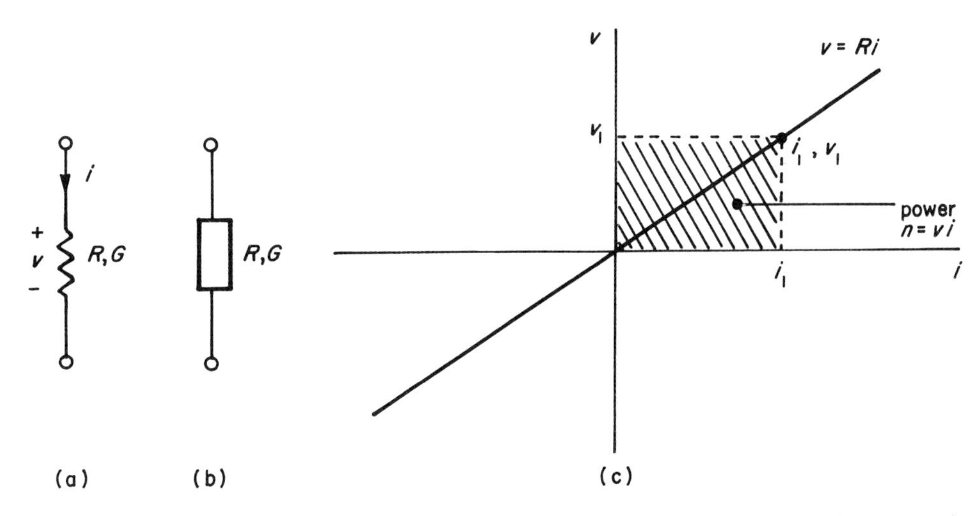

Figure 1.2 The linear resistive element (ohmic resistor): **(a)** preferred symbol with associated reference directions; **(b)** alternative symbol prevalent in the European literature; **(c)** element characteristic expressing Ohm's law. R = resistance parameter, $G = 1/R$ = conductance parameter for the same element.

current. If voltage and current are given in dimensions V and A, respectively (see Table 1.1), then we have

> Dimension for Resistance: $[R]$ = ohm (symbolized Ω)
> Dimension for Conductance: $[G]$ = siemens (symbolized S)

The dimensional relations between the variables and parameters introduced thus far are evident if we transform Eqs. (1.1) and (1.2) into dimensional equations:

$$[v]=[R]\times[i]$$

or

$$\text{volts}=\text{ohms}\times\text{amperes}$$

or

$$\mathrm{V}=\Omega\times\mathrm{A}$$

hence

$$\Omega=\frac{\mathrm{V}}{\mathrm{A}} \tag{1.3}$$

and

$$[i]=[G]\times[v]$$

or

$$\mathrm{A}=\mathrm{S}\times\mathrm{V}$$

hence

$$S = \frac{A}{V} = \frac{1}{\Omega} \tag{1.4}$$

The scaling prefixes of Table 1.2 apply also to parameter units.

Example 1.2. Determination of resistance and conductance parameters with Ohm's law.

(A)

$v = 10\text{ V}, i = 1\text{ A}, R = ?, G = ?$

By Eqs. (1.1) and (1.3)

$$R = \frac{10\text{ V}}{1\text{ A}} = 10\ \Omega$$

and by Eqs. (1.2) and (1.4)

$$G = \frac{1\text{ A}}{10\text{ V}} = 0.1\text{ S}$$

or

$$G = \frac{1}{R} = \frac{1}{10}\text{ S}$$

(B)

$v = 100\text{ mV}, i = 5\text{ nA}$

$$R = \frac{100\text{ mV}}{5\text{ nA}} = \frac{100 \times 10^{-3}\text{ V}}{5 \times 10^{-9}\text{ A}} = \frac{1}{5} \times 10^{8}\ \Omega = 20\text{ M}\Omega$$

Next let us reconsider energy dissipation in a resistor. The amount of charge dq transported through a resistor during an infinitesimally short time interval dt equals the number of charges moved per unit time (=current) multiplied by the interval dt:

$$dq = i\,dt \tag{1.5}$$

From field theory it follows that the amount of energy de dissipated during dt is the product of voltage (energy change per unit charge) and the number of charges:

$$de = v\,dq \tag{1.6}$$

With Eqs. (1.5) and (1.6) we have

$$de = vi\,dt$$

Hence the *rate* of energy dissipation or *power* is energy change per unit time:

$$n = \frac{de}{dt} = vi \tag{1.7}$$

Equation (1.7) expresses formally what has been asserted earlier. The units of the power quantity are given in Table 1.1.

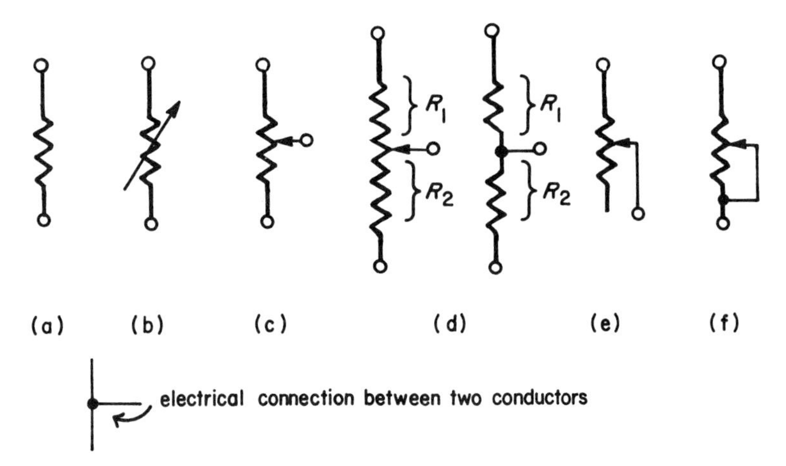

Figure 1.3 Symbols for fixed and variable resistance elements: **(a)** fixed resistor; **(b)** variable resistor; **(c)** resistor with a third movable contact (wiper) or tap, usually called a potentiometer; **(d)** equivalence of a tapped resistor to two connected fixed value resistors; **(e)** and **(f)** modifications of a tapped resistor or potentiometer for action as an adjustable resistor.

We can modify the expression for the power further by replacing either the voltage or the current term with the appropriate form of Ohm's law. From Eqs. (1.1) and (1.7) follows

$$n = Ri^2 \tag{1.8}$$

and from Eqs. (1.2) and (1.8) we get

$$n = Gv^2 = \frac{v^2}{R}$$

We should note that power is the *product* of two electric variables. The appropriate graphical representation of power in the voltage–current plot of Fig. 1.2c is, therefore, two-dimensional, i.e., the *area* of the rectangle with sides v and i is equivalent to the power dissipated.

In an actual circuit, a resistor may either be of fixed value or it may be continuously variable. The two cases are distinguished by the two symbols given in Figs. 1.3a and 1.3b. A common type of adjustable resistor has a third terminal in the form of a slider which can be moved across the resistor material to make contact at different locations. Such adjustable resistors are called potentiometers (or "pots"); the symbol is given in Fig. 1.3c.

The potentiometer is really the equivalent of a series combination of two resistors (Fig. 1.3d). It can also be connected as a two-terminal variable resistor as seen in Figs. 1.3e and 1.3f.

From the standpoint of network analysis one would always prefer linear resistors. Nonlinear element laws lead to analytical difficulties. But there are network elements which are useful precisely because of their nonlinear characteristics. The next section considers two such elements.

1.1.1.2 Nonlinear Resistive Elements

The common attribute of the nonlinear elements to be discussed in this section is that they are made of materials which belong to the class of *semiconductors*. Semiconductive materials occupy an intermediate position between metallic conductors and insulators in that their *resistivity* or *specific resistance* (=resistance of a body of unit cross-sectional area and unit length at 0°C) is one or several orders of magnitude higher than that of metallic conductors but still less than that of insulators. The resistance of semiconductive materials depends strongly on temperature and the presence of impurities in the material. Specific nonlinearities are often obtained by deliberately "doping" the semiconductor material with certain impurities and setting up electric *junctions* between differently doped materials. A more extensive discussion of semiconductors is given in Chapter 2. The following examples were chosen because they serve well to illustrate a point and they are also among the most useful nonlinear elements.

The Diode or Rectifier. This is the most important nonlinear two-terminal element. Modern diodes are semiconductor *junction diodes*, i.e., effects at the junction between two differently doped regions of a semiconductor substrate bring about the diode effect. Silicon is the most common semiconductor material used. The older term "rectifier" refers to the preferential conduction of current of one polarity as seen in Fig. 1.4b. Unlike ordinary resistors, which are *bidirectional*, diodes are *undirectional elements*. This means that the connection of the terminals is not reversible, because current flows preferentially from the *anode* terminal A to the *cathode* terminal K. If the applied voltage has the polarity indicated in Fig. 1.4a, then the diode is *forward biased* and conducting. The diode is *reverse biased* and practically nonconducting if the voltage has the opposite polarity. Only a small *leakage* current flows during reverse bias (third quadrant in Fig. 1.4b). We may think of a diode as the electric equivalent of a one-way valve.

A good approximation of the element characteristic in Fig. 1.4b is

$$i = I_s(e^{qv/kT} - 1) \tag{1.9}$$

where I_s is the leakage or *saturation* current, $q = 1.59 \times 10^{-19}$ C is the charge of an electron, $k = 1.38 \times 10^{-23}$ J/°K is the Boltzmann constant, and T is the absolute temperature at the diode junction in degrees Kelvin.

For a silicon diode, I_s is rather small, so that Eq. (1.9) can be simplified to

$$i \approx I_s e^{qv/kT} = I_s e^{v/\sigma} \tag{1.10}$$

where $\sigma = kT/q = 26$ mV for $T = 300$°K. We also have

$$v = \sigma \ln\left(\frac{i}{I_s}\right) \tag{1.10a}$$

Equations (1.10) and (1.10a) show that diodes may be used to compute exponentials or logarithms electronically (see Chapter 4). Most of the time, however, we are only interested in the one-way valve action of diodes; the details of the element law are then of no interest. Therefore, the simple straight line approximations in Fig. 1.4c suffice. In one approximation we assume that the conductance of the foward-biased diode is infinite, i.e., the characteristic coincides with the positive current axis and the negative voltage axis and thus simulates an open circuit for $v < 0$ and a short

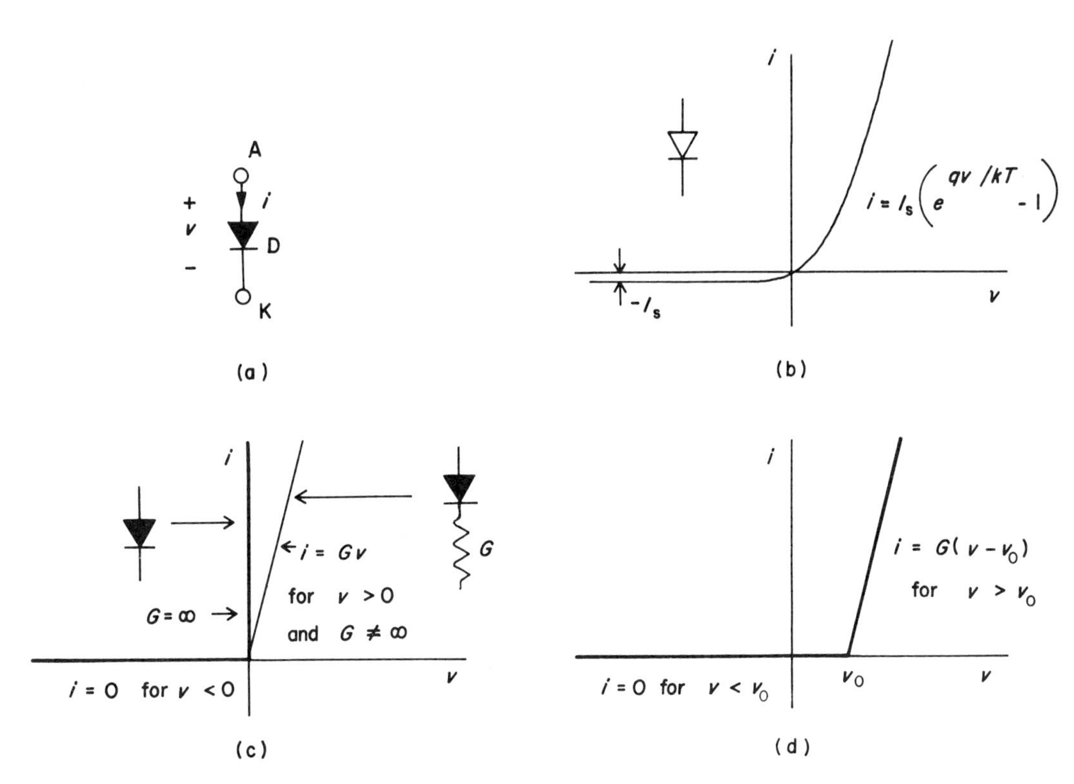

Figure 1.4 The diode or rectifier: **(a)** symbol for the ideal diode element (solid symbol) and reference directions (A, anode; K, cathode); **(b)** characteristic of a practical semiconductor diode (open symbol; see text); **(c)** idealized diode characteristic with infinite $(G = \infty)$ and finite $(G \neq \infty)$ forward conductance; **(d)** idealized diode characteristic with "knee" voltage v_0.

circuit for $v > 0$. For this *ideal diode* we will use the filled-in diode symbol. In the other approximation we take into account that the conductance of the forward-biased diode is finite. The appropriate model is the series combination of an ideal diode and a resistor (G) as shown in Fig. 1.4c.

In some instances the "knee" voltage due to the strong curvature of the $i-v$ characteristic cannot be ignored. In that case the straight line approximation of Fig. 1.4d is useful. "Knee" voltages are about 600 mV for silicon diodes and 200 mV for germanium diodes.

Constant Voltage Diodes: Zener Diodes. These devices can be used to establish voltage references and to shift or limit voltage levels by exploiting the *Zener effect*. This effect refers to a specific breakdown phenomenon arising at reverse-biased semiconductor junctions: A reverse-biased zener diode acts as a closed "valve" until the *breakdown voltage*, or *zener voltage*, v_z, is reached; at that voltage the diode becomes highly conductive and the current increases rapidly. With the reference directions in Fig. 1.5a, this part of the $i-v$ characteristic is in the third quadrant of Fig. 1.5b. Under forward-bias the zener diode behaves like an ordinary diode (first quadrant, Fig. 1.5b).

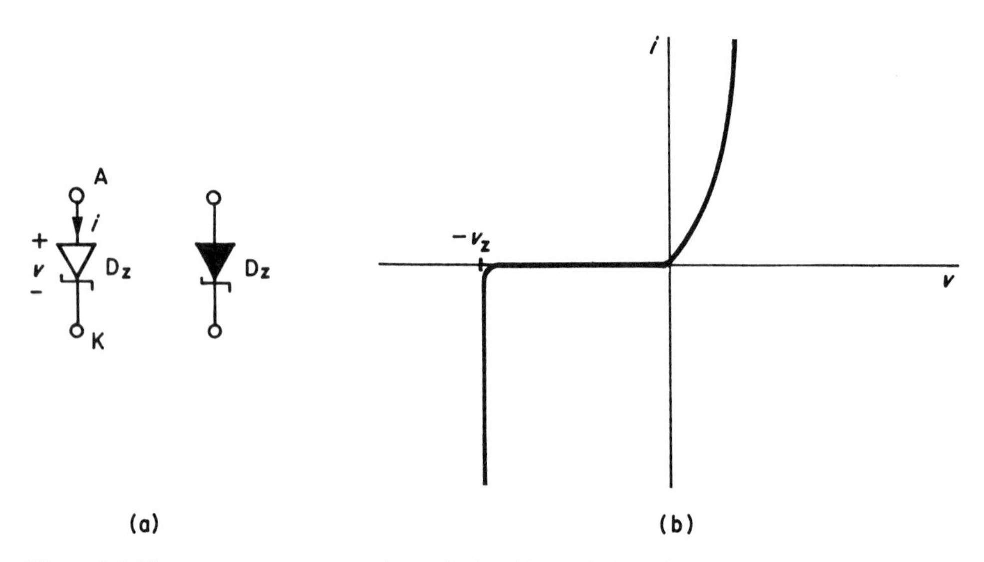

Figure 1.5 The zener or constant voltage diode: **(a)** symbols and reference directions (A, anode; K, cathode); **(b)** element characteristic. The constant voltage region (with zener voltage v_z) is in the third quadrant under backward bias conditions.

1.1.2 Storage of Electric Energy: Capacitors

In a simple *capacitor* two conductors, e.g., two metal plates, are positioned in close proximity. Between the conductors there is either a vacuum or, more commonly, an insulating material or *dielectric* so that no conductive pathway exists in the space between the two conductors. The symbols for the capacitor indicates this by a gap between the lines representing the two conductors (Fig. 1.6a).

The storage function of the capacitor may be explained as follows. Assume that a battery is connected to the two conductors of the capacitor so that one conductor takes on positive and the other negative polarity (Fig. 1.6b). This procedure *charges* the capacitor as the battery supplies *equal* amounts of charge of *opposite polarity* ($q+, q-$) to the two conductors of the capacitor. As long as the charging process persists a current i flows *into* the positive capacitor terminal, depositing positive charges on (or removing negative charges from) the conductor of positive polarity. This current is matched by a current of equal magnitude flowing *out* of the negative terminal, removing positive charges from (or depositing negative charges on) the other conductor. Once the terminals of the capacitor have a potential difference equal to the battery voltage, the charging process is finished, and the current is zero. If the battery is now removed from the capacitor, the capacitor remains charged, i.e., the charge separation persists indefinitely. The stored charges represent a source of electric energy, which becomes available when the capacitor terminals are connected to another device. The capacitor can be discharged by connecting a wire or a resistor across the capacitor terminals. The discharge current flows in the opposite direction and the charge separation is abolished.

During a charge or discharge maneuver the current flowing into one capacitor

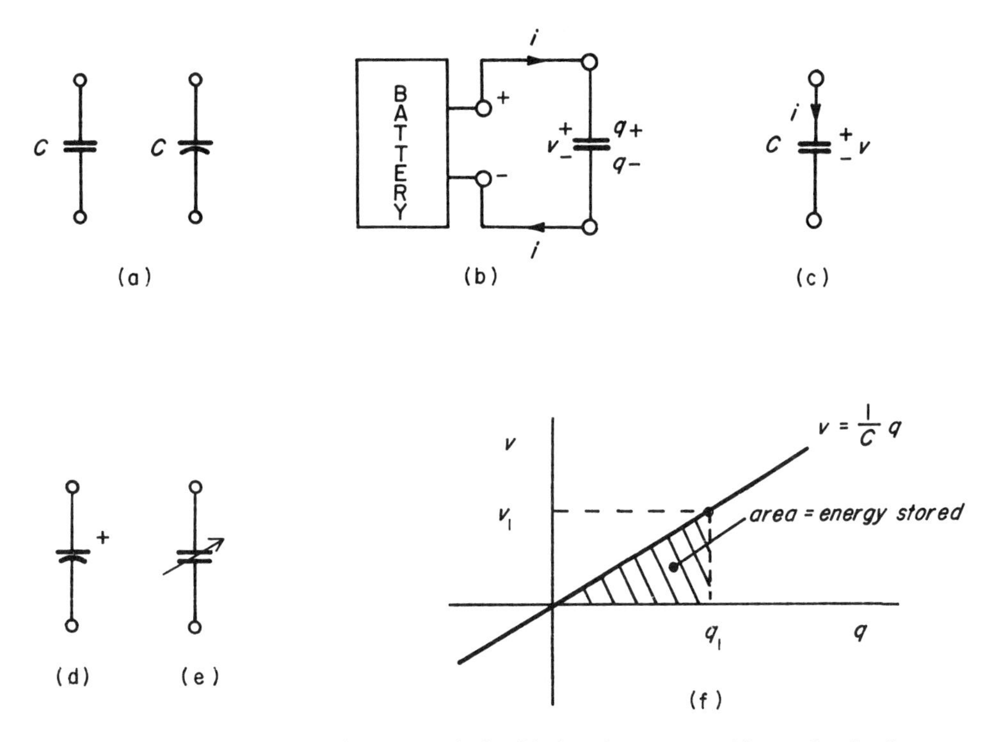

Figure 1.6 The capacitor. **(a)** element symbols; **(b)** charging process; **(c)** associated reference directions for the capacitor; **(d)** polarity indication for an electrolytic capacitor; **(e)** symbol for the variable capacitor; **(f)** element characteristic (voltage–charge curve) of the capacitor.

terminal is matched by a current of equal magnitude flowing out of the other terminal. This gives the appearance of a current flowing "through" the capacitor in spite of the insulating gap between the conductors. Consequently we can define associated reference directions of current and voltage as for any other two-terminal device (Figure 1.6c).

As long as charges are transferred, the capacitor voltage changes as a function of time. A fixed capacitor voltage indicates that charge transfer has ceased, hence current is zero. Now the capacitor acts like an insulator or *open circuit*; this means that, as a coupling element, the capacitor transfers only *changes* in capacitor voltage.

The charge separation required for a given voltage depends on the geometry of the conductors making up the capacitor and on the nature of the insulating medium between the conductors. The geometric and material parameters are combined into a single parameter for a given capacitor: the *capacitance*, C. In general the following hold: (1) The larger the conductive surfaces facing each other, the larger the capacitance. (2) The capacitance is inversely proportional to the distance between the conductors. (3) Presence of a dielectric other than vacuum increases the capacitance. The factor by which the capacitance with a dielectric increases over that with vacuum is called the *dielectric coefficient* K, *the relative dielectric constant* ε_r, or the *relative permittivity* κ. (The nomenclature is, unfortunately, not uniform.) (4) The larger the capacitance, the more charges a capacitor can hold for a given voltage.

The latter statement can be formalized as the algebraic form of the

Element Law of the Capacitor

$$q = Cv \tag{1.11}$$

or

$$v = \frac{1}{C}q \tag{1.12}$$

where C = capacitance

From Equation (1.11) it is obvious that the element law for the capacitor is linear as long as C is a constant. The law expresses a simple proportionality between the potential difference across the two conductor terminals and the number of charges on *either* of the two conductors *without* regard for the sign. (The *total* charge, i.e., the net sum of the charges on the two conductors, is, of course, zero). The resulting characteristic for the capacitor in the voltage–charge plot is a straight line through the origin with slope $1/C$ (Fig. 1.6f).

The element law in terms of voltage and charge is somewhat inconvenient for network analysis, where it is customary to express all equations in terms of voltage and current. Current is the rate at which charges are transferred, and, conversely, charge at time t is the definite integral of the current. From Eqs. (1.11) and (1.12) we derive then two additional ways of expressing the element law:

Differential Form of the Element Law of the Capacitor

$$i = \frac{dq}{dt} = \frac{d}{dt}(Cv) = C\frac{dv}{dt} \tag{1.13}$$

Integral Form of the Element Law of the Capacitor

$$\int_0^t i(\tau)\,d\tau = q(t) - q(0) = C[v(t) - v(0)]$$

or

$$v(t) = \frac{1}{C}\int_0^t i(\tau)\,d\tau + v(0) \tag{1.14}$$

where $q(0)$ and $v(0)$ are the initial capacitor charge and voltage, respectively

The unit of capacitance is the *farad* (F), named after Michael Faraday. A capacitor of one farad capacitance stores one coulomb of charge per volt potential difference. One coulomb, however, represents a very large number of charges. Practical capacitances are much smaller than one farad; typically they are between a few picofarads (1 pF $= 10^{-12}$ F) and a few thousand microfarads (1 μF $= 10^{-6}$ F; see Table 1.2).

The relation between capacitance units and those for charge, voltage, and current can be derived from the equations for the element law, Eqs. (1.11)–(1.14):

Dimensions for Capacitance

$$[C] = \text{farad (F)}$$

$$[C] = \frac{[q]}{[v]}$$

$$\text{farad} = \text{coulomb} \times \text{volt}^{-1}$$

or

$$[C] = \frac{\left[\int i\,dt\right]}{[v]}$$

$$\text{farad} = \text{ampere} \times \text{second} \times \text{volt}^{-1}$$

Next we compute the amount of energy stored in a charged capacitor. The work performed in raising a charge increment dq to the potential v equals the energy increment de (Eq. 1.6):

$$de = v\,dq$$

With Eq. (1.11) we write

$$de = \frac{1}{C} q\,dq$$

Hence the *total energy stored in the linear capacitor* is the sum of the incremental energy changes as expressed by the integral

$$e = \int_0^e de = \frac{1}{C}\int_0^q q\,dq = \frac{1}{2C} q^2 \tag{1.15}$$

Alternative forms of the energy equation can be derived by substituting Eq. (1.11) into (1.15):

$$e = \tfrac{1}{2} vq \tag{1.16}$$

and

$$e = \tfrac{1}{2} Cv^2 \tag{1.17}$$

From Eq. (1.16) we see that the graphic representation of the stored energy corresponds to the area between the capacitor characteristic and the q axis in Fig. 1.6f.

In practice it is not possible to convert all work performed in charging a capacitor into stored electric energy. Some of the energy is dissipated into heat in the dielectric medium separating the capacitor plates. And once a practical capacitor has been charged it will not hold its charge indefinitely, but will slowly discharge through its imperfectly insulating dielectric.

Such capacitors are referred to as leaky or lossy. Depending on the dielectric, a practical capacitor may retain its charge for hours or days or just a few minutes. For network analysis we can usually assume that a capacitor is ideal (not leaky). If it does become necessary to include the dissipative properties of the dielectric, the capacitor can be modeled by combining a resistor with the ideal capacitor element (typically in parallel).

1.1.3 Storage of Magnetic Energy: Inductors

Inductors are the complementary or *dual* elements to capacitors. In an ideal capacitor, with a constant voltage across the terminals and no current flowing, a constant electric field is set up between the plates. In an ideal inductor, with a constant *current* flowing, a constant *magnetic field* is established, but the voltage difference across the terminals of the inductor is zero. *Magnetic energy* is stored as long as the current through the inductor keeps flowing.

In its simplest form, an ideal inductor is made out of a piece of (ideal) wire through which a current flows, with the magnetic field set up in concentric fashion around the wire axis. From physics we know that the orientation of the magnetic field is related to the direction of current flow; reversal of the current polarity reverses the field orientation. The magnetic field can be enhanced either by increasing the current in the conductor or by closely aggregating several parallel conductors carrying currents of the same polarity. The same effect is achieved by winding a long

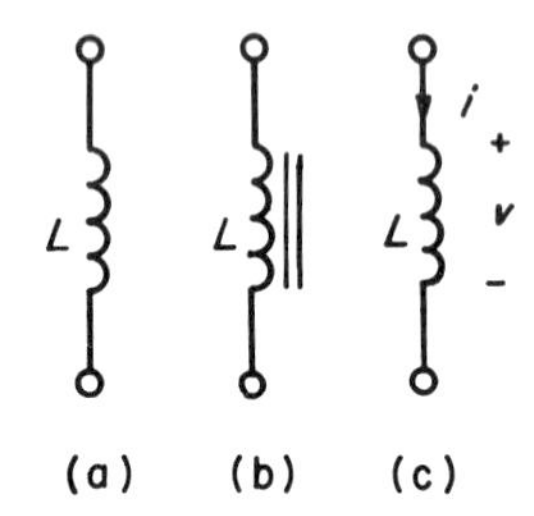

Figure 1.7 The inductor: **(a)** element symbol of an inductor without ferromagnetic core; **(b)** symbol for an inductor with ferromagnetic core; **(c)** associated reference directions for the inductor; **(d)** element characteristic (current–flux curve) of a linear inductor.

wire on a spool or reel. In the resulting *inductor coil* the individual turns or windings form a system of parallel conductors. With all windings having the same orientation and carrying the same current, a stronger magnetic field results. This efficient method of constructing an inductor is reflected in the element symbol of the inductor (Fig. 1.7a).

The magnetic storage capability of a given inductor is a function of the inductor geometry and the presence of ferromagnetic materials. (We shall neglect the effects of materials other than ferromagnetic). Geometric and material properties are combined into a single element parameter, the *inductance* or *self-inductance* L, of an inductor. In general, we can say: (1) The inductance parameter is a complicated function of geometry depending on whether the inductor is a straight wire or one of the various types of coils (single- or multilayered, cylindrical, toroidal, spiral, etc.). (2) In the case of coils, inductance usually increases with the number of windings *squared*. (3) Ferromagnetic materials increase the inductance. The factor by which ferromagnetic materials enhance the magnetic flux in a uniform field (e.g., in the "core" of a toroidal coil) over that in vacuum is called the *relative permeability* (μ_r, κ_m, K_m). (4) The larger the inductance, the larger is the magnetic flux ϕ for a given current i flowing through the inductor.

We now state the algebraic form of the element law:

Element Law of the Inductor

$$\phi = Li \tag{1.18}$$

or

$$i = \frac{1}{L}\phi \tag{1.19}$$

where L is the inductance and ϕ is the magnetic flux

Equations (1.18) and (1.19) define a linear element law if the inductance L is a constant. Generally, L is not a constant if the inductor contains ferromagnetic material. Depending on the type of ferromagnetic material used, the characteristic curves of such inductors show more or less severe nonlinearities, such as hysteresis or saturation. Unless the nonlinearities are expressly desired for a particular function, we shall assume that a ferromagnetic material has been selected that minimizes nonlinear effects.

Like Eq. (1.11), the algebraic inductor law is not particularly suitable for network analysis because it is not expressed in terms of a voltage–current pair. From *Faraday's law* we know that at the terminals of an inductor we can measure an *induced* voltage equal to the rate of change of the magnetic flux:

$$v = \frac{d\phi}{dt} \tag{1.20}$$

From Eqs. (1.20) and (1.18) follows then

$$v = \frac{d\phi}{dt} = \frac{d}{dt}(Li) \tag{1.21}$$

With the assumption that L is constant we can derive from Eq. (1.21) two forms of the linear element law suitable for network analysis:

Differential Form of the Element Law of the Inductor

$$v = L\frac{di}{dt} \tag{1.22}$$

Integral Form of the Element Law of the Inductor

$$i = \frac{1}{L}\int_0^t v(\tau)\,d\tau + i(0) \tag{1.23}$$

where $i(0)$ is the initial current in the inductor

The voltage–current pair in Eqs. (1.22) and (1.23) has the polarities indicated in Fig. 1.7c. This convention is compatible with Lenz's rule. According to this rule, for the given polarities, an increasing current (and therefore increasing magnetic flux) would give rise to an induced voltage that *opposes* the current (flux) change.

The unit of the inductance parameter is the *henry* (H) in honor of Joseph Henry. An inductor of one henry has a magnetic flux of one weber for each ampere of current flowing through it. The factors and unit prefixes of Table 1.2 apply also to inductance units. With Eqs. (1.18) and (1.22) we can state the relations between the inductance parameter unit and the dimensions of the basic variables flux, current, and voltage:

Dimensions for Inductance

$$[L] = \text{henry (H)}$$

$$[L] = \frac{[\phi]}{[i]}$$

$$\text{henry} = \text{weber} \times \text{ampere}^{-1}$$

or

$$[L] = \frac{[v]}{[di/dt]}$$

$$\text{henry} = \text{volt} \times \text{second} \times \text{ampere}^{-1}$$

Energy in the form of a magnetic field is stored by the inductor as long as a constant current i is maintained. Once the current reaches a constant level no energy is consumed in maintaining the field, because the inductor voltage v is zero and hence the power is zero, too. Before this equilibrium state is reached, voltage and current are not zero during the "charging" process. The rate of energy change (power) in establishing the magnetic field is, with Faraday's law [Eq. (1.20)],

$$p = iv = i\frac{d\phi}{dt}$$

$$n = iv = i\frac{d\phi}{dt}$$

Hence the energy increment expended in the interval dt is

$$de = n\,dt = i\,d\phi$$

Assuming that L is constant, we can write with Eq. (1.19) for the total energy stored in the magnetic field

$$e = \frac{1}{L}\int_0^{\phi} \phi\,d\phi = \frac{1}{2L}\phi^2 \tag{1.24}$$

From Eqs. (1.19) and (1.24) we obtain these expressions for the energy:

$$e = \tfrac{1}{2} i\phi \tag{1.25}$$

and

$$e = \frac{L}{2} i^2 \tag{1.26}$$

According to Eq. (1.25) the amount of magnetic energy stored in an inductor subjected to a constant current i is represented by the area under the element characteristic as indicated in Fig. 1.7d.

In practical inductors the voltage across the inductor terminals is not zero when the charging process is completed because the constant current maintaining the field causes an ohmic voltage drop due to the resistance of the nonideal windings. This winding resistance can be modeled as a resistor in series with an ideal inductor. In addition to these "copper losses" there is further energy dissipation when a ferromagnetic material is present: eddy current losses (in an electrically conductive ferromagnetic core), hysteresis losses, and still not quite specified "residual" losses which limit high-frequency performance.

The following section on *mutual inductance* discusses the magnetic interaction between adjacent inductors. It may be skipped in a first reading.

Faraday's Law [Eq. (1.20)] has a much broader meaning than that implied in the preceding discussion. If the induced voltage is only due to magnetic flux changes linked with the current through the inductor itself, then the voltage is *self-induced* and the parameter L in Eq. (1.22) is properly referred to as the *self-inductance*. However, other sources such as permanent magnets and additional inductors may contribute to the magnetic field to which an inductor is exposed. Faraday's law states that the induced voltage equals the rate of change of magnetic flux *without* specifying the source of the magnetic field. An induced voltage can be observed under a number of conditions: (1) self-induction as presented in the previous section; (2) motion of the inductor in a time invariant field (e.g., one set up by stationary permanent magnets); (3) motion of a permanent magnet relative to a stationary inductor; (4) exposure of an inductor to a time-varying field created by other (stationary) inductors through which time-varying currents flow; and (5) a combination of all the above possibilities.

The effect described by Faraday's law in its general interpretation describes, then, a way of *coupling* electrically isolated devices through the effects of the magnetic field. If the effect is undesirable, we refer to the voltages induced by moving magnetic fields and currents in neighboring conductors or inductors as *magnetic interference*. But we might deliberately ensure magnetic coupling, for example, by winding two coils on the same ferromagnetic core as in a *transformer* (Section 1.6).

Whenever a network contains inductors which are magnetically coupled (intentionally or because of inadequate magnetic shielding), Eqs. (1.18) and (1.22) need to be modified to include the effects of currents flowing on other inductors. Hence the element law of inductor j which is linked magnetically to the inductors k, l, etc., is

$$\phi_j = L_{jj} i_j + M_{jk} i_k + M_{jl} i_l + \cdots \tag{1.27}$$

where

ϕ_j is the magnetic flux for inductor j,

$i_j, i_k, i_l, \ldots$ are the currents flowing through the inductors $j, k, l, \ldots$,

L_{jj} is the *self-inductance* (equivalent to the conventional inductance L) of the inductor j,

$M_{jk}, M_{jl}, \ldots$ are the *mutual inductances* representing the coupling between conductor j and conductors $k, l, \ldots$.

From Eqs. (1.20) and (1.27) follows then for the total induced voltage in inductor j

$$v_j = L_{jj} \frac{di_j}{dt} + M_{jk} \frac{di_k}{dt} + M_{jl} \frac{di_l}{dt} + \cdots \tag{1.28}$$

Fortunately, in most instances we can assume, as a first approximation, that ordinary inductors are magnetically isolated from each other. Equation (1.28) reduces then to Eq. (1.22). This makes the resulting network equations much simpler.

The preceding section concludes the introductory discussion of passive two-terminal elements. The analysis of a circuit containing these components will be straightforward, if we can assume that the components can be represented by linear, constant parameter model elements. For this case, Table 1.3 summarizes the classifications of these model elements according to element laws and model parameters.

Table 1.3 Linear, Time Invariant, Passive Model Elements

	Resistor	Capacitor	Inductor
Algebraic element law	$v = Ri = \frac{1}{G} i$ $i = \frac{1}{R} v = Gv$	$q = Cv$ $v = \frac{1}{C} q$	$\phi = Li$ $i = \frac{1}{L} \phi$
Differential element law	———	$i = \frac{dq}{dt} = \frac{d}{dt}(Cv)$ $i = C\frac{dv}{dt}$	$v = \frac{d\phi}{dt} = \frac{d}{dt}(Li)$ $v = L\frac{di}{dt}$
Integral element law	———	$v = \frac{1}{C}\int_0^t i\, dt + v(0)$	$i = \frac{1}{L}\int_0^t v\, dt + i(0)$
Parameters	R, resistance $G = \frac{1}{R}$, conductance	C, capacitance	L, inductance
Parameter units	$[R] = \Omega$ (ohm) $\Omega = V \cdot A^{-1}$ $[G] = S$ (siemens) $S = \Omega^{-1} = V^{-1} \cdot A$	$[C] = F$ (farad) $F = C \cdot V^{-1} = A \cdot s \cdot V^{-1}$	$[L] = H$ (henry) $H = Wb \cdot A^{-1}$ $H = V \cdot s \cdot A^{-1}$

Other major assumptions justifying the use of ideal model elements are (1) *a single physical property*, e.g., dissipation, *dominates* (that excludes, e.g., the "parasitic" inductance of a wire-wound resistor), and (2) the dominant physical property of a component is *locally concentrated*, so that we can use "*lumped*" or *discrete* model elements.

1.1.4 Active Two-Terminal Elements: Sources

Active model elements represent sources of energy. We shall distinguish between two types of active two-terminal elements: *independent voltage sources* and *independent current sources*. These sources are called independent because current and voltage are independent and not linked by an element law. Accordingly, an independent voltage source maintains a specified voltage across its terminals which is independent of the current load, a short circuit being excluded. And an independent current source circulates a specified current through its terminals independent of the terminal voltage, an open circuit being excluded.

The term "independent" is broader than indicated above: the source variable (voltage or current) is not only independent of its complementary variable at the terminals (current or voltage) but is also independent of any *other* voltages or currents associated with circuit components sharing the circuit. However, there is an important class of sources having source variables which *are* a function of a voltage or current elsewhere in the circuit but are independent of the complementary variable at the source terminals. Such sources are called *controlled sources*. Their discussion is left to Section 1.6 because they are usually modelled as three- or four-terminal elements.

Symbols for independent voltage and current sources are given in Fig. 1.8. The preferred symbols are those of Figs. 1.8a and 1.8c. It is customary to reserve capital letters for sources of constant voltage or current. The battery symbol of Fig. 1.8e is used both for ideal, i.e., independent constant voltage sources and practical batteries. Terminal voltages are usually designated by the letter v, but it is also common to use the letter e for voltage sources. The symbol in Fig. 1.8f is occasionally used to indicate a source supplying sinusoidally varying voltages.

If the source variables are constant, we speak of *constant* voltage or current

Figure 1.8 Independent sources: **(a)** preferred symbol and reference polarity of an independent voltage source; **(b)** alternative polarity indication; **(c)** preferred symbol and reference polarity of an independent current source; **(d)** alternative symbol of an independent current source; **(e)** battery (fixed voltage source) symbol; **(f)** common symbol for ac or sinusoidal voltage sources.

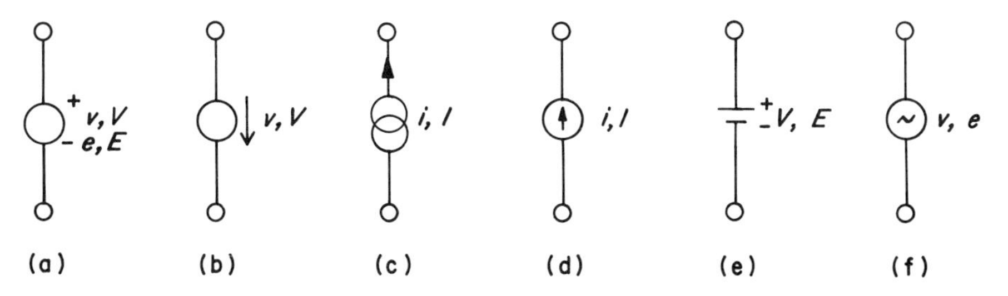

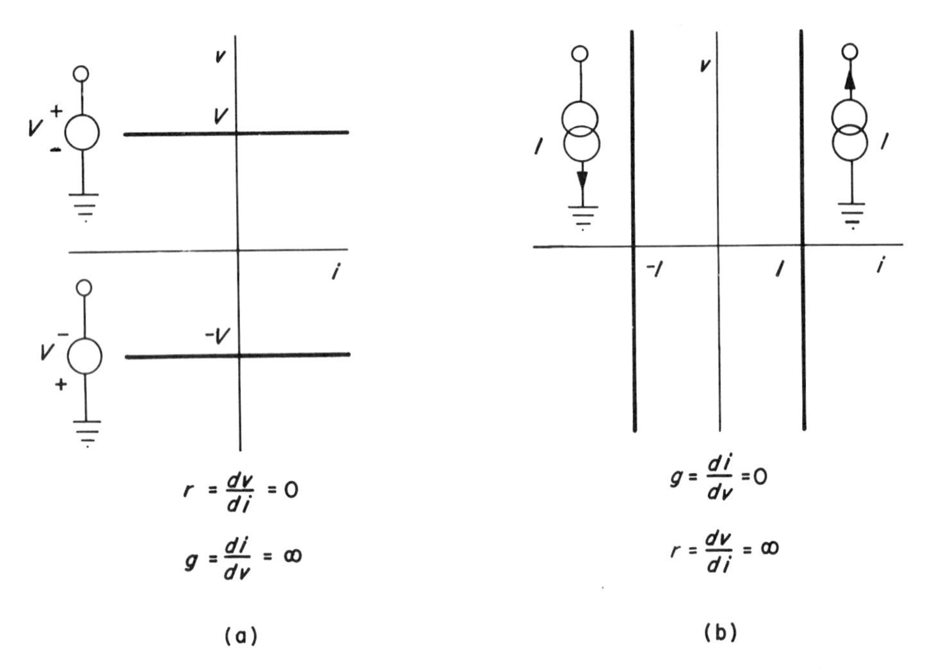

Figure 1.9 Element characteristics for independent sources: **(a)** voltage source; **(b)** current source. (See text.)

sources. Figure 1.9 illustrates the characteristics of such elements. By keeping one source terminal at *zero* or *ground* potential we can create positive or negative sources according to the polarity of the free terminal. The slopes of the characteristics, which represent the incremental resistance or conductance parameters in the case of passive elements, are here zero (or infinity) because of the independence of the sources. In general we can associate zero resistance (infinite conductance) with ideal voltage sources and zero conductance (infinite resistance) with ideal current sources.

The reference directions for active elements differ from those defined for passive elements. The appropriate polarities follow naturally if we consider a combination of active and passive elements in the role of "source" and "load." Figure 1.10a shows such an arrangement of a voltage source and a resistor. Note that the terminal voltages have the same polarity but not the currents. The situation is further complicated in the circuit in Fig. 1.10b, where the current at the positive source terminal is leaving one source but entering the other. The source with the lower terminal voltage (v_2) has the associated reference directions of a passive element; it receives energy, according to Fig. 1.10c. The source with the higher terminal voltage (v_1) supplies energy; its reference polarities for current and voltage are opposite to each other, i.e., they are not "associated." Hence the power is *negative* (Fig. 1.10d). With this convention we assure that the net power, i.e., the sum of power delivered and absorbed, is zero. For example, in the case of the simple circuit in Fig. 1.10a we have

$$iv - iv = 0$$

Practical voltage and current sources are not truly independent sources. The slopes of the characteristics of practical sources differ from zero (or infinity). This

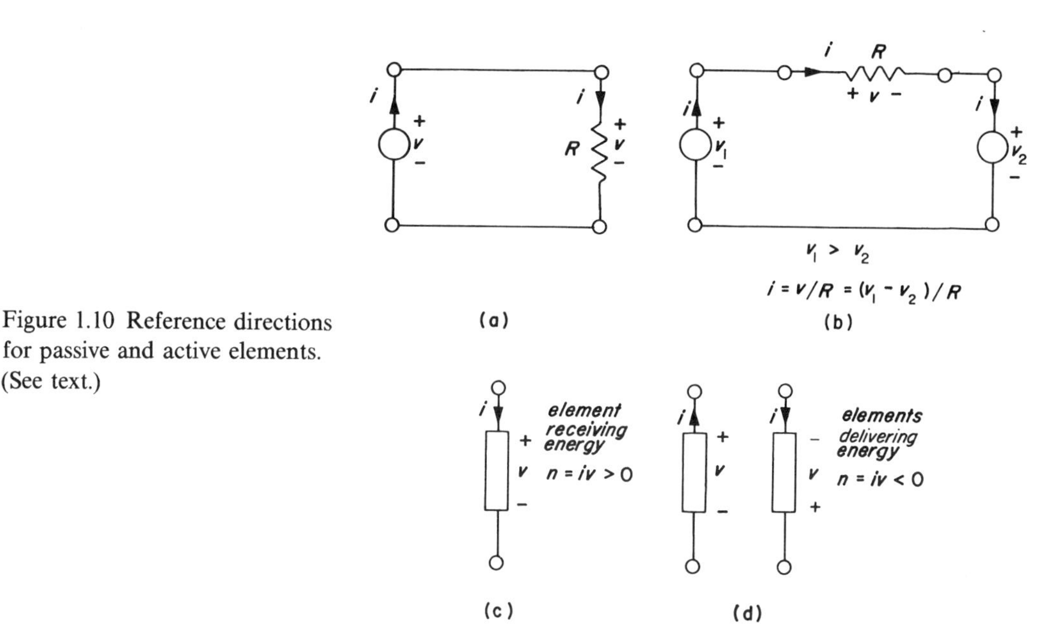

Figure 1.10 Reference directions for passive and active elements. (See text.)

indicates that "inner" resistances or conductances are associated with them. Thus they must be modeled by combinations of ideal sources and passive elements as discussed in Section 1.3.

1.2 The Generalization of Ohm's Law: Impedance and Admittance

The element laws for capacitors and inductors contain differential quotients or integrals if they are expressed in terms of voltage and current. But it is possible to reduce these equations to algebraic relations analogous to Ohm's law, if, for example, voltage and current are sinusoidal functions of time. The result is, as we shall see in this section, a generalized form of Ohm's law which defines, by analogy with resistance and conductance, the *frequency-dependent* parameters *impedance* and *admittance*. These new parameters for the storage elements have the same dimensions (ohm and siemens) as the dissipative ones; though this does *not* imply a functional change from energy storage to dissipation.

For some readers it will be necessary to review some properties of sinusoids and establish various conventions.

1.2.1 Sinusoids and Phasors

Sinusoidal signals are described by the trigonometric functions (Figs. 1.11a and 1.11b):

$$A\cos\omega t = A\cos(2\pi ft) = A\cos\frac{2\pi t}{T}$$

or

$$A\sin\omega t = A\sin(2\pi ft) = A\sin\frac{2\pi t}{T}$$

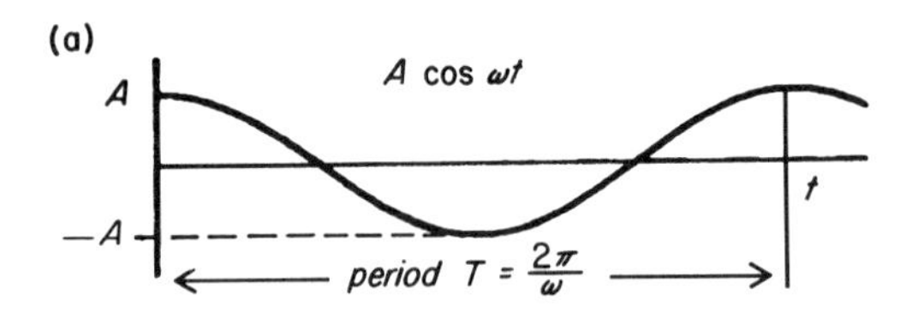

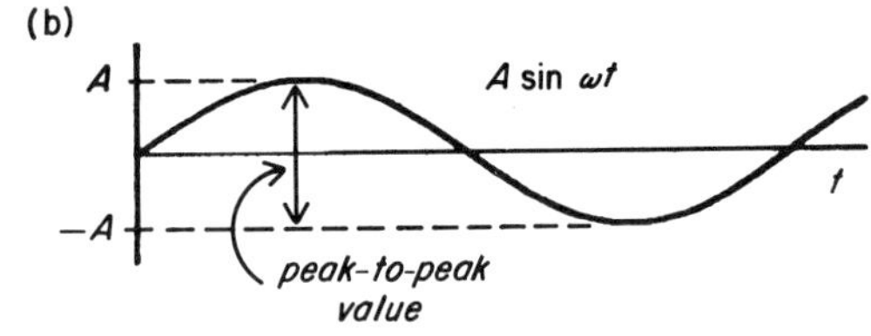

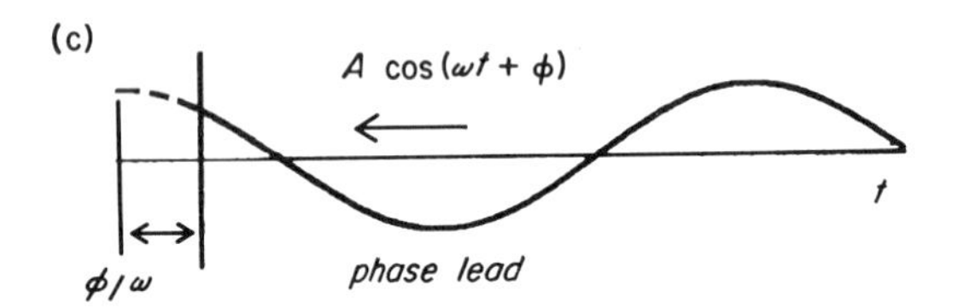

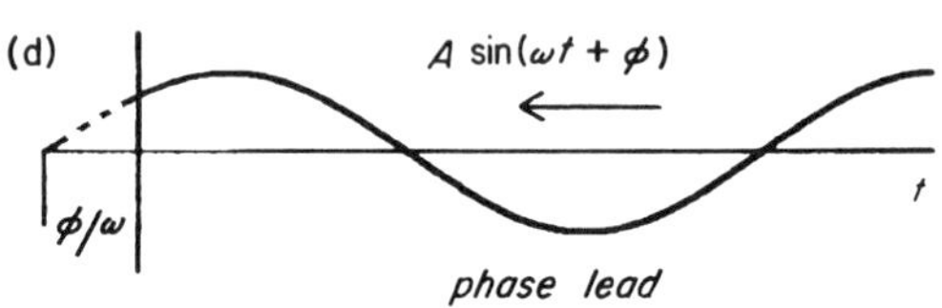

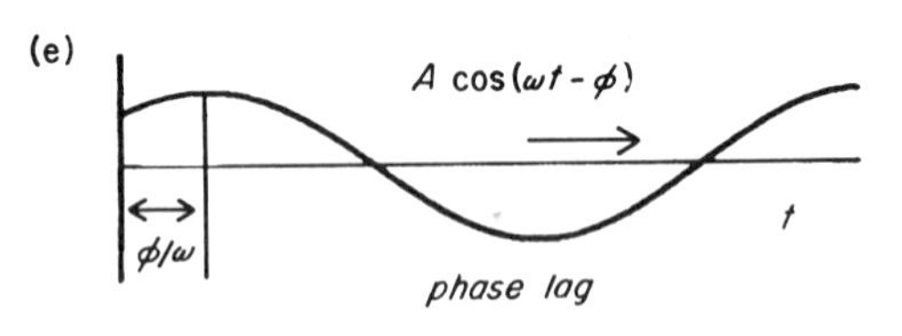

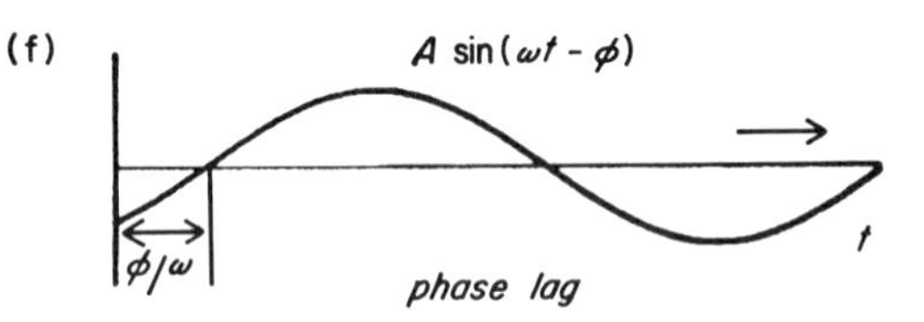

Figure 1.11 Sinusoidal signals. (See text.)

where the following definitions hold:

$A =$ amplitude (or magnitude) in volts or amperes,
$2A =$ peak-to-peak value (Fig. 1.11b),
$\omega =$ angular, radian or circular frequency in radians per second (rad/s),
$T =$ period or cycle length in seconds (s),
$f = 1/T = \omega/2\pi =$ frequency signifying the number of periods or cycles per second in hertz (Hz), which is dimensionally equivalent to reciprocal seconds (1/s).

The combination of two sinusoidal signals of the *same* frequency results again in a sinusoidal signal of *identical* frequency which is usually shifted in time. For example,

$$A_1 \cos \omega t - A_2 \sin \omega t = A \cos(\omega t + \phi)$$

where

$$A = \left(A_1^2 + A_2^2\right)^{1/2}$$

$$\phi = \tan^{-1}\left(\frac{A_2}{A_1}\right) \text{ in radians (rad) or degrees (°)}$$

Figure 1.11c shows that the result is a cosine of the same frequency but shifted to the *left* with respect to the time scale. The shift is indicated by the new parameter ϕ, the *phase* or *phase angle*. The waveform has a *phase lead* because corresponding values of the cosine function occur *earlier* (Fig. 1.11c) than in the case which has no *phase shift* (Fig. 1.11a). The *lead time* equals ϕ/ω. The sine function with phase lead is shown in Fig. 1.11d. In general, a *positive* phase angle (i.e., a plus sign precedes ϕ in the argument) is associated with a phase lead, and a *negative* one with a *phase lag* or time shift to the *right* (Figs. 1.11e and 11.f).

The mathematical operations involving sinusoids are often much simpler, if we make use of *Euler's identity* relating cosine and sine functions to the exponential function:

$$e^{\pm j\phi} = \cos\phi \pm j\sin\phi \tag{1.29}$$

or in the case of an arbitrary amplitude

$$Ae^{\pm j\phi} = A\cos\phi \pm Aj\sin\phi \tag{1.30}$$

where $j = \sqrt{-1}$. (Mathematicians use $i = \sqrt{-1}$; electrical engineers use j in order to avoid confusion with the current symbol i.)

Equations (1.29) and (1.30) describe *complex numbers* (see Appendix 1.A) which can be plotted in the *complex plane* or *Argand diagram*. The cosine and sine components represent the *real* and *imaginary parts*, respectively. It is common to draw a radius vector from the origin of the Argand diagram to the coordinate pair $(A\cos\phi, A\sin\phi)$ and call this visual representation a *phasor of magnitude A*. (Fig. 1.12a). Any complex number can be described and plotted in such a fashion. To specify a complex number we need two quantities, either the real (Re) and imaginary (Im) parts of the rectangular representation or *magnitude A* and *argument*, *angle*, or *phase*, ϕ of the equivalent phasor in the *polar* representation.

All phasors of magnitude A will lie on a circle of radius A concentric to the origin in the Argand diagram. The precise location on the magnitude circle for a particular phasor is given by its argument ϕ, which is the angle between the positive real half-axis and the phasor. Equation (1.29) describes a *unit phasor* of magnitude one. From Fig. 1.12a it is then evident that the argument ϕ represents the arc of the unit circle between the positive real half-axis and the phasor. Positive arguments are measured in *counterclockwise* fashion from the point $+1$ on the real axis. Clockwise measurement represents a negative angle (Fig. 1.12c). Accordingly, the intersections of the unit circle with the real and imaginary axes are equivalent to the following phasors:

$$e^{j0} = \cos 0 + j\sin 0 = +1$$

$$e^{j(\pi/2)} = \cos\left(\frac{\pi}{2}\right) + j\sin\left(\frac{\pi}{2}\right) = e^{-j(3\pi/2)} = +j$$

$$e^{\pm j\pi} = \cos\pi \pm j\sin\pi = -1$$

$$e^{-j(\pi/2)} = \cos\left(\frac{\pi}{2}\right) - j\sin\left(\frac{\pi}{2}\right) = e^{j(3\pi/2)} = -j$$

With Eq. (1.30) we introduce the short notation for a phasor

$$\mathbf{A} = Ae^{j\phi} \tag{1.31}$$

Magnitude and argument may be indicated in the following manner:

Magnitude: $|\mathbf{A}| = A$

Argument: $\sphericalangle\mathbf{A} = \phi$

For a negative argument we obtain the *conjugate* complex number (see Appendix 1.A and Fig. 1.12c):

$$\mathbf{A}^* = Ae^{-j\phi} \tag{1.32}$$

The argument of a phasor may change as a function of time. In the special case where the argument is a linear function of time, such as $(\omega t + \phi)$, we may visualize a *rotating phasor*

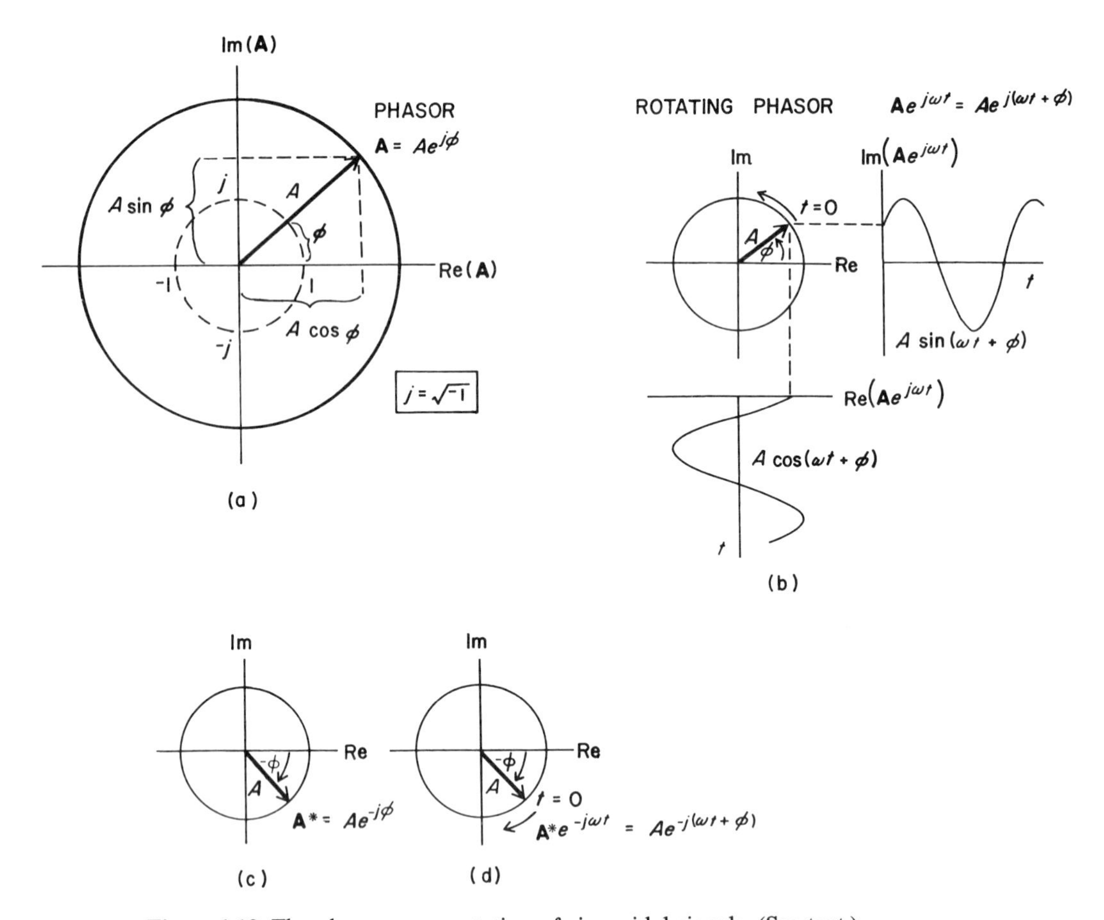

Figure 1.12 The phasor representation of sinusoidal signals. (See text.)

(Fig. 1.12b). For time $t=0$, such a phasor has the phase ϕ; thereafter its position in the Argand diagram changes like that of a pointer rotating counterclockwise with constant *angular velocity* ω.

With Euler's identity we have

$$e^{j\omega t}=\cos\omega t+j\sin\omega t \tag{1.33}$$

and, more generally,

$$\mathbf{A}e^{j\omega t}=Ae^{j(\omega t+\phi)}=A\cos(\omega t+\phi)+jA\sin(\omega t+\phi) \tag{1.34}$$

and

$$\mathbf{A}^{*}e^{-j\omega t}=Ae^{-j(\omega t+\phi)}=A\cos(\omega t+\phi)-jA\sin(\omega t+\phi) \tag{1.35}$$

where the plus sign stands for counterclockwise rotation (Fig. 1.12b) and the minus sign for clockwise rotation.

From Fig. 1.12b and Eq. (1.34) we conclude that the cosine and sine are the time-varying projections of the rotating phasor $\mathbf{A}e^{j\omega t}$ on the real and imaginary axis, respectively:

$$A\cos(\omega t+\phi)=\mathrm{Re}(\mathbf{A}e^{j\omega t})=A\,\mathrm{Re}(e^{j(\omega t+\phi)}) \tag{1.36}$$

and

$$A\sin(\omega t+\phi)=\mathrm{Im}(\mathbf{A}e^{j\omega t})=A\,\mathrm{Im}(e^{j(\omega t+\phi)}) \tag{1.37}$$

From Eqs. (1.36) and (1.37) follows another way of expressing sinusoidal signals:

$$\cos\omega t=\frac{e^{j\omega t}+e^{-j\omega t}}{2} \tag{1.38}$$

and

$$\sin\omega t=\frac{e^{j\omega t}-e^{-j\omega t}}{2j} \tag{1.39}$$

The point of introducing the exponential formulation of sinusoidal functions according to Eqs. (1.36)–(1.39) is that the mathematical analysis of networks is much easier in terms of the exponential function. The simplest strategy is to arrive at a solution for the exponential function first and then determine the solution for the particular sinusoid using the preceding equations. This procedure is justified by the superposition property of linear systems. The principle will be illustrated in subsequent sections. In most instances, however, it will not even be necessary to make the transformation from phasor to sinusoid because a solution giving magnitude and phase will be sufficient to describe the *steady state sinusoidal response* of a system. (In the steady state, all switching transients have disappeared, leaving only the response to the sinusoid proper.)

1.2.2 Passive Element Laws for Sinusoidal Excitation: Impedance and Admittance

1.2.2.1 Resistors

Let us assume that the current $i=I\cos\omega t$ flows through a resistor, R. From Ohm's law we have for the voltage

$$v=Ri=RI\cos\omega t=V\cos\omega t \tag{1.40}$$

Obviously, the voltage is also a sinusoid of amplitude $V=RI$ and angular frequency ω. In general, the current could be either a cosine or a sine function, and could have a phase angle, too. We can include all these possibilities using the rotating current phasor concept:

$$i=\mathbf{I}e^{j\omega t}=Ie^{j(\omega t+\phi)}=I\cos(\omega t+\phi)+jI\sin(\omega t+\phi)$$

From this the cosine or sine response can be recovered by taking the real or imaginary part. Equation (1.40) can now be made more general with the phasor notation:

$$v=Ri=R\mathbf{I}e^{j\omega t}=\mathbf{V}e^{j\omega t}=Ve^{j(\omega t+\phi)} \tag{1.41}$$

If the current is $i=\mathrm{Re}(\mathbf{I}e^{j\omega t})=I\cos(\omega t+\phi)$, then the corresponding voltage will be, according to Eq. (1.41),

$$v=\mathrm{Re}(\mathbf{V}e^{j\omega t})=V\cos(\omega t+\phi)=RI\cos(\omega t+\phi)$$

And, conversely, for the current $i=\mathrm{Im}(\mathbf{I}e^{j\omega t})=I\sin(\omega t+\phi)$, the voltage is

$$v=\mathrm{Im}(\mathbf{V}e^{j\omega t})=V\sin(\omega t+\phi)=RI\sin(\omega t+\phi)$$

In both instances, current and voltage differ only in amplitude. As scaled versions of each other, the two waveforms have the same frequency and phase angle: they are *coherent* and *in phase*. In general, however, we can expect that linear elements do not only cause an amplitude change but also a phase shift. The frequency, though, and the sinusoidal wave shape will remain unchanged as long as the system is linear. Because this fact is known to us beforehand, we can simply use the phasors **V** and **I** instead of $\mathbf{I}e^{j\omega t}$, as we see next.

From Eq. (1.41) we know that, for the current $\mathbf{I}e^{j\omega t}$, the voltage will also be a sinusoid, namely, $\mathbf{V}e^{j\omega t}$. (This is true for *all* linear elements, not only for resistors.) We can then modify Ohm's law with the substitutions $i=\mathbf{I}e^{j\omega t}$ and $v=\mathbf{V}e^{j\omega t}$:

$$v = Ri$$

$$\mathbf{V}e^{j\omega t} = R\mathbf{I}e^{j\omega t}$$

Cancelling the term $e^{j\omega t}$ on both sides (it contains information we already know) we obtain *Ohm's law in phasor notation*:

$$\mathbf{V} = R\mathbf{I} \qquad \text{and} \qquad R = \frac{\mathbf{V}}{\mathbf{I}} \tag{1.42}$$

This is obviously a much simpler expression than either (1.40) or (1.41). The phasor form of the element law contains both amplitude and phase information:

$$|\mathbf{V}| = V = R|\mathbf{I}| = RI$$

and

$$\measuredangle\mathbf{V} = \measuredangle\mathbf{I} = \phi$$

1.2.2.2 Capacitors. Impedance, Admittance

The advantages of the phasor method are particularly evident when we apply it to the case of sinusoidal excitation of storage elements. We will assume here that a *sinusoidal steady state* exists. This means that we wait until all transients from switching on the system have died out, and that the circuit was initially completely relaxed (zero initial conditions). In the case of the capacitor, this is equivalent to dropping the initial voltage term $v(0)$ and changing the lower integration limit to $-\infty$ in the element law given by Equation (1.14):

$$v(t) = \frac{1}{C}\int_{-\infty}^{t} i(t)\,dt = \frac{1}{C}\int i(t)\,dt \tag{1.43}$$

where the definite integral can be replaced by the indefinite one because $v(-\infty)=0$.

Proceeding as before with the resistor, we have for the application of a current

$$i = \mathbf{I}e^{j\omega t}$$

From Equation (1.43)

$$v(t) = \frac{1}{C}\int \mathbf{I}e^{j\omega t}\,dt = \frac{\mathbf{I}}{j\omega C}e^{j\omega t}$$

with

$$v(t) = \mathbf{V}e^{j\omega t}$$

we have

$$\mathbf{V}e^{j\omega t}=\frac{\mathbf{I}}{j\omega C}e^{j\omega t} \tag{1.44}$$

and after cancellation of the exponential factor we obtain

$$\mathbf{V}=\frac{1}{j\omega C}\mathbf{I} \tag{1.45}$$

and

$$\mathbf{I}=j\omega C\mathbf{V} \tag{1.46}$$

Equations (1.45) and (1.46) are formally equivalent to Ohm's law if we define new, generally *frequency-dependent* parameters analogous to resistance and conductance.

We define the *impedance* $Z(\omega)$ of an arbitrary linear two-terminal element under sinusoidal excitation as the voltage phasor/current phasor ratio:

$$Z(\omega)=\frac{\mathbf{V}}{\mathbf{I}} \tag{1.47}$$

$$Z(\omega)=|Z|e^{j\sphericalangle Z}=\frac{|\mathbf{V}|}{|\mathbf{I}|}e^{j(\sphericalangle\mathbf{V}-\sphericalangle\mathbf{I})} \tag{1.48}$$

$$|Z|=\frac{|\mathbf{V}|}{|\mathbf{I}|}=\frac{V}{I} \tag{1.49}$$

$$\sphericalangle Z=\sphericalangle\mathbf{V}-\sphericalangle\mathbf{I} \tag{1.50}$$

$$[Z]=\Omega$$

We define the *admittance* $Y(\omega)$ of an arbitrary linear two-terminal element under sinusoidal excitation as the reciprocal of the impedance or the ratio current phasor/voltage phasor:

$$Y(\omega)=\frac{1}{Z(\omega)}=\frac{\mathbf{I}}{\mathbf{V}} \tag{1.51}$$

$$Y(\omega)=|Y|e^{j\sphericalangle Y}=\frac{|\mathbf{I}|}{|\mathbf{V}|}e^{j(\sphericalangle\mathbf{I}-\sphericalangle\mathbf{V})} \tag{1.52}$$

$$|Y|=\frac{|\mathbf{I}|}{|\mathbf{V}|}=\frac{I}{V} \tag{1.53}$$

$$\sphericalangle Y=-\sphericalangle Z=\sphericalangle\mathbf{I}-\sphericalangle\mathbf{V} \tag{1.54}$$

$$[Y]=\mathrm{S}$$

These definitions are entirely general and applicable to any linear two-terminal element. If a device has several terminal pairs, we can define for each pair an impedance or admittance with conditions at the remaining terminal pairs specified (e.g., open or short circuited). To distinguish these particular phasor ratios from transfer ratios between terminal pairs we designate them as *driving point* impedances or admittances.

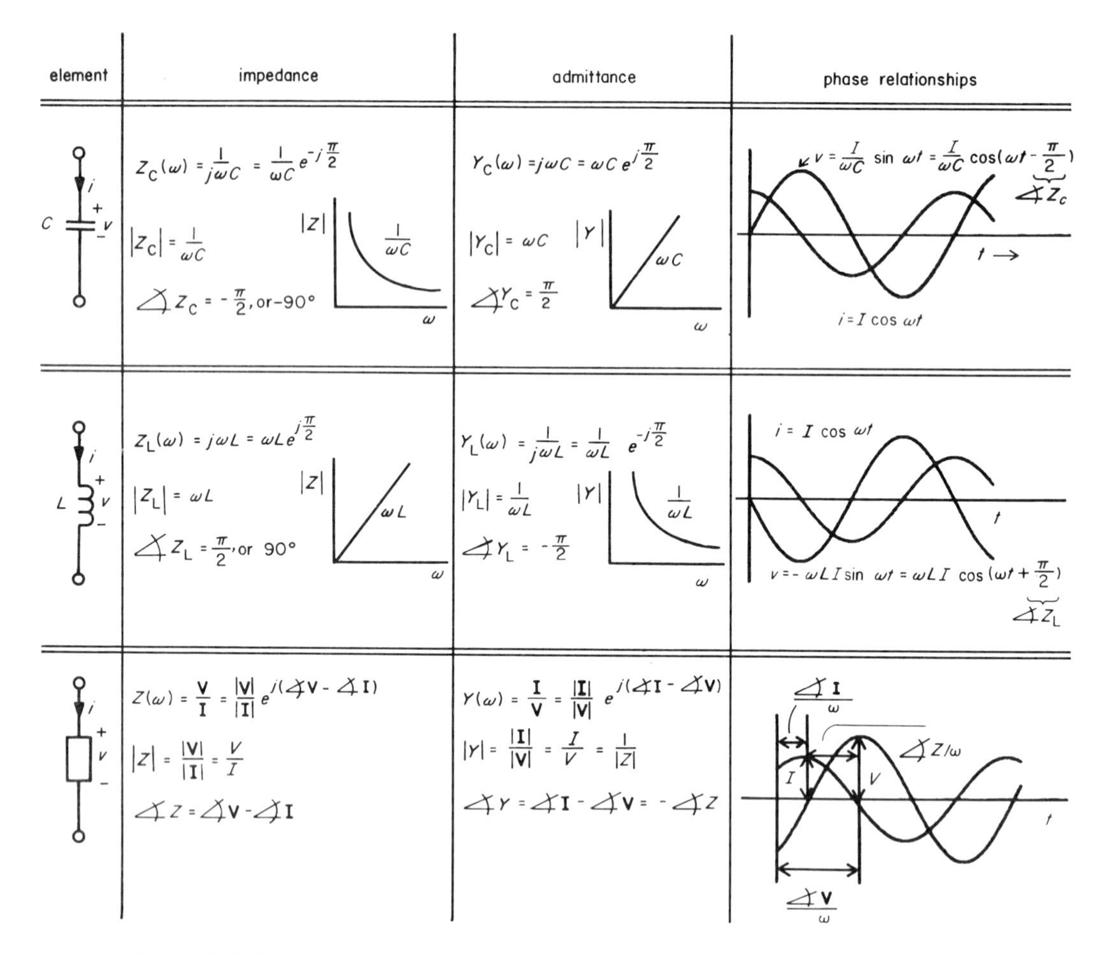

Figure 1.13 Impedance, admittance, and sinusoidal response properties of storage elements and general linear one-ports. (See text.)

Applying these definitions to the resistor element [Eq. (1.42)] we see that the impedance and admittance of a resistor equal its resistance R and conductance G, respectively.

The impedance and admittance of a capacitor (Fig. 1.13) are obtained from Eqs. (1.45) and (1.47), (1.46), and (1.51):

$$Z_C(\omega)=\frac{1}{j\omega C} \tag{1.55}$$

and

$$Y_C(\omega)=j\omega C \tag{1.56}$$

Impedance and admittance have the dimensions *ohm* (Ω) and *siemens* (S), respectively, but this does *not* imply that, for example, the function of a capacitor has been changed from energy storage to dissipation. The new parameters only indicate how much the passage of a sinusoidal current of a given frequency is "impeded" (impedance) or facilitated (admittance).

For instance, from Eq. (1.55) we see that the impedance goes to zero as the angular frequency increases; the capacitor acts like a short circuit at very high signal frequencies. And from Eq. (1.56) it follows that the admittance becomes zero as the signal frequency approaches zero; the capacitor acts like an open circuit for zero-frequency (*dc*) signals.

Example 1.3.

Compute the impedance of a 1-μF capacitor for the typical house current frequencies 50 and 60 Hz, and for the radio frequencies 100 kHz and 100 MHz.

$$Z_C=\frac{1}{j\omega C}, \quad \text{or} \quad |Z_C|=\frac{1}{\omega C}=\frac{1}{2\pi f C}$$

1. $f=50$ Hz:

$$|Z_C|=\frac{1}{2\times\pi\times 50\times 10^{-6}}\,\Omega=\frac{10^6}{100\times\pi}\,\Omega=3183\ \Omega$$

2. $f=60$ Hz:

$$|Z_C|=\frac{1}{2\times\pi\times 60\times 10^{-6}}\,\Omega=\frac{10^6}{120\times\pi}=2653\ \Omega$$

3. $f=100$ kHz:

$$|Z_C|=\frac{1}{2\times\pi\times 10^5\times 10^{-6}}\,\Omega=\frac{10}{2\pi}=1.6\ \Omega$$

4. $f=100$ MHz:

$$|Z_C|=\frac{1}{2\times\pi\times 10^8\times 10^{-2}}=\frac{1}{2\times\pi\times 100}\,\Omega=1.6\times 10^{-3}\ \Omega$$

Whether we use the impedance or the admittance parameter is entirely a matter of convenience in any given situation. They both represent the same physical properties. If we need not make a distinction between the two we refer to *im*pedances and ad*mittances* collectively, as *immittances*.

The immittance definitions, Eqs. (1.47) and (1.51), are, of course, restatements of Ohm's law in phasor notation. Because **V** and **I** are complex phasors, it is obvious that, in general, Z and Y are also complex quantities. Thus the *magnitudes of the immittances*, Eqs. (1.49) and (1.53), are given by the *amplitude* ratios of the voltage and current pairs and *not* the instantaneous ratios at a given time t (Fig. 1.13).

The positive peak values, representing the amplitudes of voltage and current, will generally not occur at the same instant. This shift on the time axis between current and voltage is represented by the *phase angle of the immitances*, Eqs. (1.50) and (1.54), and Fig. 1.13. For example, for the capacitor we find

$$Z_C(\omega)=\frac{1}{j\omega C}=\frac{-j}{\omega C}=\frac{1}{\omega C}e^{-j(\pi/2)}$$

and

$$\sphericalangle Z_C=-\frac{\pi}{2}, \quad \text{or} \quad -90^\circ$$

In this particular instance the phase angle is independent of the frequency; the capacitor voltage always *lags* the current by 90°.

Finally, we can point to other conveniences the phasor method offers. The transition from Eq. (1.43) to (1.45) demonstrates that an *integration* of a sinusoidal signal can be reduced to a division of the corresponding phasor by $j\omega$:

$$\int_{-\infty}^{t} \to \int \to \frac{1}{j\omega}$$

As we shall see next, the *differentiation* of a sinusoidal signal amounts to the *multiplication* of the phasor by $j\omega$:

$$\frac{d}{dt} \to j\omega$$

We can show this by deriving Eq. (1.56) from the differential form of the capacitor element law, Eq. (1.13). With the voltage

$$v(t) = \mathbf{V}e^{j\omega t}$$

we obtain

$$i(t) = \mathbf{I}e^{j\omega t} = \frac{d}{dt}(Cv) = \frac{d}{dt}(C\mathbf{V}e^{j\omega t})$$

and

$$\mathbf{I}e^{j\omega t} = j\omega C\mathbf{V}e^{j\omega t}$$

or

$$\mathbf{I} = j\omega C\mathbf{V}$$

These rules, which are very useful for the sinusoidal steady state analysis of networks, can be generalized to multiple integrations and differentiations.

The n-fold integration or differentiation of sinusoidal signals corresponds to n-fold division or multiplication respectively, by $j\omega$: with $x(t) = \mathbf{X}e^{j\omega t}$,

$$\underbrace{\int_{-\infty}^{t}\int_{-\infty}^{t}\cdots\int_{-\infty}^{t}}_{n \text{ times}} x(t)\,dt \leftrightarrow \left(\frac{1}{j\omega}\right)^{n}\mathbf{X}$$

and

$$\frac{d^{n}}{dt^{n}}x(t) \leftrightarrow (j\omega)^{n}\mathbf{X}$$

The rules have the effect of reducing integral and differential equations to *algebraic* phasor equations representing the sinusoidal steady state of a system. We shall apply these rules in later sections.

1.2.2.3 Inductors

With the methods just developed it is elementary to obtain the impedance and admittance of an inductor subjected to sinusoidal excitation. From the differential form of the inductor element law, Eq. (1.22), follows

$$\mathbf{V} = j\omega L \mathbf{I}$$

or

$$Z_{\mathrm{L}} = \frac{\mathbf{V}}{\mathbf{I}} = j\omega L \tag{1.57}$$

and

$$Y_{\mathrm{L}} = \frac{1}{Z_{\mathrm{L}}} = \frac{1}{j\omega L} \tag{1.58}$$

The voltage across the inductor *leads* the current with a constant phase angle of 90° because the argument of the impedance is $\pi/2$ (Fig. 1.13):

$$Z_{\mathrm{L}} = |Z_{\mathrm{L}}| e^{j \measuredangle Z_{\mathrm{L}}} = \omega L e^{j(\pi/2)}$$
$$\measuredangle Z_{\mathrm{L}} = \measuredangle \mathbf{V} - \measuredangle \mathbf{I} = \pi/2$$

As the frequency becomes infinite, the magnitude of the impedance becomes infinite according to Eq. (1.57); the inductor blocks very high signal frequencies. Conversely, for *dc* signals ($\omega = 0$) the impedance goes to zero; an ideal inductor acts as a short circuit or bypass for such signals.

REMARK: The immittances of ideal capacitors and inductors are imaginary numbers. If one of the phasors of the voltage–current pair is a real number, then the multiplication with the appropriate immittance will make the other phasor an imaginary number. *This does not mean that the resultant voltages or currents are "imaginary" in the physical sense.* Physically, they are quite real, but they have a 90° phase shift, which is *mathematically symbolized* by the factors j or $-j$.

1.2.3 Passive Element Laws for Exponential Excitation: A More General Immittance Concept

In this section we shall consider, mostly for the sake of convenience, a generalization of the phasor and immittance concepts. This approach is best justified by the application of the Laplace transformation to the element laws as set forth in Chapter 3. Here we simply introduce these ideas without concern for mathematical rigor.

1.2.3.1 Exponential Excitation: Complex Frequency

The phasors introduced in Section 1.2.1 represented sinusoids of constant amplitude. The generalized phasors encompass not only constant amplitude sinusoids but also exponentially decaying and rising sinusoids and aperiodic exponential transients. This is accomplished by substituting for the constant amplitude of a rotating phasor

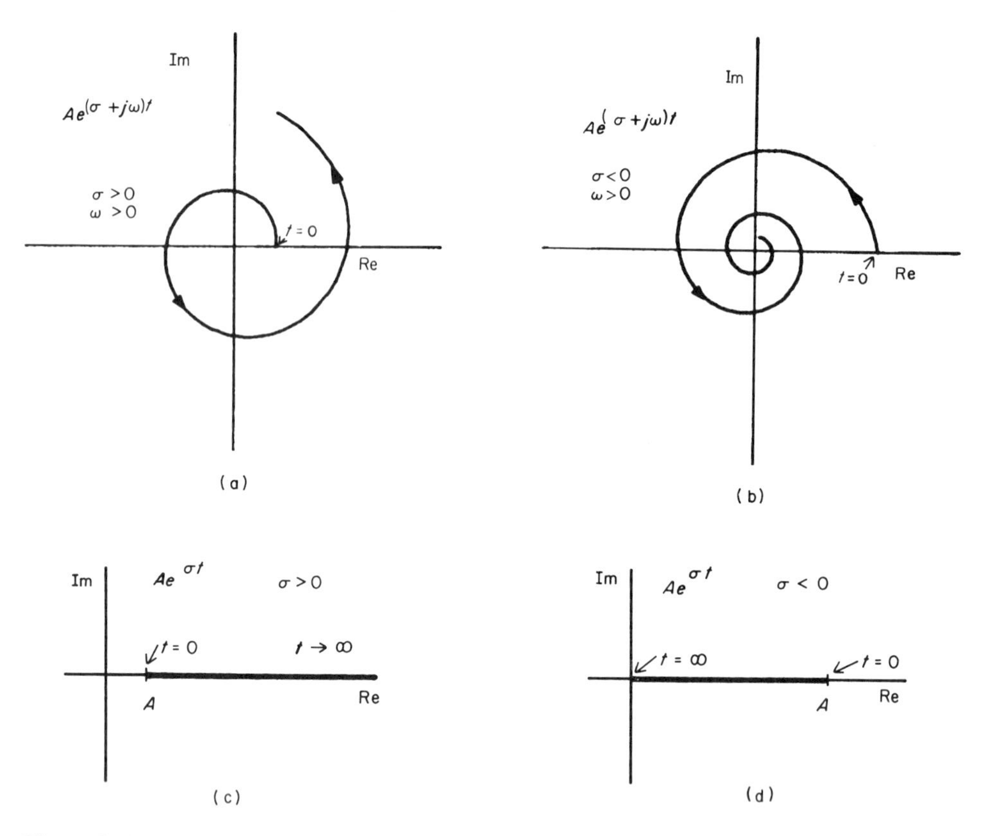

Figure 1.14 Exponential excitation. Trajectories in the complex plane. (See text.)

an "amplitude" which is an exponential function of time:

$$\mathbf{A}e^{j\omega t} \to (\mathbf{A}e^{\sigma t})e^{j\omega t} = \mathbf{A}e^{(\sigma+j\omega)t} \tag{1.59}$$

The complex number in the exponent is commonly called the *complex frequency* s:

$$s = \sigma + j\omega \tag{1.60}$$

In Eq. (1.60), ω is the radian frequency as introduced previously, and σ is the *damping constant* indicating how fast the exponential term rises ($\sigma>0$) or decays ($\sigma<0$).

The new generalized phasor

$$\mathbf{A}e^{st} = \mathbf{A}e^{(\sigma+j\omega)t} \tag{1.61}$$

can be plotted as a rotating phasor in the Argand diagram with its "tip" tracing an exponential spiral away from ($\sigma>0$) or toward ($\sigma<0$) the origin (Figs. 1.14a and 1.14b). These spiral trajectories degenerate into circles if $\sigma=0$ (Fig. 1.12) or into straight lines in the case of aperiodic functions ($\omega=0$) as seen in Figs. 1.14c and 1.14d for $\phi=0$.

The amplitude factor A provides the scaling, while the complex frequency s specifies the *type* of waveform represented by the phasor. The real and imaginary components of s, σ, and ω can be zero, positive, or negative. Setting $\phi=0$ we can

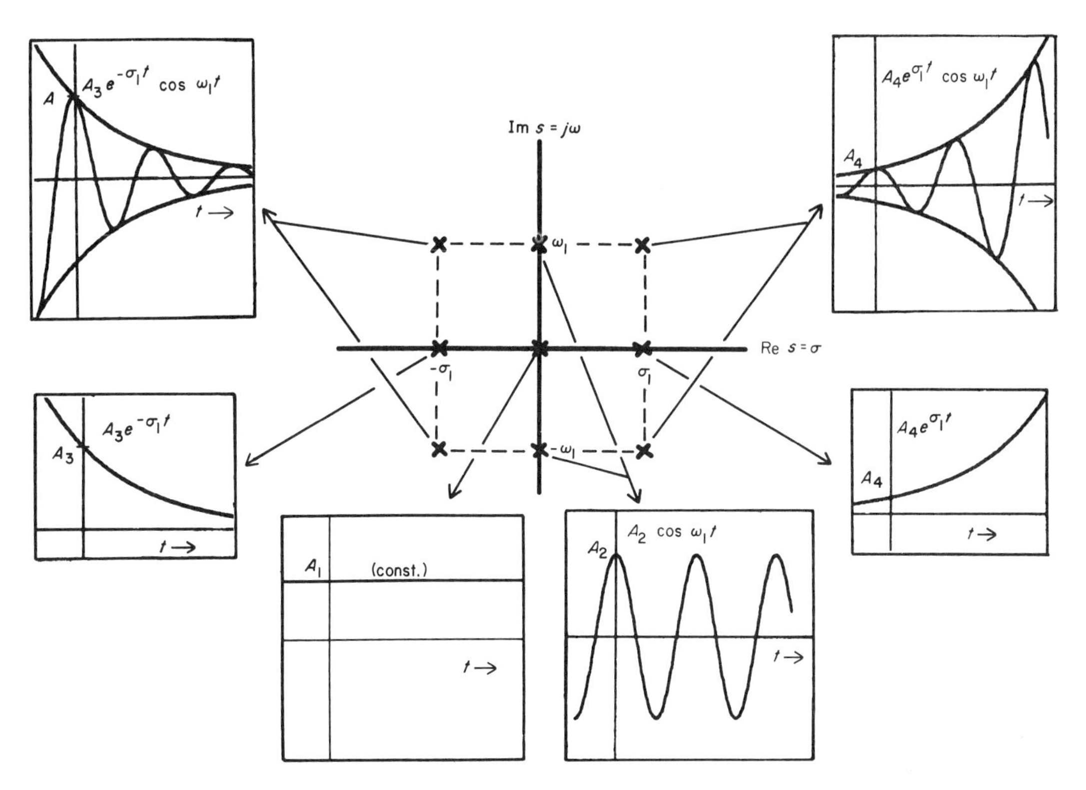

Figure 1.15 Signal patterns associated with complex exponents of exponential signals. (See text.)

readily distinguish the following cases with the aid of Eqs. (1.38), (1.39), (1.59), and (1.60):

$$\text{Re}(s)=0, \quad \text{Im}(s)=0 \rightarrow f(t)=A=\text{constant}$$

$$\text{Re}(s)<0, \quad \text{Im}(s)=0 \rightarrow f(t)=Ae^{-\sigma t}$$

$$\text{Re}(s)>0, \quad \text{Im}(s)=0 \rightarrow f(t)=Ae^{\sigma t}$$

$$\text{Re}(s)=0, \quad \text{Im}(s)\neq 0 \rightarrow f(t)=A\frac{e^{j\omega t}+e^{-j\omega t}}{2}=A\cos\omega t$$

$$\text{Re}(s)<0, \quad \text{Im}(s)\neq 0 \rightarrow f(t)=A\frac{e^{(-\sigma+j\omega)t}+e^{(-\sigma-j\omega)t}}{2}$$

$$=Ae^{-\sigma t}\cos\omega t$$

$$\text{Re}(s)>, \quad \text{Im}(s)\neq 0 \rightarrow f(t)=A\frac{e^{(\sigma+j\omega)t}+e^{(\sigma-j\omega)t}}{2}$$

$$=Ae^{\sigma t}\cos\omega t$$

Figure 1.15 correlates the position of σ and ω in the s plane with the associated waveforms in the time domain. Note that the waveforms containing sinusoidal terms are always the consequence of conjugate complex pairs

$$s=\sigma+j\omega \qquad \text{and} \qquad s^*=\sigma-j\omega$$

1.2.3.2 Impedance and Admittance in Terms of Complex Frequency

The generalized phasor of Eq. (1.61) will be used to derive element laws and immittances for the case where all voltages, currents, charges, and magnetic fluxes before application of the signal are zero (zero initial conditions). A rigorous derivation is possible with the Laplace transformation (Chapter 3).

With the current given as

$$i=\mathbf{I}e^{st} \tag{1.62}$$

we can fortunately assume that the voltage can be represented as

$$v=\mathbf{V}e^{st} \tag{1.63}$$

and vice versa because of the linearity of the elements.

For the element laws we have then the following equations.

Resistor:

$$v(t)=Ri(t)\rightarrow\mathbf{V}e^{st}=R\mathbf{I}e^{st} \tag{1.64}$$

or

$$\mathbf{V}=R\mathbf{I}$$

Capacitor:

$$v(t)=\frac{1}{C}\int_0^t i(t)\,dt\rightarrow\mathbf{V}e^{st}=\frac{1}{C}\int_0^t \mathbf{I}e^{st}\,dt\rightarrow\mathbf{V}e^{st}=\frac{1}{sC}\mathbf{I}e^{st}$$

or

$$\mathbf{V}=\frac{1}{sC}\mathbf{I} \tag{1.65}$$

Inductor:

$$v(t)=L\frac{d}{dt}i(t)\rightarrow\mathbf{V}e^{st}=L\frac{d}{dt}\mathbf{I}e^{st}\rightarrow\mathbf{V}e^{st}=sL\mathbf{I}e^{st}$$

or

$$\mathbf{V}=sL\mathbf{I} \tag{1.66}$$

By analogy with Eqs. (1.47) and (1.51) we define then the immittances in terms of the complex frequency s:

The *generalized impedance* $Z(s)$ of an arbitrary linear two-terminal element is given as

$$Z(s)=\frac{\mathbf{V}e^{st}}{\mathbf{I}e^{st}}=\frac{\mathbf{V}}{\mathbf{I}} \tag{1.67}$$

The *generalized admittance* $Y(s)$ is given as

$$Y(s)=\frac{\mathbf{I}e^{st}}{\mathbf{V}e^{st}}=\frac{\mathbf{I}}{\mathbf{V}} \tag{1.68}$$

Applying these definitions to Eqs. (1.64)–(1.66), we specifically obtain the immittances for the basic passive elements.

Resistor:

$$Z_R(s)=R; \qquad Y_R(s)=G=\frac{1}{R}$$

Capacitor:

$$Z_C(s)=\frac{1}{sC}; \qquad Y_C(s)=sC$$

Inductor:

$$Z_L(s)=sL; \qquad Y_L(s)=\frac{1}{sL}$$

Obviously, these immittances can be obtained from the expression derived for sinusoidal excitation by substituting s for $j\omega$. Conversely, the immittances for the sinusoidal steady state are a special case of the generalized immittances with s replaced by $j\omega$, i.e., $\sigma=0$.

We can also extend the integration and differentiation rules for the $j\omega$ "operator" (Section 1.2.2.2) to the complex frequency "operator" s. The following rules should only be applied to a *fully relaxed system* (zero state, zero initial conditions) in order to avoid problems which are better handled by the Laplace transformation (Chapter 3).

The n-fold integration or differentiation of a generalized phasor signal corresponds to n-fold division or multiplication, respectively, by the complex frequency s: with $x(t)=\mathbf{X}e^{st}$

$$\underbrace{\int_0^t\int_0^t\cdots\int_0^t}_{n\text{ times}} x(t)(dt)^n \leftrightarrow \frac{1}{s^n}\mathbf{X}$$

and

$$\frac{d^n}{dt^n}x(t)\leftrightarrow s^n\mathbf{X}$$

This concludes the discussion of individual circuit elements. The next step is to formulate the equations governing the interconnection of individual circuit elements into circuits.

1.3 The Laws of Kirchhoff

1.3.1 Definitions for Circuits and Networks

In an electric circuit individual network elements are connected to each other according to some *schematic* or *wiring plan*. The representation of this plan on paper is the *circuit diagram*. An example is given in Fig. 1.16a. Whenever a conductive connection between the terminals of circuit elements is to be made we draw a line

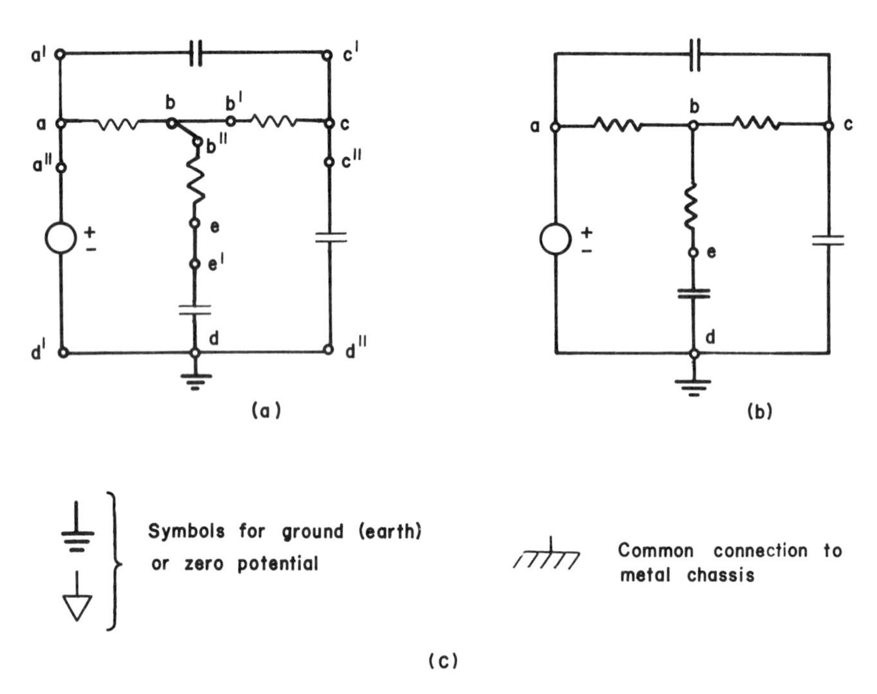

Figure 1.16 **(a)** Wiring schematic of a circuit; **(b)** Corresponding circuit diagram; **(c)** Symbols for connections to zero or reference potential.

connecting the terminals (labeled by lower case letters in Fig. 1.16a). The lines between terminals represent *ideal* conductors so that all terminals connected by such lines (e.g., a′-a-a″) have the same electric potential. (Any parasitical resistance, inductance, etc., present in practical connecting wires and solder junctions are either neglected or represented by separate model elements.) All connected terminals of the same potential are combined into a single junction or *node* in the circuit diagram. In our example the number of nodes is thus reduced to five (a, b, c, d, and e) as shown in Fig. 1.16b.

It is useful to designate one node as the *reference* or *datum node*. This is done so that the potentials at all the other nodes can be measured with respect to the potential at the reference node. We can pick any node as the reference node, but if one node is connected to *ground* or *earth* (representing *zero potential*) or to a large metal body such as the mounting chassis (representing "*common*" potential, which may be separate from ground potential) we attach one of the symbols of Fig. 1.16c to that node and designate it as the reference node. In our example node d is the reference node.

A more general form of our example is given in Fig. 1.17a, where we have left unspecified the nature of the passive elements. In addition, we have added the *reference directions* for each circuit element. The polarities for currents and voltages are entered so that they form *associated* pairs for the passive elements.

It is necessary to assign reference directions *before* we analyze the circuit or network. But the initial choice of current directions is entirely arbitrary; only at the end of the analysis will we know whether the choice was correct. If our initial choice

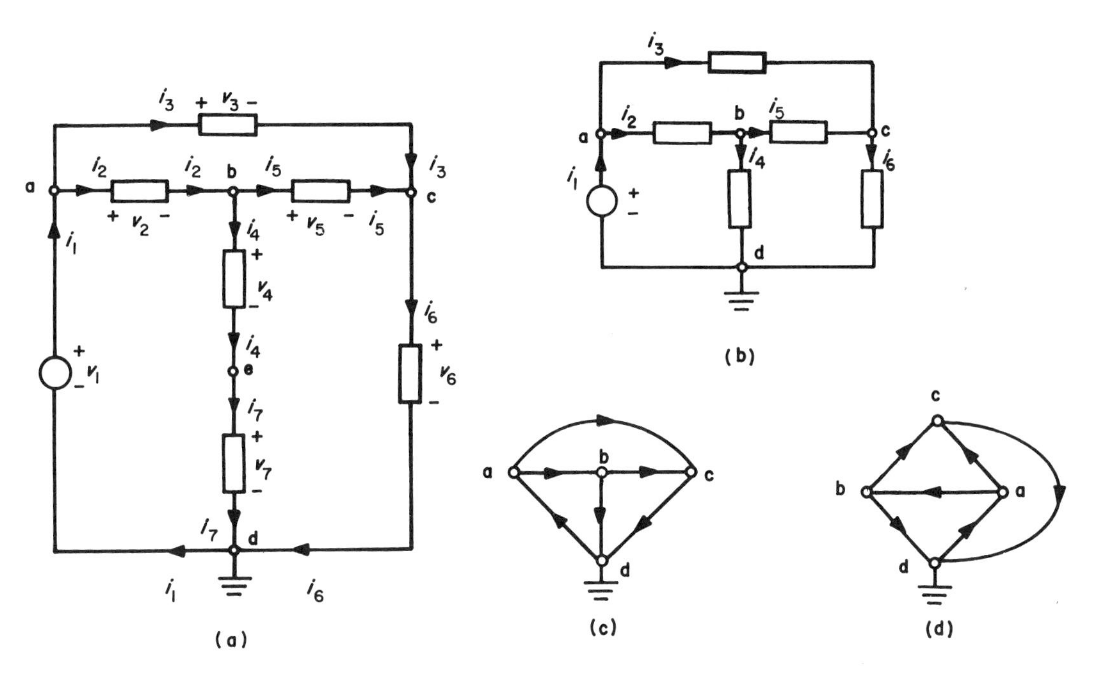

Figure 1.17 Derivation of circuit topology. (See text.)

of polarity for a particular current–voltage pair is incorrect, the analysis will yield a negative sign for the corresponding variables. We can change the polarity to the correct value at that time.

A junction of more than two network elements is designated as a *complex node* (e.g., nodes a, b, c, and d in Fig. 1.17). At a *simple node* only two elements connect. Node e in Fig. 1.17a is an example of a simple node. For reasons of continuity, the current entering that node, i_4, must equal the current leaving it, i_7. So if we know either of the two currents we will be able to compute the corresponding voltages, v_4 and v_7, from the element laws. For this reason, simple nodes are often eliminated from a network by combining the passive model elements involved into a new one of equivalent properties. How this is done will be evident later.

The result of eliminating the simple node e in our example is given in Fig. 1.17b. The model element between nodes b and d combines the properties of the two elements it replaces. In this figure only the current polarities are indicated; the voltage polarities follow automatically from the use of associated references.

The way in which the individual elements are connected is crucial. The same elements connected in a pattern different from that in Fig. 1.17b constitute a new circuit of different function. The new circuit has a different *topology*. For example, it might have more or fewer nodes; or nodes b and c might not be connected by an element. Neglecting the particular nature of the model elements, we can abstract the pattern of connection by replacing the elements with lines connecting the nodes. The result is the network *graph*. A graph representing our circuit example is given in Fig. 1.17c. In it we have retained the reference directions of the currents; we therefore call it an *oriented graph*.

Two graphs are said to be *topologically equivalent* if a deformation involving stretching, bending, flipping, rotating, etc., but *not* cutting or joining transforms one graph into the other. Hence the graph in Fig. 1.17d is topologically equivalent to that in Fig. 1.17c in spite of its different appearance. Of course, topological equivalence must also exist between the circuit diagram and the actual wiring layout of a circuit. A lack of equivalence means that we have made a wiring mistake.

For the efficient application of the rules of network analysis we need to concern ourselves with some basic properties of network graphs. We have already introduced the concepts of node and reference node. We consider several other terms next.

Independent node pair: The combination of the reference node with any other node of the network.

Node voltage: The potential difference of an independent node pair, i.e., the potential of a node with respect to the *reference* node.

Number of independent node pairs or node voltages: For N = total number of nodes (including reference node), the total number of independent node pairs or node voltages is $N-1$.

Branch (or edge): The line segment in a graph connecting two nodes directly.

Loop: A subgraph (part of a graph) forming a closed contour. Examples of loops are given in Figs. 1.18a and 1.18b for the graph considered previously. For the purpose of network analysis, loops are given a reference direction as indicated by the arrows and are usually numbered.

Mesh: A loop containing no inner branches. The loops 1, 2, and 3 in Fig. 1.18a are meshes; those indicated in Fig. 1.18b (4 and 5) are not. In some texts, however, the terms loop and mesh are used indiscriminately.

Tree: A subgraph of *connected* branches containing *all* the nodes but *no* loops. Two examples for the same graph are given in Figs. 1.18c and 1.18d (heavy lines). The number of tree branches equals the number of independent node voltages, i.e., $N-1$.

Link or chord: Branches *not* belonging to a tree (light lines in Figs. 1.18c and 1.18d). The number of links is $L = B-(N-1)$, where B is the total number of branches.

Number of independent loops: Starting with a tree, we form a loop each time we add a link to the tree. Hence, if there are L links, there must be L independent loops. (This is independent of the choice of a particular tree.)

Planar graph: A graph which can be drawn on a 2-dimensional surface without crossing branches.

Branch current: The current flowing through a branch.

Branch voltage: The potential difference between nodes connected by a branch. The term voltage drop is also used.

We are now ready to take the next important step in the procedure to analyze a network. Two sets of equations are necessary to describe network behavior mathematically. The first set comprises the element laws of all the constituent parts of the network. The other set is based on the connection pattern or the graph properties of the circuit. This second set of equations can be written down by inspection of the

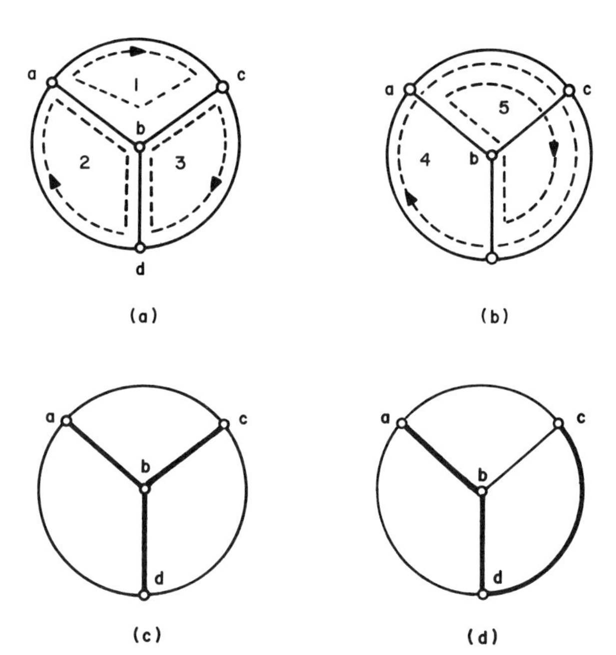

Figure 1.18 Loops, meshes, and trees: **(a)** meshes which are also loops; **(b)** loops which are not meshes; **(c)** and **(d)** two alternative sets of trees for the same network. (See text.)

oriented graph according to the rules formulated by *Kirchhoff* for branch currents and branch voltages. These rules will be introduced and applied in the next few sections. Combining the two sets of equations leads to the *network equations* describing network behavior.

1.3.2. The Kirchhoff Current Law (KCL)

Electric charges arriving at a network node cannot be stored there because, as an ideal entity, a node cannot have capacitive storage properties. Furthermore, a node cannot act as a source or sink for electric charges. We can therefore state a principle of continuity or charge conservation: At any time the number of charges arriving at the node must equal the number leaving. This principle is stated in terms of instantaneous branch currents by

> Kirchhoff's Current Law (KCL)
>
> The sum of the branch currents leaving *any* node must equal the sum of the branch currents entering that node at any time.
>
> Or, alternatively, the sum of all branch currents incident at *any* node is zero at any time.

In the latter version of the Kirchhoff current law (henceforth abbreviated KCL), it is customary to count as *positive* all currents *leaving* the node and as negative all currents *entering* the node.

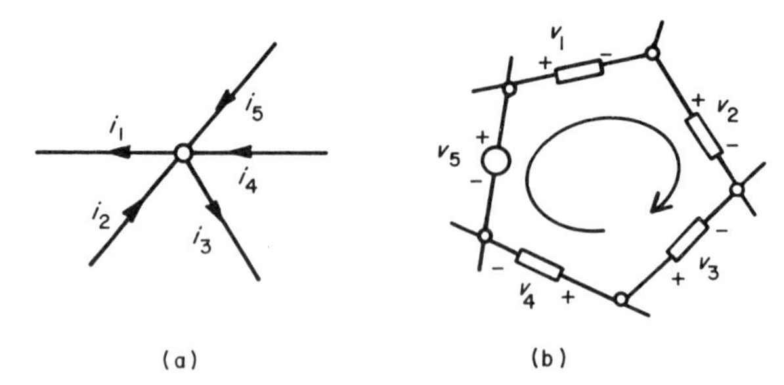

Figure 1.19 **(a)** Branch currents for the formulation of Kirchhoff's current law (KCL) at a common node. **(b)** Branch voltages for the formulation of Kirchhoff's voltage law (KVL) in a common loop.

Applying the KCL to the example in Fig. 1.19a we obtain

(i_1+i_3)	$=$	$(i_2+i_4+i_5)$
currents leaving the node		currents entering the node

or

$$(i_1+i_3-i_2-i_4-i_5)=0$$

sum of currents incident at the node

For the case where the currents in Fig. 1.19a are sinusoidal currents, the KCL postulates

$$\mathbf{I}_1 e^{j\omega t}+\mathbf{I}_3 e^{j\omega t}-\mathbf{I}_2 e^{j\omega t}-\mathbf{I}_4 e^{j\omega t}-\mathbf{I}_5 e^{j\omega t}=0$$

And after cancellation of the common exponential factor the equation reduces to the phasor sum

$$\mathbf{I}_1+\mathbf{I}_3-\mathbf{I}_2-\mathbf{I}_4-\mathbf{I}_5=0$$

From the foregoing example we can draw the conclusion that *the Kirchhoff current law applies also to current phasors*. This is a very important point, because from Section 1.2 we know that the phasor method reduces the element laws containing differential quotients and integrals to simple algebraic equations. Ultimately, this means that the network equations themselves will be algebraic.

It is necessary here to remind ourselves that a current phasor contains both the amplitude and the phase of sinusoidal current. Thus, in general, the KCL does *not* apply to *amplitudes* alone. Only in the case of a network where no phase shifts are possible would the amplitudes satisfy the KCL. However, it is possible to represent each current phasor as the sum of its real and imaginary parts. *The sums of the real parts and the sums of the imaginary parts satisfy the KCL separately.*

Returning to the example in Fig. 1.19a, we can restate the phasor version of the KCL as follows.

$$\begin{aligned}\mathrm{Re}(\mathbf{I}_1)+j\mathrm{Im}(\mathbf{I}_1)+\mathrm{Re}(\mathbf{I}_3)+j\mathrm{Im}(\mathbf{I}_3)-\mathrm{Re}(\mathbf{I}_2)-j\mathrm{Im}(\mathbf{I}_2)\\ -\mathrm{Re}(\mathbf{I}_4)-j\mathrm{Im}(\mathbf{I}_4)-\mathrm{Re}(\mathbf{I}_5)-j\mathrm{Im}(\mathbf{I}_5)=0\end{aligned}$$

Collecting like terms we obtain

$$\mathrm{Re}(\mathbf{I}_1)+\mathrm{Re}(\mathbf{I}_3)-\mathrm{Re}(\mathbf{I}_2)-\mathrm{Re}(\mathbf{I}_4)-\mathrm{Re}(\mathbf{I}_5)=0$$

and

$$\mathrm{Im}(\mathbf{I}_1)+\mathrm{Im}(\mathbf{I}_3)-\mathrm{Im}(\mathbf{I}_2)-\mathrm{Im}(\mathbf{I}_4)-\mathrm{Im}(\mathbf{I}_5)=0$$

as asserted earlier.

The KCL is of extreme generality. The only restriction we wish to place on its application is that the network consist of spatially concentrated or *lumped* elements. Otherwise, it is entirely immaterial whether the branch elements are dissipators, energy stores, or sources; they may even be time varying and nonlinear. In fact, the KCL can be applied to other, nonelectric systems where the corresponding continuity or equilibrium equations can be established for "through" variables (e.g., forces, heat fluxes, hydraulic flows) analogous to current.

1.3.3 The Kirchhoff Voltage Law (KVL)

The dual postulate to the KCL is the Kirchhoff voltage law (henceforth abbreviated KVL) for branch voltages. If we consider the node voltages of the nodes in a closed loop, we realize that they represent the potential energy levels which a unit charge "circulating" through the closed loop would encounter. The *branch* voltages represent the *voltage* (potential) *drops* or *rises* the unit charge would experience on this journey. It is customary to count a branch voltage equivalent to a *voltage drop* as *positive*, because the associated voltage polarity for a passive element is positive. A *negative* sign is associated with a voltage rise.

The rule is easily demonstrated with the example in Fig. 1.19b, where in addition to the associated reference directions of the branches a circular arrow is drawn inside the loop. It represents the path of the unit charge traversing the loop. Later we shall see that it is advantageous to think of this arrow as the *loop* or *mesh current*. It should always be drawn in whenever we wish to analyze a network loop or mesh. It provides the reference direction for classifying the branch voltages either as voltage drops $(+)$ or rises $(-)$. Whenever the loop current and the branch current of a *passive* element have the same direction the associated voltage polarity is positive and physically a voltage drop occurs (e.g., v_1 in Fig. 1.19b). The opposite is true for *active* elements; coincident directions for loop and branch currents amount to a potential rise and the corresponding voltage polarity is negative because the polarities of the current–voltage pair of a potential source are not associated (e.g., v_5 in Fig. 1.19a). In its simplest form the rule amounts to this: *Whenever the loop or mesh current "sees" the plus sign first, the branch voltage is considered positive (voltage drop); if the loop or mesh current "sees" the minus sign first, the branch voltage is negative (voltage rise).*

Clearly, for the loop current shown, the branch voltages in Fig. 1.19b are classified as follows: v_1, v_2, and v_4 are voltage drops $(+)$; v_3 and v_5 are voltage rises $(-)$. Of course, had we chosen a loop current of opposite orientation, the classification of the branch voltages would be the reverse.

The reference directions following the broken lines in Figs. 1.18a and 1.18b would be appropriate pathways for loop currents. The loop currents in Fig. 1.18a are also mesh currents because they do not cut across any branches. The distinction is important for analytical techniques discussed later on.

As a unit charge traverses the closed loop once and returns to the node it started from, the net energy gain will be zero as it returns to the same node potential. It is thus intuitively obvious that the voltage drops and rises encountered during one complete traversal of the loop must cancel each other. This conclusion is generally true for lumped networks and is stated as

> Kirchhoff's Voltage Law (KVL)
>
> The sum of the branch voltages for any closed loop is zero at any time.
>
> Or, alternatively, for any closed loop, the sum of the voltage drops equals the sum of the voltage rises at any time.

For the loop in Fig. 1.19b, KVL leads to these equations:

$$(v_1 + v_2 - v_3 + v_4 - v_5) = 0$$

sum of the branch voltages

or

$$(v_1 + v_2 + v_4) = (v_3 + v_5)$$

sum of the voltage drops	sum of the voltage rises

The KVL is also true for branch voltage phasors and their real and imaginary components. In the case of our example we have

$$\mathbf{V}_1 e^{j\omega t} + \mathbf{V}_2 e^{j\omega t} - \mathbf{V}_3 e^{j\omega t} + \mathbf{V}_4 e^{j\omega t} - \mathbf{V}_5 e^{j\omega t} = 0$$

from which follows

$$\mathbf{V}_1 + \mathbf{V}_2 - \mathbf{V}_3 + \mathbf{V}_4 - \mathbf{V}_5 = 0$$

$$\mathrm{Re}(\mathbf{V}_1) + \mathrm{Re}(\mathbf{V}_2) - \mathrm{Re}(\mathbf{V}_3) + \mathrm{Re}(\mathbf{V}_4) - \mathrm{Re}(\mathbf{V}_5) = 0$$

and

$$\mathrm{Im}(\mathbf{V}_1) + \mathrm{Im}(\mathbf{V}_2) - \mathrm{Im}(\mathbf{V}_3) + \mathrm{Im}(\mathbf{V}_4) - \mathrm{Im}(\mathbf{V}_5) = 0$$

Like the current law, the Kirchhoff voltage law is of great generality. It only requires that the network consist of lumped elements. The physical nature of the branch elements is of no consequence; they need *not* be linear or time invariant. For nonelectric networks, the KVL can be applied to "across" variables (e.g., velocity, temperature difference, pressure head difference) which correspond to voltage.

1.3.4 Sufficiency of Kirchhoff Laws; Number of Independent Equations

For the complete description of network behavior we need to know the branch currents and branch voltages at all times. Consider a network with N nodes (for simplicity assume all simple nodes removed by the insertion of equivalent branches as discussed for Figs. 1.17a and 1.17b) and B branches. Altogether we have then B branch currents and B branch voltages, or a total of $2B$ unknowns.

If we are given either a branch current or a branch voltage, we can always use the appropriate element laws to compute one from the other, i.e., we can write B equations based on the branch element laws alone. We need another set of B equations in order to determine the $2B$ unknowns. This second set of equations we get from the application of the Kirchhoff laws.

After designating one node as the reference node, we are left with $N-1$ independent nodes. For each of these we can write $N-1$ independent KCL equations. This leaves $B-(N-1)$ equations yet to be found.

All nodes can be connected by a tree containing $N-1$ tree branches. The number of links removed to generate the tree is $L=B-(N-1)$, which equals the number of equations needed. As we reinsert the links one by one, we generate an independent loop for which we can write a KVL equation. Hence we establish a set of $B-(N-1)$ independent equations based on KVL. So we have altogether a sufficient set of independent equations to solve for the unknown branch variables.

We should point out, that the preceding "proof" ignores a variety of complications, some of which arise from the presence of independent sources in a network.

Actually, through the *implicit* application of the Kirchhoff and element laws it is possible to characterize a network with fewer than B equations. Later sections will make this clear.

1.3.5 Series Combinations of Passive Elements; Complex Impedances

Networks containing many nodes and loops can often be reduced to much simpler networks through the elimination of simple nodes (representing series combinations of passive elements) and of meshes with only two nodes (representing parallel combinations of passive elements). For this purpose we must study the rules for series and parallel combinations of passive elements. First, we apply the Kirchhoff laws to series combinations.

The simple circuit in Fig. 1.20a has one mesh and three simple nodes (A, B, C).

Figure 1.20 Series combinations of linear resistors. (See text.)

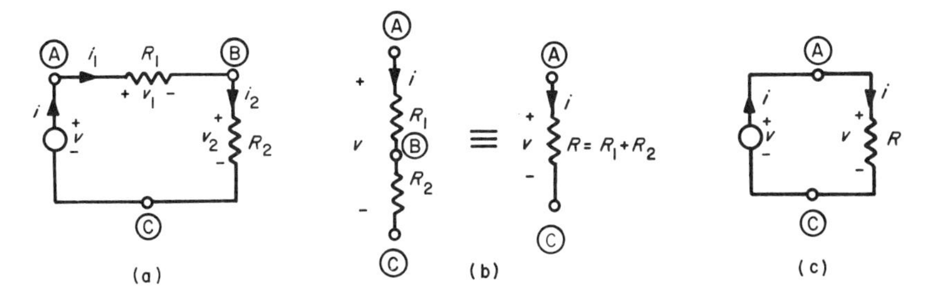

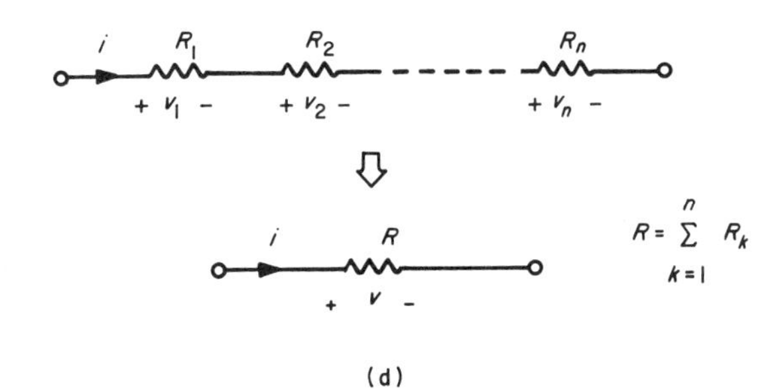

Node C could serve as the reference node, though in this particular case that is not an important consideration. Node B serves as the junction between two linear resistors connected in series; this is the node we wish to eliminate.

From KCL follows

$$i_1 = i_2 = i \tag{1.69}$$

From KVL follows

$$v_1 + v_2 - v = 0$$

or

$$v = v_1 + v_2 \tag{1.70}$$

The branch currents and branch voltages for the two elements are related by Ohm's law, hence

$$v_1 = R_1 i_1 \tag{1.71}$$

and

$$v_2 = R_2 i_2 \tag{1.72}$$

Combining Eqs. (1.69)–(1.72) we obtain

$$v = v_1 + v_2 = (R_1 + R_2)i = Ri$$

or

$$R = R_1 + R_2 \tag{1.73}$$

Equation (1.73) states that the series combination of two *linear* resistors is equivalent to a resistor representing the *sum* of the two resistances (Fig. 1.20b). Inserting the equivalent resistor into the original network leads to the network with only two nodes in Fig. 1.20c.

The summation rule for series resistors can be extended to an arbitrary number of *linear* resistors.

> The total resistance of a *series* combination of *n linear* resistors equals the sum of the resistances:
>
> $$R_{total} = R_1 + R_2 + \cdots + R_n \tag{1.74}$$

This summation rule for resistances can be generalized to impedances by using phasors in the Kirchhoff and element laws and replacing the resistance terms with impedance terms:

> The total impedance of *n linear* elements (or branches) in *series* is the sum of the individual impedances:
>
> $$Z(s)_{total} = Z_1(s) + Z_2(s) + \cdots + Z_n(s) \tag{1.75}$$
>
> and
>
> $$Z(j\omega)_{total} = Z_1(j\omega) + Z_2(j\omega) + \cdots + Z_n(j\omega) \tag{1.76}$$

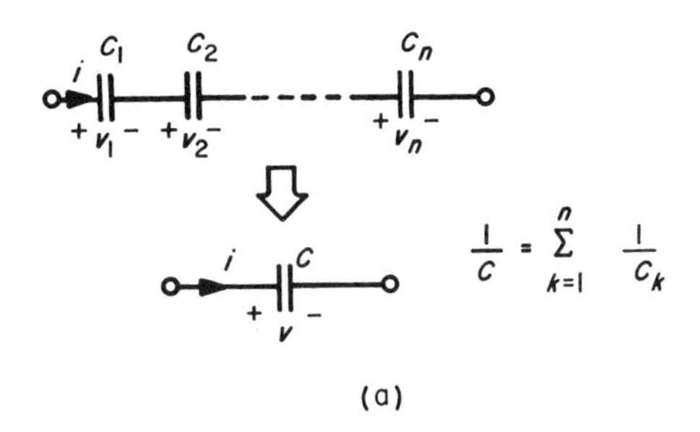

Figure 1.21 Series combinations of linear capacitors and inductors. (See text.)

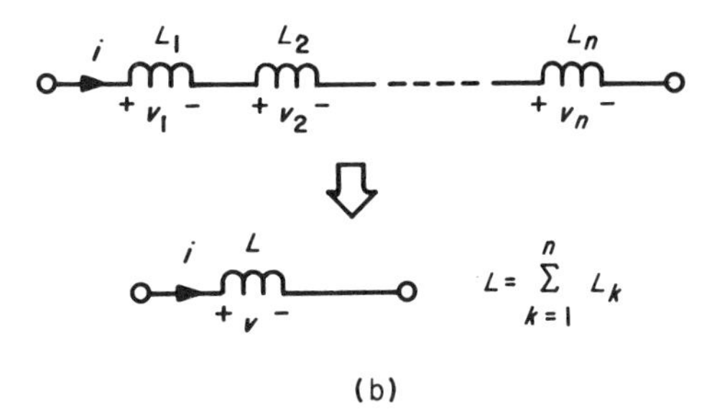

The application of these summation rules to the series combinations in the next three examples will demonstrate that (1) the total capacitance of a series combination of capacitors is less than that of any individual element because the summation extends over *reciprocal* capacitances; (2) for inductors, the summation is analogous to resistors; and (3) a total impedance can be defined for mixed series combinations of resistors and storage elements.

Example 1.4. Series combination of capacitors

(See Fig. 1.21a.) According to Section 1.2.3.2, the impedance of an individual capacitor C_i is $Z_i = 1/sC_i$. The total impedance is then, by Eq. (1.75),

$$Z(s)_{\text{total}} = \frac{1}{sC_1} + \frac{1}{sC_2} + \cdots + \frac{1}{sC_n} = \frac{1}{s} \sum_{i=1}^{n} \frac{1}{C_i} = \frac{1}{sC_{\text{total}}}$$

From this expression follows immediately the *series summation rule for reciprocal capacitances*:

$$\frac{1}{C_{\text{total}}} = \sum_{i=1}^{n} \frac{1}{C_i} \tag{1.77}$$

Because of this relation the total capacitance of series-connected capacitors is always *less* than any individual capacitance.

Example 1.5. Series combination of inductors

(See Fig. 1.21b.) The impedance of an inductor L_i is $Z_i = sL_i$. Equation (1.75) gives the total impedance as

$$Z(s)_{\text{total}} = sL_1 + sL_2 + \cdots + sL_n = s \sum_{i=1}^{n} L_i = sL_{\text{total}}$$

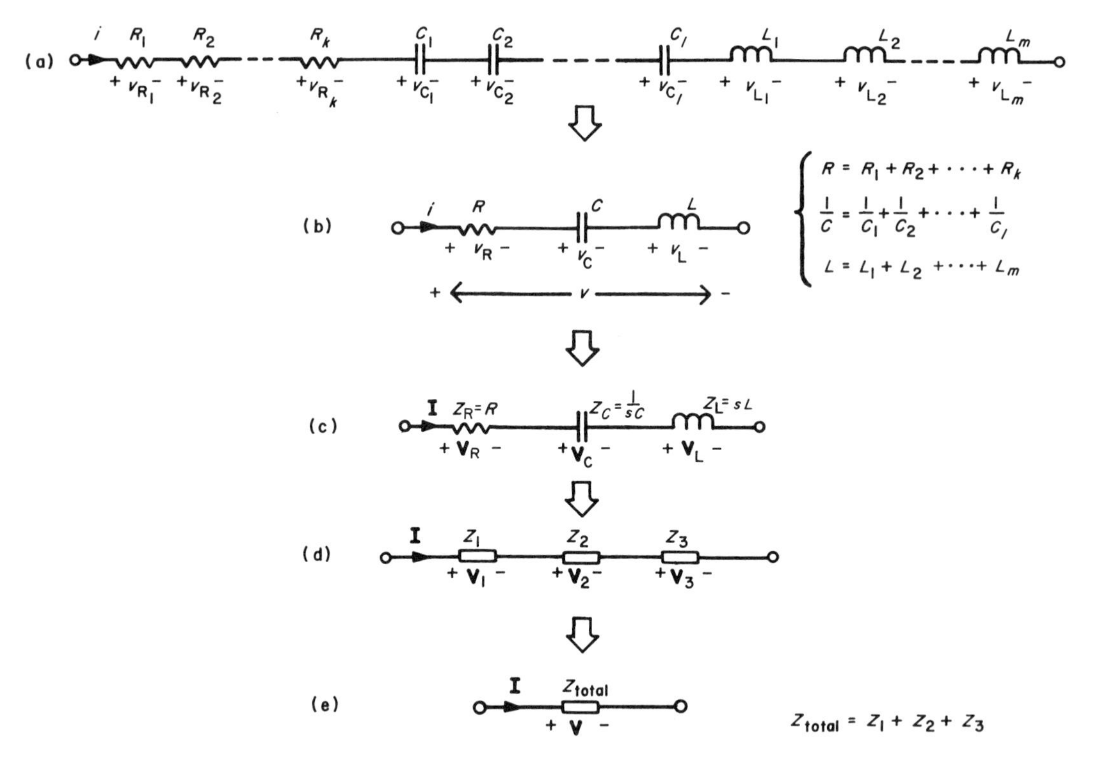

Figure 1.22 Generalized passive element branch consisting of the series combination of an equivalent resistor, capacitor, and inductor. (See text.)

The *series summation rule for inductances* is, therefore,

$$L_{\text{total}} = \sum_{i=1}^{n} L_i \tag{1.78}$$

Example 1.6. General multielement branch

(See Fig. 1.22a.) We use the series combination rules for resistances, capacitances, and inductances in order to determine the equivalent R, C, and L in Fig. 1.22b. The corresponding impedances for this three-element branch are indicated in Fig. 1.22c. According to Eq. (1.75) we can combine these impedances into a simple equivalent impedance (Figs. 1.22c and 1.22d):

$$Z_{\text{total}}(s) = Z_1 + Z_2 + Z_3 = R + \frac{1}{sC} + sL \tag{1.79}$$

The same result is obtained by changing from time domain quantities (Fig. 1.22b) to phasors (Fig. 1.22c) and applying KVL:

$$\mathbf{V} = \mathbf{V}_{\text{R}} + \mathbf{V}_{\text{C}} + \mathbf{V}_{\text{L}} = R\mathbf{I} + \frac{\mathbf{I}}{sC} + sL\mathbf{I} = \left(R + \frac{1}{sC} + sL\right)\mathbf{I} \tag{1.80}$$

and

$$Z_{\text{total}} = \frac{\mathbf{V}}{\mathbf{I}} = R + \frac{1}{sC} + sL$$

Equation (1.80) can be treated as the "element" law of a generalized series branch. It is the phasor form of the following integrodifferential equation which relates the branch voltage to the branch current, namely,

$$v = v_R + v_C + v_L = Ri + \frac{1}{C}\int_0^t i\,dt + L\frac{di}{dt} \tag{1.81}$$

The total impedance for combinations of passive elements generally is a complex quantity. For an ac impedance we can write

$$Z(j\omega) = R + jX \tag{1.82}$$

$$|Z(j\omega)| = (R^2 + X^2)^{1/2} \tag{1.83}$$

$$\sphericalangle Z(j\omega) = \tan^{-1}\left(\frac{X}{R}\right) = \phi \tag{1.84}$$

$$R = \mathrm{Re}(Z) = |Z|\cos\phi \tag{1.85}$$

$$X = \mathrm{Im}(Z) = |Z|\sin\phi \tag{1.86}$$

These equations, describing impedance as a complex number, complement the polar coordinate approach used in Eqs. (1.48)–(1.50).

The imaginary part of the impedance, X, is called the *reactance*; it represents the *storage* properties of the element(s) represented by the impedance. The real part, R, represents the *dissipative* properties. *In general, both the real and the imaginary part of an impedance may be frequency dependent.* From Eqs. (1.83) and (1.84) it is obvious that *magnitude and phase angle of an impedance are generally frequency dependent* too.

The properties of an impedance representing dissipation and energy storage will be explained next by way of a simple circuit.

Example 1.7.

The *R–L series combination* in Fig. 1.23a could be taken as the equivalent circuit of a practical air coil inductor with R representing the resistance of the coil wire.

The total ac impedance of this circuit is

$$Z = R + j\omega L = R\left(1 + j\omega\frac{L}{R}\right) \tag{1.87}$$

In Eq. (1.87), the ratio R/L is called the *characteristic frequency*, ω_0. With this new parameter, the equation reduces to

$$Z = R\left(1 + j\frac{\omega}{\omega_0}\right)$$

We can now define a normalized impedance z, which is dimensionless and contains only the parameter ω_0:

$$z = \frac{Z}{R} = 1 + j(\omega/\omega_0) \tag{1.88}$$

(a) R, L $\equiv$ Z

$Z = R + j\omega L$

$z = \frac{Z}{R} = (1 + j\omega\frac{L}{R})$

$\omega_0 = \frac{R}{L}$

(b)

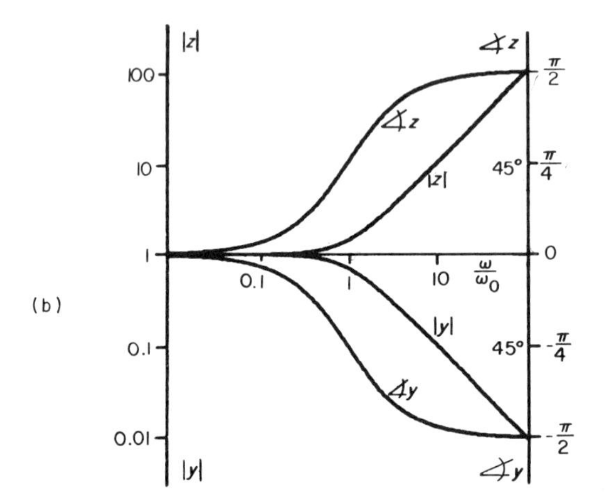

Figure 1.23 Impedance and admittance properties of the series $R-L$ combination. (See text.)

For the magnitude and phase of this normalized impedance we obtain from Eq. (1.88)

$$|z| = \left[1 + \left(\frac{\omega}{\omega_0}\right)^2\right]^{1/2}$$

$$\sphericalangle z = \tan^{-1}[\text{Im}(z)/\text{Re}(z)] = \tan^{-1}(\omega/\omega_0)$$

The magnitude and phase values for these and other frequencies are plotted with *logarithmic* scales in the *magnitude–frequency* and *phase–frequency diagram* in the upper half of Fig. 1.23b. The influence of the parasitic wire resistance is evident. We note the following:

1. At frequencies much smaller than the characteristic frequency ω_0 $(= R/L)$, the *ohmic* properties are dominant because magnitude and phase are very close to the values for the pure resistor R.
2. At frequencies much larger than ω_0, the inductor dominates, i.e., the magnitude rises linearly with frequency, and the phase asymptotically approaches 90°, the value for an ideal inductor. (We have neglected any parasitic capacitance.)

It is instructive to consider the *admittance* of this simple $R-L$ combination too. From Eq. (1.88) we write for the *normalized* admittance

$$y = \frac{1}{z} = \frac{R}{Z} = \frac{Y}{G} = \frac{1}{1 + j(\omega/\omega_0)} \tag{1.89}$$

For magnitude and phase we have then

$$\log|y| = \log\left|\frac{1}{z}\right| = -\log|z|$$

and

$$\measuredangle y = \measuredangle\left(\frac{1}{z}\right) = -\measuredangle z$$

These quantities are plotted in the lower half of Fig. 1.23b.

1.3.6 Parallel Combinations of Passive Elements; Complex Admittances

We start with the simple circuit of Fig. 1.24a. We may treat node B as the reference node. Applying KCL to node A gives us

$$i_1 + i_2 - i = 0$$

or

$$i = i_1 + i_2 \tag{1.90}$$

For the two meshes we have by KVL

$$v_1 - v = 0$$

and

$$v_2 - v_1 = 0$$

or

$$v = v_1 = v_2 \tag{1.91}$$

Figure 1.24 Parallel combinations of linear resistors. (See text.)

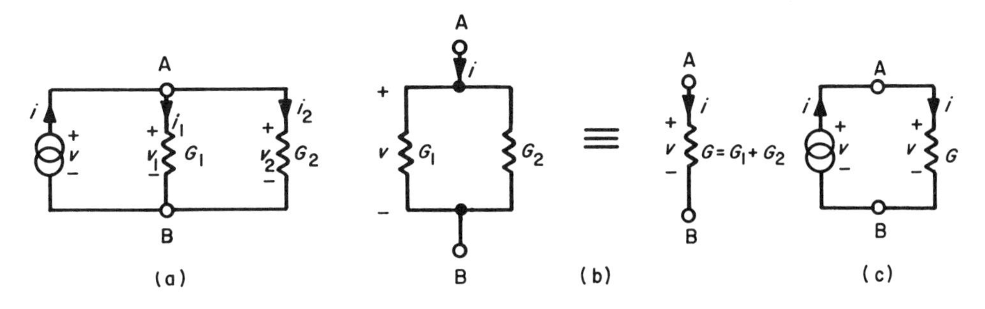

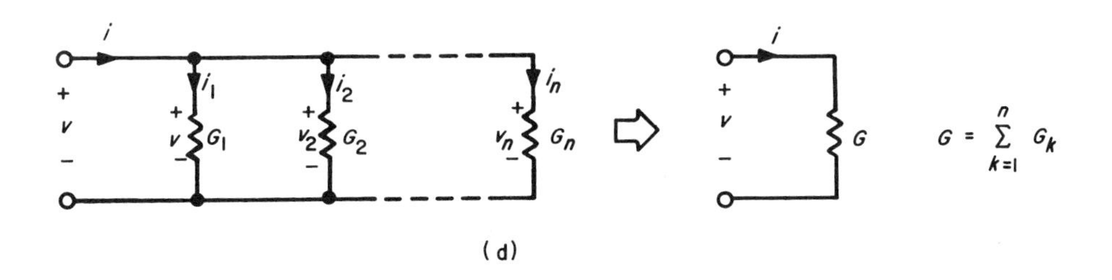

The element laws of the passive branch elements are

$$i_1 = G_1 v_1 \tag{1.92}$$

and

$$i_2 = G_2 v_2 \tag{1.93}$$

Substituting Eqs. (1.92) and (1.93) into (1.90), we obtain with (1.91)

$$i = G_1 v + G_2 v = (G_1 + G_2) v = Gv$$

or

$$G = G_1 + G_2 \tag{1.94}$$

Equation (1.94) states that the total *conductance* of the parallel combination of two *linear* resistors equals the sum of the individual conductances.

Instead of conductance parameters we could use resistance parameters in Eq. (1.94):

$$\frac{1}{R} = \frac{1}{R_1} + \frac{1}{R_2} = \frac{R_1 + R_2}{R_1 R_2}$$

or

$$R = \frac{R_1 R_2}{R_1 + R_2} \tag{1.95}$$

Though more complicated than Eq. (1.94), Eq. (1.95) is more commonly used if only two parallel elements have to be considered.

From Eq. (1.94) it is obvious that the total *conductance* of a parallel combination is always *more* than any of the individual conductances. Conversely, the total *resistance* of a parallel arrangement is always *less* than any of the individual resistances.

On the basis of Eq. (1.94) we can eliminate the mesh formed by the two resistors in Fig. 1.24a and replace it by a single equivalent branch (Fig. 1.24b). The resulting network has only one mesh with a single passive branch (Fig. 1.24c).

In the general case, there will be n resistors branches forming $n-1$ meshes which share a single node pair. (It is not necessary that one node be the reference node.) The n resistors and $n-1$ meshes can be replaced by a single equivalent branch representing the sum of the individual conductances as shown in Fig. 1.24d:

> The total conductance of a *parallel* combination of n *linear* resistors equals the sum of the conductances:
>
> $$G_{\text{total}} = G_1 + G_2 + \cdots + G_n \tag{1.96}$$
>
> or
>
> $$\frac{1}{R_{\text{total}}} = \frac{1}{R_1} + \frac{1}{R_2} + \cdots + \frac{1}{R_n} \tag{1.97}$$

We can generalize the parallel combination rule for conductances to admittances by using phasor quantities in the Kirchhoff and element laws:

The total admittance of *n linear* elements (or branches) in *parallel* equals the sum of the individual admittances:

$$Y(s)_{\text{total}} = Y_1(s) + Y_2(s) + \cdots + Y_n(s) \tag{1.98}$$

and

$$Y(j\omega)_{\text{total}} = Y_1(j\omega) + Y_2(j\omega) + \cdots + Y_n(j\omega) \tag{1.99}$$

Let us apply this rule to the parallel combination of capacitors, inductors, and mixed elements.

Example 1.8. Parallel combination of capacitors

(See Fig. 1.25a.) With the admittance of an individual capacitor being $Y_i(s) = sC_i$, we obtain for the total admittance

$$Y_{\text{total}}(s) = \sum_{i=1}^{n} sC_i = s\sum_{i=1}^{n} C_i = sC_{\text{total}} \tag{1.100}$$

The *parallel summation rule for capacitances* (Fig. 1.25a) follows directly from Eq. (1.100):

$$C_{\text{total}} = \sum_{i=1}^{n} C_i \tag{1.101}$$

Figure 1.25 Parallel combinations of linear capacitors and inductors. (See text.)

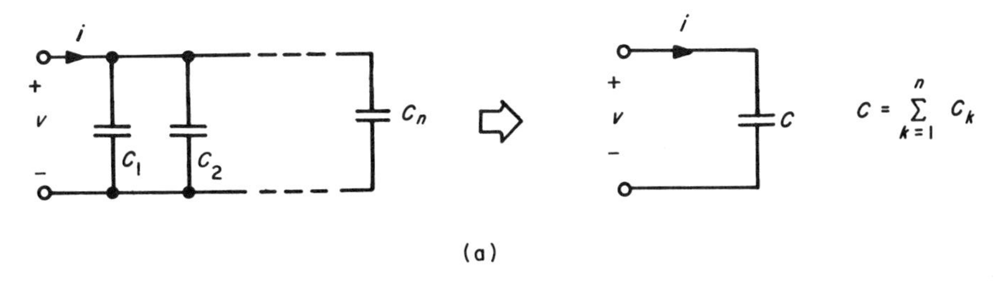

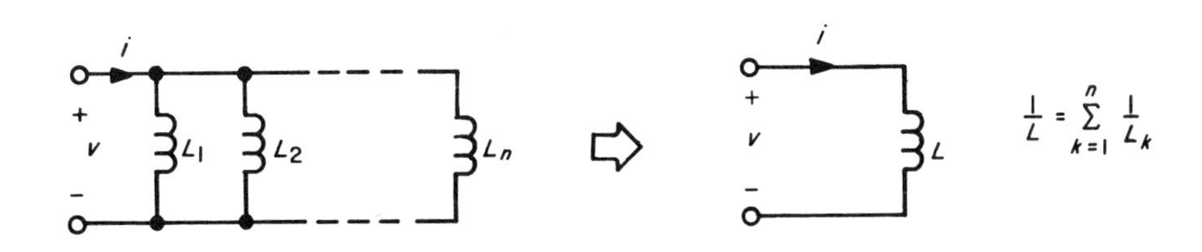

Example 1.9. Parallel combination of inductors

(See Fig. 1.25b.) An individual inductor admittance is $Y_i = 1/sL_i$. From Eq. (1.98) we have for the total admittance

$$Y_{\text{total}}(s) = \sum_{i=1}^{n} \frac{1}{sL_i} = \frac{1}{s} \sum_{i=1}^{n} \frac{1}{L_i} = \frac{1}{sL_{\text{total}}} \tag{1.102}$$

The *parallel summation rule for reciprocal inductances* is, therefore, from Eq. (1.102)

$$\frac{1}{L_{\text{total}}} = \sum_{i=1}^{n} \frac{1}{L_i} \tag{1.103}$$

Example 1.10. Generalized parallel branches

(See Fig. 1.26a.) According to the parallel combination rules for conductances, capacitances, and inductances, the multibranch circuit of Fig. 1.26a can be reduced to the three-branch circuit of Fig. 1.26b. A change to phasor notation (Fig. 1.26c) allows us to combine the admittances of the three equivalent branches into a single generalized admittance (Figs. 1.26d and 1.26e) according to Eq. (1.98):

$$Y_{\text{total}}(s) = Y_1 + Y_2 + Y_3 = G + sC + \frac{1}{sL} \tag{1.104}$$

If we combine this expression with the phasor form of KCL for this circuit, we obtain the "element" law of this generalized three-branch circuit in phasor terms:

$$\mathbf{I} = \mathbf{I}_G + \mathbf{I}_C + \mathbf{I}_L = G\mathbf{V} + sC\mathbf{V} + \frac{\mathbf{V}}{sL} = \left(G + sC + \frac{1}{sL}\right)\mathbf{V} \tag{1.105}$$

Figure 1.26 Generalized passive element branch consisting of the parallel combination of an equivalent resistor, capacitor, and inductor. (See text.)

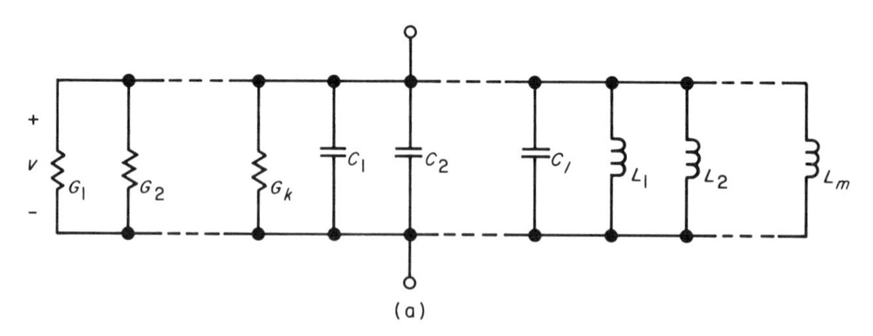

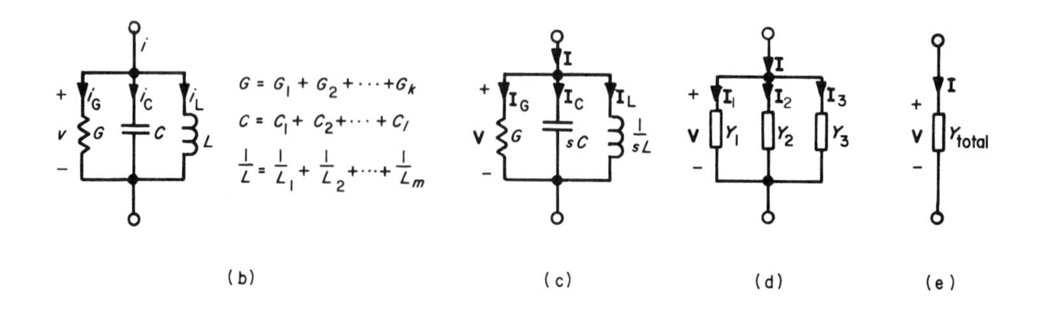

In the time domain this is equivalent to the integrodifferential equation

$$i = i_G + i_C + i_L = Gv + C\frac{dv}{dt} + \frac{1}{L}\int_0^t v\,dt \tag{1.106}$$

The admittances derived from the consolidation of network branches are in general complex numbers. By analogy with Eqs. (1.82)–(1.86) we can write

$$Y(j\omega) = G + jB \tag{1.107}$$

$$|Y(j\omega)| = [G^2 + B^2]^{1/2} \tag{1.108}$$

$$\measuredangle Y(j\omega) = \tan^{-1}\left(\frac{B}{G}\right) = \psi \tag{1.109}$$

$$G = \mathrm{Re}(Y) = |Y|\cos\psi \tag{1.110}$$

$$B = \mathrm{Im}(Y) = |Y|\sin\psi \tag{1.111}$$

Equations (1.107)–(1.111) should be compared to Eqs. (1.51)–(1.54). The real part of the admittance, G, representing the *dissipative* properties of the equivalent one-port, is *generally frequency dependent*. The imaginary part of the admittance, B, is called the *susceptance*. It represents the *storage* properties of the equivalent one-port and is *frequency dependent*. Because of the frequency dependence of the real and imaginary parts, *magnitude and phase of an admittance should generally be frequency dependent, too*. The following example demonstrates this.

Example 1.11. R–C parallel combinations

In this specific instance we may consider the combination to represent a *practical* capacitor with the resistor standing for the losses in the nonideal dielectric.

The total ac admittance of this circuit must be

$$Y = G + j\omega C = G\left(1 + j\omega\frac{C}{G}\right) \tag{1.112}$$

The ratio G/C $(=1/RC)$ is defined as the *characteristic frequency* ω_0 of the circuit. (The reciprocal, RC, is the time constant T.) With the new parameter ω_0, Eq. (1.112) becomes

$$Y = G\left(1 + j\frac{\omega}{\omega_0}\right)$$

Normalizing this expression, we obtain the dimensionless *normalized admittance* y, which contains only the parameter ω_0:

$$y = \frac{Y}{G} = 1 + (j\omega/\omega_0) \tag{1.113}$$

Equation (1.113) is on the right side exactly the same as Eq. (1.88) for the impedance of the $R-L$ series combination. The frequency response plot of Eq. (1.113) is, therefore, identical to that in Fig. 1.23, *except* we must relabel all axes from z to y.

The *normalized impedance* for the $R-C$ parallel combination is then

$$z = \frac{1}{y} = \frac{1}{1 + j(\omega/\omega_0)} \tag{1.114}$$

The right side of Eq. (1.114) is analogous to that of Eq. (1.89) for the admittance of the $R-L$ series circuit.

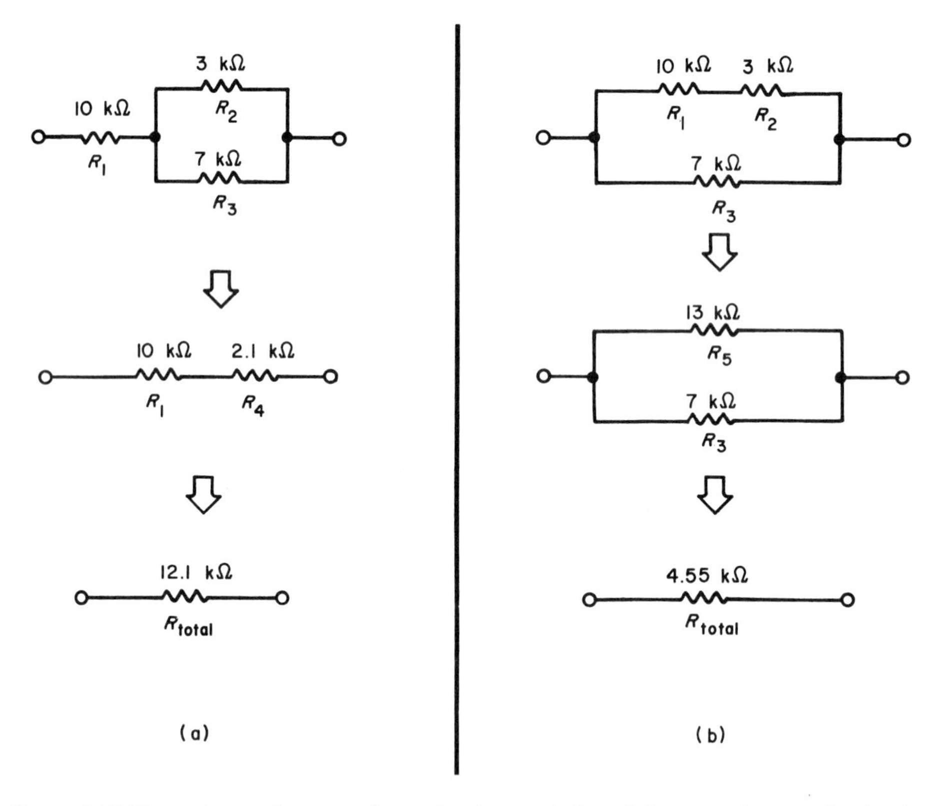

Figure 1.27 Equivalent resistances of two circuits consisting of the same elements but having different topology.

In general, circuits will have both parallel and series arrangements within them. The next three examples illustrate this. The first two examples also emphasize the importance of circuit topology. The third example simply demonstrates how the series–parallel combination rules can be used to reduce a fairly complicated circuit to a simple equivalent one-port.

Example 1.12.

Derive the equivalent resistance R_{total} of the series–parallel arrangement in Fig. 1.27a.

Applying Eq. (1.95) to R_2 and R_3, we can replace this parallel circuit with R_4 (the vertical double bar indicates parallel arrangement):

$$R_4 = R_2 \| R_3 = \frac{R_2 R_3}{R_2 + R_3} = \frac{21}{10}\,\text{k}\Omega = 2.1\,\text{k}\Omega$$

For the series arrangement R_1–R_4 we obtain from Eq. (1.74)

$$R_{\text{total}} = R_1 + R_4 = (10 + 2.1)\,\text{k}\Omega = 12.1\,\text{k}\Omega$$

Example 1.13.

Derive the equivalent resistance for the circuit in Fig. 1.27b.

According to Eq. (1.74), we replace the series circuit $R_1 - R_2$ with the equivalent resistor R_5:

$$R_5 = R_1 + R_2 = (10+3)\,\mathrm{k}\Omega = 13\,\mathrm{k}\Omega$$

The remaining parallel circuit has, with Eq. (1.95), the total resistance

$$R_{\text{total}} = \frac{R_3 R_5}{R_3 + R_5} = \frac{91}{20}\,\mathrm{k}\Omega = 4.55\,\mathrm{k}\Omega$$

Note that the *equivalent* or *total* resistances of the circuits in Figs. 1.27a and 1.27b *differ* from each other, even though the circuits are constructed of the same basic elements. The difference in circuit topology is the reason for this.

Example 1.14.

The network in Fig. 1.28a has an externally accessible node pair A–B. We treat the whole circuit as a one-port. The problem is to determine the $i - v$ relation at the terminal pair A–B.

In order to obtain the solution we choose the impedance/admittance method of eliminating series and parallel connections under the assumption that the elements are linear and all initial conditions are zero.

Figure 1.28 Reduction of a passive circuit to a single equivalent impedance by application of the rules for series and parallel combinations.

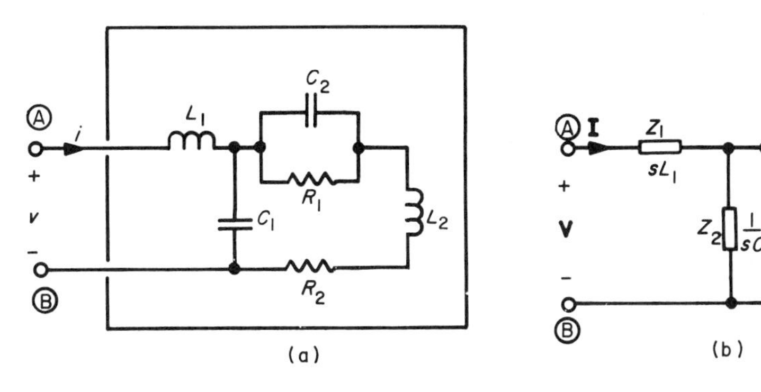

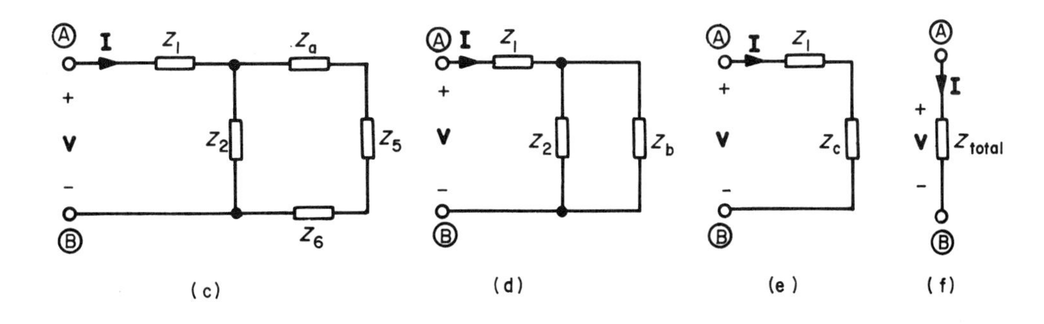

The first step is to transform the circuit into its phasor–impedance form (Fig. 1.28b). (The admittance form would serve equally well.) Next let us eliminate the parallel combination of Z_3 and Z_4. Choosing the *admittance* form, we obtain from Eq. (1.98)

$$Y_a = Y_3 + Y_4 = G_1 + sC_2 = \frac{1}{R_1} + sC_2 = \frac{1 + sR_1C_2}{R_1}$$

or

$$Z_a = \frac{1}{Y_a} = \frac{R_1}{1 + sR_1C_2}$$

The equivalent impedance, Z_a, is then inserted in place of the parallel combination as shown in Fig. 1.28c. We now see that Z_a, Z_5, and Z_6 form a series circuit which can be replaced by the equivalent impedance Z_b according to Eq. (1.75):

$$Z_b = Z_a + Z_5 + Z_6 = \frac{R_1}{1 + sR_1C_2} + sL_2 + R_2$$

$$= \frac{(R_1 + R_2) + s(R_1R_2C_2 + L_2) + s^2R_1C_2L_2}{1 + sR_1C_2}$$

or

$$Z_b = \frac{k_1 + sk_2 + sk_3}{1 + sk_4}$$

where the coefficients k_1 to k_4 are used to simplify the expression. With this new equivalent impedance the circuit takes on the form in Fig. 1.28d.

The next obvious step is to replace the parallel connection of Z_2 and Z_b with an equivalent impedance Z_c. In terms of admittances we can write

$$Y_c = Y_2 + Y_b = sC_1 + \frac{1 + sk_4}{k_1 + sk_2 + s^2k_3}$$

$$= \frac{1 + s(k_4 + k_1C_1) + s^2k_2C_1 + s^3k_3C_1}{k_1 + sk_2 + s^2k_3}$$

and with new coefficients we have

$$Y_c = \frac{1 + sk_5 + s^2k_6 + s^3k_7}{k_1 + sk_2 + s^2k_3}$$

or

$$Z_c = \frac{1}{Y_c} = \frac{k_1 + sk_2 + s^2k_3}{1 + sk_5 + s^2k_6 + s^3k_7}$$

With Z_c in place, we are left with one more series combination, Z_1 and Z_c (Fig. 1.28e), which can be reduced to the total driving point impedance of the one-port (Fig. 1.28f):

$$Z_{\text{total}} = Z_1 + Z_c = sL_1 + \frac{k_1 + sk_2 + s^2k_3}{1 + sk_5 + s^2k_6 + s^3k_7}$$

With a new set of coefficients, the last equation can be brought into the form

$$Z(s)_{\text{total}} = \frac{\mathbf{V}}{\mathbf{I}} = \frac{a_0 + a_1s + a_2s^2 + a_3s^3 + a_4s^4}{b_0 + b_1s + b_2s^2 + b_3s^3} \tag{1.115}$$

Equation (1.115) represents already the solution of the problem in terms of the generalized phasor voltage–current pair. For *ac* phasors, the ac driving point impedance is obtained by replacing s with $j\omega$.

1.3.7 Nonideal Sources: Combinations of Active and Passive Elements

An ideal voltage source provides a specified voltage at its terminals, regardless of the magnitude and polarity of the current drawn from it. Because of this, the characteristic of an ideal constant voltage source is a line parallel to the current axis in Fig. 1.9a. And that of an ideal constant current source is parallel to the voltage axis (Fig. 1.9b). "Real" or practical sources at best approximate these conditions because they combine in themselves the properties of *both* active and passive elements.

We can determine the voltage–current characteristics of practical or *nonideal* sources by connecting various load resistors across the source terminals and measuring the resulting voltage–current pairs in each case.

An example of a characteristic we might obtain in this manner for a nonideal source of voltage v_s is given in Fig. 1.29a. The characteristic of an ideal source would follow the broken line labeled v_s. But the actual characteristic intersects the voltage and current axes at v_{oc} and i_{sc}, respectively. The straight line between these two points has the slope

$$-R_s = -\frac{v_{oc}}{i_{sc}} = -\frac{v_s}{i_{sc}} \tag{1.116}$$

The equation for the source characteristic is then

$$v = v_{oc} - iR_s = v_s - iR_s \tag{1.117}$$

Figure 1.29 Source–resistor models for practical sources: short circuit and open circuit variables. (See text.)

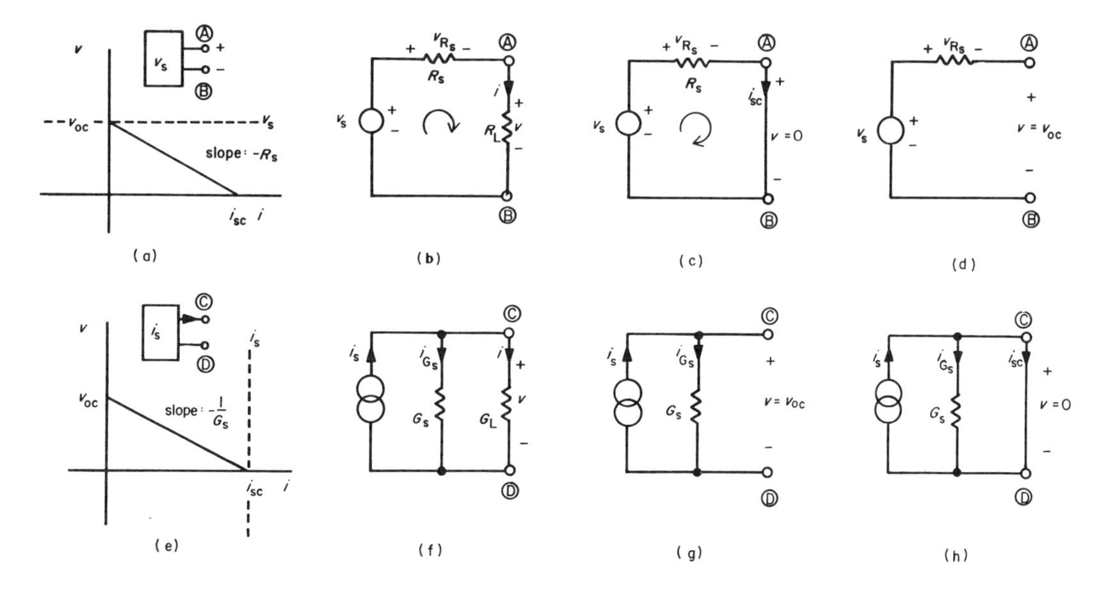

However, we obtain the same equation if we apply KVL to the circuit in Fig. 1.29b. The portion of the circuit to the left of terminals A and B serves as an *equivalent circuit for the nonideal voltage source*. The resistor R_s, in series with the ideal source v_s, is called the *source* or *inner resistance*; it signifies the nonideal behavior of the practical source.

All the information necessary to set up the equivalent circuit is given by the intersections of the characteristic with the voltage and current axes. These two points represent the extremes of possible loading conditions. A load of zero resistance amounts to a short circuit across terminals A and B (Fig. 1.29c). The resulting *short circuit current* i_{sc} equals the *maximum* current which can be drawn from this source. Whereas an ideal voltage source would "resist" the short-circuiting procedure and push an *infinite* current through the shorting connection, a practical voltage source will only supply a *finite* current at zero terminal voltage. With a load of infinite resistance, no current flows; so we simply leave terminals A and B free (Fig. 1.29d). Under this condition, we can measure the *open circuit voltage* v_{oc}. It represents the *maximum* voltage the source can supply at zero current. The open circuit voltage v_{oc} equals the nominal *source* voltage v_s.

With KVL we obtain for the circuit in Fig. 1.29c

$$v_s = v_{R_s} = i_{sc} R_s \tag{1.118}$$

And for the circuit in Fig. 1.29d we have

$$v = v_{oc} = v_s \tag{1.119}$$

because, with no current flowing, there can be no voltage drop across R_s. Equations (1.118) and (1.119) give us all the information we need for the description of the source characteristic according to Eqs. (1.116) and (1.117).

The dual problem is posed by the nonideal current source with the characteristic shown in Fig. 1.29e. The broken line indicates the characteristic of an ideal current source. For the actual characteristic we have the equation

$$i = i_{sc} - G_s v = i_s - G_s v \tag{1.120}$$

where

$$G_s = \frac{i_{sc}}{v_{oc}} = \frac{i_s}{v_{oc}} \tag{1.121}$$

Equation (1.120) represents the KCL equation for the circuit in Fig. 1.29f. The circuit to the left of terminals C and D is now the *equivalent circuit for the nonideal current source*. The nonideal aspect of the source is represented by the *source* or *inner conductance* G_s.

As in the previous case, determination of circuit performance under open circuit (Fig. 1.29g) and short circuit conditions (Fig. 1.29h) gives us all the necessary information to set up the equivalent circuit. The open circuit voltage v_{oc} equals the *maximum* voltage across the current source terminals. The short circuit current i_{sc} equals the *nominal* source current i_s.

From Fig. 1.29g we have

$$v = v_{oc} = \frac{i_s}{G_s} \tag{1.122}$$

and from Fig. 1.29h we have

$$i_{sc} = i_s \tag{1.123}$$

From Eqs. (1.116)–(1.123) it is obvious that a nonideal voltage source with a source resistance

$$R_s = \frac{1}{G_s}$$

could be described either by the series combination of an ideal voltage source ($v_s = v_{oc}$) with a source resistance R_s (Figs. 1.29c and 1.29d) or by the parallel combination of an ideal *current* source ($i_s = i_{sc}$) with a source *conductance* $G_s = 1/R_s$ (Figs. 1.29g and 1.29h). The same is true for nonideal current sources. In order to emphasize this point the *same* nonideal characteristic was drawn in Figs. 1.29a and 1.29e. Whether a practical voltage or current source is represented by the voltage source/source resistance model or the current source/source conductance model is entirely a matter of convenience. It is customary though, to designate a nonideal source as a voltage source when the source resistance is low (conductance high) *in comparison* with the load resistance, and as a current source when the source resistance is high (conductance low) in comparison with the load resistance.

1.3.8 The "Black Box" Problem: The Theorems of Helmholtz, Thévenin, and Norton

1.3.8.1 Driving Point Impedance

Consider a network S containing linear passive elements and sources. If a terminal pair provides access to S (Fig. 1.30), we can treat the network as a one-port and determine its driving point impedance Z_s or admittance Y_s. We may not know the details of the circuit representing S, just as if it were a mysterious "black box" with two terminals. But we still will be able to obtain the *equivalent driving point or "inner" immittances* simply by following the same procedures we established to obtain the equivalent circuits of nonideal sources. The steps as outlined in Figs. 1.30b–1.30d are

1. determine the open circuit voltage of the one-port,
2. determine the short circuit current of the one-port,
3. by analogy with Eqs. (1.116) and (1.121) use phasors to obtain the *driving point* or *inner impedance*

$$Z_s(s) = \frac{\mathbf{V}_{oc}(s)}{\mathbf{I}_{sc}(s)}$$

or the *driving point* or *inner admittance*

$$Y_s(s) = \frac{1}{Z_s(s)} = \frac{\mathbf{I}_{sc}(s)}{\mathbf{V}_{oc}(s)}$$

In the *special* case, where the network S has *no dependent* sources, we may use this alternative procedure:

1. *Replace* all independent voltage sources by short circuits.

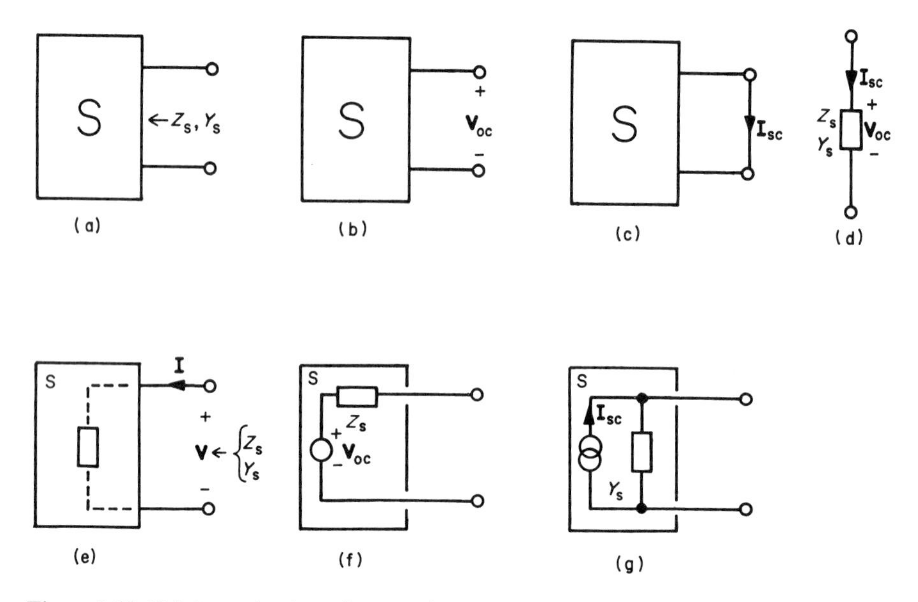

Figure 1.30 Driving point impedance (admittance) and the equivalent circuits of Thévenin and Norton for a linear network. (See text.)

2. *Remove* all independent current sources, leaving open circuits.
3. Determine the impedance or admittance of the remaining passive network as "seen" from the terminals (Fig. 1.30e).

The problem of Fig. 1.28 is a good example for the last step in the alternative procedure.

These two procedures were presented without proof. Their justification lies in the *linearity* of the network. Which of the two methods is used is a matter of convenience as long as the presence of dependent sources does not restrict us to the first one.

1.3.8.2 Theorem of Helmholtz and Thévenin

So far, the driving point immittances represent only the passive properties of the one-port S. For a *complete* description of network *behavior* (*not* the details of the circuit) we need only one other variable, the open circuit voltage.

Without proof we state the following theorem:

> Theorem of Helmholtz and Thévenin
>
> The behavior of a one-port containing linear passive elements and independent or dependent sources is equivalent to that of an independent voltage source of equal open-circuit voltage in series with the driving point impedance of the one-port (Fig. 1.30f).

The circuit derived according to this theorem is called the *Thévenin equivalent (circuit)*.

1.3.8.3 Theorem of Norton

This theorem is the dual form of the preceding one. In this case the driving point *admittance* stands for the passive properties of the network S, and an equivalent *current* source is needed for a complete description of network behavior. The complete equivalent circuit according to this approach is shown in Fig. 1.30g as proposed in this theorem:

> Theorem of Norton
>
> The behavior of a one-port containing linear passive elements and independent or dependent sources is equivalent to that of an independent current source of equal short circuit current in parallel with the driving point admittance of the one-port.

The equivalent circuit derived by the application of Norton's Theorem is called the *Norton equivalent (circuit)*.

Example 1.15.

Obtain the Norton and Thévenin equivalents at the terminal pair A and B of the circuit in Fig. 1.31a. The voltage difference across R_2 is fixed by sources v_1 and v_2. The rest of the network is, therefore, not affected by the presence of this parallel R_2. Accordingly, we remove R_2 and replace sources v_1 and v_2 by a source of voltage $v_3 = v_1 - v_2$. Inductor L_1 in series with current source i_2 can also be removed because the current supplied through it to node A is not affected by the presence of L_1.

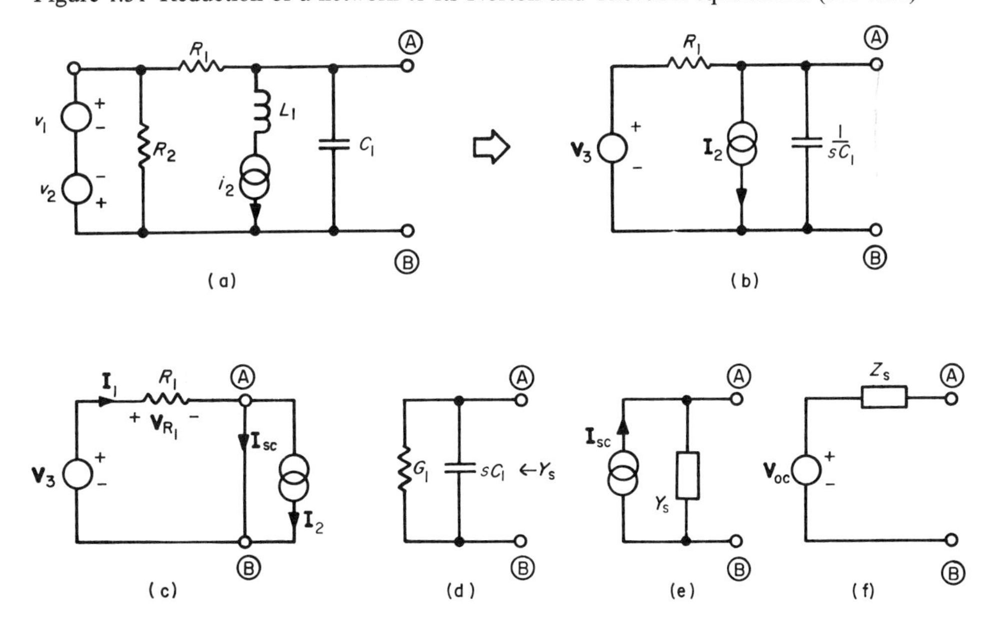

Figure 1.31 Reduction of a network to its Norton and Thévenin equivalents. (See text.)

The reduced network in generalized phasor notation is given in Fig. 1.31b. Next we determine the short circuit current across the terminals A and B. According to Fig. 1.31c we obtain

$$\mathbf{I}_{sc} = \mathbf{I}_1 - \mathbf{I}_2 = \frac{\mathbf{V}_3}{R_1} - \mathbf{I}_2 = \frac{\mathbf{V}_1 - \mathbf{V}_2}{R_1} - \mathbf{I}_2 \tag{1.124}$$

Since there are no dependent sources, we may determine the equivalent driving point admittance by reducing the network to its passive components only. Hence we replace voltage source v_3 with a short circuit and the current source i_2 with an open circuit (Section 1.3.8.1). From the resulting network (Fig. 1.31d) we obtain the driving point admittance by inspection:

$$Y_s(s) = G_1 + sC_1 \tag{1.125}$$

Equations (1.124) and (1.125) provide all that is needed to set up the equivalent circuit according to Norton's theorem (Fig. 1.31e).

The driving point impedance is the reciprocal of the admittance. From Eq. (1.125) we obtain then

$$Z_s(s) = \frac{1}{Y_s} = \frac{1}{G_1 + sC_1} \tag{1.126}$$

The open circuit voltage can be determined from Eqs. (1.124) and (1.126):

$$\mathbf{V}_{oc}(s) = \mathbf{I}_{sc}(s) Z_s(s) = \frac{\mathbf{V}_1 - \mathbf{V}_2 - R_1 \mathbf{I}_2}{1 + sR_1C_1} \tag{1.127}$$

This completes the solution, because Eqs. (1.126) and (1.127) specify the equivalent circuit according to the theorem of Helmholtz and Thévenin (Fig. 1.31f).

1.4 Analysis of Simple Circuits

In this section we shall analyze a variety of simple but useful circuits. This exercise will not only serve to illustrate the methods developed so far, but also provide the basis for new concepts. In many instances, complicated circuits can be reduced to or approximated by the circuits presented in this section.

1.4.1 The Voltage Divider

One of the simplest, yet most useful, circuits is the voltage divider. As its name implies it serves to "divide" down or lower the level of a voltage. In its most common form it is made up of two resistors.

Two resistors in series constitute the resistive voltage divider shown in Fig. 1.32a. The input voltage v_i is applied across the series combination. The output voltage v at the node pair A and B equals the voltage drop v_2 across R_2. Applying KCL and KVL we obtain the equations

$$i_1 = i_2$$

and

$$v_1 + v_2 = v_i$$

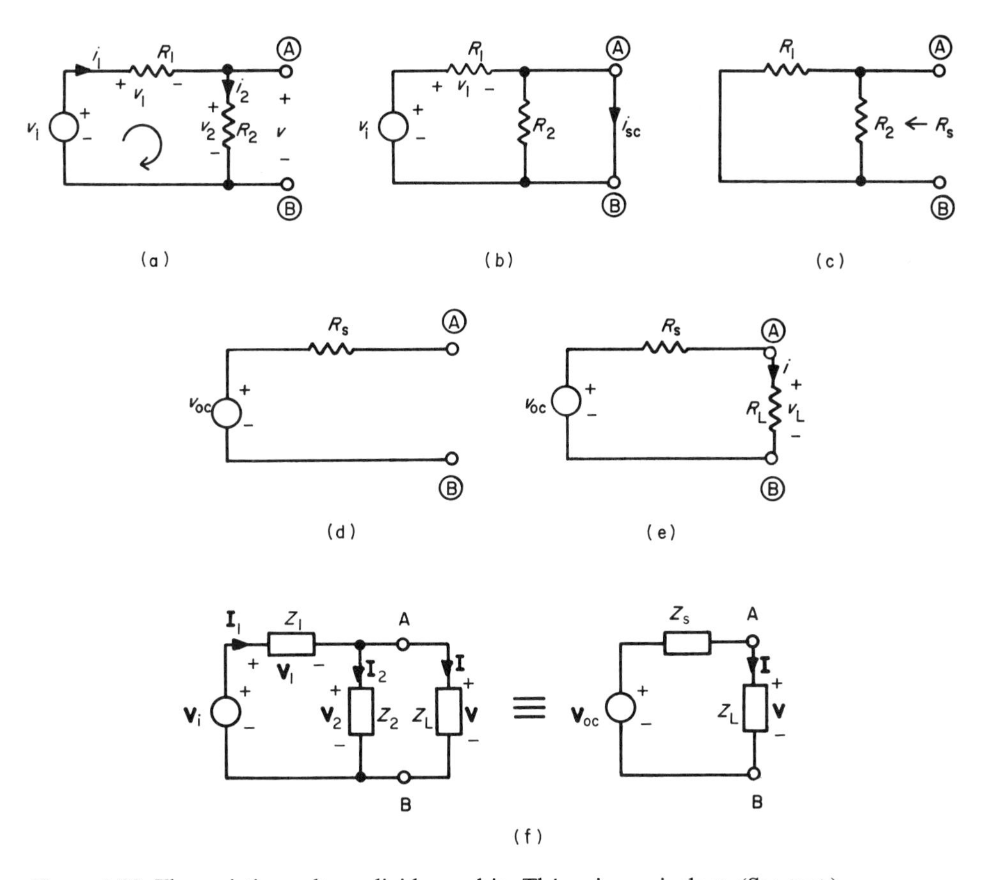

Figure 1.32 The resistive voltage divider and its Thévenin equivalent. (See text.)

With Ohm's law we have

$$R_1 i_1 + R_2 i_2 = (R_1 + R_2) i_2 = v_i$$

or

$$i_2 = \frac{v_i}{R_1 + R_2}$$

and finally

$$v = v_2 = i_2 R_2 = \frac{R_2}{R_1 + R_2} v_i$$

We could have achieved the same result with less effort by the method of equivalent circuits. According to Fig. 1.32b, the short circuit current is

$$i_{sc} = \frac{v_i}{R_1} = v_i G_1$$

The equivalent inner conductance and resistance are (Fig. 1.32c)

$$G_s = G_1 + G_2 = \frac{1}{R_s}$$

and

$$R_s = \frac{1}{G_1 + G_2} = \frac{R_1 R_2}{R_1 + R_2}$$

The node voltage v, however, is the open circuit voltage. So we obtain from Eqs. (1.118) and (1.121)

$$v = v_{oc} = i_{sc} R_s$$

and

$$v = v_{oc} = \frac{i_{sc}}{G_s}$$

Inserting the expressions for the inner resistance and conductance we can state the

Voltage Divider Equations

$$v = \frac{R_2}{R_1 + R_2} v_i \tag{1.128}$$

or

$$v = \frac{G_1}{G_1 + G_2} v_i \tag{1.129}$$

Divider ratio:

$$\alpha = \frac{R_2}{R_1 + R_2} = \frac{G_1}{G_1 + G_2} \leqslant 1$$

The Thévenin equivalent of the voltage divider is given in Fig. 1.32d, where v_{oc} is established according to Eqs. (1.128) or (1.129) and R_s represents the parallel combination of the divider resistances R_1 and R_2. The preceding equations apply to the *load-free* voltage divider. In order to determine the voltage across the output terminals A and B in the presence of a load resistance R_L we make again use of the Thévenin equivalent. Once more the problem is reduced to that of the load-free voltage divider, as comparison between Figs. 1.32a and 1.32e shows. With the labels of Fig. 1.32e we obtain from Eq. (1.129) the output voltage of the *loaded* voltage divider:

$$v_L = \frac{G_s}{G_s + G_L} v_{oc} = \left(\frac{G_1 + G_2}{G_1 + G_2 + G_L} \right) \left(\frac{G_1}{G_1 + G_2} \right) v_i$$

or

$$v_L = \frac{G_1}{G_1 + G_2 + G_L} v_i \tag{1.130}$$

By replacing the conductances with the reciprocal resistances we derive the more complicated resistance form of the loaded divider equation:

$$V_L = \frac{R_2 R_L}{R_1 R_2 + R_1 R_L + R_2 R_L} v_i \tag{1.131}$$

Equations (1.128)–(1.131) constitute a powerful set of analytical tools. Many network problems can be reduced to these equations.

The topology of the voltage divider network remains unchanged if we replace the resistances with generalized impedances. The complete impedance voltage divider with load impedance and equivalent circuit is shown in Fig. 1.32f. Because of the topological equivalence to the resistive case, we can rewrite Eqs. (1.128)–(1.131) in phasor–immittance notation:

Generalized Voltage Divider Equations

1. Load-free ($Z_L = \infty$ in Fig. 1.32f)

$$\mathbf{V}(s) = \frac{Z_2(s)}{Z_1(s) + Z_2(s)} \mathbf{V}_i(s) \tag{1.132}$$

or

$$\mathbf{V}(s) = \frac{Y_1(s)}{Y_1(s) + Y_2(s)} \mathbf{V}_i(s) \tag{1.133}$$

2. With load

$$\mathbf{V}(s) = \frac{Y_1(s)}{Y_1(s) + Y_2(s) + Y_L(s)} \mathbf{V}_i(s) \tag{1.134}$$

or

$$\mathbf{V}(s) = \frac{Z_2(s) Z_L(s)}{Z_1(s) Z_2(s) + Z_1(s) Z_L(s) + Z_2(s) Z_L(s)} \mathbf{V}_i(s) \tag{1.135}$$

Divider ratio:

$$\alpha(s) = \frac{Z_2(s)}{Z_1(s) + Z_2(s)} = \frac{Y_1(s)}{Y_1(s) + Y_2(s)} \tag{1.136}$$

The divider ratio is a dimensionless quantity but, as Eq. (1.136) shows, generally a *function of frequency*. As it relates one branch voltage to another, it does not belong to the class of immittances. It is one of the various *network functions* we can define for a circuit. Network functions relating voltages of different branches are often called *transfer functions* (in terms of generalized frequency s) or *frequency response functions* (in terms of ω for the sinusoidal steady state). These terms, however, are also used for other pairings of input–output variables.

1.4.2 First-Order Circuits; Bode Plots

The circuits discussed in this section are of *first order* because they can be described by first-order differential equations. In their *minimum form* they consist of only *two* elements (excluding sources): a *storage element* and a *dissipator*. We impose the further restriction that we consider here only circuits described by *linear* first-order differential equations.

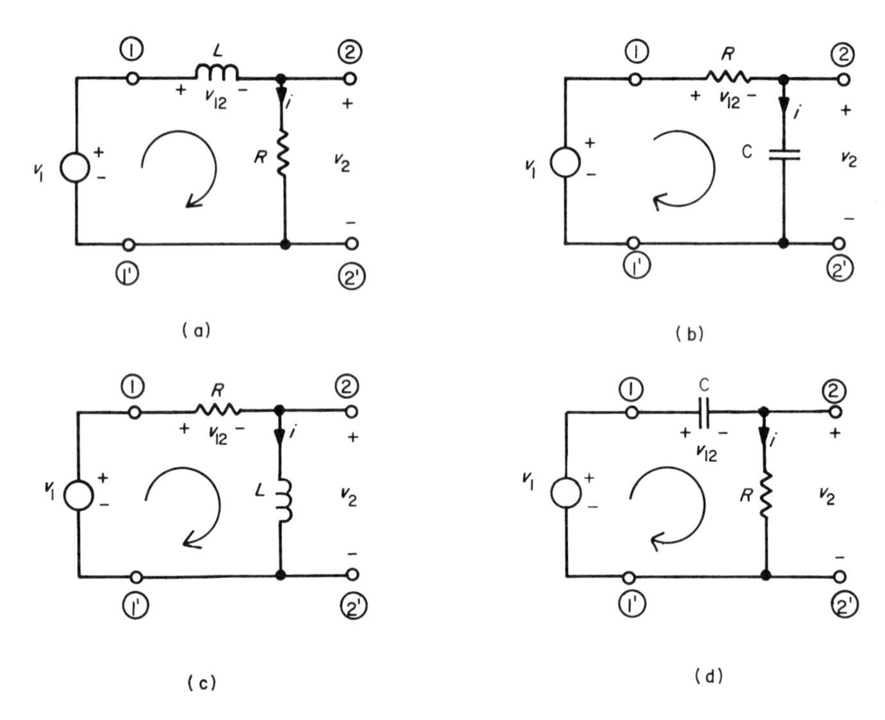

Figure 1.33 First-order low-pass **(a, b)** and high-pass **(c, d)** filters. (See text.)

1.4.2.1 Simple First-Order Circuits

Typical first-order circuits of great practical importance are shown in Fig. 1.33. In all four circuits, the elements are arranged in series to form a single mesh.

Analysis of the circuits in Fig. 1.33 is straightforward; they all have the same topology. The following KVL equation applies to all four circuits:

$$v_{12} + v_2 = v_1. \tag{1.137}$$

Inserting the element laws for the circuit of Fig. 1.33a into Eq. (1.137) we obtain the differential equation for the current i as a function of the source voltage v_1:

$$L\frac{di}{dt} + Ri = v_1.$$

Division of both sides by R gives

$$\frac{L}{R}\frac{di}{dt} + i = \frac{v_1}{R}$$

where

$$\frac{L}{R} = T_L$$

is the *time constant* of the circuit. It is in units of seconds if L is in henrys and R is in ohms. Inserting the time constant, we standardize the left side of the differential equation:

$$T_L\frac{di}{dt} + i = \frac{v_1}{R} \tag{1.138}$$

This first-order differential equation can be solved for any input v_1 by the standard solution methods for linear differential equations or by the transform methods in Chapter 3.

The current i is not the only possible output variable for the circuit. In most practical applications of these circuits, in fact, we are usually more interested to obtain an equation relating the output voltage v_2 to the input voltage v_1. For the circuit in Fig. 1.33a, the output voltage is

$$v_2 = Ri$$

Thus we can substitute $i = v_2/R$ in Eq. (1.138) in order to get the desired relation

$$T_L \frac{dv_2}{dt} + v_2 = v_1 \tag{1.139}$$

Once again we get a first-order differential equation.

From Eq. (1.137) and the element laws we can immediately write an equation for the current in the circuit of Fig. 1.33b:

$$Ri + \frac{1}{C}\int_0^t i\,dt = v_1$$

After differentiation and multiplication by C we obtain

$$RC\frac{di}{dt} + i = C\frac{dv_1}{dt}$$

We define as the *time constant* of the circuit

$$T_C = RC$$

Hence, in terms of the time constant, the differential equation is now

$$T_C\frac{di}{dt} + i = C\frac{dv_1}{dt} \tag{1.140}$$

Note that, in this instance, the forcing term (right side of the differential equation) is the first *derivative* of the input voltage, provided v_1 is differentiable.

The mesh current can also be expressed in terms of the node voltages by this equation:

$$i = \frac{v_{12}}{R} = \frac{v_1 - v_2}{R} \tag{1.141}$$

If we now combine Eqs. (1.140) and (1.141) we generate a differential equation relating v_2 to v_1:

$$T_C\frac{dv_2}{dt} + v_2 = v_1 \tag{1.142}$$

As we compare Eqs. (1.139) and (1.142) we see that the two circuits in Figs. 1.33a and 1.33b are *functionally equivalent* for the input–output *voltage* pair (v_1, v_2). The differential equations for the mesh currents, Eqs. (1.138) and (1.140) are different, however.

The differential equation for the mesh current for the circuit in Fig. 1.33c must be the same as Eq. (1.138). The equation relating the output voltage v_2 to the input voltage v_1 is derived by combining Eqs. (1.138) and (1.141):

$$T_L\frac{dv_2}{dt} + v_2 = T_L\frac{dv_1}{dt} \tag{1.143}$$

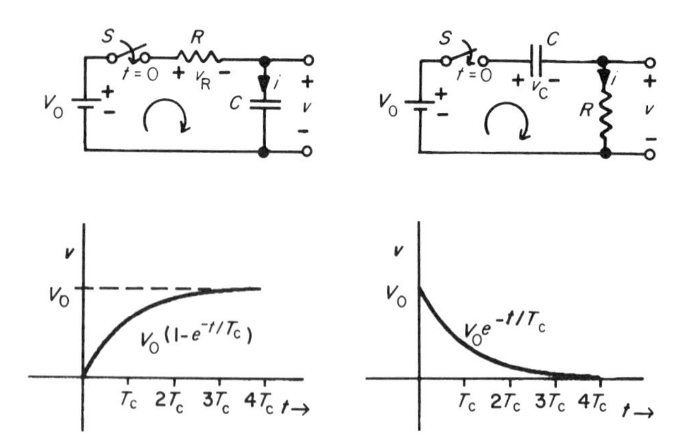

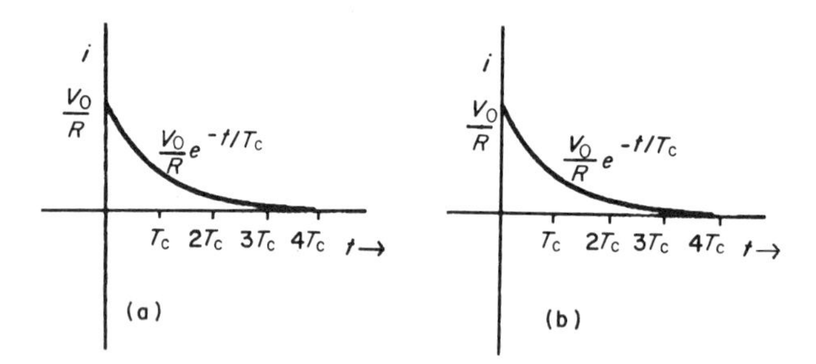

Figure 1.34 The transient behavior of two first-order circuits. (See text.)

In contrast to Eqs. (1.139) and (1.142), the output voltage is now a function of the *derivative* of the input voltage.

Finally, for the circuit in Fig. 1.33d we can write the same mesh current equation as Eq. (1.140). In order to derive the input–output relation in terms of the node voltages v_1 and v_2, we substitute $i = v_2/R$ in Eq. (1.140). The result is

$$T_C \frac{dv_2}{dt} + v_2 = T_C \frac{dv_1}{dt} \tag{1.144}$$

Equation (1.144) is obviously equivalent to Eq. (1.143). This means that in this sense the circuit in Fig. 1.33d is functionally equivalent to that in Fig. 1.33c.

The prototypical first-order differential equations of this section [Eqs. (1.138) or (1.140) and (1.143) or (1.144)] can be reduced to algebraic equations by the phasor method in the case of sinusoidal or exponential signals and zero initial conditions. For the circuits of Fig. 1.33 this amounts to a reduction to the voltage divider case, as we shall see in the next section.

The general solution of these differential equations with arbitrary initial conditions can be obtained with the standard methods outlined in differential equation texts or with the transform methods presented in Chapter 3. If these methods are used in the case where a fixed voltage $v_1 = V_0$ is suddenly applied to the circuits in Figs. 1.33b and 1.33d by closing switch S, then the results of Fig. 1.34 are obtained. We see that the circuit in Fig. 1.33b will, in the *steady state*, transmit the input voltage, while the circuit of Fig. 1.33d will only produce a switching transient. As we see in the next section, the steady state transmission of dc voltages is typical of so-called *low-pass* filters (e.g., Figs. 1.33a and 1.33b) while dc blockade is typical of *high-pass* filters (e.g., Figs. 1.33c and 1.33d).

The mesh current is the same for both circuits (bottom panels in Figs. 1.34a and 1.34b). It represents of course the charging current for the capacitor.

1.4.2.2 Frequency Response: Bode Plots

In order to assess the function of the simple first-order circuits of Fig. 1.33 as *filters* it is advantageous to consider their behavior in the *sinusoidal steady state*. The sinusoidal steady state response is of fundamental importance because arbitrary waveforms can be decomposed into sinusoidal components as shown in Chapter 3.

Network behavior as a function of sinusoidal excitation frequency can be determined by the ac phasor method. The result is the so-called *frequency response*.

We demonstrate the method on the circuit of Fig. 1.33a. In terms of ac phasors we write the KVL equation

$$\mathbf{V}_{12}+\mathbf{V}_2=\mathbf{V}_1$$

and the element laws

$$\mathbf{V}_{12}=j\omega L\mathbf{I},$$

$$\mathbf{V}_2=R\mathbf{I}$$

Substitution of the element laws into the KVL equation gives us

$$j\omega L\mathbf{I}+R\mathbf{I}=\mathbf{V}_1$$

or, by analogy with Eq. (1.138),

$$T_{\mathrm{L}}\,j\omega\mathbf{I}+\mathbf{I}=\frac{\mathbf{V}_1}{R}$$

Factoring out the current phasor $\mathbf{I}$, we obtain

$$\mathbf{I}(1+j\omega T_{\mathrm{L}})=\frac{\mathbf{V}_1}{R}$$

or

$$\mathbf{I}=\frac{\mathbf{V}_1}{R(1+j\omega T_{\mathrm{L}})}$$

This solution for the current phasor as a function of the input voltage phasor can be used to determine the transfer ratio of output voltage phasor/input voltage phasor:

$$\mathbf{V}_2=R\mathbf{I}=\frac{\mathbf{V}_1}{1+j\omega T_{\mathrm{L}}}$$

and

$$\frac{\mathbf{V}_2}{\mathbf{V}_1}=\frac{1}{1+j\omega T_{\mathrm{L}}}$$

The last result could have been obtained in a different manner because, topologically, all four networks in Fig. 1.33 belong to the class of voltage dividers. The output phasor $\mathbf{V}_2$ is then related to the input phasor $\mathbf{V}_1$ by the voltage divider ratio

$\alpha(j\omega)$ according to Eq. (1.136), provided we replace the generalized frequency s with $j\omega$.

For the low-pass filters in Figs. 1.33a and 1.33b we have then with time constant T and characteristic frequency $\omega_0 = 1/T$

$$\alpha(j\omega) = \frac{Z_2(j\omega)}{Z_1(j\omega) + Z_2(j\omega)} = \frac{1}{1 + j\omega T} = \frac{1}{1 + j(\omega/\omega_0)}$$

This voltage divider ratio represents the relation between a sinusoidal input voltage at one terminal pair and a sinusoidal output voltage at another terminal pair. It belongs to the general class of *frequency response functions*, which are often symbolized by the letter G:

$$G(j\omega) = \frac{\mathbf{V}_2}{\mathbf{V}_1} = \frac{1}{1 + j\omega T} = \frac{1}{1 + j(\omega/\omega_0)} \qquad (1.145)$$

Application of Eq. (1.136) to the high-pass filter circuits in Figs. 1.33c and 1.33d gives this frequency response function:

$$G(j\omega) = \frac{\mathbf{V}_2}{\mathbf{V}_1} = \frac{j\omega T}{1 + j\omega T} = \frac{j(\omega/\omega_0)}{1 + j(\omega/\omega_0)} \qquad (1.146)$$

Equations (1.145) and (1.146) are specific examples of frequency response functions, which, in general, have the form

$$G(j\omega) = \frac{a_0 + a_1(j\omega) + a_2(j\omega)^2 + \cdots + a_n(j\omega)^n}{b_0 + b_1(j\omega) + b_2(j\omega)^2 + \cdots + b_m(j\omega)^m} \qquad (1.147)$$

where $m \geqq n$. The roots of the numerator and denominator polynomials can be used to factor the polynomials; hence

$$G(j\omega) = k_0 \frac{(j\omega - z_1)(j\omega - z_2)\cdots(j\omega - z_n)}{(j\omega - p_1)(j\omega - p_2)\cdots(j\omega - p_m)}$$

where $k_0 = a_n/b_m$. The roots of the denominator polynomial are the *poles* of $G(j\omega)$; they are designated by the letter p. The roots of the numerator z are the zeros of $G(j\omega)$. We define now the *negative reciprocals of the zeros* and *poles* of $G(j\omega)$ as the *time constants* of the frequency response function such that

$$\frac{-1}{z_1} = T_a, \qquad \frac{-1}{z_2} = T_b, \qquad \cdots$$

and

$$\frac{-1}{p_1} = T_1, \qquad \frac{-1}{p_2} = T_2, \qquad \cdots$$

The substitution of these terms into the last equation gives us

$$G(j\omega) = G_0 \frac{(1 + j\omega T_a)(1 + j\omega T_b)\cdots(1 + j\omega T_n)}{(1 + j\omega T_1)(1 + j\omega T_2)\cdots(1 + j\omega T_m)} \qquad (1.148)$$

where the constant G_0 is given as

$$G_0 = \frac{a_n T_1 T_2 \cdots T_m}{b_m T_a T_b \cdots T_n}$$

Equation (1.148) represents actually the special case for which all poles and zeros are distinct, but there may be multiple poles and zeros. Some of the poles and zeros may also vanish ($p=0$, $z=0$) or be complex numbers. The latter case is frequently avoided by not splitting the quadratic factors which give rise to the complex poles or zeros. So the most general form of $G(j\omega)$ is then

$$G(j\omega)=G_0\frac{(j\omega)^s(1+j\omega T_a)(1+j\omega T_b)^t\left[1+j\omega\zeta_c T_c+(j\omega T_c)^2\right]\cdots}{(j\omega)^u(1+j\omega T_1)(1+j\omega T_2)^v\left[1+j\omega\zeta_3 T_3+(j\omega T_3)^2\right]\cdots} \tag{1.149}$$

The frequency response function described by Eq. (1.149) has a multiple zero of order s at $\omega=0$, a multiple zero of order t at $\omega=-1/T_b$, a multiple pole of order u at $\omega=0$, a multiple pole of order v at $\omega=-1/T_2$, and quadratic factors indicating the presence of complex zeros and poles. (The latter case will be discussed in Section 1.4.3.)

The rationale for using the factored forms of Eqs. (1.148) and (1.149) rather than Eq. (1.147) is that double-logarithmic plots of the frequency response are more easily constructed from the factored forms. Designating the factors in the numerator of either Eq. (1.148) or (1.149) by $N(j\omega)$ and the factors in the denominator by $D(j\omega)$, we can write

$$G(j\omega)=G_0\frac{N_a(j\omega)N_b(j\omega)\cdots N_n(j\omega)}{D_1(j\omega)D_2(j\omega)\cdots D_m(j\omega)}$$

In terms of magnitude and phase, this equation can be expressed as follows:

$$\begin{aligned}G(j\omega)&=|G(j\omega)|e^{j\phi(\omega)}\\&=G_0\frac{|N_a|\cdot|N_b|\cdots|N_n|}{|D_1|\cdot|D_2|\cdots|D_m|}\\&\quad\cdot\exp\{j[(\sphericalangle N_a+\sphericalangle N_b+\cdots)-(\sphericalangle D_1+\sphericalangle D_2+\cdots)]\}\end{aligned}$$

Taking the natural logarithm of this expression, we obtain

$$\begin{aligned}\ln G(j\omega)&=\ln|G(\omega)|+j\phi(\omega)\\&=\ln G_0+\ln|N_a|+\ln|N_b|+\cdots\\&\quad+\ln|N_n|-\ln|D_1|-\ln|D_2|-\cdots\\&\quad-\ln|D_m|+j[(\sphericalangle N_a+\sphericalangle N_b+\cdots)-(\sphericalangle D_1+\sphericalangle D_2+\cdots)]\end{aligned} \tag{1.150}$$

Equation (1.150) is a very useful expression because it relates in a simple way the magnitude and phase of $G(j\omega)$ to the magnitude and phase of its constituent factors. First, according to this equation, the *logarithm of the magnitude of* $G(j\omega)$ *equals the sum of the logarithms of the factor magnitudes*; terms corresponding to the denominator are negative. Second, *the phase of* $G(j\omega)$, $\phi(j\omega)$, *equals the sum of the phase terms in the numerator minus the sum of the phase terms in the denominator.*

In practice, we express the magnitude as the *log-magnitude* of $G(j\omega)$, abbreviated $\mathrm{Lm}\,G$, which is defined as

$$\mathrm{Lm}\,G=20\log_{10}|G(\omega)| \tag{1.151}$$

Although $G(j\omega)$ typically is a dimensionless number, $\mathrm{Lm}\,G$ is usually expressed in the units *decibel* (abbreviated dB). For $|G|=10$, the log-magnitude would be 20 dB.

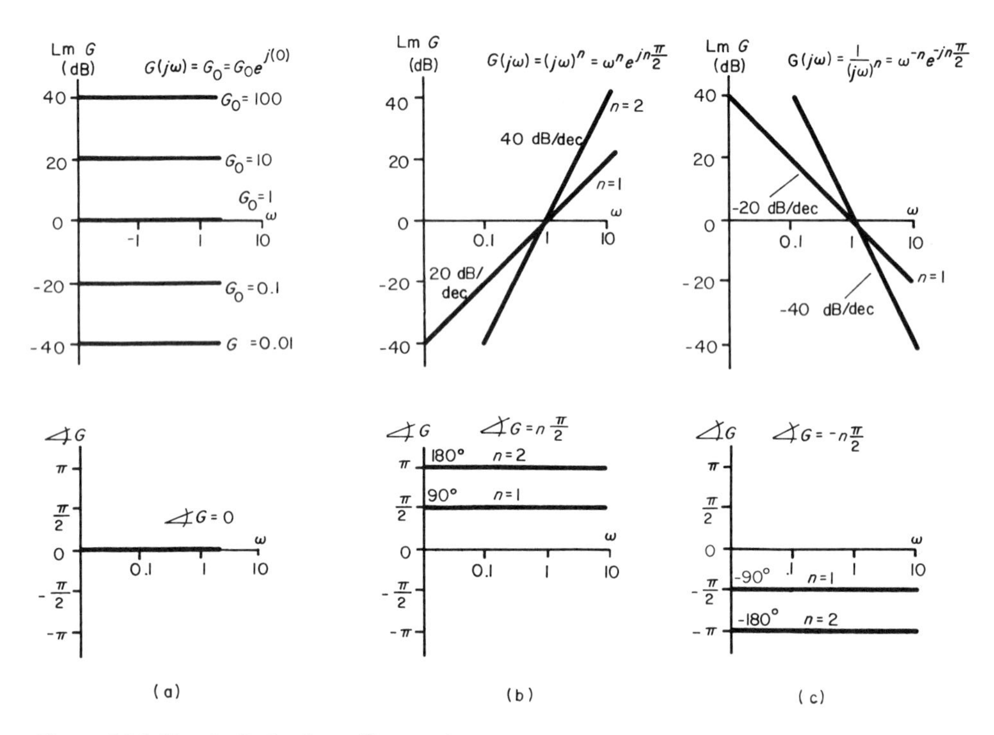

Figure 1.35 Simple Bode plots. (See text.)

From Eqs. (1.150) and (1.151), it follows that

$$\begin{aligned}\operatorname{Lm} G = {} & \operatorname{Lm} G_0 + \operatorname{Lm} N_a + \operatorname{Lm} N_b + \cdots + \operatorname{Lm} N_n \\ & - (\operatorname{Lm} D_1 + \operatorname{Lm} D_2 + \cdots + \operatorname{Lm} D_m)\end{aligned} \tag{1.152}$$

and for the phase we can write

$$\phi(\omega) = (\sphericalangle N_a + \sphericalangle N_b + \cdots + \sphericalangle N_n) - (\sphericalangle D_1 + \sphericalangle D_2 + \cdots + \sphericalangle D_m) \tag{1.153}$$

A plot of the log-magnitude, Eq. (1.152), and the phase, Eq. (1.153), against frequency (on a *logarithmic* scale) is called a *Bode plot* in honor of H. W. Bode, who was one of the pioneers of frequency response analysis.

Bode plots for the simplest factors in Eq. (1.149) are shown in Fig. 1.35. The Bode plots of quadratic factors are discussed in Section 1.4.3. This leaves us with first-order factors, which we shall consider next as we establish the frequency response plots for first-order filters.

The frequency response function for the first-order low-pass filters of Figs. 1.33a and 1.33b is, according to Eq. (1.145),

$$G(j\omega) = \frac{1}{1 + j\omega T} = \frac{1}{1 + j(\omega/\omega_0)}$$

The magnitude of $G(j\omega)$ is given by

$$|G| = \left[\frac{1}{1 + (\omega/\omega_0)^2}\right]^{1/2} \tag{1.154}$$

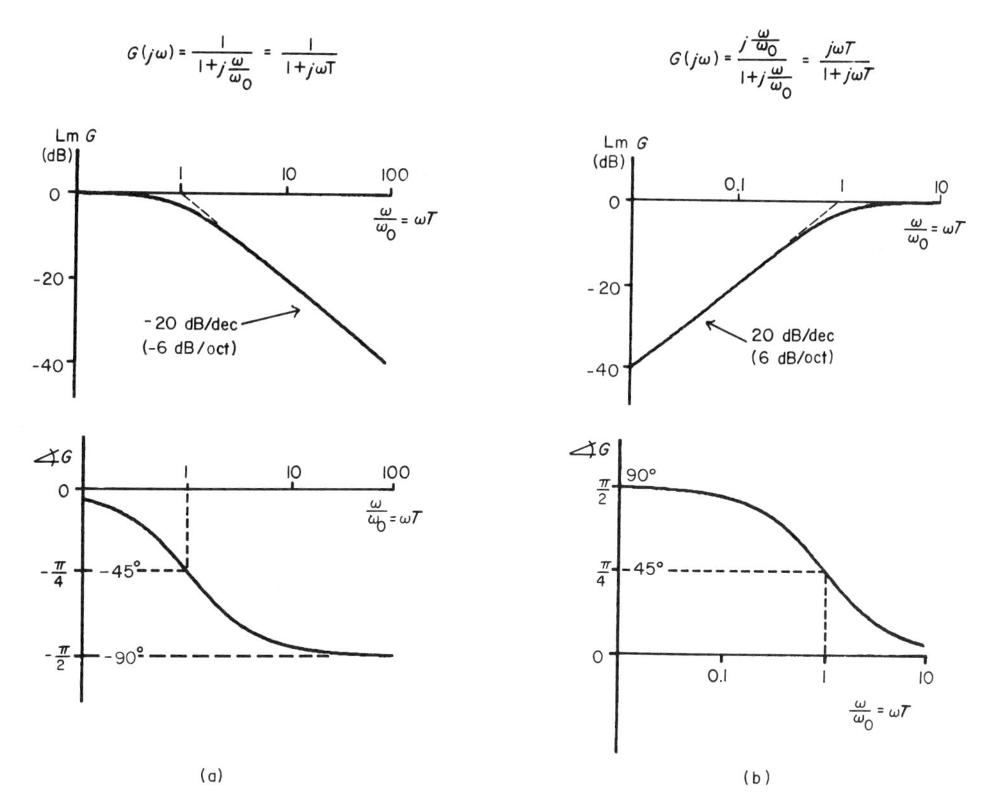

Figure 1.36 First-order Bode plots: **(a)** low-pass filter; **(b)** high-pass filter. (See text.)

and the phase by

$$\sphericalangle G = -\tan^{-1}\left(\frac{\omega}{\omega_0}\right) \tag{1.155}$$

From Eq. (1.154) we obtain the log-magnitude

$$\mathrm{Lm}\,G = 20\log\left[1+\left(\frac{\omega}{\omega_0}\right)^2\right]^{-1/2} \tag{1.156}$$

For $\omega \ll \omega_0$, Eq. (1.156) reduces to

$$\mathrm{Lm}\,G \approx 20\log(1) = 0$$

This, however, is the equation for the horizontal axis, $\mathrm{Lm}\,G = 0$. It represents the low-frequency *asymptote* of the log-magnitude plot in Fig. 1.36a.

For high frequencies, $\omega \gg \omega_0$, Eq. (1.156) is approximated by

$$\mathrm{Lm}\,G \approx 20\log\left(\frac{\omega}{\omega_0}\right)$$

This *high-frequency asymptote* is drawn as a broken line in Fig. 1.36a with a slope of -20 dB/decade (-6 dB/octave).

The two asymptotes intersect at the point $\omega = \omega_0$. If we use the asymptotes as approximations to the actual LmG curve, the "break" or transition from one asymptote to the other occurs at the characteristic frequency ω_0. This frequency is, therefore, called the *break* or *corner* frequency of the filter.

The *low-pass* character of the filter is evident. From $\omega=0$ (dc) to the corner or break frequency, signals are passed with little attenuation; thereafter attenuation increases. For frequencies $\omega>10\omega_0$, signal magnitudes are attenuated at -20 dB/decade.

The deviations of the actual LmG curve from the asymptotes are insignificant for frequencies one decade below and above the characteristic frequency. The deviation is largest at the characteristic frequency. For $\omega=\omega_0$, we have from Eq. (1.156)

$$\mathrm{Lm}G=20\log(2)^{-1/2}=-10\log 2\approx-3\text{ dB}$$

The corner frequency is, therefore, also called the *3-dB frequency*. At this point the magnitude has decreased by a factor of $1/\sqrt{2}$, i.e., it is down to 70.7% of its value at very low frequencies. The frequency interval from $f=0$ to $f_0=\omega_0/2\pi$, corresponding to the 3-dB frequency, is called the *bandwidth* of the low-pass filter.

The phase lag at the corner frequency is, from Eq. (1.155),

$$\measuredangle G(\omega_0)=-\tan^{-1}(1)=-\frac{\pi}{4}(45°)$$

From the Bode plot, Fig. 1.36a, we see that the phase curve has its steepest slope in this region, i.e., phase change as a function of frequency becomes most pronounced in the vicinity of the corner frequency. For very low frequencies, the phase asymptotically approaches 0, and for very high frequencies it approaches the *maximum phase shift for a first-order system*, namely, $-\pi/2$ ($-90°$). Note that the phase curve is rotationally symmetric around the $-\pi/4$ point.

As corner frequencies are changed, the log-magnitude and phase curves are simply shifted in parallel to the frequency axis. Bode plots for a particular case are quickly constructed with the aid of templates shifted along the frequency axis or by using the values given in Table 1.4. For rough calculations it is sufficient to use the asymptotes only.

Next we consider the first-order factor

$$(1+j\omega T)=(1+j(\omega/\omega_0))$$

Table 1.4 Log-Magnitude (Lm) and Phase Values for $G(j\omega)=\left(1+j\dfrac{\omega}{\omega_0}\right)^{\pm 1}$

ω	Asymptote of LmG (dB)	Actual LmG (dB)	Difference between LmG and Asymptote (dB)	Phase rad/degree
$0.1\omega_0$	0	0	0	$\pm0.1/\pm6°$
$0.5\omega_0$	0	±1	±1	$\pm0.47/\pm27°$
$0.75\omega_0$	0	±2	±2	$\pm0.65/\pm37°$
ω_0	0	±3	±3	$\pm\frac{\pi}{4}/\pm45°$
$1.33\omega_0$	±2.5	±4.5	±2	$\pm0.93/\pm53°$
$2\omega_0$	±6	±7	±1	$\pm1.1/\pm63°$
$10\omega_0$	±20	±20	0	$\pm1.5/\pm84°$

This frequency response function is the reciprocal of the preceding one. The curves in the Bode plot would then be the *mirror images* (reflected along the frequency axes) of the low-pass curves in Fig. 1.36a. The high-frequency asymptote has, therefore, a positive slope of 20 dB/decade, and the phase angle is positive (phase lead). Although such a frequency response function exhibits enhancement of frequencies higher than the break frequency, it is not a high-pass filter in the strict sense because frequencies below the break frequency are not attenuated.

Finally, let us consider the Bode plot for the first-order high-pass filters of Figs. 1.33c and 1.33d. Their frequency response function, according to Eq. (1.146), is

$$G(j\omega)=\frac{j\omega T}{1+j\omega T}=\frac{j\omega/\omega_0}{1+j(\omega/\omega_0)}$$

The log-magnitudes and phases of numerator and denominator are additive:

$$\operatorname{Lm} G=\operatorname{Lm}(j\omega T)+\operatorname{Lm}(1+j\omega T)^{-1}$$

and

$$\sphericalangle G=\sphericalangle(j\omega T)+\sphericalangle\left[(1+j\omega T)^{-1}\right]$$

Therefore, we can synthesize the corresponding Bode plot from Figs. 1.35b and 1.36a. The result is shown in Fig. 1.36b.

A comparison of the Bode plots in Figs. 1.36a and 1.36b shows why the high-pass filter is the exact opposite of the low-pass filter. Frequencies well above the corner frequency ($\omega>10\omega_0$) are passed with no attenuation and minimum phase shift. Below the corner frequency, attenuation approaches a rate of -20 dB/decade and there is a maximum phase *lead* of $\pi/2$ for very low frequencies.

This concludes the discussion of single first-order circuits.

1.4.3 Second-Order Circuits

Second-order networks contain at least two storage elements that cannot be reduced to a single one by equivalent replacement of series or parallel arrangements, etc. A typical case would be a network containing an inductor and a capacitor. In any event, *the differential equations describing such networks will be of second order.* A special class of second-order circuits are the series and parallel combinations of an inductor and a capacitor. We shall explore such circuits next.

1.4.3.1 Ideal Resonance Circuits

The differential equation for the *LC* series circuit in Fig. 1.37a may be derived as follows. From KVL we have

$$L\frac{di}{dt}+\frac{1}{C}\int_0^t i\,dt+\frac{q(0)}{C}=v(t)$$

where $q(0)$ is the initial charge on the capacitor. We obtain the necessary second-order differential equation by differentiating both sides of the equation:

$$L\frac{d^2i}{dt^2}+\frac{i}{C}=\frac{dv}{dt}$$

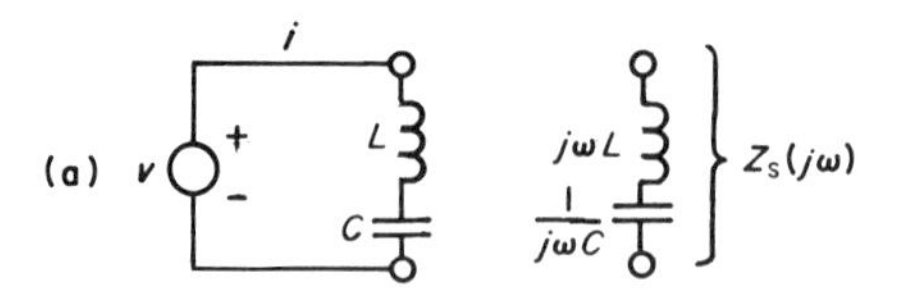

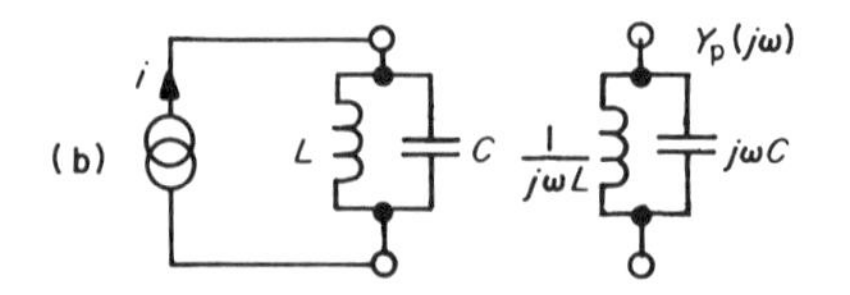

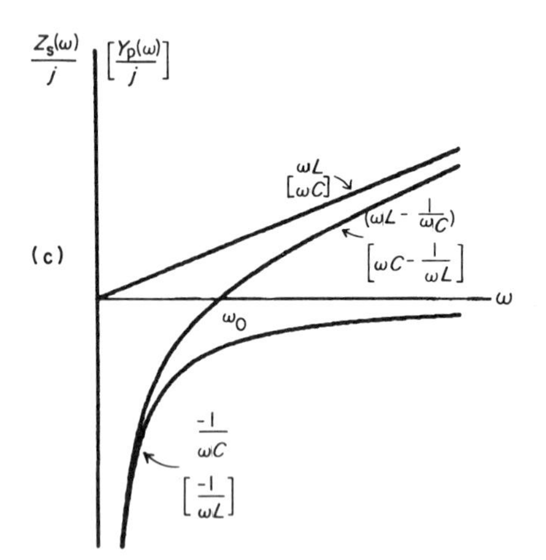

Figure 1.37 Second-order circuits: ideal LC resonance circuits. (See text.)

The case of special interest to us is the sinusoidal excitation of this circuit. But instead of solving the differential equation for the sinusoidal steady state, we simply write down the ac impedance of the LC combination by inspection, because this quantity gives us all we need to know for the relation between voltage and current in the sinusoidal steady state. For the series impedance we have

$$Z_s(j\omega) = \frac{\mathbf{V}(j\omega)}{\mathbf{I}(j\omega)} = j\omega L + \frac{1}{j\omega C} = j\left(\omega L - \frac{1}{\omega C}\right) \tag{1.157}$$

We see that the impedance becomes zero whenever $\omega L = 1/\omega C$. Correspondingly, the current should become infinite because this represents a short circuit. This is the condition of *resonance*, and the excitation frequency for which it occurs is the *resonance frequency* f_0. This characteristic frequency is related to the circuit parameters by the following expression:

$$\omega_0 = 2\pi f_0 = \frac{1}{\sqrt{LC}} \tag{1.158}$$

Substituting Eq. (1.158) for ω in Eq. (1.157) verifies that resonance occurs at this frequency.

With Eq. (1.158) we can write Eq. (1.157) in frequency-normalized forms:

$$Z_s(j\omega) = j\sqrt{\frac{L}{C}}\left(\frac{\omega}{\omega_0} - \frac{\omega_0}{\omega}\right) \tag{1.159}$$

$$Z_s(jf) = j\sqrt{\frac{L}{C}}\left(\frac{f}{f_0} - \frac{f_0}{f}\right) \tag{1.160}$$

The series impedance, exclusive of the factor j, is plotted as a function of the radian frequency ω in Fig. 1.37c. At a frequency much below ω_0 the total impedance approaches that of the capacitor alone ($-1/\omega C$). At frequencies much higher than the resonance frequency, the impedance is approximately that of the inductor alone (ωL). At resonance ($\omega = \omega_0$), the component impedances "cancel" each other in the sense that *the voltage drops across inductor and capacitor are of equal magnitude, but of opposite sign, so that the total voltage drop is zero.* (Note that, individually, the voltage drops across the storage elements can be at dangerously high levels.)

The impedance, Eqs. (1.159) and (1.160), is an imaginary number. Its phase, therefore, is either $+\pi/2$ or $-\pi/2$. From the above equations it is evident that the phase is $-\pi/2$ for all frequencies below resonance and $+\pi/2$ for all frequencies above resonance. At the resonance frequency itself, the phase "jumps" from $-\pi/2$ to $+\pi/2$, the total phase *change*, therefore, is π (180°).

The dual case to the preceding one is the *parallel* combination of an inductor and a capacitor, Fig. 1.37b. We write the *admittance* by inspection as

$$Y_p(j\omega) = j\omega C + \frac{1}{j\omega L} = j\left(\omega C - \frac{1}{\omega L}\right)$$

For the resonance frequency, as defined by Eq. (1.158), the total admittance of the parallel circuit will be zero, i.e., the network behaves like an open circuit, at resonance. With a finite excitation current, the voltage drop across the parallel circuit would be infinite. The inductor and capacitor currents are of opposite sign because a 90° phase lead and a 90° phase lag add up to a total of 180°. At resonance, these two currents are also of *equal* magnitude. The resulting circular current pattern represents the loss-free exchange of energy between the two storage elements without the need for *external* current excitation.

The equation for the admittance of the parallel circuit has the same general form as Eq. (1.157), except that the roles of L and C are interchanged. The parallel admittance $Y_p(\omega)$, therefore, has the same general frequency dependence as the series impedance $Z_s(\omega)$ (bracket terms in Fig. 1.37c).

The frequency-normalized expressions for the parallel admittance differ from those of the series impedance, Eqs. (1.159) and (1.160), only by a factor:

$$Y_p(j\omega) = j\sqrt{\frac{C}{L}}\left(\frac{\omega}{\omega_0} - \frac{\omega_0}{\omega}\right) \tag{1.161}$$

$$Y_p(jf) = j\sqrt{\frac{C}{L}}\left(\frac{f}{f_0} - \frac{f_0}{f}\right) \tag{1.162}$$

Practical *LC* resonance circuits always contain parasitic or deliberately added dissipative components. Such circuits are also called *damped* resonance circuits in contrast to the dissipation free, *undamped* ones discussed in this section. If the

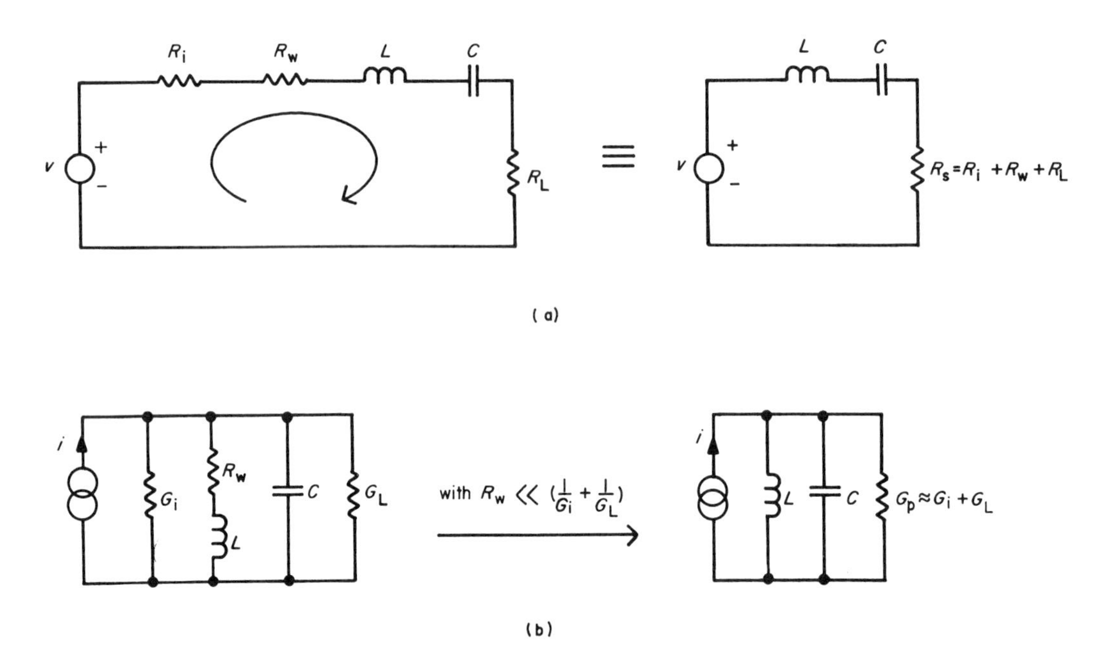

Figure 1.38 Second-order circuits with dissipation: *RLC* series and parallel circuits.

dissipation cannot be ignored, we must apply the methods developed in the next section.

1.4.3.2 Damped Resonance Circuits: Transient Behavior

A series *LC* resonance circuit may contain several dissipative components. For example, the circuit on the left of Fig. 1.38a includes the inner resistance of the source R_i, a resistance representing the wire losses in the inductor R_w, and a load resistance R_L. (The dielectric losses are assumed to be negligible). All resistive components are combined in an equivalent series resistance R_s. With this simplification we obtain the *series RLC circuit* on the right of Fig. 1.38a. The dual, *parallel RLC (or GLC) circuit* on the right of Fig. 1.38b is an approximate model of the realistic circuit on the left.

Before we derive the immittances of the damped resonance circuits, we shall develop their differential equations and consider their transient response to stepwise excitations.

Consider the series *RLC* circuit in Fig. 1.39a. With the switch closed at time $t=0$, the KVL equation is

$$v_L + v_R + v_C = V_0$$

Insertion of the element laws into this equation gives us

$$L\frac{di_s}{dt} + i_s R_s + \frac{1}{C}\int_0^t i_s\,dt + \frac{q(0)}{C} = V_0 \tag{1.163}$$

where $q(0)$ is the initial charge on the capacitor. Differentiation on both sides leads

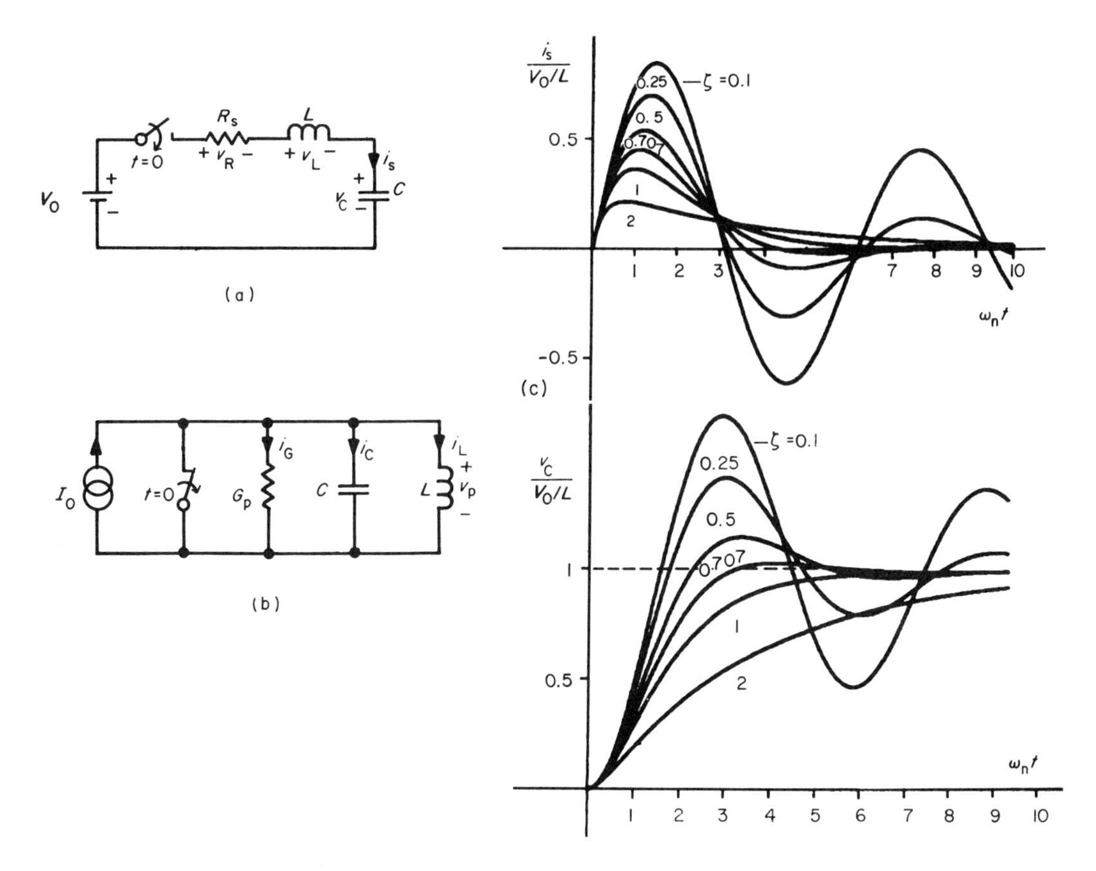

Figure 1.39 Transient response of second-order *RLC* circuits. (See text.)

to the *homogeneous second-order differential equation*

$$\frac{d^2 i_s}{dt^2}+\frac{R_s}{L}\frac{di_s}{dt}+\frac{1}{LC}i_s=0 \tag{1.164}$$

Equation (1.164) is often written in the *standard form*

$$\frac{d^2 i_s}{dt^2}+2\zeta\omega_n\frac{di_s}{dt}+\omega_n^2 i_s=0 \tag{1.165}$$

where $\omega_n=1/\sqrt{LC}$ is the *undamped natural frequency*, corresponding to the radian frequency at resonance for the loss-free case according to Eq. (1.158), and ζ is the *damping ratio*, defined as

$$\zeta=R_s/2\omega_n L$$

From the theory of linear differential equations with constant coefficients we know that Eq. (1.165) has solutions of the form

$$i_s=Ae^{st}$$

Inserting this solution into Eq. (1.165) we obtain

$$s^2Ae^{st}+2\zeta\omega_n sAe^{st}+\omega_n^2 Ae^{st}=0$$

and after cancellation of the common factor Ae^{st}, we are left with

$$s^2 + 2\zeta\omega_n s + \omega_n^2 = 0 \tag{1.166}$$

Equation (1.166) is the *characteristic equation* of the differential equations, Eq. (1.165). The roots of the characteristic polynomial are called the *characteristic values* or *natural modes* of the system. They specify the time course of the transient behavior.

As the characteristic equation is of second order, there must be two roots, namely,

$$s_1, s_2 = -\zeta\omega_n \pm \omega_n\sqrt{\zeta^2 - 1} = -\sigma \pm \omega_d \tag{1.167}$$

where $\sigma = \zeta\omega_n$ is the damping constant and $\omega_d = \omega_n\sqrt{1-\zeta^2}$ is the *damped natural frequency*. The damped natural frequency ω_d approaches the undamped natural frequency ω_n when the damping ratio ζ becomes very small. On the other hand, ω_d approaches zero as the damping ratio ζ approaches unity.

The damping ratio is a function of the ratio of the resistive component R_s to the impedance magnitude of either of the two storage elements at frequency ω_n. At that frequency the impedance of the inductor is $Z_L(\omega_n) = \omega_n L$, and the impedance of the capacitor is $Z_C(\omega_n) = 1/\omega_n C$. With these quantities we obtain from the earlier definition

$$\zeta = \frac{1}{2}\frac{R_s}{Z_L} = \frac{1}{2}\frac{R_s}{Z_C} = \frac{1}{2Q} \tag{1.168}$$

where $Q = \omega_n L/R_s = 1/\omega_n C R_s$ is the *circuit Q* of the *series RLC* circuit. Like the damping ratio, the circuit Q is a figure of merit. As Q increases to infinity, the condition of loss-free resonance is approached. High Q values are possible with very good circuit components. Frequently, however, the dissipative component R_s is rather sizable, and a low Q results. The effects of different Q values will become apparent before this section is finished.

Depending on whether the characteristic equation has two distinct real roots, a repeated real root, or a pair of complex conjugate roots, the system is defined as *overdamped*, *critically damped*, or *underdamped*, respectively.

Let us consider the three cases.

1. The *overdamped* system—two distinct real roots. From Eqs. (1.167) and (1.168) we know that the following conditions must be satisfied:

   $$\zeta > 1, \qquad Q < \tfrac{1}{2}$$

   The general solution of the homogeneous differential equation, Eq (1.165), has then the form

   $$i_s(t) = A_1 e^{s_1 t} + A_2 e^{s_2 t} \tag{1.169}$$

   where

   $$s_1 = -\zeta\omega_n + \omega_n\sqrt{\zeta^2 - 1} = -\sigma + \omega_d$$

   $$s_2 = -\zeta\omega_n - \omega_n\sqrt{\zeta^2 - 1} = -\sigma - \omega_d$$

   Note that s_1 and s_2 are *negative* and real; the solution is, therefore, the superposition of two *decaying* exponentials.

The application of the transform methods outlined in Chapter 3 to Eq. (1.165) for the initial conditions $q(0)=0$ and $i_s(0)=0$ (which lead to $di_s/dt = V_0/L$ at $t=0$) gives the overdamped solution as

$$i_s(t)=\frac{V_0}{2\omega_n L\sqrt{\zeta^2-1}}\left[e^{-\left(\zeta-\sqrt{\zeta^2-1}\right)\omega_n t}-e^{-\left(\zeta+\sqrt{\zeta^2-1}\right)\omega_n t}\right] \tag{1.170}$$

This solution is plotted in the upper panel of Fig. 1.39c for $\zeta=2$.

2. The *critically damped* system—a repeated real root. The appropriate parameter conditions are

$$\zeta=1, \qquad Q=\tfrac{1}{2}$$

In this case the previous solution model, Eq. (1.169), must be modified because $s_1=s_2=-\omega_n$. The theory of differential equations specifies that, in addition to the exponential solution, there must be a solution of the type $te^{s_1 t}$. The appropriate model for the total solution is now

$$i_s(t)=A_1e^{s_1 t}+A_2te^{s_1 t}$$

If we take the initial values to be the same as in the previous case, the final form of the solution can be shown to be

$$i_s(t)=\frac{V_0}{L}te^{-\omega_n t} \tag{1.171}$$

This solution, too, is plotted in the upper panel of Fig. 1.39c ($\zeta=1$). Note: in this case the undamped natural "frequency" ω_n is mathematically equivalent to the reciprocal of a time constant. As in the preceding case, its presence in the equations representing the solutions, obviously, does *not* imply that the time course of the current is oscillatory. The reason for the absence of oscillations lies in the fact that in the cases considered thus far, damping is either critical or more than critical.

3. The *underdamped* system—a pair of complex conjugate roots. The corresponding parameter conditions are

$$\zeta<1, \qquad Q>\tfrac{1}{2}$$

The two complex roots are in this case

$$s_1=-\zeta\omega_n+j\omega_n\sqrt{1-\zeta^2}=-\sigma+j\omega_d$$

$$s_2=-\zeta\omega_n-j\omega_n\sqrt{1-\zeta^2}=-\sigma-j\omega_d$$

With the same initial conditions as before, the solution is

$$i_s(t)=\frac{V_0}{\omega_d L}e^{-\sigma t}\sin\omega_d t \tag{1.172}$$

or

$$i_s(t)=\frac{V_0e^{-\zeta\omega_n t}}{\omega_n L\sqrt{1-\zeta^2}}\sin\left(\omega_n\sqrt{1-\zeta^2}\,t\right) \tag{1.173}$$

Equation (1.172) or (1.173) describes a decaying or damped sinusoidal oscillation. The *envelope* of the oscillating function is given by the exponential damping factor

$$\frac{\pm V_0}{\omega_d L} e^{-\sigma t} = \frac{\pm V_0}{\omega_n L\sqrt{1-\zeta^2}} e^{-\zeta\omega_n t}$$

The solution is plotted for several values of ζ(0.1, 0.25, 15, and $1/\sqrt{2}=0.707$) in the upper panel of Fig. 1.39c.

At this point, the reader may wish to review Fig. 1.15 and correlate the generalized frequency concept with the various types of roots of the characteristic equation, Eq. (1.166).

In the cases discussed so far in this section, the current always decayed to zero as long as the damping ratio ζ was not zero. Thus in the *steady state*, as t approaches infinity, the current vanishes. The various solutions plotted in the upper panel of Fig. 1.39c, therefore, represent *transient responses* to the switching process. They should be compared to the first-order plots of Figs. 1.34a and 1.34b (bottom panels). The steady state response could have been nonzero if the network topology had been different or if we had chosen a variable other than the current. We shall briefly consider the latter case, and discuss the behavior of the capacitor voltage for the network in Fig. 1.39a.

According to the element law for the capacitor we have

$$v_C(t) = \frac{1}{C}\int_0^t i_s\,dt + \frac{q(0)}{C}$$

We can now insert the various solutions for i_s into this equation and solve for v_C by integration. The solutions for various damping ratios are plotted in the bottom panel of Fig. 1.39c. They represent *prototypical responses of a second-order system in transition from one steady state level* (here $v_C=0$ for $t<0$) *to another* (here $v_C(\infty)=V_0$). The manner in which the new steady state level is approached is a function of the damping ratio ζ. The voltage curves in Fig. 1.39c show this clearly.

For critical ($\zeta=1$) or more than critical ($\zeta>1$) damping, the new steady state level is approached in a monotonic *non*oscillatory fashion. In general, the larger the damping ratio ζ, the longer it will take the overdamped system to come close to the new level. For strongly underdamped systems (e.g., $\zeta=0.1$), the new level is reached *more quickly*, but the system response *overshoots* and finally settles in an *oscillatory* fashion to the new level. This oscillatory transient is often called *ringing*. In general, the lower the damping ratio, the larger are the oscillatory overshoots and undershoots, and the more cycles it will take for the system to settle down.

The parallel circuit of Fig. 1.39b can be analyzed in analogous fashion to the series circuit just discussed. KCL is chosen instead of KVL. The resulting differential equation for the voltage v_p is exactly analogous to that for i_s, Eq. (1.165). The solution for v_p after opening the switch at $t=0$ has the same form as that shown for i_s in the upper panel of Fig. 1.39c, while that for the inductor current i_L corresponds to the plot in the lower panel of Fig. 1.39c.

1.4.3.3 Damped Resonance Circuits: Immittances

The sinusoidal steady state behavior of the series and parallel *RLC* circuits is completely described by their ac immittances. The derivation of these immittances is straightforward according to the rules established in Sections 1.3.5 and 1.3.6.

For the circuit on the right of Fig. 1.38a we can write by inspection

$$Z_s(j\omega) = R_s + j\omega L + \frac{1}{j\omega C} = R_s + j\left(\omega L - \frac{1}{\omega C}\right) \tag{1.174}$$

The imaginary part of the impedance vanishes if the radian frequency ω equals the undamped natural frequency, which is defined as before:

$$\omega_n = 2\pi f_n = \frac{1}{\sqrt{LC}} \tag{1.175}$$

With Eq. (1.175) we can change Eq. (1.174) into the three frequency-normalized forms:

$$Z_s(j\omega) = R_s + j\sqrt{\frac{L}{C}}\left(\frac{\omega}{\omega_n} - \frac{\omega_n}{\omega}\right) = R_s\left[1 + jQ\left(\frac{\omega}{\omega_n} - \frac{\omega_n}{\omega}\right)\right] \tag{1.176}$$

$$Z_s(jf) = R_s + j\sqrt{\frac{L}{C}}\left(\frac{f}{f_n} - \frac{f_n}{f}\right) = R_s\left[1 + jQ\left(\frac{f}{f_n} - \frac{f_n}{f}\right)\right] \tag{1.177}$$

$$Z_s(j\Omega) = R_s + j\sqrt{\frac{L}{C}}\left(\Omega - \frac{1}{\Omega}\right) = R_s\left[1 + jQ\left(\Omega - \frac{1}{\Omega}\right)\right] \tag{1.178}$$

where $\Omega = \omega/\omega_n = f/f_n$ is the normalized frequency and $Q = G_s\sqrt{L/C} = \omega_n L/R_s = 1/\omega_n C R_s$ is the circuit Q of the series RLC circuit.

In order to plot the impedance independently of the particular value R_s, it is advantageous to normalize once more and obtain the *dimensionless* impedance:

$$z_s(j\Omega) = \frac{Z_s(j\Omega)}{R_s} = 1 + jQ\left(\Omega - \frac{1}{\Omega}\right) \tag{1.179}$$

The magnitude and phase of this normalized impedance are

$$|z_s(j\Omega)| = \left[1 + Q^2\left(\Omega - \frac{1}{\Omega}\right)^2\right]^{1/2} \tag{1.180}$$

and

$$\sphericalangle z_s = \tan^{-1}\left[Q\left(\Omega - \frac{1}{\Omega}\right)\right] \tag{1.181}$$

The curves corresponding to Eqs. (1.180) and (1.181) are plotted in Fig. 1.40a. At the undamped natural frequency ($\Omega = 1$), the impedance magnitude is unity (its minimum) and the phase angle is zero, as long as $Q < \infty$. As Q increases, the slopes of the magnitude curves become steeper and the transition of the phase angle from $-\pi/2$ to $+\pi/2$ becomes more abrupt. As a figure of merit, the circuit Q, indicates how closely the series RLC approximates the loss-free LC circuit.

For a frequency f_1 below the resonance frequency f_n, and for a frequency f_2 above the resonance frequency, the real part of Z_s equals the absolute value of the imaginary part. Under these conditions we have

$$|Z_s(f_1)| = |Z_s(f_2)| = \sqrt{2}\,R_s$$

$$\sphericalangle Z_s(f_1) = -\frac{\pi}{4}\ (-45°)$$

$$\sphericalangle Z_s(f_2) = \frac{\pi}{4}\ (45°)$$

It can be shown that the resonance frequency f_n is the *geometric* mean of the frequencies f_1 and f_2, i.e., on a *logarithmic* scale f_1 and f_2 are equidistant from f_n.

The frequency interval from f_1 to f_2 is defined as the *bandwidth* B of the impedance. The following relations hold:

$$B = f_1 - f_2 = \frac{f_n}{Q} = df_n \tag{1.182}$$

where $d = 1/Q$ is defined as the *dissipation factor*. From Eq. (1.182) it is obvious that a narrowly "tuned" resonance circuit, i.e., a circuit with a narrow bandwidth, requires a very high Q or, conversely, a very low dissipation factor. A constant Q or d means that the *relative* bandwidth B/f_n stays unchanged as f_n is varied.

The *admittance* of the series *RLC* circuit is, of course, given by the reciprocals of the impedance equations. The dimensionless *normalized* admittance is obtained from Eq. (1.179):

$$y_s(j\Omega) = \frac{Y_s(j\Omega)}{1/R_s} = \frac{1}{1 + jQ\left(\Omega - \frac{1}{\Omega}\right)} \tag{1.183}$$

For the magnitude and phase we have then

$$|y_s(\Omega)| = \frac{1}{|z_s(\Omega)|} = \frac{1}{\left[1 + Q^2\left(\Omega - \frac{1}{\Omega}\right)^2\right]^{1/2}} \tag{1.184}$$

and

$$\sphericalangle y_s = -\sphericalangle z_s = \tan^{-1}\left[Q\left(\frac{1}{\Omega} - \Omega\right)\right] \tag{1.185}$$

The plots of Eqs. (1.184) and (1.185) appear in Fig. 1.41b.

The immittance formulas of the *parallel RLC* circuit naturally are the exact duals to the formulas derived for the series circuit. For the circuit on the right in Fig. 1.38b, we can write the admittance as

$$Y_p(j\omega) = G_p + j\omega C + \frac{1}{j\omega L} = G_p + j\left(\omega C - \frac{1}{\omega L}\right) \tag{1.186}$$

With the definition for the undamped natural frequency ω_n, according to Eq. (1.175) we obtain the equivalent expression to Eq. (1.179) for the normalized admittance, namely,

$$y_p(j\Omega) = \frac{Y_p(j\Omega)}{G_p} = 1 + jQ\left(\Omega - \frac{1}{\Omega}\right), \tag{1.187}$$

where $\Omega = \omega/\omega_n = f/f_n$ is again the normalized frequency and $Q = \sqrt{C/L}/G_p = R_p\sqrt{C/L} = \omega_n C R_p = R_p/\omega_n L$ is the circuit Q of the *parallel RLC* circuit.

The magnitude and phase of this dimensionless admittance are plotted in Fig. 1.40a.

The expression for the *impedance* of the parallel *RLC* circuit is the reciprocal of the admittance equation, Eq. (1.186). Mathematically, the impedance equations are equivalent to Eqs. (1.183)–(1.185) for the admittance of the series *RLC* circuit. The corresponding magnitude and phase plots are in Fig. 1.40b.

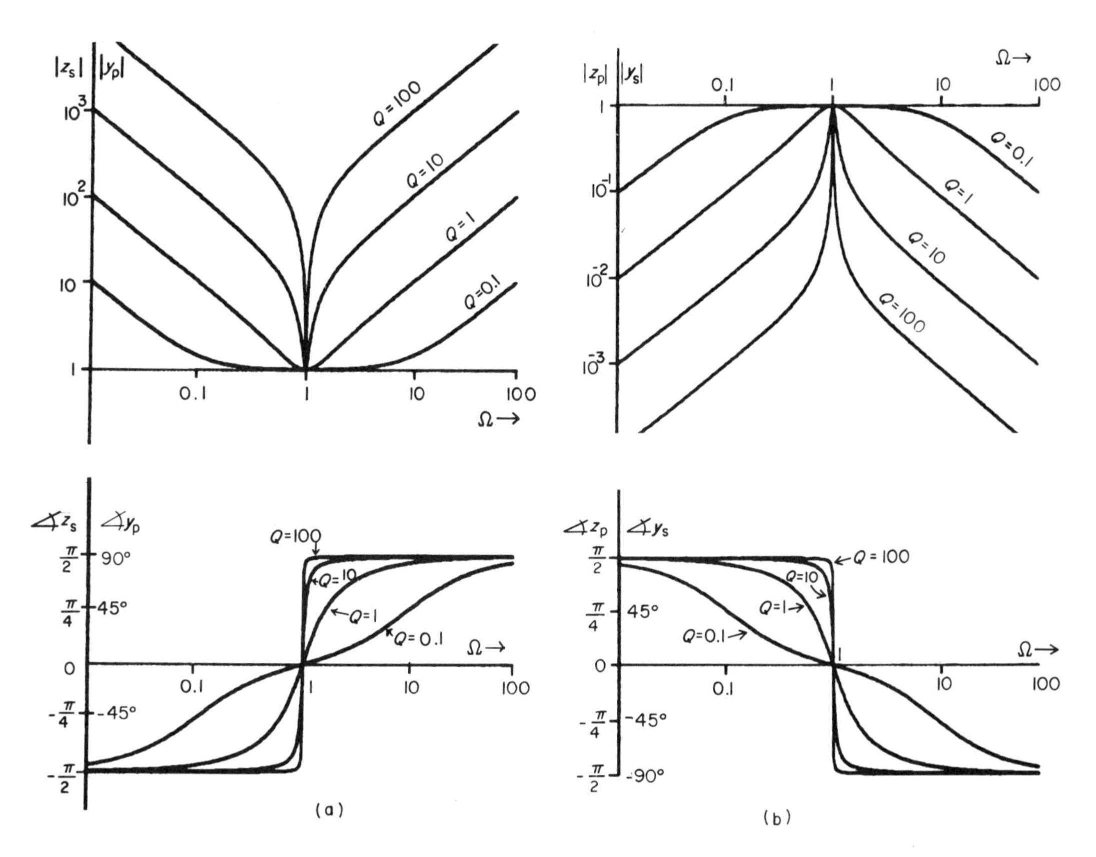

Figure 1.40 Normalized immittance plots for series and parallel *RLC* circuits: **(a)** magnitude and phase versus normalized frequency for normalized series impedance and parallel admittance; **(b)** magnitude and phase versus normalized frequency for normalized series admittance and parallel impedance.

1.4.3.4 Damped Resonance Circuits: Simple Filters

Series and parallel *RLC* circuits serve as simple filters in the form of frequency-dependent voltage dividers and current dividers, respectively. Four useful examples are given in Fig. 1.41.

For the voltage divider in Fig. 1.41a, we derive the frequency response function $G(j\omega)=\mathbf{V}_2(j\omega)/\mathbf{V}_1(j\omega)$ from Eqs. (1.136), with $s=j\omega$, and (1.176):

$$G(j\omega)=\frac{\mathbf{V}_2(j\omega)}{\mathbf{V}_1(j\omega)}=\frac{R_s}{Z_s(j\omega)}=\frac{R_s}{R_s+j\sqrt{\frac{L}{C}}\left(\frac{\omega}{\omega_n}-\frac{\omega_n}{\omega}\right)}$$

Dividing numerator and denominator by R_s, and normalizing the frequency gives

$$G(j\Omega)=\frac{1}{1+jQ[\Omega-(1/\Omega)]} \tag{1.188}$$

This, of course, is the same as the equation of the admittance of the series *RLC*

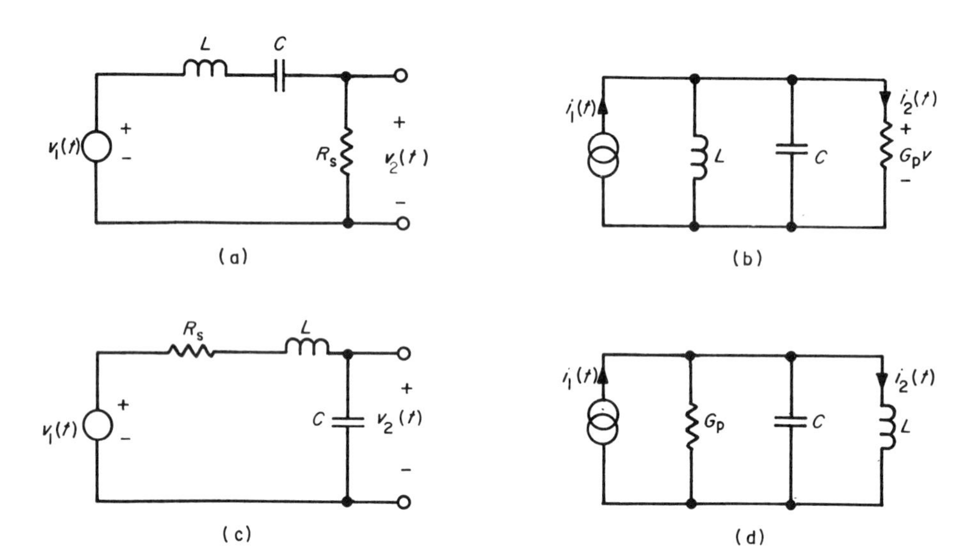

Figure 1.41 Second-order filter: **(a)** bandpass (voltage divider); **(b)** bandpass (current divider); **(c)** low pass (voltage divider); **(d)** low pass (current divider).

circuit, Eq. (1.183). The corresponding frequency response curves are the same as those in Fig. 1.40b.

From the frequency response curves it is obvious that this filter maximizes transmission at the resonance frequency f_n ($\omega_n, \Omega = 1$). Frequencies below and above f_n are attenuated. The filter is, therefore, called a *bandpass* tuned to the resonance frequency f_n. The bandwidth or *pass band* is defined as before by Eq. (1.182). The pass band ranges from frequency f_1 to frequency f_2. At these frequencies, the magnitude of $G(j\Omega)$ has dropped to $1/\sqrt{2} = 0.707$, or 71% of its maximal value at the resonance frequency.

Example 1.16. Parallel RLC as bandpass

Consider the current divider circuit in Fig. 1.41b. For the frequency response function relating the output current i_2 to the input current i_1, we can write, in phasor notation,

$$G(j\omega) = \frac{\mathbf{I}_2(j\omega)}{\mathbf{I}_1(j\omega)} = \frac{\mathbf{V}(j\omega)G_p}{\mathbf{V}(j\omega)Y_p(j\omega)} = \frac{G_p}{Y_p(j\omega)}$$

$$= \frac{G_p}{G_p + j\sqrt{\dfrac{C}{L}}\left(\dfrac{\omega}{\omega_n} - \dfrac{\omega_n}{\omega}\right)}$$

Normalization according to Eq. (1.187) gives the same frequency response function as Eq. (1.188).

Two second-order *low-pass* filters, again based on the voltage and current divider principle, are shown in Figs. 1.41c and 1.41d. The frequency response function for the circuit in Fig. 1.41c can be derived from the voltage divider equation as in the case of Fig. 1.41a:

$$G(j\omega)=\frac{\mathbf{V}_2(j\omega)}{\mathbf{V}_1(j\omega)}=\frac{1/j\omega C}{R_s+j\omega L+(1/j\omega C)}$$

or

$$G(j\omega)=\frac{1}{(1-\omega^2 CL)+j\omega CR_s}$$

With $\omega_n=1/\sqrt{CL}$ and $\zeta=\frac{1}{2}\omega_n CR_s$, as defined earlier, we obtain the frequency-normalized form of the filter transfer function:

$$G(j\omega)=\frac{1}{1-(\omega/\omega_n)^2+j2\zeta(\omega/\omega_n)} \tag{1.189}$$

Equation (1.189) is the frequency-normalized form of the quadratic factor found in the denominator of Eq. (1.149), if we set $T=1/\omega_n$.

Exactly the same frequency response function can be derived for the input–output pair i_1 and i_2 of Fig. 1.41d.

The magnitude and phase equations of $G(j\omega)$, as defined by Eq. (1.189), are

$$|G(\omega)|=\left\{\left[1-\left(\frac{\omega}{\omega_n}\right)^2\right]^2+\left(2\zeta\frac{\omega}{\omega_n}\right)^2\right\}^{-1/2} \tag{1.190}$$

$$\measuredangle G(\omega)=-\tan^{-1}\left[\frac{2\zeta(\omega/\omega_n)}{1-(\omega/\omega_n)^2}\right] \tag{1.191}$$

For frequencies very small compared to the resonance frequency, Eqs. (1.190) and (1.191) are approximated by

$$|G(\omega)|\simeq 1 \tag{1.192}$$

$$\measuredangle G(\omega)\simeq 0 \tag{1.193}$$

for

$$\omega \ll \omega_n$$

Equations (1.192) and (1.193) represent the *low-frequency* asymptotes of the frequency response.

For very large frequencies in comparison to the resonance frequency, the following approximations are applicable:

$$|G(\omega)|\simeq\frac{1}{(\omega/\omega_n)^2} \tag{1.194}$$

$$\measuredangle G(\omega)\simeq -\pi\ (-180°) \tag{1.195}$$

Equations (1.194) and (1.195) apply to the *high-frequency* asymptotes.

At resonance, $\omega = \omega_n$, the *real* part of $G(j\omega)$ vanishes, and the magnitude and phase are

$$|G(\omega_n)| = \frac{1}{2\zeta} \tag{1.196}$$

$$\sphericalangle G(\omega_n) = -\frac{\pi}{2} \; (-90°) \tag{1.197}$$

As we compare Eq. (1.196) with Eqs. (1.192) and (1.194), we see that, at resonance, the magnitude is highly dependent on the damping ratio, whereas at frequencies far away from the resonance frequency (at least by a decade), the damping ratio is of little importance.

From the notion of resonance, one is easily misled into thinking that the magnitude $|G(\omega)|$ reaches its maximum at ω_n. In fact, the maximum of $|G(\omega)|$, Eq. (1.190), occurs at the frequency

$$\omega_m = \omega_n \sqrt{1 - 2\zeta^2} \tag{1.198}$$

Note that Eq. (1.198) requires $\zeta < 1/\sqrt{2} = 0.707$; otherwise ω_m becomes an imaginary number. For damping ratios $\zeta \geqq 0.707$, we always have $|G(\omega)| \leqq 1$, according to Eq. (1.190). Magnitudes larger than unity are possible for $\zeta < 0.707$. In that case, the maximum occurs at ω_m. Inserting the expression for ω_m, Eq. (1.198), into Eq. (1.190) gives the peak magnitude as a function of the damping ratio ζ:

$$|G(\omega_m)| = \frac{1}{2\zeta\sqrt{1 - \zeta^2}} \tag{1.199}$$

for

$$\zeta < \frac{1}{\sqrt{2}} = 0.707$$

The features just discussed are easily visualized in the Bode plot, Fig. 1.42. From Eq. (1.197) we derive the log-magnitude of the low-frequency asymptote as

$$\mathrm{Lm}\, G(\omega) = 0 \text{ dB} \qquad \text{for} \quad \omega \to 0$$

The high-frequency asymptote in the Bode plot is, according to Eq. (1.194),

$$\mathrm{Lm}\, G(\omega) \simeq -40 \log\left(\frac{\omega}{\omega_n}\right) \text{dB} \qquad \text{for} \quad \omega \gg \omega_n$$

The slope of the high-frequency asymptote is, therefore, -40 dB/decade or -12 dB/octave.

Table 1.5 lists a number of useful values computed from Eqs. (1.190), (1.191), (1.193), (1.198), and (1.199). This table also applies to quadratic factors in the *numerator* of a frequency response function, if we simply multiply all log-magnitude and phase values in the table by -1. The Bode plot of such a numerator term is the *mirror* image of the plots in Fig. 1.42, reflected about the axes $\mathrm{Lm}\, G = 0$ and $\sphericalangle G = 0$.

For $\zeta = 0.707$, the maximal deviation from the asymptotic approximation is -3 dB at ω_n. This is the same as in the case of the first-order filter, Fig. 1.36a, except that frequencies much higher than ω_n are attenuated at -40 dB/decade, i.e., at twice the rate of the first-order low pass, and the phase lag is $-90°$ at ω_n instead of $-45°$.

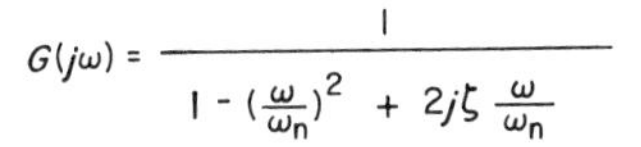

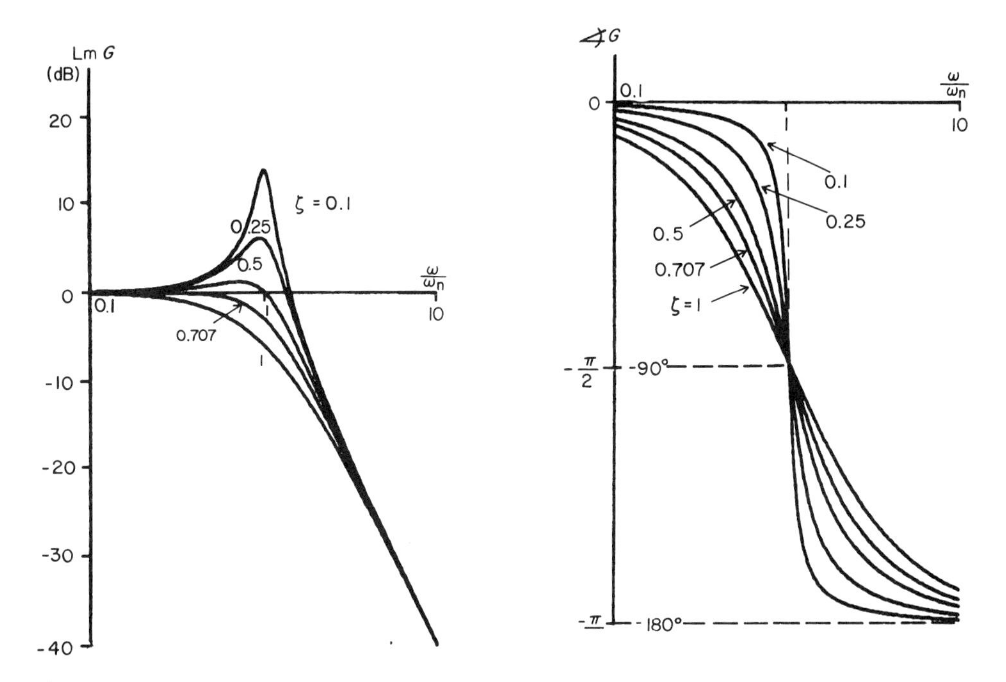

Figure 1.42 Second-order Bode plot.

The methods developed in Section 1.4 enable us to deal with a very large number of practical problems. Networks more complicated than the ones discussed in this section can often be reduced to or approximated by such simple networks. However, when such simplifications are not possible, we can use the strategies described in the next sections.

Table 1.5 Log-magnitude (Lm) and Phase Values for $G(j\omega)=\left[1-\left(\frac{\omega}{\omega_n}\right)^2+j2\zeta\left(\frac{\omega}{\omega_n}\right)\right]^{-1/2}$

	$\mathrm{Lm}\,G(\omega)$ (dB)					$\measuredangle G(\omega)$ (degrees)					Peak Frequency (ω_m) and Peak Magnitude	
ζ	$0.1\omega_n$	$0.5\omega_n$	ω_n	$2\omega_n$	$10\omega_n$	$0.1\omega_n$	$0.5\omega_n$	ω_n	$2\omega_n$	$10\omega_n$	ω_m/ω_n	$\mathrm{Lm}\,G(\omega_m)$ (db)
1	−0.09	−1.9	−6	−14	−40.09	−11	−53	−90	−127	−169	(0)	(0)
0.707	~0	−0.26	−3	−12	−40	−8	−43	−90	−137	−172	(0)	(0)
0.5	~0	+0.9	0	−11	−40	−6	−34	−90	−146	−174	0.707	+1.25
0.25	+0.08	+2	+6	−10	−39.9	−3	−18	−90	−162	−177	0.94	+6.3
0.1	+0.09	+2.4	+14	−9.6	−39.9	−1	−7.6	−90	−172.4	−179	0.99	+14

1.5 Analysis of Complex Networks

This section deals with general strategies for the analysis of networks. For the sake of brevity, we shall not discuss networks with coupled inductances (mutual inductance); they are well presented in various texts (Van Valkenburg, 1974; Desoer and Kuh, 1969). Also, the treatment of dependent or controlled sources is postponed until Section 1.6.

The first method of analysis is the simple, but often cumbersome, application of the Kirchhoff laws in combination with the element laws. This method, however, is easy to grasp and has the advantage that it is also applicable in the case of time-varying or nonlinear elements, though the solutions of the resulting equations might be difficult to obtain.

A large part of this section is devoted to a number of "shortcuts." These shortcuts greatly reduce the labor involved in the straight application of the Kirchhoff laws. The reduction in labor comes about because we can simplify the network topology or reduce the number of equations by the implicit application of the Kirchhoff laws or the superposition principle. In most instances, it is necessary that the passive network elements be linear.

The approaches outlined in this and the next section and in Chapters 2 and 3 are entirely adequate for most analysis problems of discrete networks. However, readers who wish to study the subject in depth should consult more advanced texts (Desoer and Kuh, 1969; Ley et al., 1959).

1.5.1 The Straight Application of the Kirchhoff and Element Laws

The point of the analytical procedure is to find a sufficient number of independent equations that can be solved for the voltages or currents we desire to know. In the general case, we have to determine the voltage–current pairs of the branch elements. A network with B branches presents us then, in general, with $2B$ unknowns. Section 1.3.4 introduced one method to find the $2B$ equations necessary for the complete analysis. We illustrate the procedure in this section.

In order to introduce the method we first consider a resistance network. Our demonstration object is the *lattice* or *bridge* network in Fig. 1.43. The schematics of Figs. 1.43a and 1.43b are topologically identical; the latter version is easier to label in preparation for the analysis.

The first step is to *identify the nodes* of the network and label them (A, B, C, and D; Fig. 1.43b). One node is designated as the *reference node* (D); it is often the node at zero or ground potential. The other nodes are the independent nodes (A, B, and C) of the network.

The second step is to find a complete set of *independent loops* for the network. The procedure followed here is to choose a tree connecting all nodes (branches 1, 2, and 5; heavy lines in Fig. 1.43c), and then form independent meshes as the links (3, 4, and 6; light lines in Fig. 1.43c) are added to the tree. For each mesh a reference direction is defined by a circular arrow. (Other procedures for finding independent loops or meshes can be found in, among other works, Desoer and Kuh, 1969; Ley et al. 1959).

The third preparatory step is to label all *branch currents* and *branch voltages*. At this state, it is entirely arbitrary which polarities are assigned to either the branch

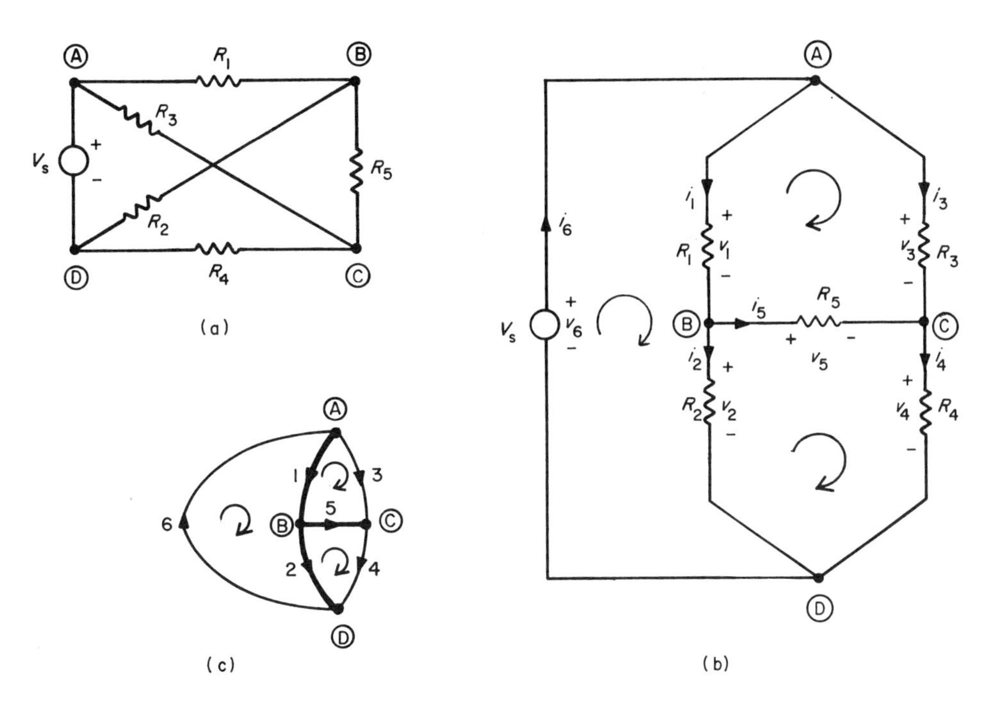

Figure 1.43 Resistive bridge or lattice circuit. (See text.)

currents or the branch voltages. It is required, however, that the reference directions for the current–voltage pair of a passive branch be associated (Fig. 1.1).

With the network properly "prepared" in this fashion, we can now write $2B$ equations for the $2B$ unknowns. The equations are then solved by any of several methods. For now, the crucial aspect is the proper formulation of the network equations. Three procedures may be used.

I. The network equations *in terms of branch currents and branch voltages* are represented by three sets of equations:
   1. The KCL equations for all the independent nodes.
   2. The KVL equations for a complete set of independent loops (meshes).
   3. The element laws for the branches.

II. The network equations *in terms of branch currents alone* are obtained by inserting the element laws (3) into the KVL equations for the independent loops (2). The resulting equations, together with the KCL equations for the independent nodes (1), form a complete set of branch current equations.

III. The network equations *in terms of branch voltages alone* are obtained by inserting the element laws (3) into the KCL equations for the independent nodes (1). The resulting equations, together with the KVL equations for the independent loops (2), form a complete set of branch voltage equations.

Example 1.17.

According to procedure I, we can write the set of network equations for the circuit in Fig. 1.43 by inspection.

1. The KCL equations for the independent nodes:

Node A: $i_1 + i_3 - i_6 = 0$ (1.200)

Node B: $-i_1 + i_2 + i_5 = 0$ (1.201)

Node C: $-i_3 + i_4 - i_5 = 0$ (1.202)

2. The KVL equations for the independent meshes according to Fig. 1.43c:

Link 3: $-v_1 + v_3 - v_5 = 0$ (1.203)

Link 4: $-v_2 + v_4 + v_5 = 0$ (1.204)

Link 6: $v_1 + v_2 = v_6$ (1.205)

3. The element laws for the six branches:

$$v_1 = R_1 i_1 = i_1/G_1 \quad (1.206)$$

$$v_2 = R_2 i_2 = i_2/G_2 \quad (1.207)$$

$$v_2 = R_3 i_3 = i_3/G_3 \quad (1.208)$$

$$v_4 = R_4 i_4 = i_4/G_4 \quad (1.209)$$

$$v_5 = R_5 i_5 = i_5/G_5 \quad (1.210)$$

$$v_6 = V_s \quad (1.211)$$

Equations (1.200)–(1.211) constitute a set of 12 equations for the 6 branch currents and the 6 branch voltages. Mathematical textbooks provide a variety of methods of solving these linear simultaneous equations. Among these are the method of determinants, matrices, and the Gauss elimination or reduction scheme. The last method is particularly effective in the case of 4 or more unknowns.

Usually, it is better to express the network equations in terms of either the branch currents or the branch voltages alone. This reduces the number of equations by 50%, and the complementary branch variables can still be determined from the element laws if we need to know them. This reduction in the number of equations is demonstrated next.

Example 1.18.

Following procedure II, we use the element laws, Eqs. (1.206)–(1.211), in the KVL expressions, Eqs. (1.203)–(1.205). With Eqs. (1.200)–(1.202) we have then a complete set of network equations in terms of the branch currents:

$$i_1 + i_3 - i_6 = 0 \quad (1.200)$$

$$-i_1 + i_2 + i_5 = 0 \quad (1.201)$$

$$-i_3 + i_4 - i_5 = 0 \quad (1.202)$$

$$-R_1 i_1 + R_3 i_3 - R_5 i_5 = 0 \quad (1.212)$$

$$-R_2 i_2 + R_4 i_4 + R_5 i_5 = 0 \quad (1.213)$$

$$R_1 i_1 + R_2 i_2 = V_s \quad (1.214)$$

This set of equations determines all branch currents. Branch current i_6, however, is the current supplied by the ideal voltage source, which supplies whatever current is needed. If we are not interested in it, we may omit Eq. (1.200), and concern ourselves only with the remaining five equations for $i_1 - i_5$.

Example 1.19.

The alternative set of network equations in terms of the unknown branch voltages is derived according to procedure III. The element laws, Eqs. (1.206)–(1.211), are used in the KCL expressions, Eqs. (1.201) and (1.202). Together with Eqs. (1.203)–(1.205), these equations constitute the desired set:

$$-v_1 + v_3 - v_5 = 0 \tag{1.203}$$

$$-v_2 + v_4 + v_5 = 0 \tag{1.204}$$

$$v_1 + v_2 = v_6 = V_s \tag{1.205}$$

$$-G_1 v_1 + G_2 v_2 + G_5 v_5 = 0 \tag{1.215}$$

$$-G_3 v_3 + G_4 v_4 - G_5 v_5 = 0 \tag{1.216}$$

Note that we omitted Eq. (1.200) because it contains the current, i_6, supplied by the independent voltage source. This current cannot be expressed in terms of an element law; i_6 is independent of $v_6 = V_s$. In any case, we do not need this additional equation; the corresponding branch voltage is the known source voltage V_s.

In most instances it is better to use either the branch current set (procedure II) or the branch voltage set (procedure III) rather than the complete set (procedure I) because we rarely need to know all branch variables. In the case of linear networks, it is possible to reduce the number of equations even more if we use variables *other* than the branch currents or branch voltages. Later in this section we shall see that we can write an equation for each independent loop (mesh) in terms of *loop (mesh) currents* or, alternatively, an equation for each independent node in terms of *node-to-datum voltages (node voltages)*. Any branch current can then be expressed in terms of loop currents, or any branch voltage in terms of node voltages.

The general strategy for setting up network equations in terms of branch variables (procedures I–III) is also suitable for networks containing storage elements. Because of the nature of the element laws of storage elements, the integrals and/or differential quotients will be introduced into the network equations. However, if phasor methods (or transform methods; Chapter 3) are applicable, the network equations will again be algebraic.

The strategies developed so far in this section are straightforward but cumbersome. A number of shortcuts make it possible to reduce the complexity of most network analysis problems. Generally the shortcuts require that the network be linear. The rest of this section is devoted to an exposition of several such shortcuts.

1.5.2 Shortcut I: Use of Equivalent Circuits

The point of this shortcut is to *reduce the number of independent nodes and loops in a network through the introduction of Thévenin and Norton equivalent circuits and through the replacement of series and parallel arrangements of passive elements by their appropriate equivalents.* The strategy is particularly advantageous if we are interested in only one or two of the branch variables of a network, as in the next example.

Example 1.20.

Let us return to the network in Fig. 1.43 and apply the shortcut to it. The network in Fig. 1.43 is also called a *Wheatstone bridge*, which is used to measure unknown resistances by the *bridge null* method. The resistors R_1-R_4 are called the "arms" of the bridge, and resistor R_5 is replaced by a sensitive ammeter which monitors the "bridge current" i_5. One of the resistors, e.g., R_4, is the unknown element. The object is to express the resistance of the unknown resistor in terms of the other known resistances, i.e., R_1, R_2, and R_3. A simple relation exists between the four resistances of the bridge arms for the null condition $i_5=0$, which is achieved by adjusting one or more of the known resistors and monitoring the galvanometer replacing R_5.

In order to derive the relation between the resistances for the null condition $i_5=0$, we need to focus our analysis only on the branch current i_5. We shall now manipulate the network in Fig. 1.43 into a simpler version suitable for the investigation of i_5.

The first step in the transformation of the bridge network is shown in Fig. 1.44a. Node A of the original network (Fig. 1.43) is "split" in order to reveal that the network contains two voltage dividers formed by R_1-R_2 and R_3-R_4, respectively. (The source, V_s, is repeated in order to facilitate the visualization of the voltage dividers.)

Next, the two voltage dividers are replaced by their Thévenin equivalents. This reduces the original network to a single-mesh circuit with the bridge current i_5, as the mesh current (Fig. 1.44b).

We now apply KVL to this equivalent circuit of the bridge and obtain from Fig. 1.44b

$$I_5(R_{s_1}+R_5+R_{s_2})=V_{oc_1}-V_{oc_2}$$

Obviously, the null condition $i_5=0$ is achieved for

$$V_{oc_1}=V_{oc_2}$$

Figure 1.44 Application of the voltage divider principle to the circuit in Fig. 1.43. (See text.)

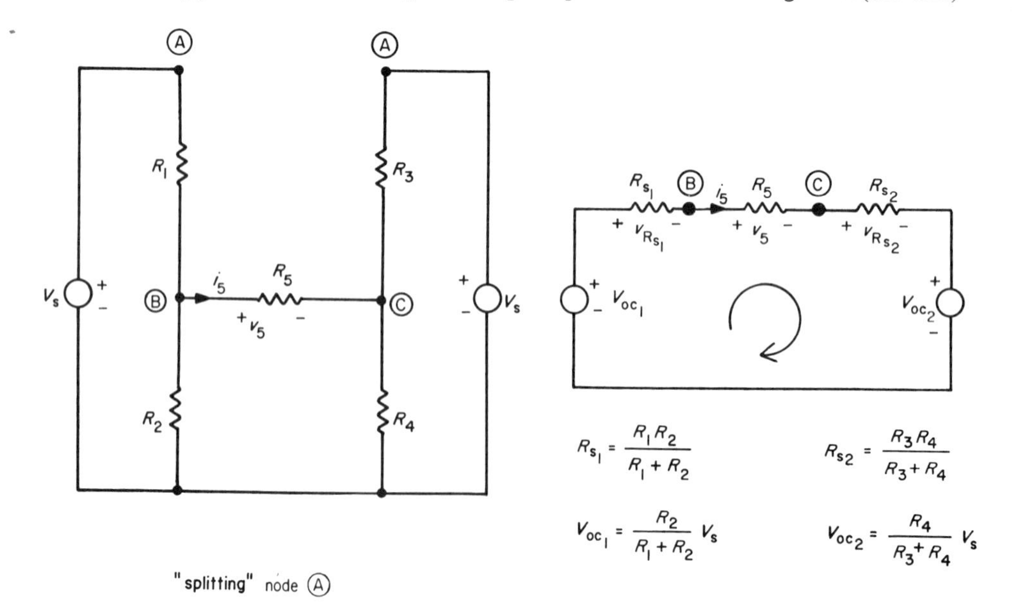

or

$$\frac{R_2}{R_1+R_2}=\frac{R_4}{R_3+R_4} \tag{1.217}$$

i.e., the voltage divider ratios for the bridge arms must be the same. Equation (1.217) can be manipulated further to give us the *null condition in terms of the bridge arm resistances* or the *bridge ratio*:

$$\frac{R_1}{R_2}=\frac{R_3}{R_4} \tag{1.218}$$

If R_4 is the unknown resistance, R_x, Eq. (1.218) becomes

$$R_x=\left(\frac{R_2}{R_1}\right)R_3 \qquad (\text{for } i_5=0)$$

This shortcut is, of course, not restricted to resistive circuits. In Fig. 1.45 it is indicated how a cascaded low-pass filter can be reduced to a simple voltage divider by using generalized phasor analysis and the method of equivalent circuits. Figure 1.45a shows that, initially, the parts to the left of the broken line a and to the right of the broken line b are replaced by a Thévenin equivalent ($\mathbf{V}_{oc}$ and Z_{s_1}) and a parallel combination (Z_L), respectively. In Fig. 1.45b, the node B is eliminated so that only the impedance voltage divider is left (Fig. 1.45c). Note that the output voltage phasor $\mathbf{V}_L$ is left as an explicit variable.

Figure 1.45 Reduction of a network. (See text.)

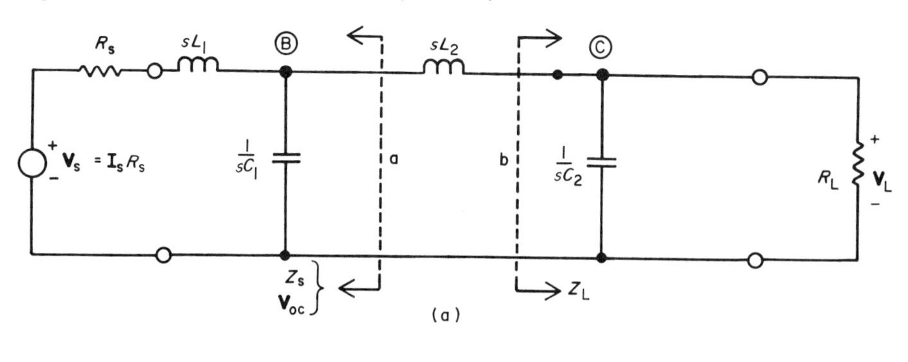

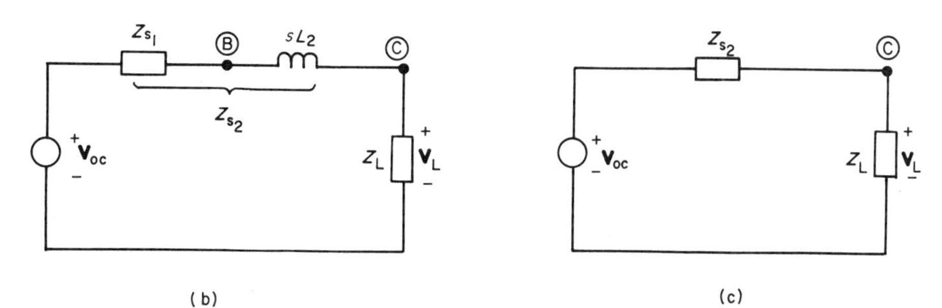

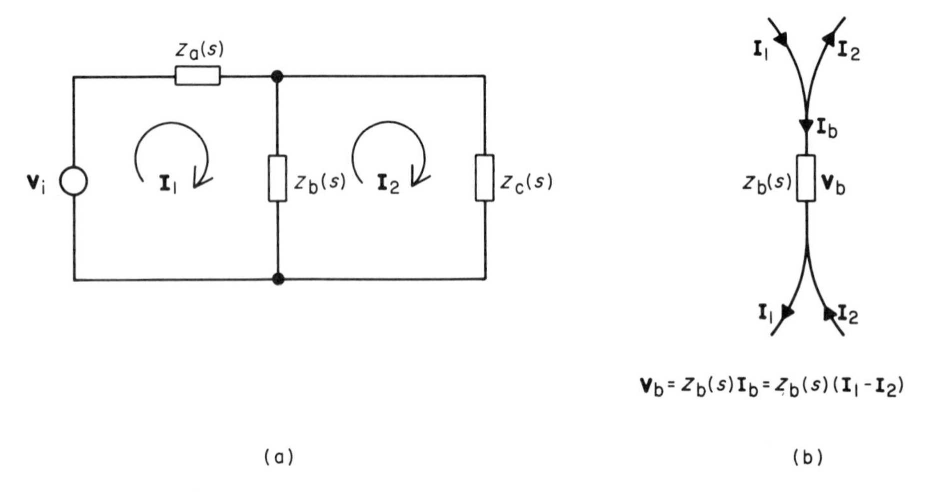

Figure 1.46 Mesh analysis: **(a)** mesh currents; **(b)** mesh and branch currents. (See text.)

1.5.3 Shortcut II: Mesh Analysis with Mesh Currents

This method differs from all previous ones in that the analysis is not based on branch variables. For example, procedure II of Section 1.5.1 gives the network equations in terms of branch currents. The shortcut to be introduced in this section *uses mesh currents in order to formulate a complete set of KVL equations*. After solving this set of equations for the mesh currents, any branch voltage or current can be determined.

For the purpose of illustrating the method, consider the circuit of Fig. 1.46a. In terms of generalized phasors we can write two *mesh equations*, i.e., KVL equations in terms of mesh current phasors. Note that the impedance $Z_b(s)$ is common to both meshes and is, therefore, traversed by *two* mesh currents circulating through it in opposite directions (Fig. 1.46b). The two mesh equations are then

$$Z_a(s)\mathbf{I}_1 + Z_b(s)(\mathbf{I}_1 - \mathbf{I}_2) = \mathbf{V}_i$$

$$Z_b(s)(\mathbf{I}_2 - \mathbf{I}_1) + Z_c(s)\mathbf{I}_2 = 0$$

After collecting terms, we get a typical set of mesh equations:

$$[Z_a(s) + Z_b(s)]\mathbf{I}_1 - Z_b(s)\mathbf{I}_2 = \mathbf{V}_i \tag{1.219}$$

$$-Z_b(s)\mathbf{I}_1 + [Z_b(s) + Z_c(s)]\mathbf{I}_2 = 0 \tag{1.220}$$

With these two equations, network behavior is completely described because branch currents are either identical to mesh currents (except, maybe, for the sign) or are simple superpositions of mesh currents (Fig. 1.46b) and branch voltages can be determined from these branch currents and the appropriate element laws.

REMARKS: (1) The bracket terms in Eqs. (1.219) and (1.220) are the *self* or total *mesh impedances*, i.e., the sum of all impedances in a mesh. (2) The terms with a negative sign represent the interaction between meshes. (3) Very often it is convenient to transform current sources into equivalent voltage sources in order to facilitate the writing of the KVL mesh equations. (4) The method is not limited to phasors. In the time domain, the mesh equations simply contain the appropriate integral and differential forms of the element laws for storage elements. (5) If several independent sources are present, the use of ac phasors requires that all sources operate at the same frequency. Otherwise the superposition principle must be used according to Section 1.5.6.

1.5.4 Shortcut III: Nodal Analysis with Node Voltages

The *node* voltages are the electric potential differences between each independent node and the datum or reference node (node–datum voltages). Each *branch* voltage is the potential difference between two independent nodes; it is, therefore, the difference of the two corresponding node–datum voltages. The branch variables are thus completely determined by the node voltages and the element laws.

The basic idea of nodal analysis is to *write the KCL equations for each independent node in terms of the nodal voltages with the aid of the element laws*. The number of equations necessary equals the number of independent nodes.

Whether we pick mesh analysis or nodal analysis is a matter of convenience; both methods require fewer equations than the branch variable method of Section 1.5.1. Generally we choose the nodal method if there are fewer independent nodes than independent meshes, and vice versa, but other factors may influence our choice.

We illustrate the general procedure with the following example.

Example 1.21.

The network in Fig. 1.47a has four independent nodes and four independent meshes (after replacing the current source i_{30} with its Thévenin equivalent). Mesh analysis would require 4 equations in terms of mesh current. But node analysis requires only 3 equations because we can eliminate one of the independent nodes by transforming the voltage sources into their Norton equivalents (Fig. 1.47b). We write the node equations according to the labels in Fig. 1.47b by inspection:

Node 1: $$(G_1+G_2)v_1+\frac{1}{L_1}\int_0^t (v_1-v_2)\,dt$$
$$=i_{10}+i_{20}-i_{L_1}(0)$$

Node 2: $$\frac{1}{L_1}\int_0^t (v_2-v_1)\,dt+\frac{1}{L_2}\int_0^t (v_2-v_3)\,dt+C_1\frac{dv_2}{dt}$$
$$=i_{30}+i_{L_1}(0)+i_{L_2}(0)$$

Node 3: $$\frac{1}{L_2}\int_0^t (v_3-v_2)\,dt+G_3v_3+C_2\frac{dv_3}{dt}$$
$$=i_{30}-i_{L_2}(0)$$

with $v_1(0)=0$, $v_2(0)=v_{C_1}(0)$, and $v_3(0)=v_{C_2}(0)$. These equations emphasize the importance of initial conditions. Similar provisions would have to be made in a formulation of mesh

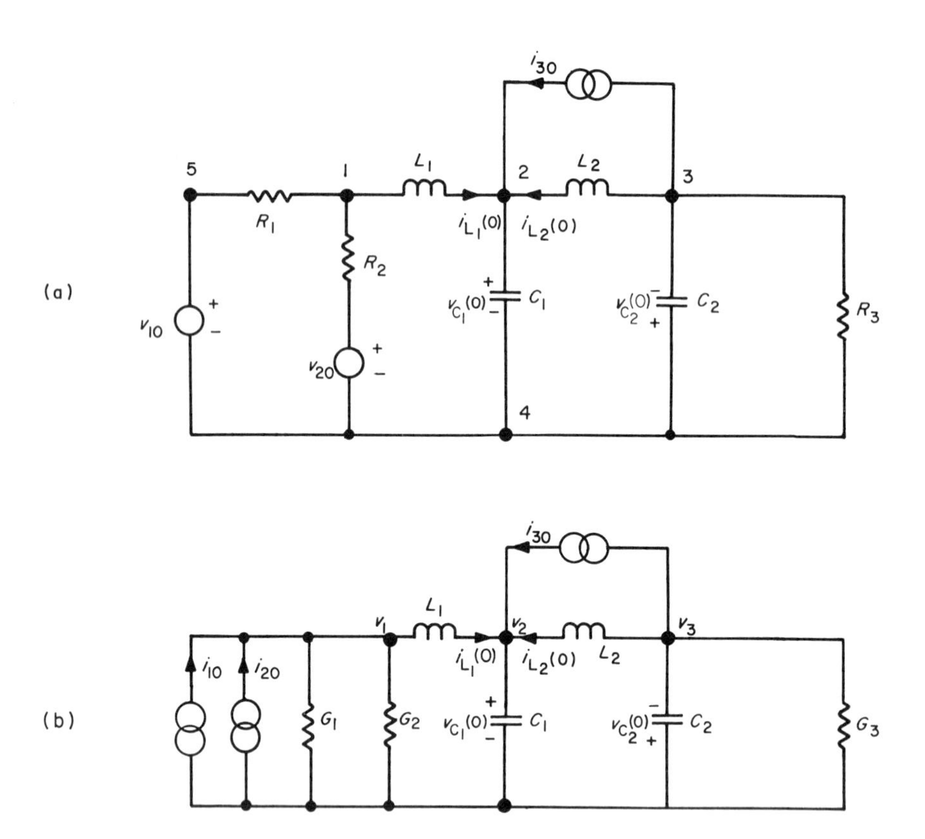

Figure 1.47 Example of a network prepared for node analysis. (See text.)

equations. In the case of zero initial conditions, introduction of phasors and rearrangement leads to the more customary form of the nodal equations:

$$\left(G_1 + G_2 + \frac{1}{sL_1}\right)\mathbf{V}_1 - \frac{1}{sL_1}\mathbf{V}_2 = \mathbf{I}_{10} + \mathbf{I}_{20} \tag{1.221}$$

$$-\frac{1}{sL_1}\mathbf{V}_1 + \left(sC_1 + \frac{1}{sL_1} + \frac{1}{sL_2}\right)\mathbf{V}_2 - \frac{1}{sL_2}\mathbf{V}_3 = \mathbf{I}_{30} \tag{1.222}$$

$$-\frac{1}{sL_2}\mathbf{V}_2 + \left(G_3 + sC_2 + \frac{1}{sL_2}\right)\mathbf{V}_3 = -\mathbf{I}_{30} \tag{1.223}$$

REMARKS: (1) The terms in parentheses in Eqs. (1.221)–(1.223) are the *self* or *total node* admittances, i.e., the sum of all admittances connected to a given node. (2) The negative terms represent the interaction between two nodes. (3) Voltage sources are usually converted to current sources as indicated in Fig. 1.47. But especially in the application of the nodal analysis to operational amplifier circuits (Chapter 4), it is customary to treat voltage sources explicitly in terms of *known* node voltages. (4) Here too, the presence of several sources requires that all of them operate at the same frequency if the ac phasor method is to be used.

1.5.5 Shortcut IV: The Superposition Principle

The superposition principle is a fundamental property of linear systems (or piece-wise linear systems with properly chosen operating points). Consider a *linear* network with several independent voltage and/or current sources, which we designate as the inputs x_1, x_2, etc. One or all of the branch variables or mesh currents or node voltages may be taken as the output(s). Let us designate one such variable by the symbol y.

As we observe the output response to one input at a time, while the other inputs are made inoperative, we obtain the "individual" responses

$$y_1 = f_1(x_1) \qquad \text{with} \qquad x_2 = 0, x_3 = 0, \ldots$$

$$y_2 = f_2(x_2) \qquad \text{with} \qquad x_1 = 0, x_3 = 0, \ldots$$

$$\vdots$$

or generally,

$$y_i = f_i(x_i), \qquad x_j = 0, \quad j \neq i$$

The *total* response is then given as the sum of individual responses. This is formally stated in

The Superposition Principle

The total response of a *linear* system to n independent inputs is the sum (superposition) of the individual responses to each input separately:

$$y = \sum_{i=1}^{n} y_i = \sum_{i=1}^{n} f_i(x_i) \tag{1.224}$$

The individual responses are obtained with respect to one input at a time while the other inputs are eliminated from the system.

The notion of "input" in this context is quite general. Obviously, if several independent sources are present, each independent source variable serves as an input variable. However, we may also apply the superposition principle where a *single* independent source provides a voltage (or current) that is *mathematically* describable as a sum of several functions. Each component function can then be regarded as a separate independent input variable. Such a "compound" source might also arise from the combination of several sources. Another example would be a source supplying an input which can be decomposed mathematically into several sinusoids of *different frequencies*. The individual sinusoidal responses for each frequency can then be computed by the phasor method giving the magnitude (amplitude) and phase for the sinusoid of that particular frequency. Thus an input

$$x(t) = A_1 \sin \omega_1 t + A_2 \sin \omega_2 t + \cdots$$

gives us the total output

$$y(t) = B_1(\omega_1) \sin[\omega_1 t + \phi_1(\omega_1)] + B_2(\omega_2) \sin[\omega_2 t + \phi_2(\omega_2)] + \cdots$$

where $B_1(\omega_1), B_2(\omega_2),\ldots$ and $\phi_1(\omega_1), \phi_2(\omega_2),\ldots$ are the phasor magnitudes and phases, respectively, for each frequency [Eq. (1.37)].

Whether we deal with a single source containing different signals or with several separate sources, the "elimination" of an input always means that, mathematically, the corresponding terms in the system or network equations must be set to zero. In this respect the two cases are indistinguishable. *Topologically*, however, differences do arise. In the case of a single source, there are no topological changes necessary. (It can be advantageous, though, to change the topology, if, for example, we wish to separate dc from ac responses, see Section 1.5.6.) When several sources are present the topology *does* change as we eliminate all sources but one because *physically* the "elimination" of sources means this: (1) a zero *voltage source* term is equivalent to the removal of the voltage source from the network and its *replacement by a short circuit* (i.e., the terminals to which the source was attached are now connected to each other); (2) a zero *current source* term is equivalent to the removal of the current source from the network and its *replacement by an open circuit* (i.e., the terminals to which the source was attached are left unconnected or "open").

The rationale for applying the superposition principle as a shortcut in the case of several sources lies precisely in the topological changes effected by the elimination of sources. The topological changes nearly always make the network simpler or reduce it to forms for which we already have solutions. The following example demonstrates this.

Example 1.22.

The circuit in Fig. 1.48a has two voltage sources. The output voltage v_1 across the terminal pair 1–1′ is a function of both source voltages. Let us designate the part of v_1 due to the source voltage v_a as v_{1a}, and that due to v_b as v_{1b}. According to the superposition principle we have then

$$v_1 = v_{1a} + v_{1b} \tag{1.225}$$

In order to determine v_{1a}, we replace the source of voltage v_b by a short circuit. For the resulting voltage divider (Fig. 1.48b) we have from Eqs. (1.128) and (1.129)

$$v_{1a} = \frac{R_2}{R_1+R_2} v_a = \frac{G_1}{G_2+G_2} v_a \tag{1.226}$$

Repeating the process in order to obtain v_{1b}, we can write for the network in Fig. 1.48c

$$v_{1b} = \frac{R_1}{R_1+R_2} v_b = \frac{G_2}{G_1+G_2} v_b \tag{1.227}$$

The complete solution follows from the addition of these partial solutions, Eqs. (1.226) and (1.227), according to Eq. (1.225):

$$v_1 = \frac{R_2 v_a + R_1 v_b}{R_1+R_2} = \frac{G_1 v_a + G_2 v_b}{G_1+G_2}$$

Another example of the topological changes resulting from the elimination of sources is shown in Figs. 1.48d–1.48f. In Fig. 1.48e the voltage source is replaced by a short circuit, whereas in Fig. 1.48f the current source is replaced by an open circuit. If the voltage drop v_3 is the desired output, we compute the partial responses v_3' and v_3'' separately from the simpler

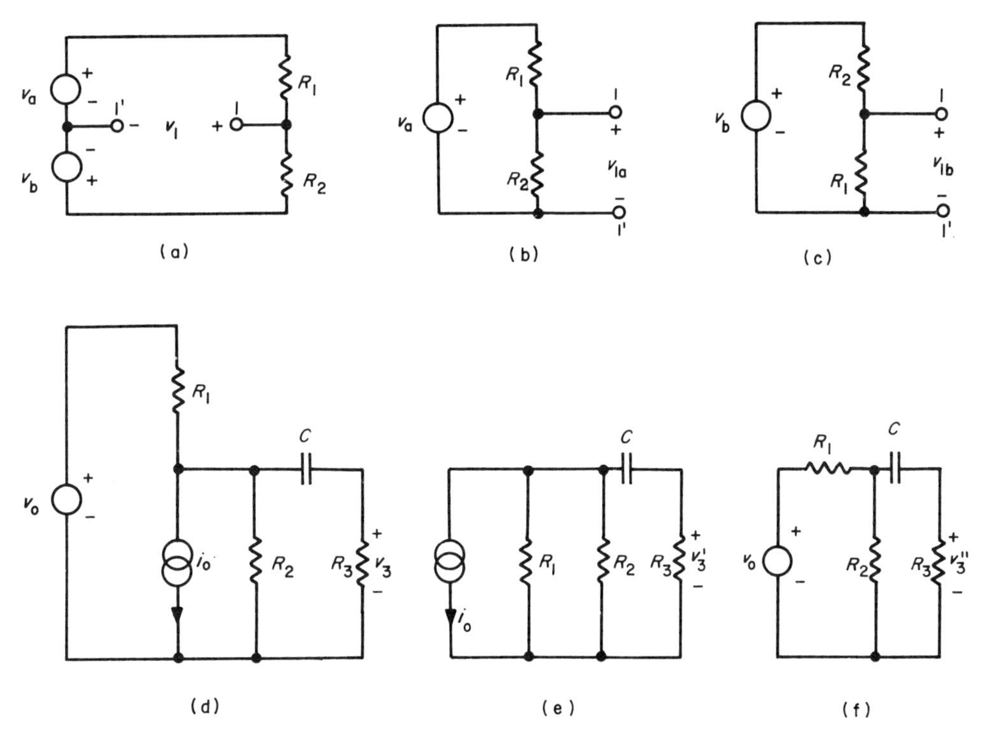

Figure 1.48 Application of the superposition principle. (See text.)

networks in Figs. 1.48e and 1.48f, respectively, and then add them for the complete response, i.e., $v_3 = v_3' + v_3''$.

1.5.6 Shortcut V: Frequency-Dependent Circuit Approximations

The point of this shortcut is to exploit the frequency dependence of the immittance of storage elements. To this end it is advantageous to divide the signals presented by the sources of a network into groups according to their frequencies. Useful subdivisions are the following: (1) dc (zero-frequency) signals; (2) low-frequency sinusoidal signals; (3) intermediate-frequency sinusoidal signals; and (4) high-frequency sinusoidal signals. (In Chapter 3 we shall discuss methods of "allocating" the appropriate frequency ranges for nonsinusoidal time-varying signals. For now it is sufficient to note that rapid changes in voltage or current levels correspond to broad frequency bands.)

Whether a signal is considered of low, intermediate, or high frequency depends on the parameter values of the storage elements of the particular circuit. *The only important question is whether for a given frequency a storage element may be removed from the circuit because its impedance equals (or approaches) either zero or infinity.* An element representing zero impedance or a very small impedance (in relation to that of other elements) is replaced by a *short circuit* in the simplified circuit diagram, whereas an element representing an infinite or a very large impedance (relative to

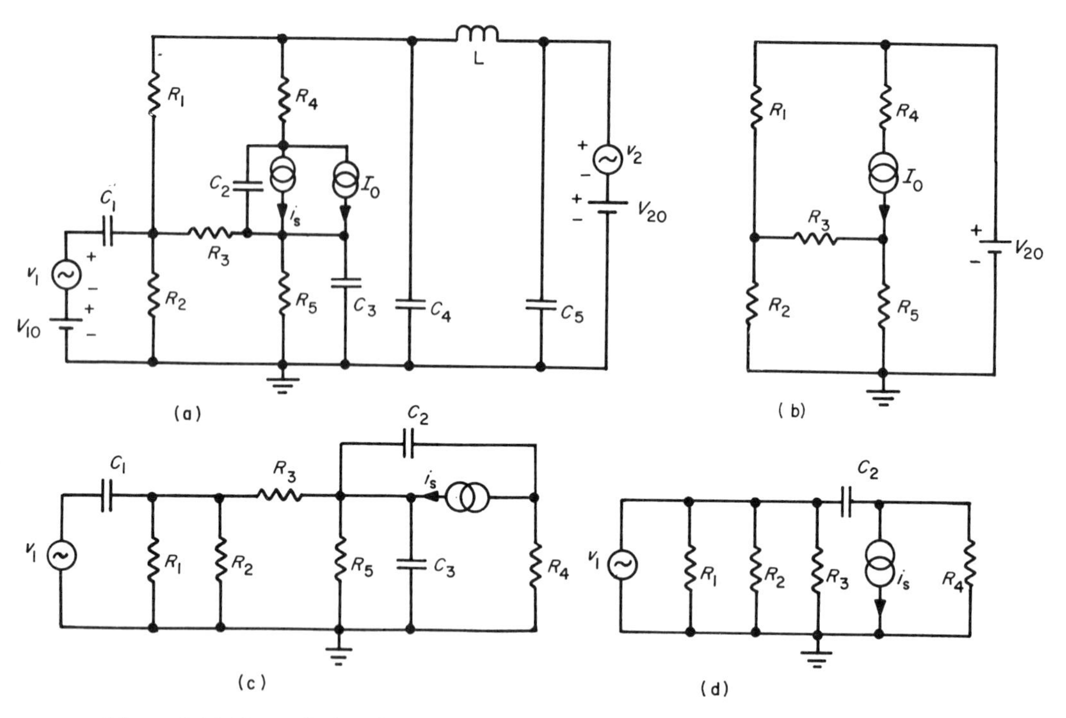

Figure 1.49 Network simplification according to frequency dependence. (See text.)

other circuit elements) is replaced by an *open circuit*. For example, at dc, capacitors are equivalent to open circuits (once the charging process is completed), and inductors can be replaced by short circuits.

The above rules for the elimination of storage elements representing extreme impedance levels are particularly useful if a network contains only storage elements of one kind, say, capacitors. Whenever both types of storage elements are present, an approximate analysis might not catch resonance effects. Hence, caution is in order.

We now demonstrate the procedure with the aid of the network in Fig. 1.49. Let us assume that the series combination $v_1 + V_{10}$ (Fig. 1.49a) represents a signal source (v_1) with a "dc bias" (V_{10}). Often we wish to prevent this dc bias from having any effect on the network. This is achieved by using "capacitor coupling" between source and network. The coupling capacitor (C_1) is also called a "blocking" capacitor because it blocks dc signals. For the series combination $v_2 + V_{20}$ (Fig. 1.49a), assume, for example, that v_2 represents an undesirable "ripple" voltage contaminating the dc supply which delivers V_{20}. The ripple voltage is kept from the circuit to the left of C_4 by the π filter C_4–L–C_5. In this low-pass filter, C_4 and C_5 are so large that their impedance is very small at the ripple frequency. The corresponding capacitors are called "bypass" capacitors because they provide a shunting bypass to ground for the undesirable ripple signal. The inductance L is chosen to present a large impedance at the ripple frequency; the corresponding inductor is called a "choke." For sufficiently large frequencies, C_3 will also act as a bypass capacitor shunting R_5.

With this discussion of the function of various storage elements, the fairly complicated network in Fig. 1.49 can now easily be broken down into simpler equivalent or approximate networks for various frequency ranges. For a given frequency range, we may first *eliminate all sources which contribute no signals in this frequency range* according to the rules set forth in Section 1.5.5. We choose three frequency ranges and obtain the following.

1. *The dc equivalent network.* The simple network in Fig. 1.49b is derived from the original by replacing the inductor with a short circuit ($Z_L(j\omega)= j\omega L=0$ for $\omega=0$) and all capacitors by open circuits ($Z_C(j\omega)=1/j\omega C=\infty$ for $\omega=0$). The time-varying but dc-free voltage sources v_1, v_2 are replaced by short circuits; the time-varying, dc-free current source i_s is removed to leave an open circuit.

   The need to analyze such dc equivalent networks arises typically in amplification circuits where the dc *operating point* must be determined. The resistors (here R_1 to R_5) are then chosen to provide the desired dc levels.
2. *The approximate low/intermediate-frequency network.* For the simplified network in Fig. 1.49c we have chosen a frequency range sufficiently high so that the bypass capacitors C_4 and C_5 can be treated as short circuits and the inductor as an open circuit, but sufficiently low so that C_1 and C_3 cannot be eliminated. All dc sources are removed according to the superposition principle. The time-varying source v_2 is eliminated because we assume that it is blocked and shunted by the π filter C_4-L-C_5 as discussed previously.
3. *The approximate high-frequency network.* The frequency is now considered so high that the blocking capacitor C_1 and the bypass capacitor C_3 act as short circuits. The remaining capacitor C_2 could, for example, be a small parasitic capacitance which becomes increasingly influential as frequency is increased.

   In this particular case, the high-frequency approximation gives us the simplest network (Fig. 1.49d), especially after the parallel combination $R_1-R_2-R_3$ is consolidated.

The shortcut method outlined in this section is extremely important for the analysis of *practical* circuits that contain many ancillary components not dedicated to the main function of the circuit. Elimination of these "extraneous" components by limiting the analysis to the frequency band of interest leads to simpler networks and hence simpler analysis.

1.6 Two-Ports, Couplers, and Controlled Sources

All the model elements discussed in the preceding sections had a single terminal pair (port). It is advantageous to broaden the class of model elements and consider *two-ports* or elements with four terminals (Fig. 1.50a). Occasionally, the two terminal pairs of the two-port share a common terminal (Fig. 1.50b). Hence a two-port must have at least three separate terminals, namely, two independent terminals and a common reference terminal (Fig. 1.50c). In general, an *n-port* has at least n independent terminals and a reference terminal so that each independent terminal forms a port or terminal pair with the common reference terminal. For the limited goals of this introductory treatment we shall limit ourselves to a simple discussion of two-ports only.

In order to incorporate two-ports into network analysis we must assign reference polarities to the port voltages and currents. The *associated reference directions* for

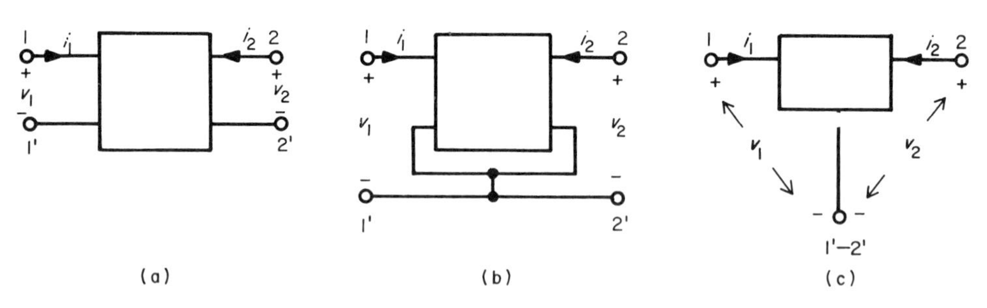

Figure 1.50 General two-ports.

two-ports are indicated in Fig. 1.50. The terminals 1′ and 2′, singly (Fig. 1.50a) or combined (Figs. 1.50b and 1.50c), serve as reference terminals.

Two-ports are often used to connect or "couple" a source (at terminals 1 and 1′) to a load (at terminals 2 and 2′) or a network to another. Therefore, they are also called *couplers*. Couplers are further classified as passive or active. *Passive* two-ports or couplers contain only passive elements. *Active* two-ports contain *dependent* sources whose output is a function of one of the port variables (in the simplest cases at least). We shall discuss now important classes of passive and active couplers.

1.6.1 A Passive Two-Port: The Ideal Transformer

This model element is a rather extreme idealization of a practical transformer. As a transformer exploits the magnetic coupling (mutual inductance) of two inductors, it is common to place two inductor coils on the *same* magnetic core in close proximity to each other. The idealization consists in assuming that (1) no energy is dissipated in the coil wire or the core, (2) the leakage flux is negligible, (3) there is no winding capacitance or other parasitic property, and (4) the self-inductance of each coil is infinite, which means that the usual element law of an inductor does *not* describe the current–voltage relations at the coil terminals. The last two assumptions also mean that the ideal transformer does not store energy.

In order to derive the laws of this coupling element let us consider Fig. 1.51a, which represents two coils of n_1 and n_2 turns, respectively, sharing a common magnetic core. In addition to the associated reference directions for voltage and current at the ports 1–1′ and 2–2′, this figure also indicates the winding sense of each coil by the *dot* convention. If both dots are either at the upper terminals (1 and 2) or the lower terminals (1′ and 2′), we can assume, according to the dot convention, that the magnetic fluxes generated by currents circulating through coils 1 and 2 will have the same polarity by the right-hand rule of electromagnetic theory. The ideal transformer symbol for the case of identical polarity is given in Fig. 1.51b. Opposite winding polarity is indicated in Fig. 1.51c. The practical consequences of equal and opposite winding polarity will be evident shortly.

Let ϕ in Fig. 1.51a stand for the magnetic flux per single winding turn. With n_1 and n_2 being the number of coil turns, we have the total fluxes for each coil as

$$\phi_1 = n_1\phi$$

and

$$\phi_2 = n_2\phi$$

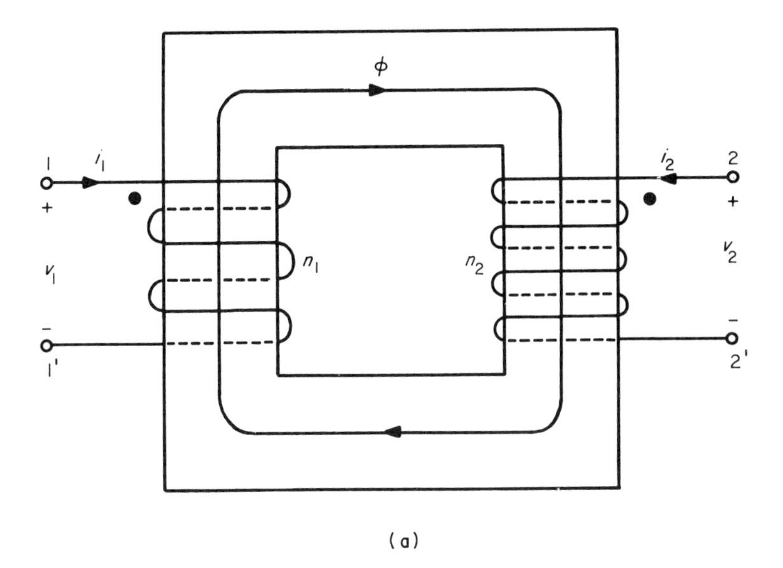

Figure 1.51 The ideal transformer. (See text.)

By Faradays's law we obtain then the port voltages

$$v_1(t) = \frac{d\phi_1}{dt} = n_1 \frac{d\phi}{dt} \tag{1.228}$$

and

$$v_2(t) = \frac{d\phi_2}{dt} = n_2 \frac{d\phi}{dt} \tag{1.229}$$

We eliminate the common flux term in Eqs. (1.228) and (1.229) and state the

Fundamental Law of the Ideal Transformer
(See Fig. 1.51b)

$$\frac{v_1(t)}{v_2(t)} = \frac{n_1}{n_2} \tag{1.230}$$

and

$$\frac{i_1(t)}{i_2(t)} = -\frac{n_2}{n_1} \tag{1.231}$$

The last equation relating the two port currents follows from our assumptions that the idealized transformer neither dissipates nor stores energy. In terms of power this means

$$v_1(t)i_1(t)+v_2(t)i_2(t)=0$$

With Eq. (1.230) we can eliminate v_2 and write

$$v_1(t)i_1(t)+\frac{n_1}{n_2}v_1(t)i_2(t)=0$$

Canceling v_1 in this expression we obtain Eq. (1.231).

If the winding sense of one of the coils is reversed (Fig. 1.51c), the fundamental law is changed to

$$\frac{v_1(t)}{v_2(t)}=-\frac{n_1}{n_2} \tag{1.232}$$

and

$$\frac{i_1(t)}{i_2(t)}=\frac{n_2}{n_1} \tag{1.233}$$

Equations (1.232) and (1.233) differ from (1.230) and (1.231) only by the sign.

REMARKS: From the preceding derivation of the fundamental transformer law we can draw several conclusions:

1. The ideal (and practical) transformer can couple only *time-varying* signals. (Alternatively, it provides "dc isolation"!) This is obvious from Faraday's law, Eqs. (1.228) and (1.229).
2. If we take one port voltage as the input, say v_1, then the other port voltage will be either an attenuated or amplified, positive or negative replica of the first depending on the *turns ratio* or *coupler ratio* (n_1/n_2) and the winding sense [Eqs. (1.230) and (1.232)]. However, amplification of a port voltage comes at the expense of attenuation of the corresponding port current and vice versa [Eqs. (1.231) and (1.233)]. This is obvious from the requirement that the power at one port equals the power observed at the other port of this passive coupler.
3. The fundamental law of the ideal transformer specifies *independent* relations for port voltages and currents, respectively. For example, the port currents might be zero because one of the ports is open circuited, but if one of the port voltages is not zero, the other will be given by Eqs. (1.230) or (1.232) no matter what the port currents are. Because of the independence of the voltage–current pairs at the ports, the element law of the ideal transformer cannot be expressed in terms of an impedance or admittance. Rather it is given as a *dimensionless transfer ratio* in terms of the turns ratio (n_1/n_2) or its reciprocal (n_2/n_1).

Now we shall consider the effects of inserting ideal transformers into circuits. As a general case, take two linear networks, S_1 and S_2, which are connected by an ideal transformer. For simplicity's sake, assume that the networks contain only sources and resistors and no storage elements. Let us represent the two linear networks by their Thévenin equivalent circuits (Fig. 1.52a). Before we proceed with the analysis we should note that the ideal transformer couples the network only magnetically, *not*

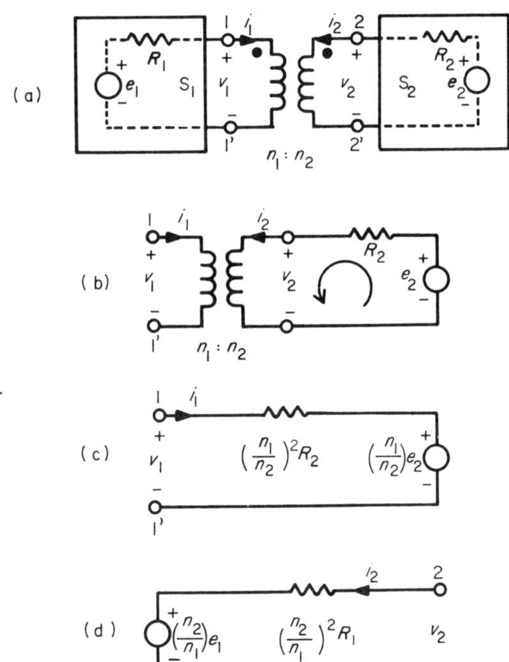

Figure 1.52 Impedance and source transformations with an ideal transformer.

galvanically (dc isolation). Hence the two networks affect each other only when *time-varying* voltages or currents are present.

The question we wish to pose is, "Looking" into port 1–1′, how does network S_1 "see" network S_2? The point here is to eliminate i_2 and v_2 by applying the coupler formulas and the appropriate network equations. We use KVL for the loop on the right in Fig. 1.52b to obtain

$$v_2 + R_2 i_2 = e_2$$

With Eqs. (1.230) and (1.231) we eliminate v_2 and i_2 and write

$$v_1 = \left(\frac{n_1}{n_2}\right)^2 R_2 i_1 + \left(\frac{n_1}{n_2}\right) e_2 \tag{1.234}$$

The equivalent circuit corresponding to Eq. (1.234) is shown in Fig. 1.52c. According to this figure, the network S_1 "sees" at the port 1–1′ an equivalent branch consisting of the *scaled* equivalent circuit of network S_2, but the scales for equivalent source and driving point resistance differ! The source voltage e_2 is multiplied by the turns ratio as expected from Eq. (1.230), whereas the equivalent resistance is multiplied by the *square* of the turns ratio.

We can generalize (on the basis of the phasor method) from resistive networks to circuits containing storage elements and say that *ideal transformers magnify driving point impedances by the square of the turns ratio.* Such impedance transformations are very useful in matching load impedances to source impedances for optimal power transfer.

Naturally, we could equally ask how network S_2 "sees" network S_1 at the port

2–2′. Writing the KVL equation for the left mesh of the circuit in Fig. 1.52a and eliminating i_1 and v_1 with the transformer equations, we obtain

$$v_2 = \left(\frac{n_2}{n_1}\right)^2 R_1 i_2 + \left(\frac{n_2}{n_1}\right) e_1$$

which is similar to Eq. (1.234) except that the *reciprocal* of the turns ratio is used. The corresponding equivalent circuit appears in Fig. 1.52d.

From the preceding analysis we conclude that the coupling action of the ideal transformer works both forward and backward: the transformer is a *bilateral* two-port.

REMARKS: The usefulness of the ideal transformer is obvious from the preceding discussion. Often, however, the scaling transformations are not the primary reason for using transformers. In fact, transformers with a turns ratio of 1 : 1 ($n_1 = n_2$) are frequently used for dc isolation or polarity reversal (Fig. 1.51c).

Before we conclude the discussion of the ideal transformer, we should stress that *practical* transformers approximate ideal transformers over a limited range of frequencies only. Resistive, capacitive, and finite inductance elements must be added to the ideal model element in order to represent actually available transformers. The reader may consult reference works such as the *Radiotron Designer's Handbook* for further details.

The ideal transformer is an example of a passive and bilateral coupler. In the next section we shall introduce a group of active and unilateral couplers.

1.6.2 Active Two-Ports I: Ideal Controlled Sources

Like the ideal transformer, the model elements discussed in this section are idealized models. However, the designer now has access to modern integrated circuit components (Chapter 4), which can be arranged into excellent approximations of these simple model elements. For a preliminary analysis, the models presented next are often quite satisfactory.

1.6.2.1 Basic Properties of Controlled Sources

We distinguish four basic types of controlled sources: voltage-controlled voltage sources (Fig. 1.53a), current-controlled current sources (Fig. 1.53b), current-controlled voltage sources (Fig. 1.53c), and voltage-controlled current sources (Fig. 1.53d). Note that diamond-shaped symbols are used to distinguish these *controlled* or *dependent* sources from independent sources (Figs. 1.8a and 1.8c). Like independent voltage (current) sources, dependent voltage (current) sources maintain the voltage (current) at the port 2–2′ *independently* of the complementary current (voltage) at this port, but the source voltage (current) *is* dependent on (or controlled by) either the voltage or the current at port 1–1′.

For the basic models introduced here, we assume that the source variable is simply proportional to the controlling variable. We can then state the

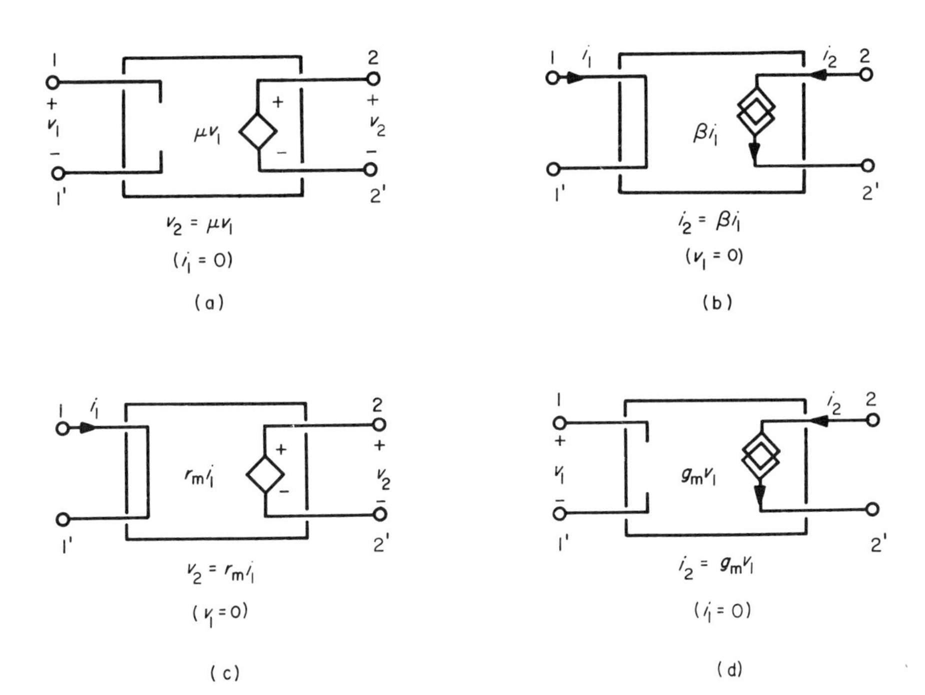

Figure 1.53 Controlled sources: **(a)** voltage-controlled voltage source; **(b)** current-controlled current source; **(c)** current-controlled voltage source; **(d)** voltage-controlled current source.

Element Laws for Ideal Controlled Sources

1. Voltage-controlled voltage source (Fig. 1.53a)

$$v_2 = \mu v_1$$
$$\mu \equiv \text{voltage ratio, voltage transfer ratio} \qquad (1.235)$$

2. Current-controlled current source (Fig. 1.53b)

$$i_2 = \beta i_1$$
$$\beta \equiv \text{current ratio, current transfer ratio} \qquad (1.236)$$

3. Current-controlled voltage source (Fig. 1.53c)

$$v_2 = r_m i_1$$
$$r_m \equiv \text{transfer resistance} \qquad (1.237)$$

4. Voltage-controlled current source (Fig. 1.53d)

$$i_2 = g_m v_1$$
$$g_m \equiv \text{transfer conductance, transconductance} \qquad (1.238)$$

Note that these element laws, Eqs. (1.235)–(1.238), do not involve all four variables from the two ports but only one each from the input port (1–1′) and the

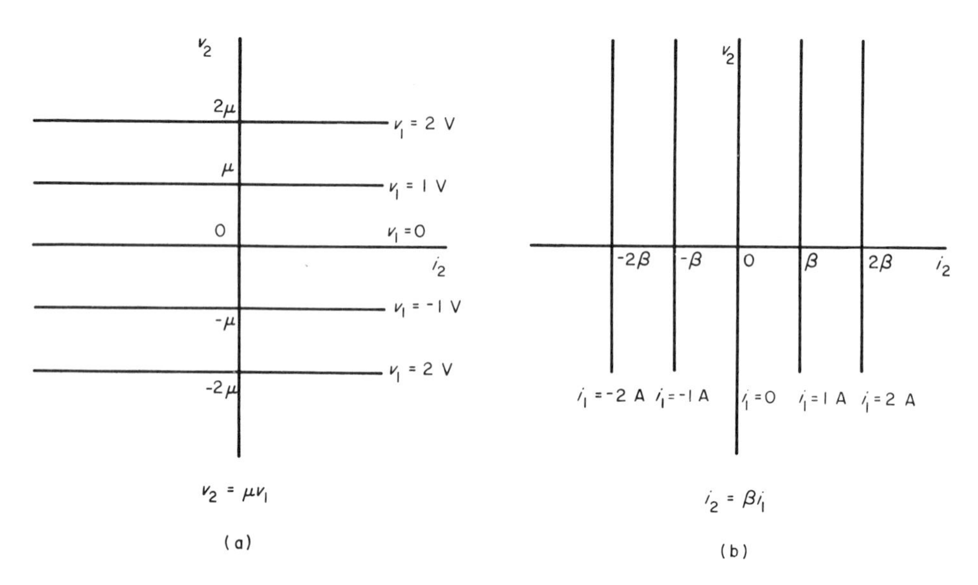

Figure 1.54 Characteristics for controlled sources: **(a)** voltage-controlled voltage source; **(b)** current-controlled current source.

output port (2–2′). The source current–voltage characteristics are, therefore, straight lines parallel to the current axis (voltage source) or to the voltage axis (current source) with the controlling variables as parameters as shown in Fig. 1.54 for two types of controlled sources. Of course, the element laws and characteristics are only valid if the voltage sources are not short circuited and the current sources are not open circuited.

The voltage and current ratios defined by Eqs. (1.235) and (1.236) are dimensionless numbers like the turns ratio of the ideal transformer. In contrast, the transfer resistance and conductance defined by Eqs. (1.237) and (1.238) have the dimensions ohm and siemens, respectively. But this dimensional equivalence to dissipative elements does not imply that these controlled sources are passive. The voltages and currents related by these parameters do not belong to the same terminal pair.

It is also important that we properly interpret the inverse forms of the element laws given by Eqs. (1.235)–(1.238). For example, we can say that Eq. (1.235) means that applying a voltage v_1 to the input port 1–1′ of the voltage-controlled voltage source *generates* the voltage $v_2 = \mu v_1$ at the output port 2–2′. But the inverse statement

$$v_1 = \frac{1}{\mu} v_2 \tag{1.239}$$

does not imply that we can apply a voltage v_2 to port 2–2′ and "generate" a voltage $v_1 = v_2/\mu$ at the input port 1–1′. Rather, the controlled sources in Fig. 1.53 are *unilateral* elements without backward action in contrast to the bilateral ideal transformer. Equation (1.239), therefore, only specifies the input voltage v_1 *necessary* to generate a given output voltage v_2.

Another interesting property of the source models in Fig. 1.53 is that the instantaneous current–voltage product is zero at the input port 1–1′ because one of

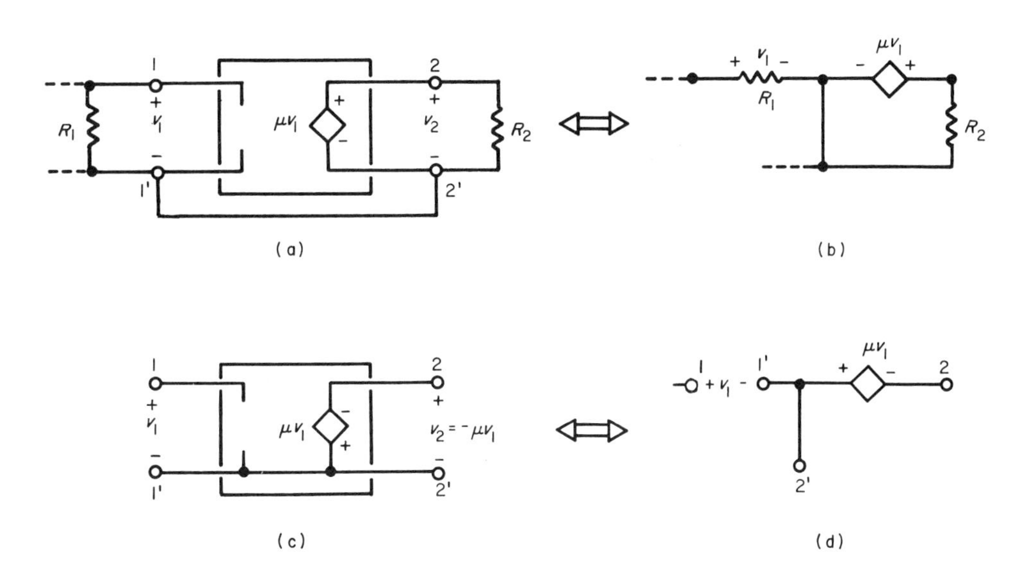

Figure 1.55 Controlled sources in networks. (See text.)

the variables is always zero. These model elements, therefore, require no input power but provide whatever power the load at the output demands.

The "boxed" representations of the controlled sources in Fig. 1.53 stress the fact that these elements are two-ports. In many instances, however, these symbols are rather cumbersome. Figure 1.55 shows how the functional relations can be indicated by simply labeling model elements and terminals.

With the basic properties of controlled sources established, we can now consider methods of analysis of circuits containing controlled sources.

1.6.2.2 Networks with Controlled Sources

In the straight application of the Kirchhoff laws to networks containing controlled sources, the model laws of controlled sources, Eqs. (1.235)–(1.238), are simply included in the set of element laws where appropriate.

Next we shall illustrate the use of mesh equations in the presence of a controlled source.

Example 1.23. Mesh equations with controlled source terms

The networks in Fig. 1.56 are identical except that one contains a controlled source. We can write the mesh equations for the network in Fig. 1.56a by inspection:

$$(R_1+R_3)i_1 - R_3 i_2 = v_1$$
$$-R_3 i_1 + (R_2+R_3)i_2 = -v_2$$

The mesh equations for the circuit containing the controlled source (Fig. 1.56b) are also written by inspection as if the controlled source were independent:

$$(R_1+R_3)i_1 - R_3 i_2 = v_1$$
$$-R_3 i_1 + (R_2+R_3)i_2 = -r_m i_1$$

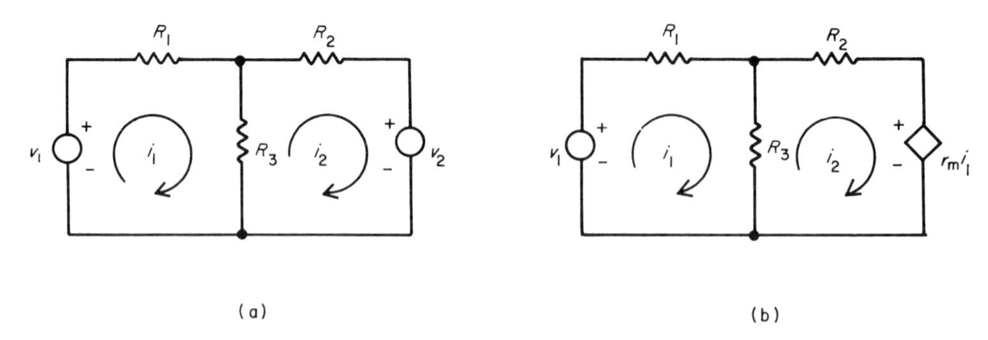

Figure 1.56 Mesh analysis with independent and controlled sources. (See text.)

On the right of the second equation, however, there is a mesh current term. Transposing this term to the left, we obtain the equation set

$$(R_1+R_3)i_1-R_3 i_2=v_1$$
$$-(R_3-r_m)i_1+(R_2+R_3)i_2=0$$

Another common shortcut is to replace networks or portions of networks by their Thévenin or Norton equivalents (Section 1.3.8). Problems arise with controlled sources especially when equivalent driving point immittances are derived without determining both the open circuit voltage and the short circuit current. *The general rule is to leave controlled sources in place* and not eliminate them for the determination of the driving point immittances.

1.6.3 Active Two-Ports II: Amplifier Models

Practical amplifiers can often be treated as functional blocks whose properties can be represented by simple two-port equivalent circuits. For example, we may use a voltage-controlled voltage source (Fig. 1.57a) as a first approximation for a practical voltage amplifier. From Fig. 1.57a we write by inspection in generalized phasor notation

$$\mathbf{V}_1'(s)=\mathbf{V}_1(s)$$

and

$$\mathbf{V}_2(s)=\mu\mathbf{V}_1(s)=\mu\mathbf{V}_1'(s)$$

so that the *voltage-to-voltage gain* or *voltage transfer ratio* is

$$G_{vv}(s)=\frac{\mathbf{V}_2(s)}{\mathbf{V}_1'(s)}=\mu \tag{1.240}$$

A more realistic approximation of an actual amplifier is the equivalent circuit in Fig. 1.57b. Typically, such amplifiers have an input impedance $Z_{in}(s)$, which causes an input current $\mathbf{I}_1(s)$ to flow. Usually there is also an output impedance $Z_{out}(s)$, so that only a fraction of the source voltage appears across the load impedance $Z_L(s)$.

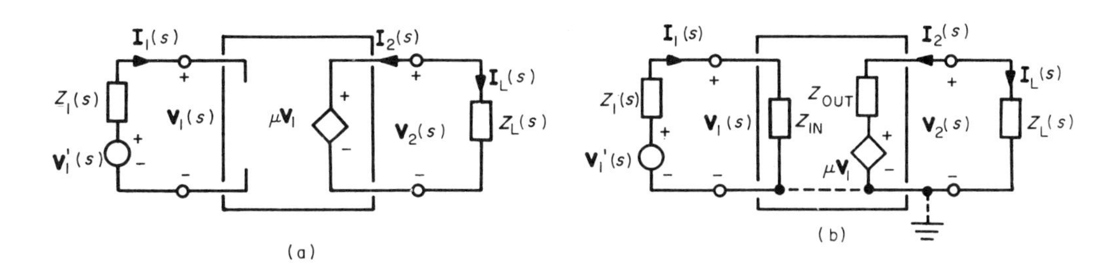

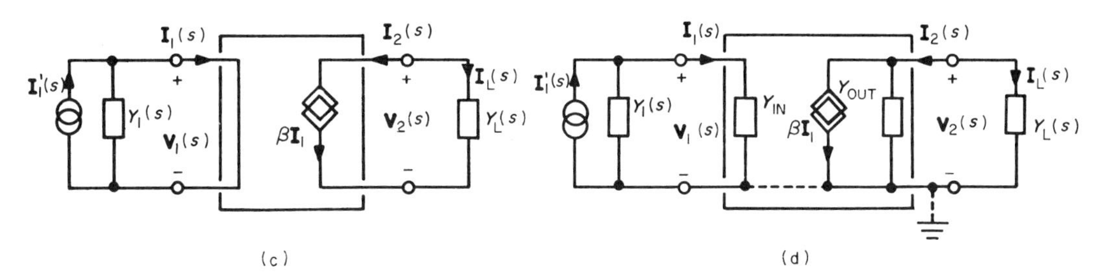

Figure 1.57 A comparison of ideal **(a, c)** and "real" **(b, d)** amplifiers. (See text.)

It is also quite typical for the input and output ports to share a common ground terminal as indicated by the broken line in Fig. 1.57b.

The analysis of the voltage amplifying properties of the circuit in Fig. 1.57b is straightforward. The two meshes in the circuit form simple voltage dividers. With Eq. (1.132) we write

$$\mathbf{V}_1(s)=\frac{Z_{\text{in}}(s)}{Z_1(s)+Z_{\text{in}}(s)}\mathbf{V}_1'(s)=\alpha_1\mathbf{V}_1'(s) \tag{1.241}$$

and

$$\mathbf{V}_2(s)=\frac{Z_{\text{L}}(s)}{Z_{\text{out}}(s)+Z_{\text{L}}(s)}\mu\mathbf{V}_1(s)=\alpha_2\mu\mathbf{V}_1(s) \tag{1.242}$$

or

$$\mathbf{V}_2(s)=\alpha_1\alpha_2\mu\mathbf{V}_1'(s)$$

where α_1 and α_2 are the input and output voltage divider ratios.

From the last equation we derive the voltage-to-voltage gain as

$$G_{\text{vv}}(s)=\frac{\mathbf{V}_2(s)}{\mathbf{V}_1'(s)}=\alpha_1\alpha_2\mu \tag{1.243}$$

In comparing Eqs. (1.240) and (1.243), we note that the more realistic amplifier model still acts like a voltage-controlled voltage source but with a reduced gain. In fact, for amplification to take place we must have $\mu>\alpha_1\alpha_2$.

The dual case of the current amplifier is modeled in Figs. 1.57c and 1.57d. The ideal current amplifier (Fig. 1.57c) has a *current-to-current gain* or *current gain*

$$G_{ii}(s)=\frac{\mathbf{I}_L(s)}{\mathbf{I}_1'(s)}=\frac{-\mathbf{I}_2(s)}{\mathbf{I}_1'(s)}=-\beta$$

The analogous formula for the more realistic model of Fig. 1.57d is

$$G_{ii}(s)=\frac{\mathbf{I}_L(s)}{\mathbf{I}_1'(s)}=\left(\frac{Y_{in}(s)}{Y_1(s)+Y_{in}(s)}\right)\left(\frac{Y_L(s)}{Y_{out}(s)+Y_L(s)}\right)(-\beta)$$

which the reader may derive from the nodal equations for the input and output ports and the expressions $\mathbf{V}_2(s)=\mathbf{I}_L(s)/Y_L(s)$ and $\mathbf{V}_1(s)=\mathbf{I}_1(s)/Y_{in}(s)$.

1.6.4 General Two-Ports

An arbitrary network with two accessible ports may serve as a coupling network connecting two circuits S_1 and S_2. In Fig. 1.58a such a situation is depicted. While the circuits S_1 and S_2 are represented by their Norton equivalents (an arbitrary choice), the coupling two-port is treated like a "black box." The only restrictions we wish to impose on the coupling network are the following. We assume that it is made up of linear, time invariant model elements, that initially all storage elements are free of energy (zero state), and that no independent sources are present.

1.6.4.1 Open Circuit Impedance Parameters

In order to describe the action of the coupling network, we simplify the analysis by assuming that the port currents supplied by the networks S_1 and S_2 in Fig. 1.58a are generated by equivalent current sources. Given the linearity of the coupling network, we apply the superposition principle and state that the port voltages should additively reflect these two external inputs (there being no internal independent

Figure 1.58 *z* parameters for a general two-port and corresponding equivalent circuits. (See text.)

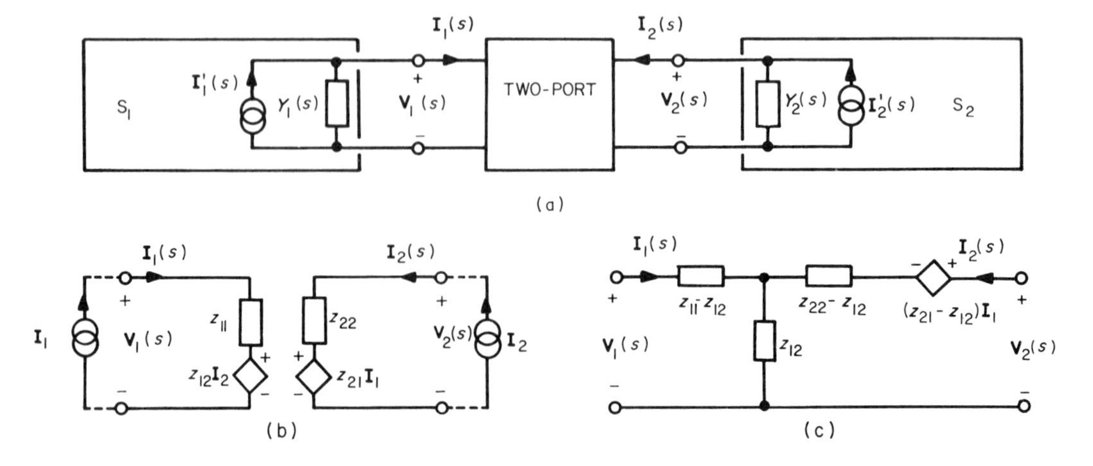

source by our assumption). In generalized phasor notation this statement is expressed as

$$\mathbf{V}_1(s)=z_{11}(s)\mathbf{I}_1(s)+z_{12}\mathbf{I}_2(s) \tag{1.244}$$

$$\mathbf{V}_2(s)=z_{21}(s)\mathbf{I}_1(s)+z_{22}(s)\mathbf{I}_2(s) \tag{1.245}$$

An equivalent circuit for the two-port satisfying Eqs. (1.244) and (1.245) is shown in Fig. 1.58b. The choice of this particular equivalent circuit means that further analysis of networks S_1 and S_2 can proceed simply by including the equivalent port branches in the formulation of the network equations. The coupling effect is evident in the controlled sources of the port branches. Of course, before the analysis can be undertaken, we must determine the coupler coefficients z_{11}, z_{12}, z_{21}, and z_{22}. If we selectively set $\mathbf{I}_2=0$ we obtain

$$z_{11}(s)=\left.\frac{\mathbf{V}_1(s)}{\mathbf{I}_1(s)}\right|_{\mathbf{I}_2(s)=0} \quad \text{and} \quad z_{21}(s)=\left.\frac{\mathbf{V}_2(s)}{\mathbf{I}_1(s)}\right|_{\mathbf{I}_2(s)=0}$$

and for $\mathbf{I}_1=0$ we have

$$z_{12}(s)=\left.\frac{\mathbf{V}_1(s)}{\mathbf{I}_2(s)}\right|_{\mathbf{I}_1(s)=0} \quad \text{and} \quad z_{22}(s)=\left.\frac{\mathbf{V}_2(s)}{\mathbf{I}_2(s)}\right|_{\mathbf{I}_1(s)=0}$$

The coefficients defined in this manner are called the *open circuit impedance parameters* of the two-port. The circuit in Fig. 1.58c is an alternative equivalent circuit for Eqs. (1.244) and (1.245). It has an entirely different topology but it is no less valid than the circuit in Fig. 1.58b. This *T-equivalent circuit* of the two-port, however, implies that the two ports have a common connection.

REMARK: At this point we should make a cautionary statement. Aside from the arbitrariness of two-port equivalent circuits just discussed, we should keep in mind that not all two-ports fit the scheme of Eqs. (1.244) and (1.245). Examples are the ideal transformer, the voltage-controlled voltage source, and the current-controlled current source, all of which cannot be described by Eqs. (1.244) and (1.245) and are characterized by dimensionless parameters (turns ratio, voltage ratio, current ratio). The current-controlled voltage source, however, does fit the scheme. Its law, Eq. (1.237), is simply a degenerate form of the impedance parameter equations. The open circuit parameters for this two-port are

$$z_{11}=0, \qquad z_{12}=0$$

$$z_{21}=r_{\mathrm{m}}, \qquad z_{22}=0$$

In order to accommodate two-ports which cannot be described by impedance parameters, a variety of other formulations have been developed. Later in this section we shall introduce two of these. More extensive discussions of the many possibilities with tabulated summaries are available in standard references (e.g., Van Valkenburg, 1974; Desoer and Kuh, 1969).

Two more comments are appropriate for the equivalent circuit of Fig. 1.58c. First, we note that the controlled source is eliminated for $z_{12}=z_{21}$. Two-ports which meet this condition are said to be reciprocal. This condition is always met by two-ports containing linear resistors, capacitors, and inductors *only*. The other point to be made is that the series combination of controlled source and impedance to the right of the T-junction in Fig. 1.58c is often replaced with a Norton equivalent. This modified form of the equivalent circuit is particularly suitable for transistors.

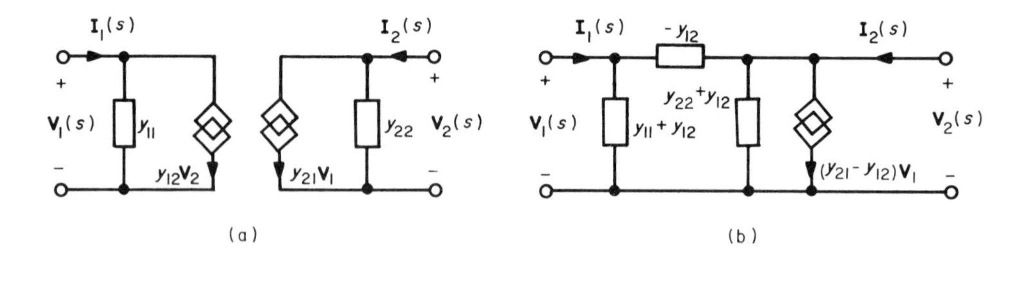

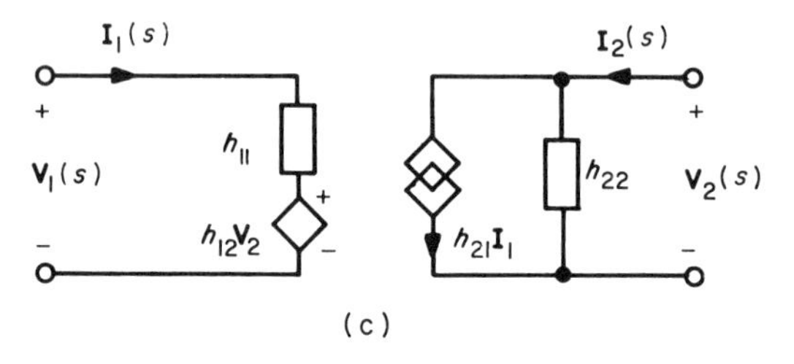

Figure 1.59 y and h parameters for a two-port. (See text.)

1.6.4.2 Short Circuit Admittance Parameters

The choice of Eqs. (1.244) and (1.245) was, of course, arbitrary. The superposition principle would have been equally satisfied had we assumed that the port *voltages* are supplied by equivalent sources, so that the port currents would be linear functions of these sources. The appropriate equations are now

$$\mathbf{I}_1(s) = y_{11}(s)\mathbf{V}_1(s) + y_{12}(s)\mathbf{V}_2(s) \tag{1.246}$$

$$\mathbf{I}_2(s) = y_{21}(s)\mathbf{V}_1(s) + y_{22}(s)\mathbf{V}_2(s) \tag{1.247}$$

Two common equivalent circuits fitting this set of coupling equations are shown in Figs. 1.59a and 1.59b. The latter is known as the *π-equivalent circuit* of a two-port.

The y coefficients are called the *short circuit admittance parameters* of the two-port because they can be determined by setting $\mathbf{V}_1 = 0$ and $\mathbf{V}_2 = 0$, respectively. The following definitions hold:

$$y_{11}(s) = \left.\frac{\mathbf{I}_1(s)}{\mathbf{V}_1(s)}\right|_{\mathbf{V}_2(s)=0} \quad \text{and} \quad y_{21}(s) = \left.\frac{\mathbf{I}_2(s)}{\mathbf{V}_1(s)}\right|_{\mathbf{V}_2(s)=0}$$

$$y_{12}(s) = \left.\frac{\mathbf{I}_1(s)}{\mathbf{V}_2(s)}\right|_{\mathbf{V}_1(s)=0} \quad \text{and} \quad y_{22}(s) = \left.\frac{\mathbf{I}_2(s)}{\mathbf{V}_2(s)}\right|_{\mathbf{V}_1(s)=0}$$

1.6.4.3 Hybrid Parameters

In this section we introduce one more alternative for the description of a two-port. We may assume that the port current arising from network S_1 in Fig. 1.58a is supplied by an equivalent current source, and the port voltage due to S_2 is supplied by an equivalent voltage source. The remaining port variables can then be determined with the superposition principle:

$$\mathbf{V}_1(s)=h_{11}(s)\mathbf{I}_1(s)+h_{12}(s)\mathbf{V}_2(s) \tag{1.248}$$

$$\mathbf{I}_2(s)=h_{21}(s)\mathbf{I}_1(s)+h_{22}(s)\mathbf{V}_2(s) \tag{1.249}$$

The *h parameters* are both short circuit and open circuit ratios as we see from the following definitions:

$$h_{11}(s)=\left.\frac{\mathbf{V}_1(s)}{\mathbf{I}_1(s)}\right|_{\mathbf{V}_2=0}=\frac{1}{y_{11}(s)} \quad \text{(short circuit input impedance)}$$

$$h_{12}(s)=\left.\frac{\mathbf{V}_1(s)}{\mathbf{V}_2(s)}\right|_{\mathbf{I}_1=0}=\frac{z_{12}(s)}{z_{21}(s)} \quad \text{(open circuit reverse voltage transfer ratio)}$$

$$h_{21}(s)=\left.\frac{\mathbf{I}_2(s)}{\mathbf{I}_1(s)}\right|_{\mathbf{V}_2=0}=\frac{y_{21}(s)}{y_{11}(s)} \quad \text{(short circuit current transfer ratio)}$$

$$h_{22}(s)=\left.\frac{\mathbf{I}_2(s)}{\mathbf{V}_2(s)}\right|_{\mathbf{I}_1=0}=\frac{1}{z_{22}(s)} \quad \text{(open circuit output admittance)}$$

The usual equivalent circuit for Eqs. (1.248) and (1.249) is presented in Fig. 1.59c. It is widely applied to transistor circuits.

The hybrid representation is suitable for most of the two-ports discussed earlier. For example, with $h_{11}=h_{12}=h_{22}=0$ and $h_{21}=\beta$ it fits the ideal current-controlled current source (Fig. 1.57c). And with $h_{11}=Y_{\text{in}}$, $h_{12}=0$, $h_{21}=\beta$, and $h_{22}=Y_{\text{out}}$ we have the representation for the current amplifier in Fig. 1.57d. The ideal transformer can also be described by setting $h_{11}=h_{22}=0$, $h_{12}=n_1/n_2$ (turns ratio), and $h_{21}=-n_1/n_2$.

Further applications of the two-port concept are found in Chapter 2.

Appendix 1.A Complex Numbers

A complex number or complex variable z is defined by a pair of numbers x and y such that

$$z=x+jy \tag{1.A1}$$

where $j=\sqrt{-1}$. (Engineers prefer j in order to avoid confusion with the current variable i. Mathematicians usually define $i=\sqrt{-1}$.)

We say that z is a *real* number if $x\neq 0$ and $y=0$, or z is an *imaginary* number if $x=0$ and $y\neq 0$, or z is a *complex* number if $x\neq 0$ and $y\neq 0$ in Eq. (1.A1). Accordingly, we define the *real part* of z:

$$\operatorname{Re}(z)=x$$

and the *imaginary part* of z:

$$\operatorname{Im}(z)=y$$

(But note that y by itself is a *real* number; it makes up the imaginary component of z after multiplication by j.)

A complex number can be graphically visualized in the *complex plane* by a point with the real part of z as the abscissa and the imaginary part as the ordinate. Fig. 1.60a gives an

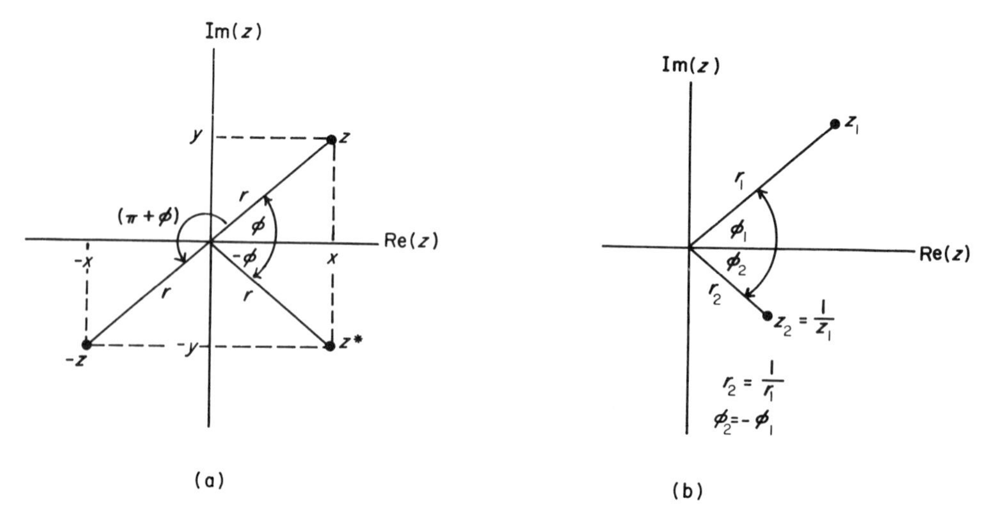

Figure 1.60 Graphic representations of a complex number. (See text.)

example. From this diagram it is also evident that z can be described by the polar coordinates r and ϕ. For the polar representation we define the *magnitude*, *absolute value*, or *modulus* of z:

$$r = |z| = (x^2 + y^2)^{1/2} \tag{1.A2}$$

and the *argument*, or *angle*, of z:

$$\phi = \sphericalangle z = \tan^{-1}\left(\frac{y}{x}\right) \tag{1.A3}$$

ϕ is positive for a counterclockwise orientation starting at the positive real (Re) half-axis.

The relations between real and imaginary parts and polar coordinates follow from the geometric relations in Fig. 1.60a. Specifically we have

$$\mathrm{Re}(z) = x = r\cos\phi \tag{1.A4}$$

$$\mathrm{Im}(z) = y = r\sin\phi \tag{1.A5}$$

Equations (1.A1)–(1.A5) can be combined to yield

$$z = x + jy = r\cos\phi + jr\sin\phi = r(\cos\phi + j\sin\phi) \tag{1.A6}$$

With Euler's identity we can deduce the polar or exponential form of z from Eq. (1.A6):

$$z = re^{j\phi} = |z|e^{j\phi} \tag{1.A7}$$

For simplification, the exponential is often symbolized by $1\angle\phi$. Thus we have the alternative form

$$z = |z|\angle\phi$$

In plots of the argument, ϕ and $\angle\phi$ are often not distinguished.

For every complex number z we can define its so-called *complex conjugate* z^*. The complex conjugate of $z = x + jy$ is defined as

$$z^* = \mathrm{Re}(z) - j\,\mathrm{Im}(z) = x - jy \tag{1.A8}$$

Plotting z^* according to Eq. (1.A8) in Fig. 1.60a reveals that z^* is the mirror image of z

reflected at the Re(z) axis. From Eqs. (1.A2), (1.A3), (1.A7), and (1.A8) we have

$$|z^*| = |z| = r$$

$$\measuredangle z^* = -(\measuredangle z) = -\phi = \tan^{-1}\left(\frac{-y}{x}\right)$$

$$z^* = re^{-j\phi}$$

In general, *functions* of a complex variable z, $f(z)$, can themselves be represented as complex variables:

$$z_2 = f(z_1) = x_2 + jy_2$$

where $x_2 = \mathrm{Re}\, f(z_1)$ and $y_2 = \mathrm{Im}\, f(z_1)$. Several examples will be given next as a part of the presentation of algebraic rules.

Algebraic Rules. We now consider several algebraic manipulations which are used in elementary circuit analysis.

1. *Scaling* For a = real number, $az = a(x + jy) = ax + jay$
2. *Addition*

$$z_3 = z_1 \pm z_2 = (x_1 \pm x_2) + j(y_1 \pm y_2)$$

$$z_1 + z_1^* = (x_1 + x_1) + j(y_1 - y_1) = 2x_1 \qquad \text{(real)}$$

$$z_1 - z_1^* = (x_1 - x_1) + j(y_1 + y_1) = j2y_1 \qquad \text{(imaginary)}$$

3. *Multiplication*

$$z_1 z_2 = (x_1 x_2 - y_1 y_2) + j(x_1 y_2 + x_2 y_1)$$

$$z_1 z_1^* = x_1^2 + y_1^2 = |z_1|^2 = r_1^2$$

$$z_1 z_2 = r_1 e^{j\phi_1} r_2 e^{j\phi_2} = r_1 r_2 e^{j(\phi_1 + \phi_2)}$$

$$z^n = r^n e^{jn\phi}$$

4. *Division*

$$\frac{z_1}{z_2} = \frac{r_1 e^{j\phi_1}}{r_2 e^{j\phi_2}} = \frac{r_1}{r_2} e^{j(\phi_1 - \phi_2)}$$

The *reciprocal* of a complex variable z_1 has the reciprocal magnitude of z_1 and the negative argument of z_1 (Fig. 1.60b):

$$z_2 = \frac{1}{z_1} = \frac{1}{r_1} e^{j(-\phi_1)}$$

Logarithms. From the polar form of a complex variable follows the expression

$$\ln z = \ln(re^{j\phi}) = \ln r + j\phi$$

With Euler's identity we can prove

$$e^{j\phi} = e^{j(\phi + 2\pi)}$$

Let ϕ_1 ($0 \leqslant \phi_1 < 2\pi$) be the *principal argument* of z, i.e., $\phi = \phi_1 + 2n\pi$; then we can write

$$\ln z = \ln|z| + j(\phi_1 + 2n\pi), \qquad n = 0, 1, 2, \ldots$$

The *principal value of the logarithm* is then defined as

$$\ln z = \ln|z| + j\phi_1$$

Problems

1. Compute the impedance of a 1 μH inductor for frequencies of 50 Hz, 60 Hz, 100 kHz, 100 MHz. Compare the results with those of Example 1.3.

2. Assume that the $R-L$ series combination of Fig. 1.23 has the parameters $R=10\ \Omega$, $L=10$ mH. (a) At what radian frequency ω is $<Z=45°$? How is this frequency related to the characteristic frequency ω_0? (b) Determine $<Z$ at $\omega_0/10$ and $10\omega_0$. (c) Determine $|Z|$ at $\omega_0/10$, ω_0, and $10\omega_0$.

3. In Fig. 1.31a, assume $v_1=10$ V, $v_2=9$ V, $i_2=1$ mA, $R_1=100\ \Omega$, $R_2=1$ kΩ, $L_1=1$ mH, and $C_1=1\ \mu$F. Determine the parameters of the Norton and Thévenin equivalent circuits of Figs. 1.31e and 1.31f.

4. In Fig. 1.32a, assume $R_1=R_2=R$. (a) Determine the driving point impedance R_s at terminals A and B. (b) Determine the transfer ratio v_L/v_i for $R_L=R/10$, $R_L=R$, and $R_L=10R$.

5. Design a low-pass filter according to Fig. 1.33 with a characteristic frequency $\omega_0=1/T=1000\ \text{s}^{-1}$ and a dc driving point impedance of 10 kΩ at terminals 2 and 2′.

6. Determine the resultant Bode diagram for a filter cascade consisting of a low-pass filter with gain G_{LP} according to Fig. 1.36a, followed by an ideal amplifier of gain $G_A=2$ which in turn is followed by a high-pass filter with a gain G_{HP} according to Fig. 1.36b. Assume that the filter time constants are identical and that the amplifier "buffers" the filters so that the low-pass filter "sees" an infinite load resistance, and the high-pass filter "sees" a zero source resistance. The overall gain is then $G=G_{LP}G_AG_{HP}$. What is the 3-dB bandwidth of the cascade?

7. The rise time T_R of a transient from one steady state level to another equals the time it takes for the transition from 10% to 90% of the difference between the two levels. For the sake of comparing the response time of first- and second-order systems, assume that the time constant T_C of the first-order circuit in Fig. 1.34a is $1/\omega_n$. Compare the rise time of the capacitor voltage of this circuit with that of the second-order circuit in Fig. 1.39a for $\zeta=0.707$ and $\zeta=1$.

8. Consider the normalized frequency response function of a second-order bandpass according to Eq. (1.188) in Section 1.4.3.4. For the normalized *deviation* x from the undamped natural frequency we can write

$$x=\frac{\Delta f}{f_n}$$

and

$$\Omega=\frac{f}{f_n}=\frac{f_n+\Delta f}{f_n}=1+x$$

Show that for small deviations ($x \ll 1$), Eq. (1.188) can be approximated by

$$G(j\Omega)\approx\frac{1}{1+j2Qx}=\frac{1}{1+2jQ(\Omega-1)}$$

[*Hint*: Use a series for $1/\Omega=(x+1)^{-1}$.]

Plot $|G(j\Omega)|$ according to this approximation and Eq. (1.188) on a *linear* Ω scale. Compare the symmetry of the two curves around $\Omega=1$. Locate the 3-dB frequencies

$\Omega_1 = f_1/f_n$ and $\Omega_2 = f_2/f_n$ for both curves. Show that f_n ($\Omega = 1$) is the *arithmetic* mean $(f_1 + f_2)/2$ for the approximation and the *geometric* mean $(f_1 f_2)^{1/2}$ for the exact curve.

9. In the Wheatstone bridge of Fig. 1.43, add a variable capacitor C_1 in parallel to R_1. Further, assume that R_4 is replaced by a series combination $R_x - L_x$ representing a practical inductor of unknown inductance L_x and unknown parasitic resistance R_x. This arrangement is called a *Maxwell* bridge. Prove that this bridge can be used to determine L_x and R_x by the null method ($i_5 = 0$). [*Hint*: Use ac impedances and the method of Example 1.20 in Section 1.5.2.] You should obtain $L_x = C_1 R_2 R_3$ and $R_x = R_2 R_3 / R_1$.

10. Cascade two identical *RL* low-pass filters (Fig. 1.33a) without buffering; i.e., the output voltage of the first filter is the input for the second. Compute the overall frequency response function by mesh analysis according to Section 1.5.3.

11. Repeat Problem 10 but use node analysis according to Section 1.5.4.

12. Apply the superposition principle to the circuit of Fig. 1.31b in order to determine the voltage at terminals A and B.

13. In the circuit of Fig. 1.48d, set $v_0 = 1$ V, $i_0 = 0$, $R_1 = R_2 = R_3 = 10$ kΩ, $C = 0.1$ μF. What is the voltage across R_2 in the dc case ($\omega = 0$) and for very high frequencies ($\omega = \infty$)? [*Hint*: Use the concepts of Section 1.5.6.]

14. Determine the open circuit impedance parameters z_{ij} (Section 1.6.4.1) for the amplifier of Fig. 1.57b in terms of μ, Z_{in}, and Z_{out}.

15. Determine the short circuit admittance parameters y_{ij} (Section 1.6.4.2) for the amplifier of Fig. 1.57d in terms of β, Y_{in}, Y_{out}.

16. Determine the hybrid parameters h_{ij} for the same amplifier (Section 1.6.4.3) in terms of β, Y_{in}, Y_{out}.

References

References for General Circuit Analysis

Desoer, C. A., and E. S. Kuh, *Basic Circuit Theory*, McGraw-Hill, New York, 1969. (Excellent.)

Ley, B. J., S. G. Lutz, and C. F. Rehberg, *Linear Circuit Analysis*, McGraw-Hill, New York, 1959.

Van Valkenburg, M. E., *Network Analysis*, 3rd ed., Prentice-Hall, Englewood Cliffs, NJ, 1974. (Clear and straightforward.)

Physics

Sears, F. W., and M. W. Zemansky, *University Physics*, 2nd ed., Addison-Wesley, Reading, MA, 1955.

Mathematics

Wylie, C. R., Jr., *Advanced Engineering Mathematics*, McGraw-Hill, New York, 1960.

Bode Plots (frequency response plots)

D'Azzo, J. J., and C. H. Houpis, *Feedback Control System Analysis and Synthesis*, 2nd ed., McGraw-Hill, New York, 1966.

Diodes

Angelo, E. J., *Electronic Circuits*, McGraw-Hill, New York, 1958.

Strauss, L., *Wave Generation and Shaping*, McGraw-Hill, New York, 1960.

Transformers

Langford-Smith, F., ed., *Radiotron Designer's Handbook*, 4th ed., Amalgamated Wireless Valve Company, Pty., Ltd., 47 York Street, Sydney, Australia, 1953. (Also distributed in the USA by Radio Corporation of America, Harrison, NJ.)

2

Controlled Devices: Transistors

Gunter N. Franz

This chapter gives a brief introduction to a class of basic electronic components which are used to approximate the characteristics of controlled sources. We restrict ourselves to three-terminal devices called *transistors*. Made of semiconductor materials, transistors are either manufactured as single components, as pairs in one housing, or in much larger numbers in miniaturized arrays. In the latter form, many transistors are typically produced on a single semiconductor chip, complete with all interconnections and passive components to form a functional *integrated circuit* (IC). Beginning with Chapter 4, this book stresses the use of integrated circuits because they facilitate the design of electronic circuits by the building block approach and thus allow us to bypass the cumbersome design and construction of electronic apparatus "from scratch" with *discrete* components. Circuits combining discrete transistors and IC components are often referred to as *hybrid* circuits.

Transistors and ICs are not the only electronic devices to be used as controlled sources or "active" devices. For example, the vacuum tube was the device of choice for many decades. Now, except for specialized applications, vacuum tubes are obsolete. For a discussion of vacuum tubes the reader may consult any of the numerous older electronics texts.

Given the orientation toward modern integrated circuits, this discussion of transistors is limited to basic function and very simple circuits.

2.1 Semiconductors

A brief discussion of the properties of semiconductor materials will facilitate our understanding of the functioning of a transistor.

Semiconductors occupy a middle ground between metallic conductors and insulators. Whereas metallic conductors have a large number of mobile or "free" electrons, insulators have only a minute number of loosely held electrons.

In the absence of an external electric field, the motion of free electrons is random and of increasing intensity as temperature increases. With the imposition of an electric field, the random motion becomes ordered so that a *directional* current flows. The current intensity depends on the externally applied electric potential gradient and on the number of free electrons available, as expressed by a temperature-dependent material property, i.e., the conductivity.

REMARK: In metallic conductors, current is carried by *electrons* so that negative elementary charges move from the low-potential point (−) to the high-potential point (+). In contrast, for the purpose of circuit analysis we always assume a current of positive polarity to flow from the high-potential point (+) to the low-potential point (−). Thus the direction of a positive current coincides with the direction of flux of positive charge carriers. From the standpoint of net charge transfer, an electron flux in the opposite direction is, of course, the equivalent of a positive current. In circuit analysis the actual polarity of the charge carrier is ignored, and only "conventional" currents are considered.

Semiconductors are materials such as pure crystalline germanium (Ge) and silicon (Si). The Ge or Si atoms have four valence electrons in the outer orbital shell and form crystal subunits so that each atom joins four neighbor atoms in a tetrahedral arrangement through electron pair bonds. In this way each atom effectively has eight electrons in the outer shell.

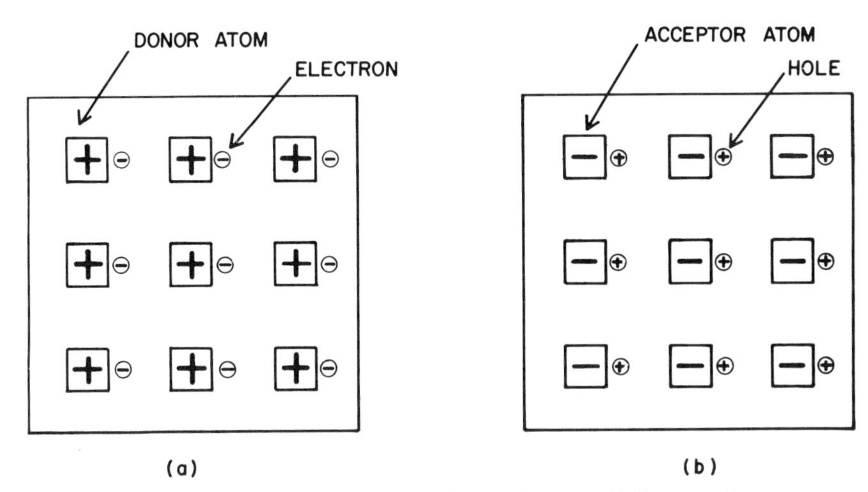

Figure 2.1 Doped semiconductor materials: **(a)** N-type; **(b)** P-type. (See text.)

In the pure crystalline state, these elements have actually very low conductivity. The interesting properties of semiconductors arise from the presence of deliberately introduced impurities in the crystal lattice. The controlled insertion of foreign atoms of different valency into the lattice structure is called *doping*.

Assume that a Ge or Si crystal has been doped with a small number of atoms with five valence electrons. Upon incorporation in the tetrahedral crystal unit, four of the five valence electrons will participate in electron pair bonds, while the fifth will become a mobile free electron. The foreign atoms are thus *donors* of free electrons. The excess of free electrons makes the doped semiconductor a so-called *N-type semiconductor*. In N-type material we have then fixed positive charges, the donor atoms, and mobile negative charges, the free electrons (Fig. 2.1a).

Had we introduced doping atoms with only three valence electrons, only three electron pairs would have been formed. That would have created a local electron deficit or a so-called *hole*. Such a hole can be "filled" by an electron from a neighboring atom breaking a pair bond and forming a new one with the previously unpaired electron. In this way a hole can in fact move to another location. Therefore, *a hole is really the equivalent of a mobile carrier of a positive elementary charge*. Now the doped semiconductor material is of the *P-type* because of the presence of mobile positive charge carriers. The impurity atoms creating this situation are called *acceptor* atoms. They accept one neighboring electron each in order to complete their outer shell. Thus they are the complementary fixed negative charges to the mobile holes (Fig. 2.1b).

Electrons and holes differ not only in their polarity, which makes them move in opposite directions in the presence of an electric field, but also in their mobility; holes are slower.

Practical N or P materials do not have doping atoms of only the donor or acceptor type, respectively. Rather they are predominantly of the N- and P-types. This means that aside from the *majority* carriers (e.g., electrons in the case of N material), there are also *minority* carriers of opposite polarity. The significance of minority carriers will become evident shortly.

Next we consider an arrangement crucial for the functioning of semiconductor diodes and transistors: the N–P junction.

2.2 The Semiconductor Junction; Junction Diodes

The mobile charge carriers in a piece of N or P silicon will, on the average, be evenly distributed throughout the material. However, the distribution becomes uneven if N material is joined to P material to form a so-called *N–P junction* (Fig. 2.2). Because of the strong concentration differences, electrons and holes will diffuse toward each other and cancel each other by combination. Left behind are uncompensated fixed charges in the areas adjacent to the junction. These fixed charges create a barrier to further carrier movement in that the donor atoms repel holes and the acceptor atoms repel electrons. The net result is that (1) complete recombination of mobile carriers is not possible, (2) a *depletion layer* or space charge region free of mobile carriers straddles the junction, and (3) a potential difference V_D, representing an energy barrier to the mobile carriers, is created by the separation of fixed charges in the depletion layer (Fig. 2.2). This potential difference does *not* depend on the presence of an external field. It amounts to a fraction of a volt at room temperature. From Fig. 2.2 it is evident that V_D represents an "uphill" barrier to the majority carriers. But for minority carriers the energy difference is "downhill." Minority carriers can, therefore, generate leakage currents across P–N junctions.

The properties of P–N junctions are of fundamental importance for the operation of transistors. But before we attack this subject we shall first consider the operation of a P–N junction as a rectifying element or diode. Diodes can be based on several physical principles, but modern diodes are typically semiconductor diodes with P–N junctions (*junction diodes*).

The function of a P–N junction as a diode is explained by Fig. 2.3. In this case the two halves of the junction are equipped with metallic terminals or leads corresponding to the anode and cathode terminals, respectively. By definition, the application of an external voltage source, such that the positive source terminal is

Figure 2.2 The P–N junction.

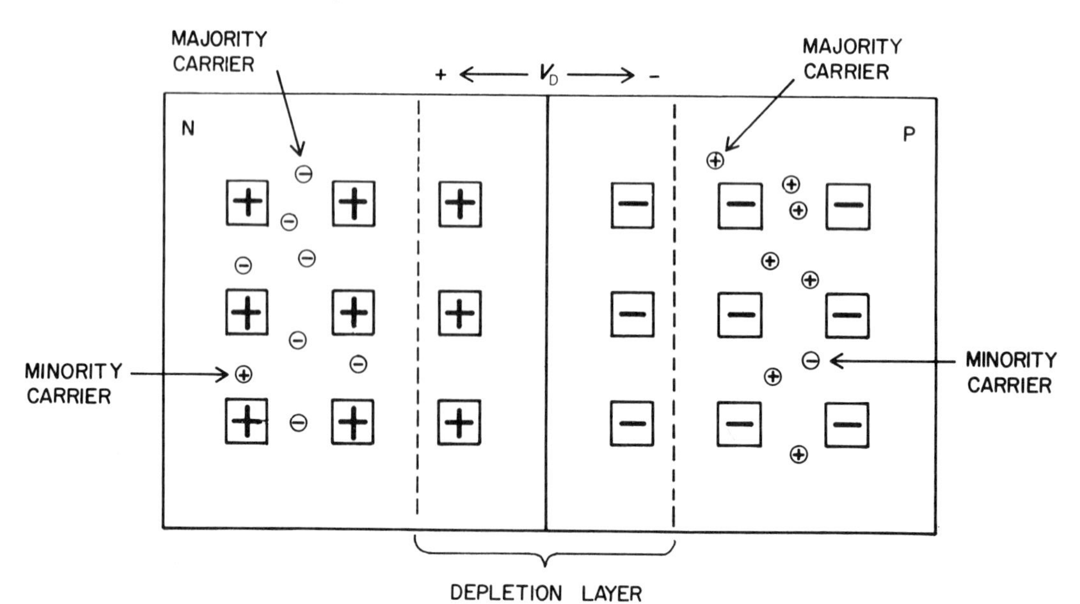

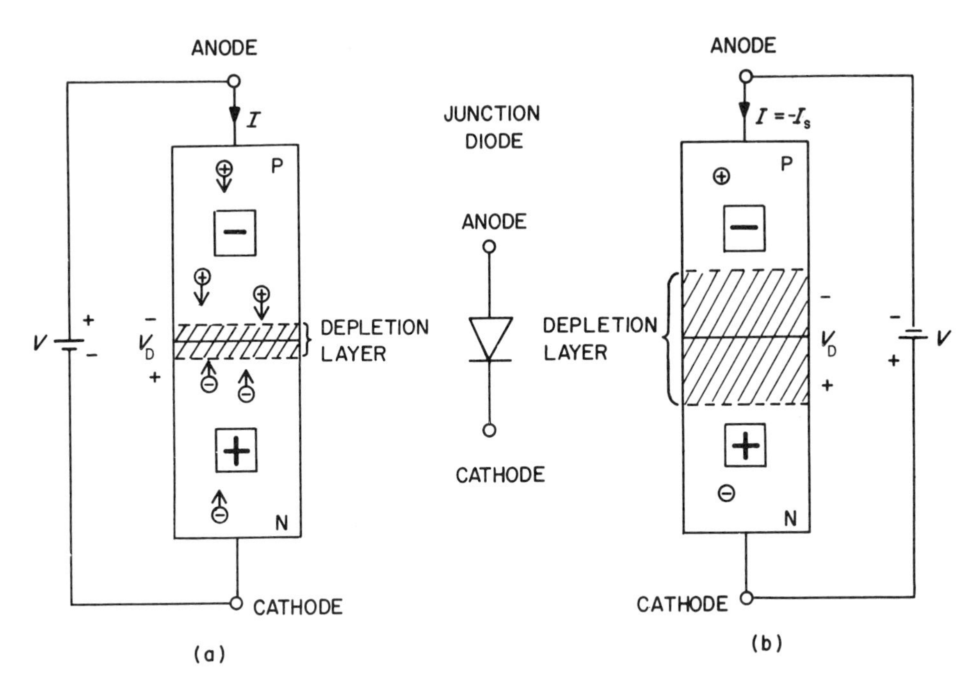

Figure 2.3 The junction diode under forward **(a)** and reverse **(b)** bias.

connected to the P half via the anode, and the negative one to the N half via the cathode, constitutes forward bias (Fig. 2.3a). In this instance, the external voltage source opposes the energy barrier, represented by V_D, reduces the width of the depletion layer, and a majority carrier current (*forward current*) will flow across the junction: Electrons released from the metallic cathode move as majority carriers through the N material and then across the barrier to recombine with holes, while at the anode electrons leave the P material and thus generate holes which move toward the depletion layer in order to recombine with electrons from the N half.

The current under forward bias is approximately an exponential function of the external voltage, namely,

$$I = I_s(e^{V/\delta} - 1) \approx I_s e^{V/\delta} \qquad (2.1)$$

where $\delta = kT/q \approx 26$ mV $\approx 1/40$ V at 300°K; $k = 1.38 \times 10^{-16}$ erg/°K, the Boltzmann constant; T is the absolute temperature in °K; $q = 1.59 \times 10^{-14}$ C, the elemental charge; and I_s is the so-called saturation current, which will be discussed next.

Under reverse bias (Fig. 2.3b), the width of the depletion layer is increased as is the magnitude of the junction potential V_D. For a reverse bias voltage several times larger than δ, the current is approximately equal to the saturation current, i.e.,

$$I \approx -I_s \qquad (2.2)$$

This current, opposite in polarity to the forward current, is the result of minority carrier drift. At room temperature this current is very small but escalates rapidly with temperature. In the case of germanium it doubles for every temperature increment of 10°K, while it doubles for every 6°K in the case of silicon. Because its value for silicon at room temperature is so much lower than that for germanium, silicon is still the preferred substrate.

The graphic representation of the diode voltage–current characteristic is given by Fig. 1.4b.

As reverse bias is increased, avalanche breakdown will occur as the zener voltage V_z is reached (Fig. 1.5b).

Unless specifically exploited, the temperature-dependent reverse bias or leakage current and the breakdown voltage of a semiconductor junction are part of the nonideal behavior of transistors and integrated circuits which must be considered in the practical application of such components.

We are now ready to discuss the two major types of transistors: field effect transistors and bipolar transistors.

2.3 Field Effect Transistors

In a simplistic way we can say that the operation of the various types of field effect transistors or FETs depends on the modulation of the conductivity of a doped semiconductor by an external control voltage. We shall distinguish here two basic types: the junction field effect transistor (JFET) and the insulated gate field effect transistor (IGFET).

2.3.1 The Junction Field Effect Transistor (JFET)

A simple model of a so-called *N-channel* JFET is shown in Fig. 2.4. (The actual geometries of the various types of JFETs are quite different from this schematic.) As the name indicates, such a transistor has a conductive channel of N material

Figure 2.4 Functional schematic for the N-channel JFET. (See text.)

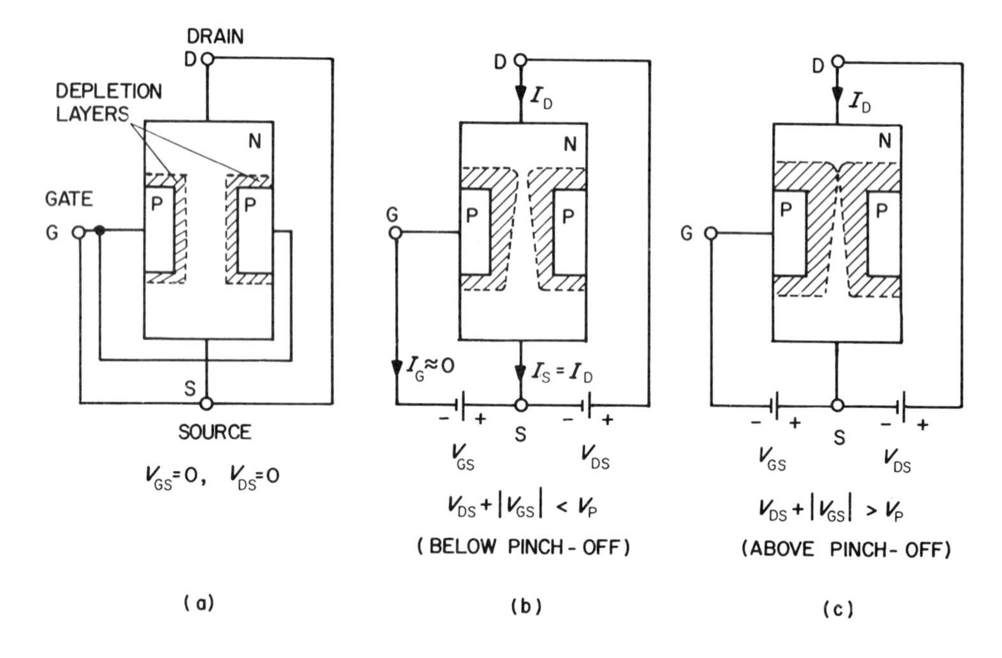

connecting two external terminals which are the *drain* (D) and the *source* (S), respectively. Diffused into the substrate are also two P regions which are connected to the *gate* terminal (G). The typical JFET is, therefore, a three-terminal device or two-port.

Let us first assume that all terminals are tied together so that no external potential differences exist, i.e., $V_{GS}=0$. Under these conditions (Fig. 2.4a), depletion layers form at the P–N junctions, and any majority carrier fluxes are balanced by minority carrier fluxes for zero net currents.

The application of an external voltage source V_{DS} between the drain and source terminals (Fig. 2.4b) makes a current flow from the drain terminal to the source terminal by way of the conductive N channel. The conductivity of this channel and hence the magnitude of the current can now be modulated by *reverse* biasing the P–N junctions (Fig. 2.4b). The negative gate–source voltage V_{GS} (with respect to the source) acts now as the control voltage of the channel conductivity by modulating the width of the depletion layers. As this voltage becomes more negative, the depletion layers widen, the channel width decreases, and the resistance of the N channel increases.

With the P–N junctions always reverse biased, the gate current I_G equals the usually very small leakage current of these junctions. Compared to the drain and source currents, this current can be neglected. Therefore, under these specific conditions, the JFET behaves like a *voltage-controlled resistor*. Actually this behavior is limited to conditions which do not induce "pinch-off." This phenomenon we shall discuss next.

For a fixed V_{GS}, the width of the depletion layers is not uniform along the conductive channel because internal ohmic voltage drops generate a spatially non-uniform potential difference across the P–N junctions (Fig. 2.4b). With the drain terminal being at a more positive potential than the source, the depletion layers are wider in the drain region. Hence, as the drain–source voltage V_{DS} is increased, the conductive channel tends to pinch off at the drain region before it does so at the

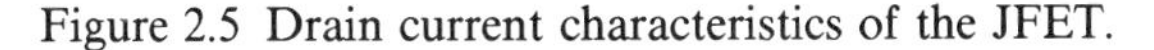

Figure 2.5 Drain current characteristics of the JFET.

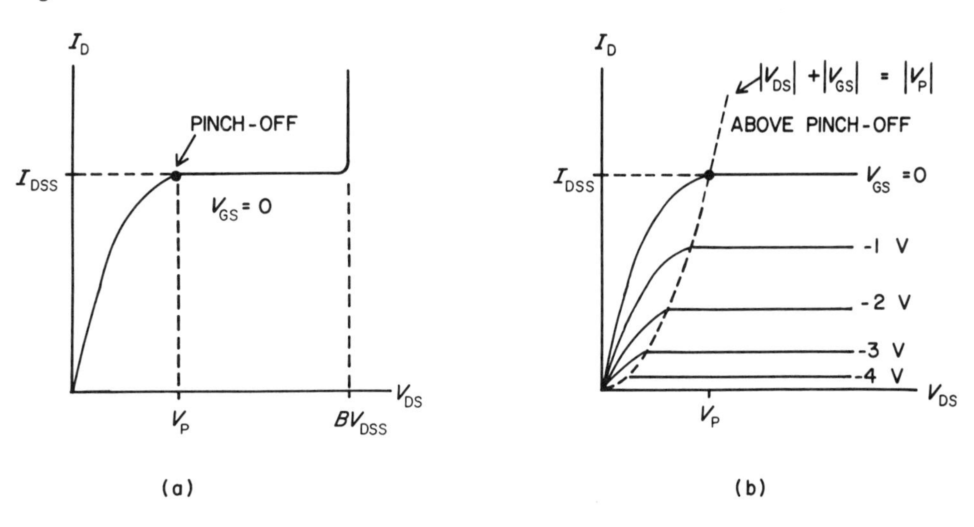

source region. The result is a nonlinear current–voltage characteristic, of which an example is given in Fig. 2.5a for zero gate–source voltage.

Note that for small V_{DS} the current rises linearly, corresponding to the behavior of a linear resistor. As we further raise V_{DS}, the nonuniformity of the depletion layers becomes more pronounced, and the drain current begins to level off. After V_{DS} passes the *pinch-off voltage* V_P, the drain current stays practically unchanged unless the breakdown voltage BV_{DSS} is reached.

In general, pinch-off occurs whenever the following condition is satisfied:

$$V_{DS} + |V_{GS}| = V_P \tag{2.3}$$

Above pinch-off, i.e., when the sum on the left of Eq. (2.3) exceeds the pinch-off voltage, the progressive reduction of the channel conductance (Fig. 2.4c) keeps the current limited. The maximal limiting current above pinch-off (and below breakdown) is designated I_{DSS}. In the subscript, DS stands for "drain–source" and the second S for "gate short circuited" (to the source), i.e., $V_{GS} = 0$.

The important point here is then that above pinch-off the JFET (in combination with the power supply, V_{DS}) is an excellent approximation of a *constant current source*.

In practical applications, the JFET is used either as an electronic switch, because it behaves like a voltage-controlled resistor *below* pinch-off, or as an amplifying element, because it represents a *voltage-controlled* current source above pinch-off, as we see next.

2.3.2 Output Characteristics of JFETs

Equation (2.3) shows that pinch-off is reached for progressively lower drain–source voltages as V_{GS} is made more negative. Hence the limiting currents above pinch-off must also become correspondingly smaller. In fact, when the absolute value of the gate–source voltage equals the pinch-off voltage V_P, pinch-off begins at $V_{DS} = 0$, and the observable current is only a very small leakage current. Operationally, we can therefore define V_P in two ways: (1) V_P equals the drain–source voltage for which the drain current reaches I_{DSS} or (2) V_P equals the absolute value of the gate–source voltage for which the drain current declines to an irreducible leakage current.

An example of a family of I_D–V_{DS} curves for various gate–source voltages is shown in Fig. 2.5b. The broken parabolic curve represents the locus at which the pinch-off condition is reached. Well to the left of this locus, the slope of the individual curves varies as a function of gate–source voltage (voltage-controlled resistance behavior). To the right of the locus, the JFET acts as a voltage-controlled current source. (By itself, the transistor cannot act as a source or active element. Source behavior is possible only in conjunction with a power supply.) Action as a *controlled* source is the crucial property which makes the transistor an amplifying element.

Above pinch-off, I_D depends on the gate–source voltage according to the following equation:

$$I_D = I_{DSS}\left(1 - \frac{|V_{GS}|}{V_P}\right)^2 \tag{2.4}$$

Equation (2.4), of course, also represents the parabolic pinch-off locus of Fig. 2.5b.

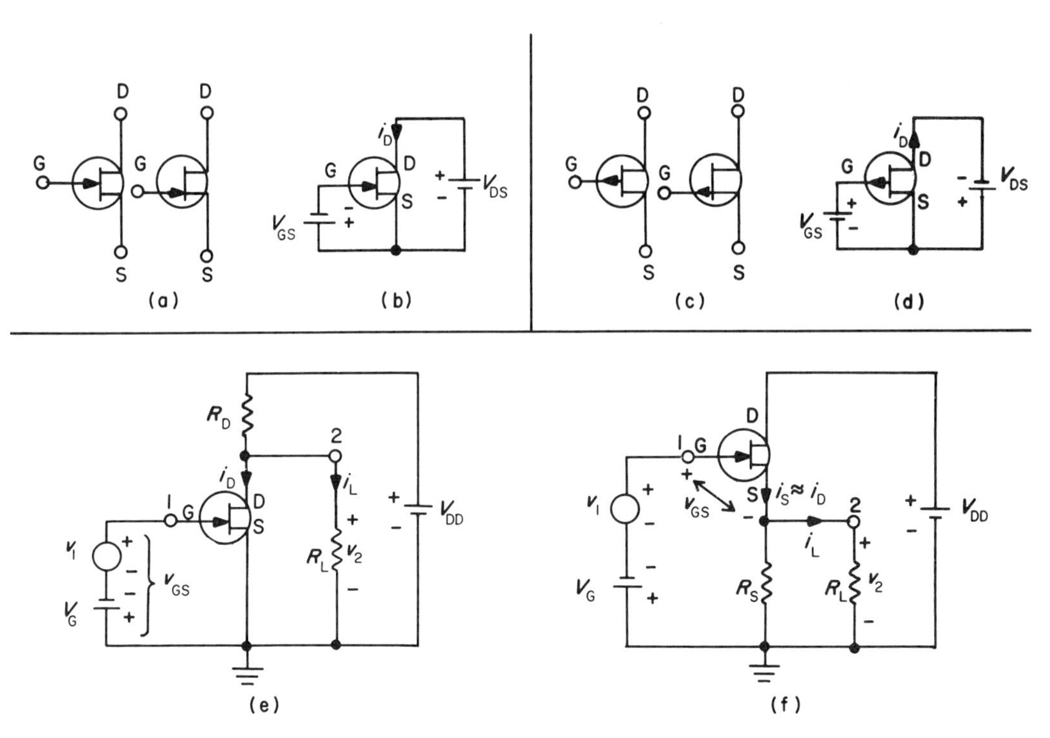

Figure 2.6 Symbols, polarities, and elementary amplifier stages for JFETs: **(a)** and **(b)** N-channel JFET. **(c)** and **(d)** P-channel JFET; **(e)** common source amplifier; **(f)** common drain amplifier (source follower).

In our discussion so far we have only treated the case of N-channel JFETs. But it is possible to reverse the roles of the N and P materials and manufacture a *P-channel* JFET. In this case the channel current is carried by holes, the control and supply voltages have the reversed polarities, i.e., $V_{GS} \geqslant 0$ and $V_{DS} \leqslant 0$, and the drain current has negative polarity because it leaves the drain terminal.

In the next section we shall consider simple amplifier stages with a single JFET.

2.3.3 Simple JFET Stages; DC Analysis

The incorporation of JFETs into circuit diagrams requires appropriate symbols. Figures 2.6a–2.6d show the usual symbols, terminal designations, and voltage source polarities for N- and P-channel JFETs. The arrow at the gate terminal indicates the direction of current flow *if* the gate junctions are forward biased. But note that proper operation of a JFET requires that the gate junction be *reverse* biased, so that the actual gate current is in most cases negligible.

Two fundamental JFET stages are shown in Figs. 2.6e and 2.6f for an N-channel JFET. (Complementary circuits with reversed polarities could be drawn for a P-channel JFET.) The *common source* (CS) amplifier stage (Fig. 2.6e) offers high input impedance ($i_G \approx 0$), moderate output impedance, and voltage amplification. The *common drain* (CD) or *source follower* stage (Figure 2.6f) offers high input impedance, low output impedance, and a small-signal voltage gain of unity; i.e., the output or source voltage "follows" the input signal v_1. This last circuit is particularly useful as a buffer stage.

The term "common" in the name of these two circuits refers to the fact that, for ac signals (of sufficiently high frequency), the source or the drain is connected to "common" or ground potential via (1) a direct connection (e.g., Fig. 2.6e), (2) a dc source of negligible internal resistance (e.g., V_{DD} in Fig. 2.6f), or (3) a capacitor of sufficiently large value.

In either case the input voltage is the sum of a signal voltage v_1 and a bias voltage V_G, which sets the operating point or quiescent state. The output voltage v_2 is the voltage drop across the load resistor R_L. V_{DD} is the power supply voltage.

For the CS stage, the control or gate–source voltage equals the input voltage, and the output voltage equals the drain–source voltage. Hence, a simple analysis gives us

$$i_L = \frac{v_2}{R_L} = \frac{v_{DS}}{R_L} \tag{2.5}$$

$$V_{DD} - v_{DS} = (i_D + i_L)R_D \tag{2.6}$$

and by combining Eqs. (2.5) and (2.6) we obtain

$$i_D = -\frac{v_{DS}}{R_{eq}} + \frac{V_{DD}}{R_D} \tag{2.7}$$

Figure 2.7 DC analysis of the common source stage. (See text.)

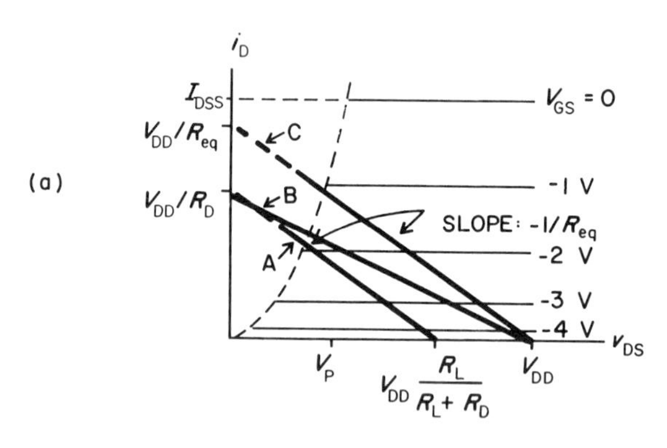

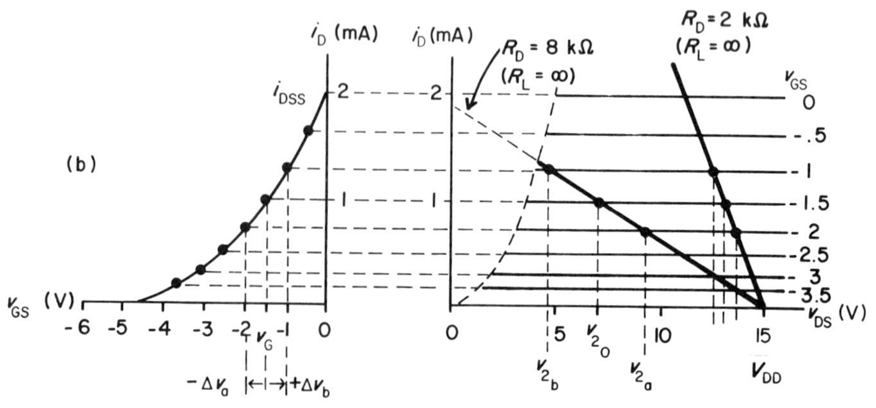

where

$$R_{eq} = \frac{R_D R_L}{R_D + R_L} \tag{2.8}$$

Equation (2.7) describes the load line in the $i_D - v_{DS}$ diagram. It establishes the constraint that the $i_D - v_{DS}$ pairs may not be just anywhere on the family of $i_D - v_{DS}$ curves but are restricted to the intersections of these curves with the load line (line A in Fig. 2.7a). Note that in the load-free state ($R_L = \infty$), the i axis intercept stays unchanged but the slope of the load line changes from $-1/R_{eq}$ to $-1/R_D$, and the range of v_{DS} increases.

In an actual design of a CS stage, v_{DS} must be limited to a range on the load lines to the right of the pinch-off locus (broken curve in Fig. 2.7a).

Before we consider the amplification action of a CS stage, let us repeat the same analysis for the CD or source follower stage (Fig. 2.6f). In this case the control or gate–source voltage is the difference between the input ($v_1 + V_G$) and output (v_2) voltages. And for the load line we have, with $i_D \approx i_S$,

$$i_D = -\frac{v_{DS}}{R_{eq}} + \frac{V_{DD}}{R_{eq}} \tag{2.9}$$

where

$$R_{eq} = \frac{R_S R_L}{R_S + R_L} \tag{2.10}$$

If R_S has the same value as R_D in the previous example, Eq. (2.9) gives us the line C in Fig. 2.7a as the load line of the CD stage.

The amplifying action of the CS stage is best explained with the aid of the two graphs in Fig. 2.7b. The basic assumption is, of course, that the transistor is operated above pinch-off.

The graph on the left side of Fig. 2.7b is a plot of Eq. (2.4). This transconductance curve demonstrates the dependence of the drain current on the control voltage v_{GS}. The usual parameter relating the source current to the control voltage is the *transconductance* g_m. In this case, however, only an *incremental* transconductance (i.e., the slope of the $i_D - v_{GS}$ curve) can be defined. For a given operating point, g_m will be approximately constant, if v_{GS} is restricted to small variations around a fixed value. Under such conditions, linear incremental analysis can be used as demonstrated in Section 2.3.4. For now we shall use only a simple graphic analysis.

For each point on the transconductance curve on the left of Fig. 2.7b we can draw a constant current curve in the $i_D - v_{DS}$ plot on the right. Thus for a signal $v_1 = 0$ and a bias voltage $V_G = -1.5$ V we have on the right the corresponding 1-mA curve for i_D on which the quiescent operating point must lie. The exact location of the operating point is given by the intersection of this current characteristic with the load line. If, for example, we choose $R_D = 8$ kΩ and $R_L = \infty$, the load line will intersect at v_{2_0}, the quiescent output voltage.

Next let us assume that the gate–source voltage is made more negative by the signal voltage $v_1 = -\Delta v_a = -0.5$ V. The corresponding current characteristic is labeled $v_{GS} = -2$ V. From its intersection with the load line for $R_D = 8$ kΩ we deduce that the output voltage is now v_{2_a}. Conversely, a signal voltage change of $+\Delta v_b$ generates the output voltage v_{2_b}.

From the graphic analysis we make two important observations: (1) The magnitude of the incremental change of the output voltage is larger than that of the input voltage ($|v_{2_a} - v_{2_0}| > |\Delta v_a|$), i.e., the input signal has been *amplified.* (2) Positive (negative) increments in the input voltage cause negative (positive) increments in the output voltage, i.e., a *polarity reversal* of the amplified signal occurs. Thus a sinusoidal input signal (v_1) will experience a *180° phase shift* in the CS stage.

From the shape of the parabolic transconductance curve on the left of Fig. 2.7b we can also deduce the following: (1) Distortion of the amplified signal can be minimized by keeping the input signal variations small and locating the operating point (via V_G) on a relatively straight segment of the transconductance curve. (2) The gate–source voltage, being the sum of signal and bias voltages, cannot be more positive than zero; otherwise the gate junctions will be forward biased. (3) The amplification factor increases with the slope (incremental transconductance) of the transconductance curve. However, the amplification factor or gain is not entirely determined by the incremental transconductance. The graph on the right shows that increasing the slope of the load line ($R_D = 2$ kΩ) reduces the gain.

Graphic analysis gives a quick overview of operating point, gain, and possible signal distortion of a transistor amplifier stage. However, when a transistor is imbedded in a more complicated network requiring the use of a set of network equations, it is advantageous to "replace" the transistor with a two-port model described by incrementally linear equations. Naturally, such models are limited to small-signal analysis. This modeling approach is the topic of the next section.

2.3.4 JFET Equivalent Circuits and Small-Signal Analysis

The JFET equivalent or model circuits in this section are based on the assumptions that the transistor is operated above pinch-off and that the incremental signal variations are small. We indicate the latter assumption by using lower-case letters as subscripts in the circuit models.

The simplest and most idealized equivalent circuit is the voltage-controlled current source model of Fig. 2.8a. It is quite satisfactory for many calculations.

The ideal current source model can be made more realistic by the addition of the output conductance g_{ds}. This simply reflects the fact that actual i_D curves, unlike the curves in Figs. 2.5b and 2.7b, are not horizontal but have a slightly positive slope. The circuit model of Fig. 2.8b is the result of this refinement.

Next, if we take into account various parasitic capacitances and the finite resistances of the reverse-biased gate junctions, the fairly complex model of Fig. 2.8c results. With the inclusion of the capacitances, it follows that the two-port parameters of the model are now frequency dependent. Fortunately, for moderately high frequencies it is often sufficient to use the simpler model of Fig. 2.8d, which contains only a single equivalent input capacitance.

The y parameters of the two-port of Fig. 2.8c are easily derived with the aid of Fig. 1.59b. Thus we have

$$-y_{12} = g_{gd} + sC_{gd} \approx sC_{gd}$$

$$y_{11} + y_{12} = g_{gs} + sC_{gs} \approx sC_{gs}$$

$$y_{22} + y_{12} = g_{ds} + sC_{ds} \approx g_{ds}$$

$$y_{21} - y_{12} = g_m$$

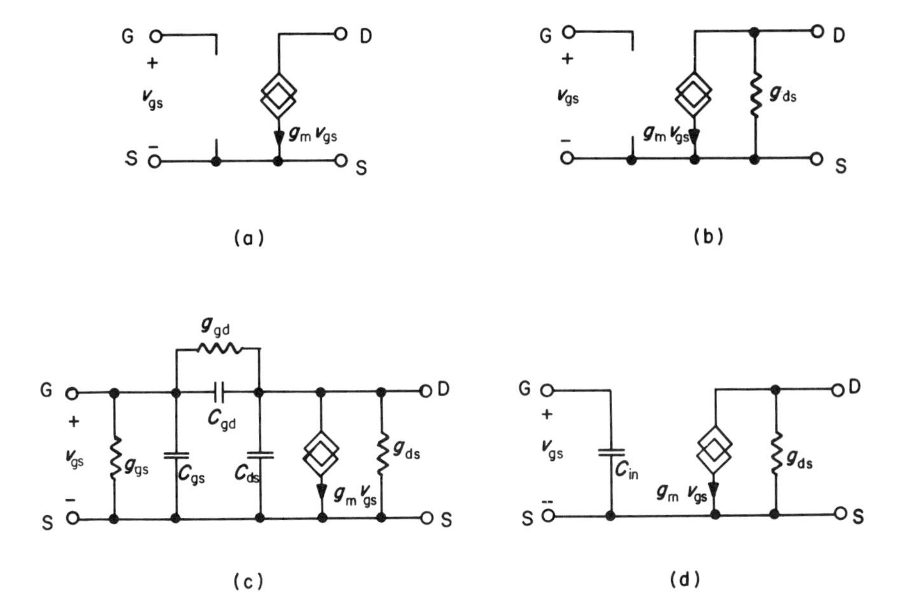

Figure 2.8 Small-signal models for FETs.

From this set of equations we derive the parameters as

$$y_{12} \approx -sC_{gd}$$

$$y_{11} = g_{gs} + s(C_{gs} + C_{gd})$$

$$y_{21} = g_m - sC_{gd}$$

$$y_{22} = g_{ds} + sC_{gd}$$

If the voltage gain of the two-port is A_V, then we can write the phasor equations

$$y_{12}\mathbf{V}_2 = y_{12}A_V\mathbf{V}_1 \approx -sC_{gd}A_V\mathbf{V}_1$$

Hence

$$\mathbf{I}_1 = y_{11}\mathbf{V}_1 + y_{12}\mathbf{V}_2 \approx s\left[C_{gs} + (1 - A_V)C_{gd}\right]\mathbf{V}_1$$

The term in square brackets is the equivalent input capacitance of the model in Fig. 2.8d. The other parameters of that model are obtained by simplification of y_{21} and y_{22}:

$$y_{21} \approx g_m$$

$$y_{22} \approx g_{ds}$$

With these models, the small-signal analysis of the prototypical circuits of Figs. 2.6e and 2.6f is now quite simple. First we reduce the circuits to their ac or small-signal equivalents by replacing all dc sources with short circuits as shown in Section 1.5.6. The resulting common source and common drain circuits are given in Figs. 2.9a and 2.9b. Upon insertion of the simple current source model of Fig. 2.8b, we obtain the equivalent circuits of Figs. 2.9c and 2.9d.

For the circuit in Fig. 2.9c we write the node equation for terminal 2:

$$g_m v_{gs} = g_m v_1 = -v_2(g_{ds} + G_D + G_L) \tag{2.11}$$

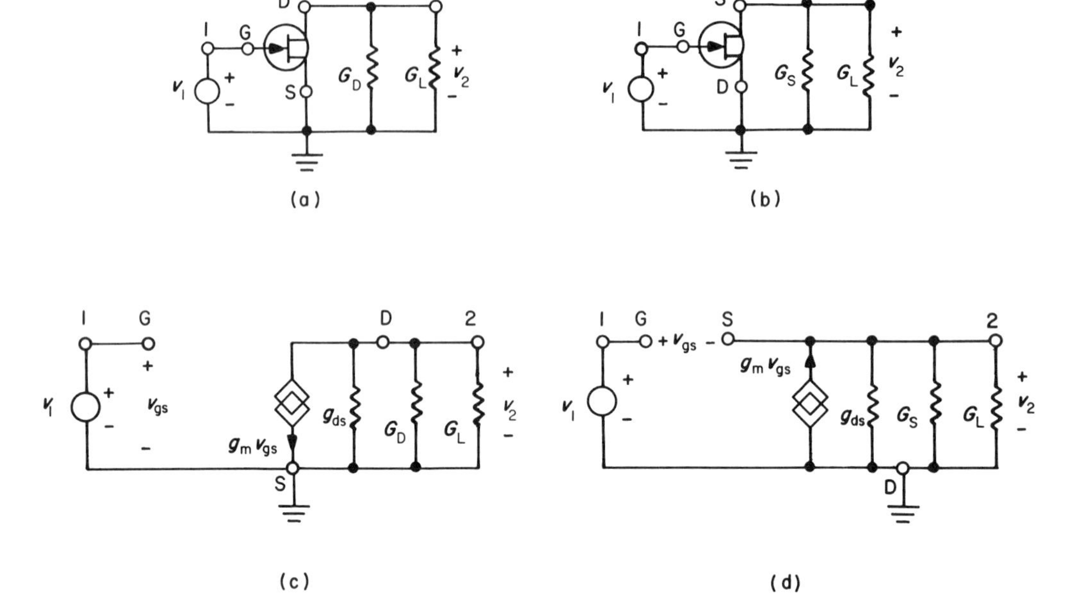

Figure 2.9 AC equivalent circuits for common source and common drain amplifier stages.

The voltage amplification factor or gain follows directly from Eq. (2.11):

$$A_v = \frac{v_2}{v_1} = \frac{-g_m}{g_{ds} + G_D + G_L} \tag{2.12}$$

For $(G_D + G_L) \gg g_{ds}$, the use of the model in Fig. 2.8a is indicated, and Eq. (2.12) reduces with Eq. (2.8) to

$$A_v = \frac{-g_m}{G_D + G_L} = -g_m \frac{R_D R_L}{R_D + R_L} = -g_m R_{eq} \tag{2.13}$$

Note: (1) The negative sign indicates polarity reversal of the signal. (2) A gain larger than unity requires $g_m > (G_D + G_L)$. (3) $-(G_D + G_L)$ or $-1/R_{eq}$ is the slope of the load line . (4) g_m is the slope of the transconductance curve.

The analysis of the CD stage is an easy matter, too, if we use the model circuit of Fig. 2.9d. For the control voltage we have

$$v_{gs} = v_1 - v_2 \tag{2.14}$$

and the node equation for terminal 2 is

$$g_m v_{gs} = v_2(g_{ds} + G_S + G_L) \tag{2.15}$$

The combination of Eqs. (2.14) and (2.15) and the assumption $g_{ds} \ll (G_S + G_L)$ yield the gain

$$A_v = \frac{v_2}{v_1} = \frac{g_m}{g_m + G_S + G_L} \leqslant 1 \tag{2.16}$$

The gain is positive: hence there is no polarity reversal of the signal. For $g_m \gg (G_S + G_L)$, the gain is approximately unity and the output "follows" the input.

The output short circuit ($v_2 = 0$) current is

$i_{sc} = v_1 g_m$

The output open circuit ($i_2 = 0$) voltage is

$v_{oc} = g_m(v_1 - v_2)/G_S = g_m(v_1 - v_{oc})/G_S$

or

$$v_{oc} = v_1 \frac{g_m}{g_m + G_S}$$

The output (driving point) admittance is then

$$g_{eq} = \frac{i_{sc}}{v_{oc}} = g_m + G_S$$

For large transconductances, the output impedance is therefore low, i.e., $r_{eq} \approx 1/g_m$. The CD stage is thus an excellent means of coupling a nonideal source with high inner impedance to a low-impedance load. In other words it is a *unity gain buffer*.

2.3.5 The Insulated Gate Field Effect Transistors (IGFET)

In a second type of FET, a metallic gate electrode is not connected to a reverse-biased P–N junction but is separated from the conductive channel by a layer of insulating material, typically silicon dioxide. This FET is called an *insulated gate field effect transistor* (IGFET). If silicon dioxide is the insulating medium, the name *metal-oxide-silicon FET* (MOSFET) is also used.

Within the class of IGFETs we distinguish two different functional types: the *depletion* IGFET and the *enhancement* IGFET. Figure 2.10 shows the simplified schematics and network symbols for the two types.

The N-channel depletion IGFET consists of a substrate of P material into which two heavily doped N regions are diffused (Fig. 2.10a). These N regions are attached to the source and drain terminals, respectively, and connected to each other by a conductive N-channel. Adjacent to the channel but separated by the insulator is the gate electrode.

With the gate terminal connected to the source ($V_{GS} = 0$), the current I_{DSS} will flow provided the drain–source voltage is large enough. If we make the gate negative with respect to the source ($V_{GS} < 0$), positive charges will be induced on the channel side opposite to the gate. This reduces the number of free electrons, and the drain current falls. From the "depletion" of these electrons follows the name for the device.

As described so far, the behavior of the IGFET is equivalent to that of the JFET. However, an important difference is that the gate–source voltage can be made positive ($V_{GS} > 0$) because there is no danger of forward biasing any gate junctions as would be the case with JFETs. A positive gate voltage, in fact, induces negative charges in the conductive channel and thus enhances the drain current. The control and output characteristics for an N-channel depletion IGFET are shown in Fig. 2.11a.

The circuit symbols for the N-channel depletion IGFET and its complementary P-channel version are shown in Fig. 2.10b. The symbol expresses the fact that the gate electrode is electrically isolated from the S–D channel. The arrow on the

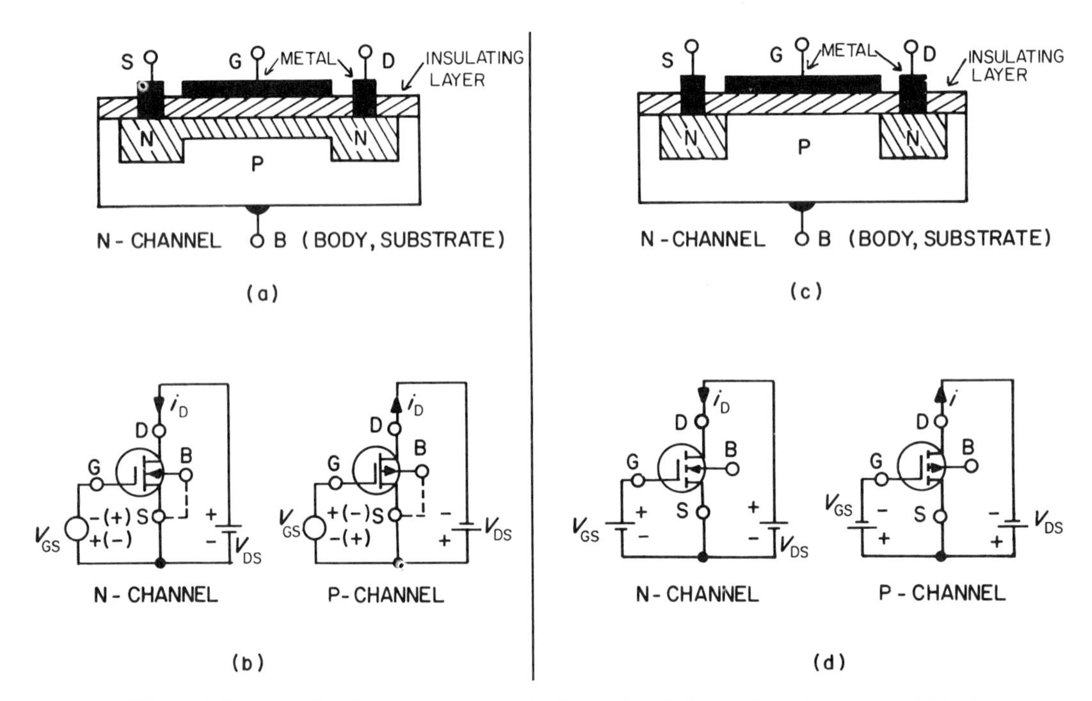

Figure 2.10 Functional schematics, symbols, and polarities of insulate gate field effect transistors: **(a)** depletion IGFET; **(b)** enhancement IGFET.

substrate terminal indicates the direction of the current across the channel–substrate junction if that junction is forward biased. The substrate is either floating, tied to the source, or connected to the gate via a zener diode in order to protect the thin gate insulation layer from electric punch-through.

In contrast to the JFET and the depletion IGFET, the *enhancement* IGFET has zero drain current for $V_{GS}=0$; i.e., no conductive channel exists between source and drain. This transistor changes from this "off" state to the "on" state if $V_{GS}>0$ for N-channel devices and $V_{GS}<0$ for P-channel devices. With these control voltage polarities the appropriate carriers are induced on the substrate side of the gate insulation. The conductive channel thus formed is further "enhanced" as the magnitude of the control voltage is increased. This is expressed clearly in the transconductance and output characteristics which are given for an N-channel device in Fig. 2.11b.

The nonconducting off state is indicated by the broken S–D path in the symbol of the enhancement IGFET (Fig. 2.10d). Also note that the polarities of V_{GS} and V_{DS} are the same in the on state (Fig. 2.10d).

We now briefly summarize and compare certain properties of JFETs and IGFETs.

The obvious difference in the gate arrangements means that IGFETs have much smaller gate currents than JFETs with their reverse-biased junctions.

For $V_{GS}=0$, enhancement IGFETs are off (nonconducting), while depletion IGFETs take on an intermediate position: control voltages of appropriate polarity swing the device into either the depletion or the enhancement mode. Correspondingly, in comparison to V_{DS}, the gate–source voltages are of the same (enhancement IGFET, enhancement mode of depletion IGFET) or of the opposite polarity (JFET,

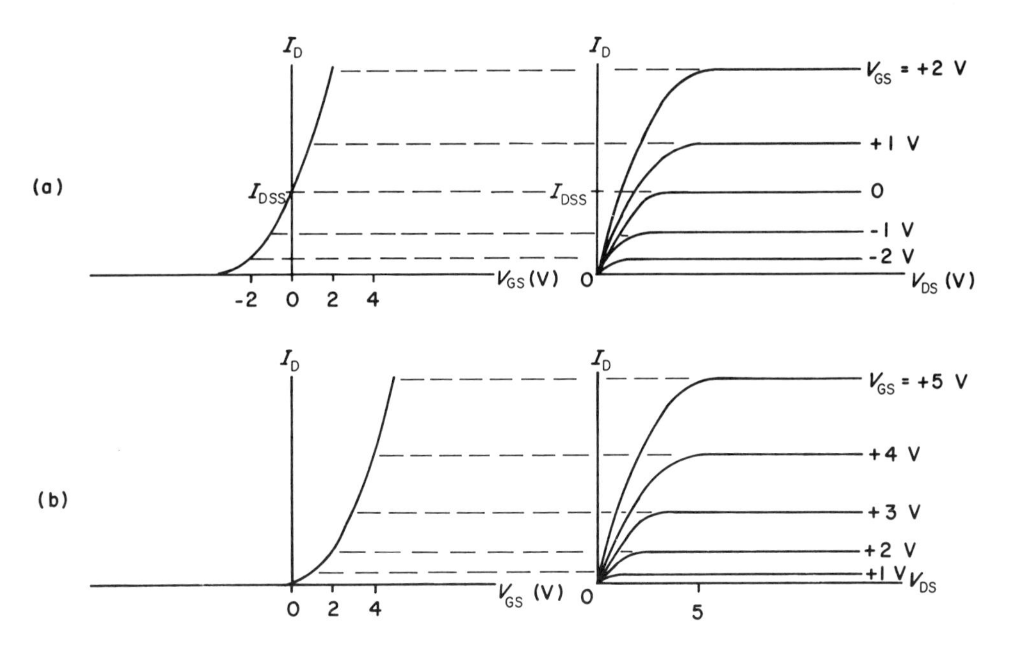

Figure 2.11 Drain current characteristics of IGFETs: **(a)** N-channel depletion IGFET; **(b)** N-channel enhancement IGFET.

depletion mode of depletion IGFET). These functional differences are particularly important in the application of FETs as electronic switches.

A discussion of "noise" (random current and voltage fluctuations superimposed on the desirable signal) arising from *within* transistor stages is beyond the scope of this chapter. Here it is sufficient to say that IGFETs are generally worse than JFETs in this respect. It is, therefore, preferable to use JFETs when signals of very low level must be amplified. (If lower input resistances are acceptable, there is also the alternative of using bipolar junction transistors, which are introduced in Section 2.3.8.)

2.3.6 Biasing Methods for FETs

In order to ensure operation in a specified region of the output characteristics, an operating point must be defined with the aid of a fixed gate bias voltage V_G (see Figs. 2.6e, 2.6f, and 2.7b).

The obvious methods of supplying bias by connecting a voltage source or a voltage divider to the gate terminal are shown in Fig. 2.12a, where R_G is the equivalent of the parallel combination of R_a and R_b, i.e., the inner resistance of the voltage divider. Device variations, however, make this method quite undesirable. This is made clear by the graph in Fig. 2.12b. There it is assumed that two JFETs of the same device type have nevertheless quite different transconductance curves—a normal outcome of the manufacturing process. A circuit designed with a fixed V_G would give greatly different drain currents (I_1 versus I_2) if the device with transconductance curve 1 were replaced with one having the transconductance curve 2.

The problem of parameter variation can be minimized with the *self-bias* method of Fig. 2.12c. In this method the transistor generates its own bias by the voltage drop

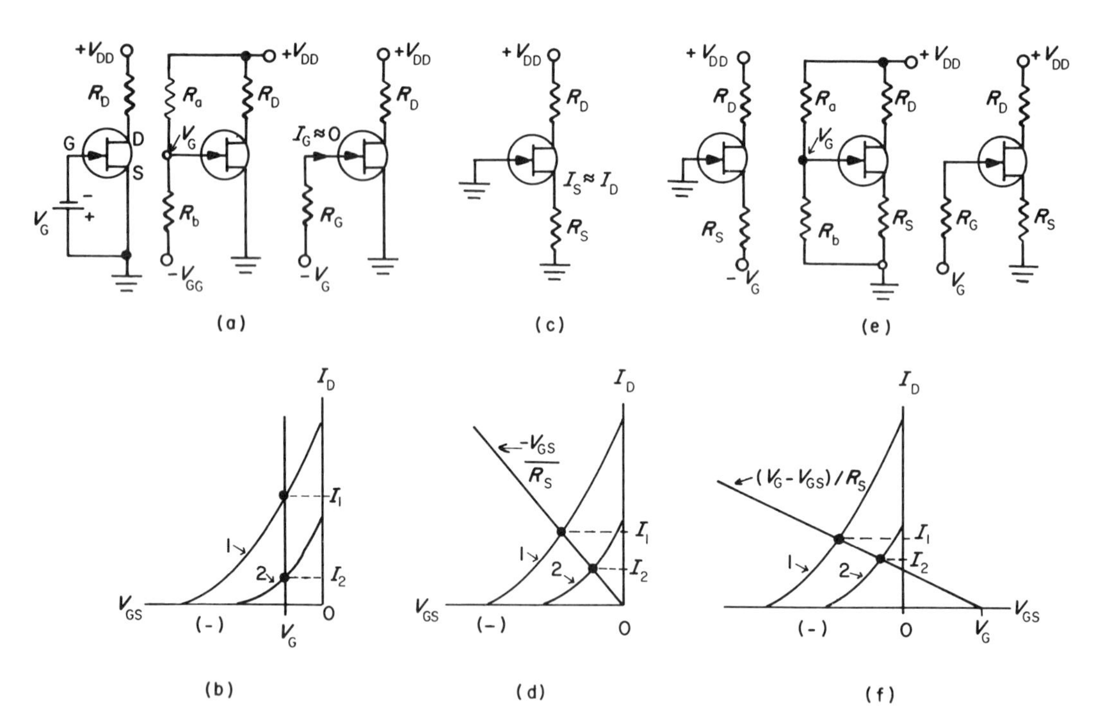

Figure 2.12 Bias methods for FETs. (See text.)

across the source resistor R_S. With the gate grounded we have

$$V_{GS}=V_G-V_S=0-V_S=-I_D R_S$$

or

$$I_D=-V_{GS}/R_S \tag{2.17}$$

We have now two relations between V_{GS} and I_D: Eq. (2.17) and the transconductance curve. The solution satisfying both constraints is the intersection of the line $-V_{GS}/R_S$ with the transconductance curve in Fig. 2.12d. Thus self-bias automatically establishes a $V_{GS}-I_D$ pair as the operating point. Self-bias also tends to *stabilize* the operating point. For example, if temperature changes were to raise I_D, V_{GS} would become more negative and thus reduce I_D.

Note that under self-bias (Fig. 2.12d) the difference between I_1 and I_2 is much less than in the case of external fixed bias (Fig. 2.12b). From a geometric point of view, this is a matter of intersecting the transconductance curves with a straight line of finite slope ($-1/R_S$) in one case and of infinite slope in the other. Obviously, the difference in currents is reduced further if the slope is made still smaller by increasing R_S. One disadvantage of doing this is that the intersection with the transconductance curve occurs in increasingly nonlinear regions of that curve at more negative values for V_{GS}. The solution to this problem is to shift the straight line away from the origin to the right so that the intersection occurs again in the quasi-linear area of the transconductance curve. The improved self-bias methods accomplishing this are shown in Fig. 2.12e. By Kirchhoff's voltage law we have with $I_G=0$

$$V_{GS}+I_D R_S-V_G=0$$

or

$$I_D = (V_G - V_{GS})/R_S \tag{2.18}$$

The graph of Eq. (2.18) in Fig. 2.12f demonstrates that for the same I_1 as in Fig. 1.12d, R_S can thus be increased in order to reduce the current variations.

In extreme cases a horizontal bias line is used. This amounts to *fixed source current bias* and requires that a constant current source (Fig. 2.12a to 2.12d) be connected to the source terminal.

2.3.7 RC-Coupled Amplifier Stages

A conventional *RC*-coupled ac amplifier stage is given in Fig. 2.13a. In this circuit a nonideal signal source with inner resistance R_1 is connected via the coupling capacitor C_c to the gate of the JFET. In this way, the ac input signal can be added to the dc bias which is established by the voltage divider $R_a - R_b$ and the source resistor R_S according to Fig. 2.12e.

Another important feature is the bypass capacitor C_S. This capacitor is so large that it effectively grounds the source terminal for all signal frequencies of interest, i.e., $R_S \gg 1/2\pi f C_S$. We shall see shortly that without this capacitor, the gain is greatly reduced.

The bias arrangement of Fig. 2.13a has the disadvantage that the equivalent resistance $R_a R_b/(R_a + R_b)$ of the voltage divider shunts the very large input resistance of the JFET. This causes an attenuation of the input signal unless R_1 is significantly smaller than this equivalent resistance. (The impedance of the coupling

Figure 2.13 The *RC*-coupled and self-biased common source amplifier stage.

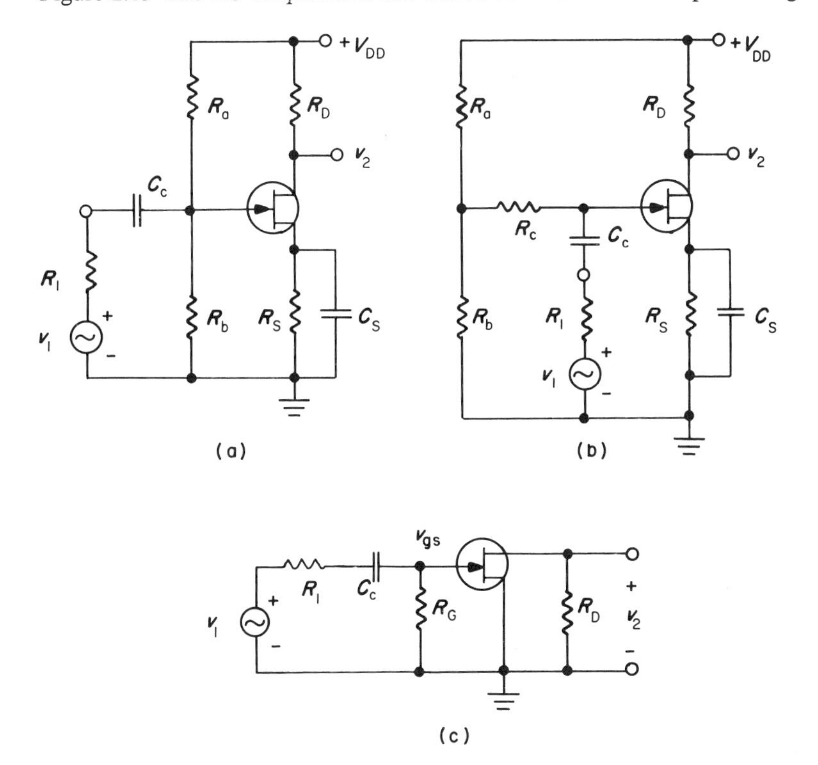

capacitor is assumed to be negligible for all frequencies of interest.) This attenuation is minimized if the gate bias is supplied through a large resistor R_c (Fig. 2.13b).

The ac equivalent circuit for the two stages just introduced is shown in Fig. 2.13c. The V_{DD} power supply represents an ac short circuit to ground. R_S is missing because of the ac shunt to ground through C_S. R_G equals either $R_a R_b/(R_a + R_b)$ (Fig. 2.13a) or $R_c + [R_a R_b/(R_a + R_b)]$ (Fig. 2.13b). The coupling capacitor C_c is included only to remind us of the frequency-dependent nature of RC-coupling. Generally, for the signal frequencies of interest C_c will be chosen so large that it can be replaced by a short circuit. With this assumption we have the voltage division

$$v_{gs} = v_1 \frac{R_G}{R_1 + R_G} \tag{2.19}$$

With the model of Fig. 2.8a and Eq. (2.19), the equation for the drain current is

$$i_d = v_{gs} g_m = v_1 g_m \frac{R_G}{R_1 + R_G} \tag{2.20}$$

The output voltage is then

$$v_2 = -i_d R_D = -v_1 g_m R_D \frac{R_G}{R_1 + R_G}$$

and the voltage gain is, therefore,

$$A_v = \frac{v_2}{v_1} = -g_m R_D \frac{R_G}{R_1 + R_G} \tag{2.21}$$

Equation (2.21) demonstrates that the bias network reduces the voltage gain by the attenuation factor $R_G/(R_1 + R_G)$. Obviously, R_G should be made as large as feasible. That is the justification for the bias arrangement of Fig. 2.13b.

At low frequencies, the coupling capacitor cannot be ignored, and the attenuation factor becomes frequency dependent:

$$\frac{R_G}{R_1 + (1/j\omega C_c) + R_G} = \frac{j\omega R_G C_c}{1 + j\omega(R_1 + R_G)C_c} \tag{2.22}$$

Equation (2.22) is the frequency response function of a high-pass filter. The expression of Eq. (2.22) must be inserted into Eq. (2.21) in place of the attenuation factor $R_G/(R_1 + R_G)$.

At high frequencies, the input capacitance of the JFET (Fig. 2.8d) must be included in the model. The reader may take it as an exercise to verify the low-pass effect of that capacitance at high signal frequencies.

Finally, we shall demonstrate that the self-bias resistor R_S reduces the ac gain if it is not bypassed by a large C_S.

Without C_S, the small-signal equations are

$$v_{gs} = v_1 \frac{R_G}{R_1 + R_G} - i_d R_S \tag{2.23}$$

and

$$i_d = v_{gs} g_m \tag{2.23a}$$

We combine Eqs. (2.23) and (2.23a) and obtain

$$i_{\mathrm{d}} = v_1 g_{\mathrm{m}} \left(\frac{R_{\mathrm{G}}}{R_1 + R_{\mathrm{G}}} \right) \left(\frac{1}{1 + g_{\mathrm{m}} R_{\mathrm{S}}} \right)$$

The output voltage is now

$$v_2 = -i_{\mathrm{d}} R_{\mathrm{D}} = -v_1 g_{\mathrm{m}} R_{\mathrm{D}} \left(\frac{R_{\mathrm{G}}}{R_1 + R_{\mathrm{G}}} \right) \left(\frac{1}{1 + g_{\mathrm{m}} R_{\mathrm{S}}} \right)$$

and the voltage gain is, therefore,

$$A_{\mathrm{v}} = \frac{v_2}{v_1} = -g_{\mathrm{m}} R_{\mathrm{D}} \left(\frac{R_{\mathrm{G}}}{R_1 + R_{\mathrm{G}}} \right) \left(\frac{1}{1 + g_{\mathrm{m}} R_{\mathrm{S}}} \right) \tag{2.24}$$

If we compare Eqs. (2.21) and (2.24) we see that without the bypass capacitor the gain is reduced by the factor $1/(1 + g_{\mathrm{m}} R_{\mathrm{S}})$. This is also true for very low signal frequencies when the impedance of the bypass capacitor is so high that it is effectively out of the circuit.

This concludes our discussion of FET amplifier circuits. Next we introduce a different type of transistor, the bipolar junction transistor. Historically, this transistor was in widespread application long before the FET. It continues to be the dominant transistor device.

2.4 Bipolar Junction Transistors

The bipolar junction transistor consists of a series arrangement of two P–N junctions either in the order N–P–N (*NPN transistor*) or in the sequence P–N–P (*PNP transistor*). Very simplified schematics of this arrangement for the two transistor types are given in Figs. 2.14a and 2.14d.

Each part of the semiconductor "sandwich" has its own electrode terminal, namely, *emitter* (E), *base* (B), and *collector* (C). In terms of circuit performance, but not physical function, emitter and collector correspond to source and drain of the FET, while the base is the control terminal analogous to the gate. In actual physical function, however, the bipolar transistor is quite different from the FET: (1) the major current path is *across* the P–N junctions; (2) one P–N junction is *forward* biased, the other reverse biased; (3) carriers of both polarities cross the forward-biased junction (hence the name *bipolar* junction transistor); (4) the junction transistor is more appropriately modeled as a *current-controlled current source* rather than a voltage-controlled one; (5) because of the forward-biased P–N junction, the controlling base current can be quite large unlike the gate leakage current of the voltage-controlled FET.

We shall now present a very simplified explanation of the operation of bipolar junction transistors. Consider the scheme of Fig. 2.14a for an NPN transistor. (The actual geometry of a manufactured transistor is much more complicated.) Note that there is an *emitter–base junction* which is *forward* biased, and a *base–collector junction* which is *reverse* biased. Compared to the unbiased state, this means that the width of the depletion layer is reduced for the emitter–base junction and enlarged for the base-collector junction.

In the case of the NPN transistor, the forward bias on the emitter–base junction represents a reduced energy barrier for electrons (majority carriers of the N material)

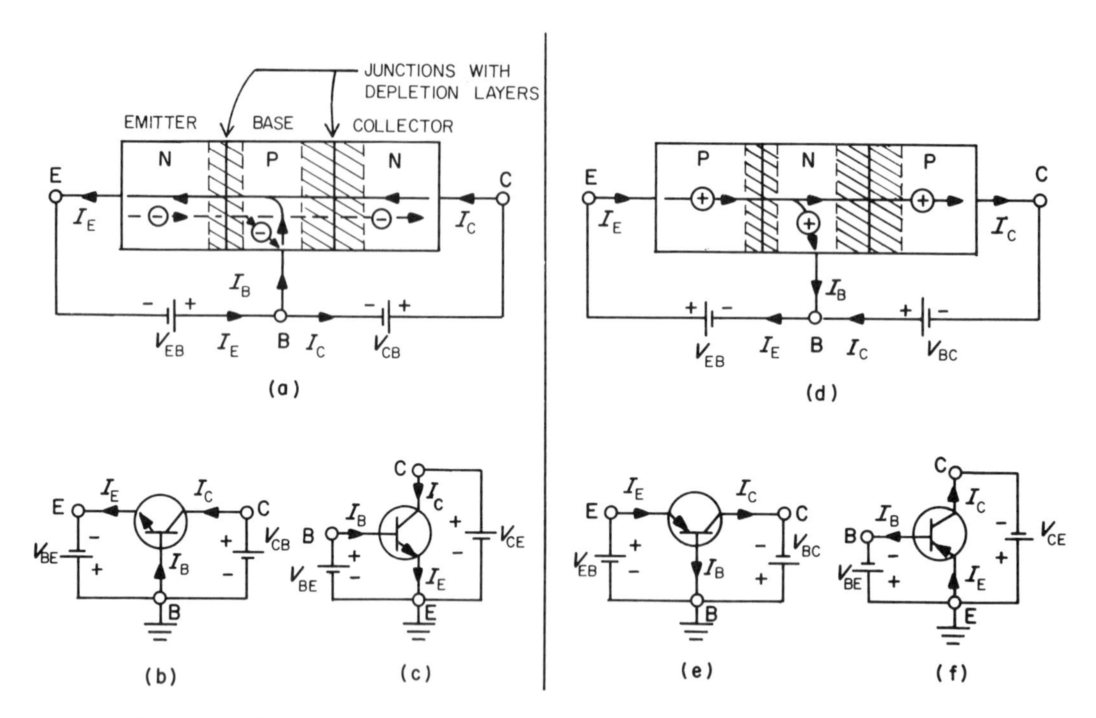

Figure 2.14 Functional schematics, symbols, and polarities of bipolar junction transistors: **(a)**, **(b)**, and **(c)** NPN junction transistors; **(d)**, **(e)**, and **(f)** PNP junction transistors.

and holes (majority carriers of the P material); thus electrons flow from N to P and holes from P to N. The reverse bias on the base–collector junction prohibits the crossing of holes but greatly *facilitates* the passage of electrons down the energy gradient from base to collector regions. The crucial aspect of NPN transistor function is that most of the electrons "emitted" from the emitter side and crossing into the usually very narrow base do not undergo recombination but are "collected" in the energetically lower (i.e., for electrons) collector region. The net electron current is indicated by the broken line in Fig. 2.14a. Anywhere from 95% to more than 99% of the emitter current I_E ends up as the collector current I_C; the remainder makes up the base current I_B. The electron current is, of course, functionally equivalent to a "conventional" current of unspecified positive charge carriers moving in the opposite direction. This is indicated by the solid line in Fig. 2.14a.

The network symbol for the NPN transistor, the positive current polarities (in terms of conventional current), and the bias arrangements are shown in Fig. 2.14b. The arrow inside the circle indicates the direction of the emitter current under forward bias of the emitter–base junction. The alternative bias arrangement (Fig. 2.14c) with $V_{CE} > V_{BE}$ is much more common because only one power supply (V_{CE}) carries the large collector–emitter current, while the second supply (V_{BE}) provides only the much smaller control current.

Functionally, and in terms of current and voltage polarities, the PNP transistor is the reverse image of the NPN transistor (Figs. 2.14d–2.14f). Net current flow is now described by the flux of holes from emitter to collector and base, respectively.

The controlled source function of the bipolar junction transistor can be described by two modes of current control: control by emitter current or by base current. (It is also possible to use a voltage-controlled current source model, but the relation is much more nonlinear). We introduce these concepts next with the aid of NPN transistors.

To understand the first mode we plot the collector current I_C as a function of the controlling emitter current I_E. This relation is approximately linear as shown in Fig. 2.15a. An example of output characteristics, Fig. 2.15b, demonstrates that the transistor in this mode is an excellent approximation of a current-controlled current source as long as the collector–base voltage is large enough to maintain sufficient backward bias. Note that a collector current I_{CB0} flows even if there is no emitter current (emitter electrode floating). This current is the temperature-dependent leakage current across the reverse-biased base–collector junction. A simple nonlinear model approximating this behavior consists of two idealized diodes for the emitter–base and base–collector junctions and two controlled current sources arranged as in Fig. 2.15c. Although certain applications require that transistors be operated in this mode, the fact that the controlling current I_E is actually a bit larger than the output current I_C is often a drawback. The controlling current is much smaller in the mode discussed next.

Figure 2.15 Collector current characteristics and large-signal models of junction transistors.

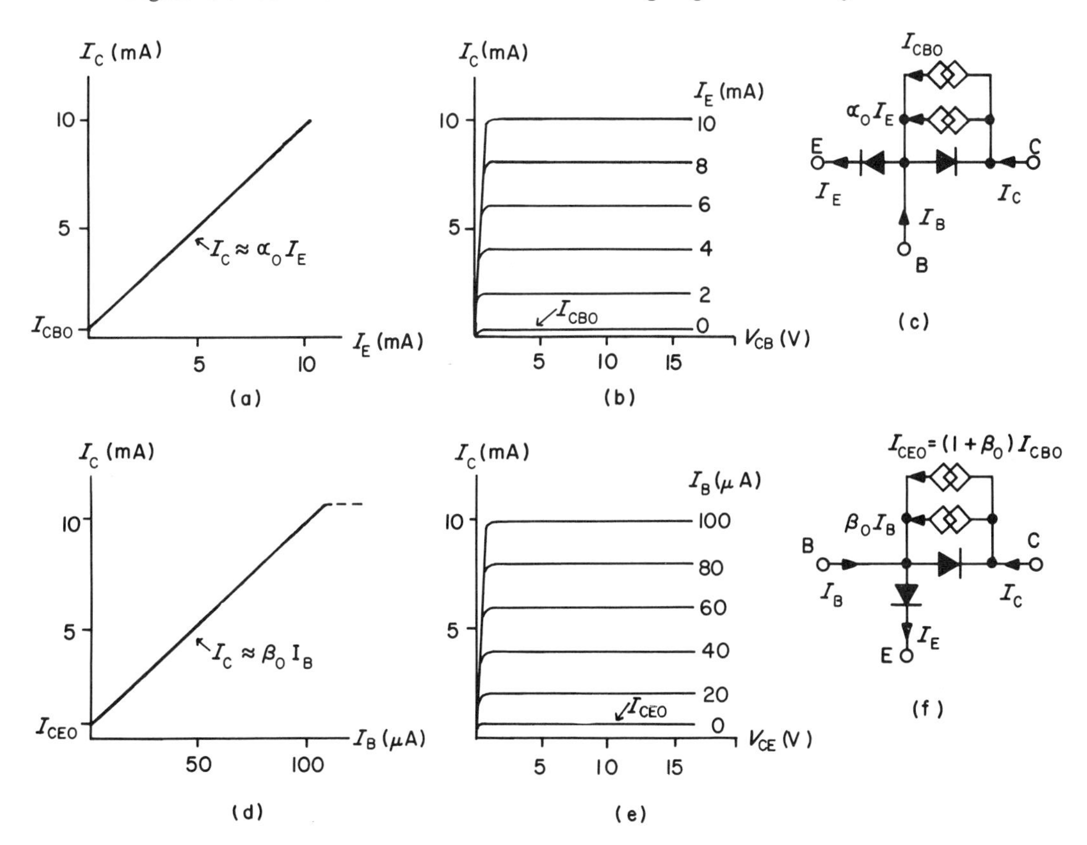

For base current control, we assume than an external current source imposes a base current. With a power supply V_{CE} of appropriate polarity (Fig. 2.14c) across the collector and emitter terminals, the collector current will be approximately proportional to the base current. In the example of Fig. 2.15d, the collector current is about a 100 times the base current. The family of idealized output characteristics is shown in Fig. 2.15e. Note that in this case, too, an output current I_{CE0} flows even if the controlling current is zero (base electrode floating). Now the appropriate nonlinear device model is that of Fig. 2.15f.

By reversing polarities, exactly complementary curves and models could be obtained for PNP transistors.

The control relations just developed and Kirchhoff's current law provide us with the means to state important relations between the various transistor currents. They are valid for either NPN or PNP transistors.

According to Fig. 2.14 and Kirchhoff's current law we write

$$I_E = I_B + I_C \tag{2.25}$$

From the characteristics of Fig. 2.15, we obtain for the model in Fig. 2.15c the control law

$$I_C = \alpha_0 I_E + I_{CB0}, \tag{2.26}$$

and for the model in Fig. 2.15f

$$I_C = \beta_0 I_B + I_{CE0} \tag{2.27}$$

In Eqs. (2.26) and (2.27), α_0 and β_0 are the dc or large-signal *forward current gains*. And the leakage or *cut-off* currents are defined as

$$I_{CB0} = I_C|_{I_E=0} = -I_B|_{I_E=0} \tag{2.28}$$

and

$$I_{CE0} = I_C|_{I_B=0} = I_E|_{I_B=0} \tag{2.29}$$

The minus sign in Eq. (2.28) indicates that the base leakage current flows *out* of the base terminal. In order to derive the relation between the two leakage currents we impose the condition of Eq. (2.29) on Eq. (2.26). This gives us

$$I_{CE0} = \alpha_0 I_{CE0} + I_{CB0} = I_{CB0}/(1-\alpha_0)$$

An alternative expression follows from combining Eqs. (2.27) and (2.28), namely,

$$I_{CB0} = -\beta_0 I_{CB0} + I_{CE0} = I_{CE0}/(1+\beta_0)$$

At room temperature, a silicon transistor might have a cut-off current $I_{CB0} \approx 100$ nA and a current gain $\beta_0 \approx 100$; therefore, I_{CE0} is about 10 μA.

2.4.1 Equivalent Circuits for Junction Transistors

For small-signal operation, say, in the middle of the characteristics of Figs. 2.15a and 2.15d, Eqs. (2.26) and (2.27) reduce to

$$i_C = \alpha i_E = h_{fb} i_E \tag{2.30}$$

and

$$i_C = \beta i_B = h_{fe} i_B \tag{2.31}$$

respectively. The gain factors α and β are now the *small-signal short circuit current gains*. In transistor data sheets, α and β are often replaced by the h parameters h_{fb} and h_{fe}, respectively. The h parameters in Eqs. (2.30) and (2.31) have subscripts which indicate that these are *forward* current gains (f), determined either with the base (b) or the emitter (e) connected at ac common. We leave it to the reader to prove with the aid of Eqs. (2.25), (2.30), and (2.31) that the current gains are related as follows:

$$\alpha = \beta/(1+\beta) < 1 \tag{2.32}$$

$$\beta = \alpha/(1-\alpha) > 1 \tag{2.33}$$

Typical values for modern junction transistors are $\alpha = 0.93$–0.995, and $\beta = 50$–250.

Equations (2.30)–(2.33) are very important for setting up circuit models of junction transistors. A collection of such small-signal model circuits is given in Figs. 2.16 and 2.17 for NPN transistors. For PNP transistors, the source polarities are simply reversed.

The simplest of the T-equivalent models, Fig. 2.16a, is derived from the large-signal model of Fig. 2.15c; i.e., the forward-biased emitter diode is replaced by a short circuit, the reverse-biased collector diode is taken as an open circuit, and (as in all other small-signal models) the cut-off currents are ignored.

Figure 2.16 Small-signal T models of junction transistors.

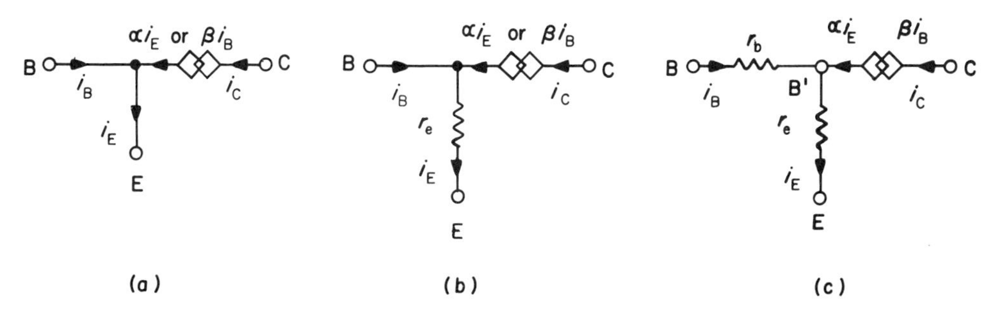

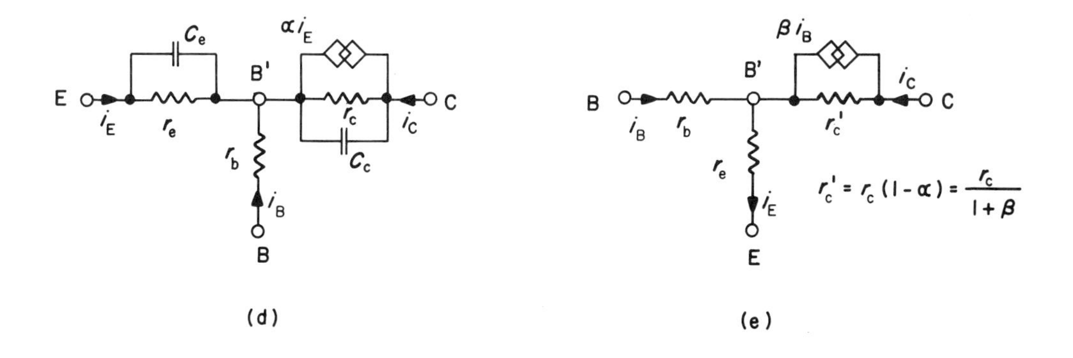

The next refinement (Fig. 2.16b) is to include the effective resistance of the emitter diode r_e, which is typically about 25–30 Ω. There is also an access resistance from the external base terminal B to the "inner" base region B′. The corresponding resistance r_b in Fig. 2.16c is about 100 Ω or more.

The parasitic junctional capacitances C_e and C_c and the finite resistance of the collector junction r_c make the model of Fig. 2.16d more realistic. The capacitances are of the order of a few pF, while r_c is about 1–1.5 MΩ. Note that $r_c' = (1-\alpha)r_c$ is the appropriate collector resistance if the current source βi_B is used (Fig. 2.16e).

The finite input resistance of a transistor justifies also the use of *voltage-controlled* current source models. A rather complicated one for high-frequency modeling is the so-called *hybrid-π* of Fig. 2.17a. The parameter r_μ, indicating the "backward" action of the collector voltage, is usually rather large and can be ignored. For the other parameters we have in comparison with Fig. 2.16d

$$r_x = r_b$$
$$r_\pi = (1+\beta)r_e$$
$$g_o = 1/r_c$$
$$C_\pi = C_e$$
$$C_\mu = C_c$$
$$g_m = \alpha/r_e = \beta/r_\pi$$

Figure 2.17 Hybrid-π and *h* parameter models of junction transistors.

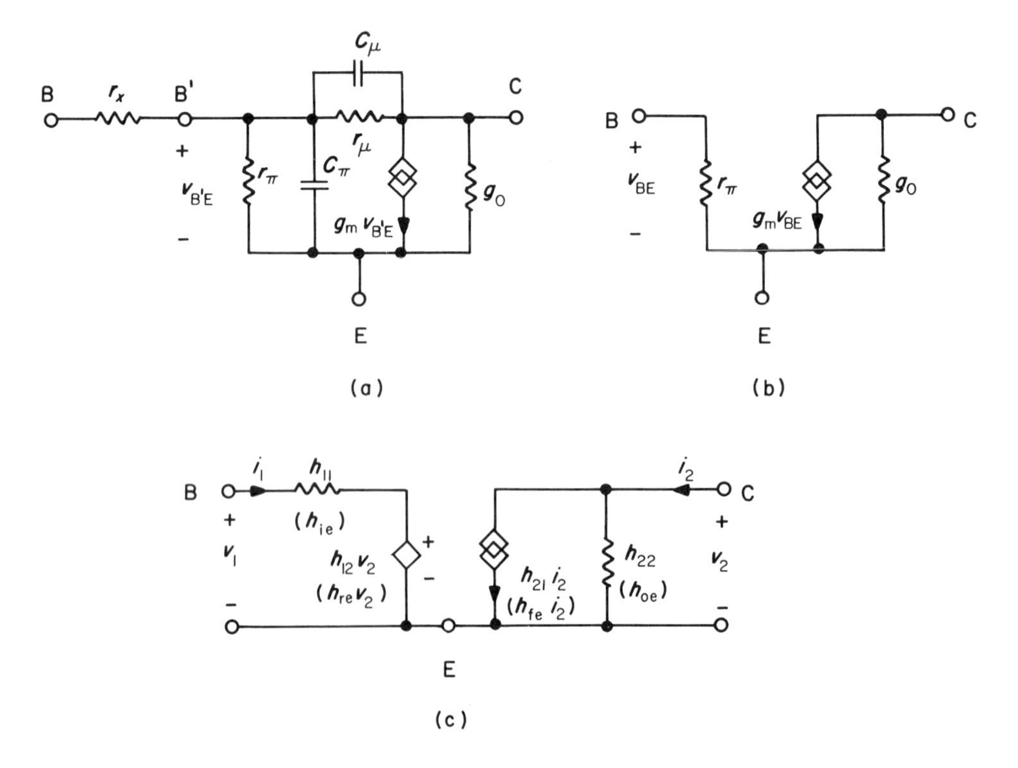

A simplified hybrid-π model that is adequate for approximate calculations is shown in Fig. 2.17b. Typical values would be $r_\pi = 3.6\ \text{k}\Omega$, $g_m = 33$ mS, $g_o = 0.7\ \mu$S.

Finally, another model explicitly expressing the "backward" action of the output voltage with a controlled voltage source is the h parameter equivalent circuit of Fig. 2.17c. The parameters in parentheses are the ones appropriate for transistors with the emitter connected to ac ground. Transistor data sheets often contain tabulations and curves for h parameters as functions of operating conditions.

The h parameters of Fig. 2.17c are related to the parameters of the T-equivalent of Fig. 2.16e as follows:

$$h_{fe} = \beta$$

$$h_{oe} = 1/r_c' = (1+\beta)/r_c$$

$$h_{re} = r_e/r_c' = (1+\beta)r_e/r_c$$

$$h_{ie} = r_b + (1+\beta)r_e$$

We shall now apply some of the equivalent circuits in a simple analysis of the three basic amplifier configurations.

2.4.2 Three Basic Amplifier Configurations of Junction Transistors

In this section we shall discuss the *common base* (CB), the *common emitter* (CE), and the *common collector* (CC) or *emitter follower* configurations. In each instance, "common" means that the appropriate terminal is at ground or common reference potential.

From the example exercises to follow we can conclude that, in relative terms,

1. the *voltage gain* is high for the CB and CE stage and low (about 1) for the CC or emitter follower stage;
2. the *current gain* is high for the CE and CC (emitter follower) stage and low (about 1) for the CB stage;
3. the *input resistance* is high for the CC (emitter follower) stage, medium for the CE stage, and low for the CB stage;
4. the *output resistance* is high for the CB stage, medium for the CE stage, and low for the CC (emitter follower) stage.

2.4.2.1 The Common Base (CB) Amplifier

This circuit is the least important configuration except in high-frequency applications where the CB connection minimizes the effect of parasitic capacitances. The CB configuration is, however, the most obvious one based on the scheme of Fig. 2.14a.

Figure 2.18a (top) shows a biased CB stage, where i_1 is the input current source, I_B the dc bias current source, v_2 the output voltage across the load R_L, and V_{CC} the collector power supply. For ac or *small-signal* analysis we replace the bias current source with an open circuit, and the power supply with a short circuit. This gives us the circuit in the middle of Fig. 2.18a. Using the simple T-equivalent of Fig. 2.16b with the addition of r_c, we obtain the model circuit at the bottom of Fig. 2.18a.

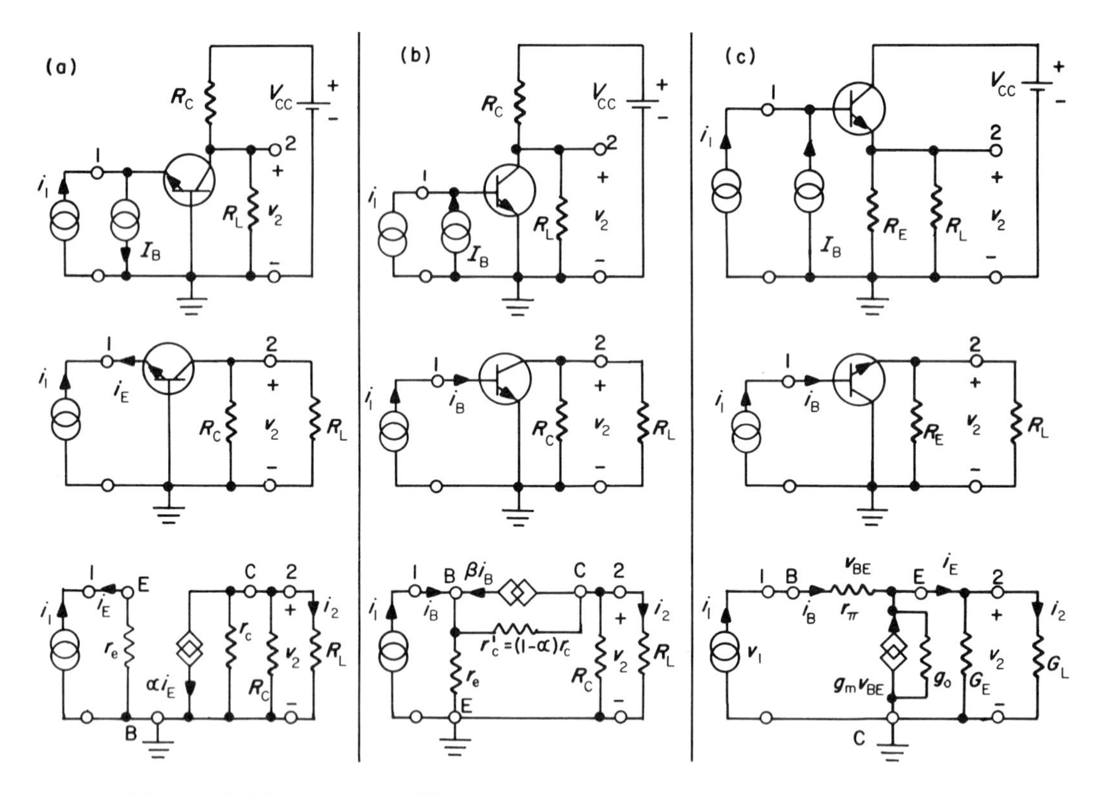

Figure 2.18 Three basic amplifier stages and their ac equivalent circuits: **(a)** common base (CB) stage; **(b)** common emitter (CE) stage; **(c)** common collector (CC) stage.

Example 2.1. Analysis of the CB stage

By inspection of the circuit of Fig. 2.18a (bottom) we can write for the *input resistance*

$$R_{\text{in}} = v_1/i_1 = r_e \tag{2.34}$$

The short circuit output current ($R_L = 0$) is

$$i_{2sc} = -\alpha i_E$$

The open circuit output voltage ($R_L = \infty$) is

$$v_{2oc} = -\alpha i_E r_c R_C/(r_c + R_C)$$

The driving point or *output resistance* is, therefore,

$$R_{\text{out}} = \frac{v_{2oc}}{i_{2sc}} = \frac{r_c R_C}{r_c + R_C} = \frac{1}{g_c + G_C} \tag{2.35}$$

The input current is

$$i_1 = -i_E$$

The output voltage is, therefore,

$$v_2 = -\alpha i_E/(g_c + G_C + G_L) = \alpha i_1/(g_c + G_C + G_L) \tag{2.36}$$

From Eq. (2.36) we derive the output current

$$i_2 = v_2 G_L = \alpha G_L i_1/(g_c + G_C + G_L)$$

and the *current gain*

$$A_i = \frac{i_2}{i_1} = \frac{\alpha G_L}{g_c + G_C + G_L} \approx \frac{\alpha G_L}{G_C + G_L} \tag{2.37}$$

By inspection we write for the input current

$$i_1 = v_1 / r_e = v_1 g_e$$

With this expression, the *voltage gain* follows from Eq. (2.36):

$$A_v = \frac{v_2}{v_1} = \frac{\alpha g_e}{g_c + G_C + G_L} \approx \frac{\alpha g_e}{G_C + G_L} \tag{2.38}$$

Note that the voltage gain is positive; i.e., there is *no* sign reversal or 180° phase shift between input and output voltage. With $\alpha \approx 1$ and $g_e \gg (G_C + G_L)$, the gain can also be made quite large.

2.4.2.2 The Common Emitter (CE) Amplifier

This is the most useful configuration. In comparison to the CB stage, this stage not only has a higher input resistance, lower output resistance, and a higher current amplification, but also a 180° phase shift (sign reversal) between input and output. Furthermore, the high current and voltage amplifications combined result in the highest power amplification of the three stage types.

The dc, ac, and equivalent circuits of the CE stage are given in Fig. 2.18b. With the T-equivalent of Fig. 2.16b augmented by r_c', the analysis is quite simple, as the following example shows.

Example 2.2. Analysis of the CE stage

The node equation for the input terminal is

$$v_1(g_e + g_c') = v_2 g_c' + (1+\beta) i_1 \approx (1+\beta) i_1$$

The *input resistance* is then approximately

$$R_{in} = \frac{v_1}{i_1} \approx \frac{1+\beta}{g_e + g_c'} \approx \frac{1+\beta}{g_e} = (1+\beta) r_e \tag{2.39}$$

A comparison of Eqs. (2.34) and (2.39) proves that the input impedance of the CE stage is much larger than that of the CB stage.

For the calculation of the output impedance we make the assumption $r_e \ll r_c'$. The output short circuit current ($R_L = 0$) is then

$$i_{2sc} \approx -\beta i_B$$

and the output open circuit voltage ($R_L = \infty$) is

$$v_{2oc} \approx -\beta i_B r_c' R_C / (r_c' + R_C)$$

Hence the *output resistance* is approximately

$$R_{out} = \frac{v_{2oc}}{i_{2sc}} \approx \frac{r_c' R_C}{r_c' + R_C} = \frac{1}{g_c' + G_C} = \frac{1}{(1+\beta) g_c + G_C} \tag{2.40}$$

The impedance according to Eq. (2.40) is smaller than that of the CB stage, Eq. (2.35).

If we continue to consider r_e to be small compared to the other resistances, we obtain simple node equations for terminals 1 and 2. Thus

$$v_1 g_e \approx (1+\beta) i_B \tag{2.41}$$

and

$$v_2(g_c' + G_C + G_L) \approx -\beta i_B \tag{2.42}$$

We combine Eqs. (2.41) and (2.42) to obtain the *voltage gain*

$$A_v = \frac{-\beta g_e/(1+\beta)}{g_c' + G_C + G_L} = \frac{-\alpha g_e}{g_c' + G_C + G_L} \approx \frac{-\alpha g_e}{G_C + G_L} \tag{2.43}$$

Compare Eq. (2.43) to Eq. (2.38) and note the negative sign indicating a 180° phase shift.

With Eq. (2.42) and $i_B = i_1$, the output current is approximately

$$i_2 = v_2 G_L \approx \frac{-\beta G_L i_1}{g_c' + G_C + G_L}$$

The *current gain* is, therefore,

$$A_i = \frac{i_2}{i_1} = \frac{-\beta G_L}{g_c' + G_C + G_L} \approx \frac{-\beta G_L}{G_C + G_L} = \frac{-\alpha G_L}{(1-\alpha)(G_C + G_L)} \tag{2.44}$$

which is larger than the corresponding gain for the CB stage, Eq. (2.37).

2.4.2.3 The Common Collector (CC) Amplifier or Emitter Follower

This stage is used mostly as a buffer circuit between a high-impedance source and a low-impedance load because it has unity voltage gain, a high input impedance, and a low output impedance. With the voltage gain very close to unity, the emitter "follows" the base input; hence the name "emitter follower."

In order to analyze the circuit we shall use the simplified hybrid-π model of Fig. 2.17b for the sake of variety. This is done in the following example.

Example 2.3. Analysis of the CC stage

Kirchhoff's voltage law gives us

$$v_1 = v_{BE} + v_2 = i_B r_\pi + v_2 \tag{2.45}$$

With $g_o \ll (G_E + G_L)$ the output voltage is approximately

$$v_2 \approx (i_1 + g_m v_{BE})/(G_E + G_L) = i_1(1 + g_m r_\pi)/(G_E + G_L) \tag{2.46}$$

In Section 2.4.1 we stated that $r_\pi = (1+\beta) r_e$. The reader may also prove that $g_m r_\pi = \beta$. With these two expressions, Eqs. (2.45) and (2.46) can be combined for the derivation of the *input resistance* to yield

$$R_{in} = (1+\beta)\left(r_e + \frac{R_E R_L}{R_E + R_L} \right) \tag{2.47}$$

Obviously, of the three basic stages, the CC stage has the largest input resistance [cf. Eqs. (2.34), (2.39), and (2.47)].

The output short circuit current ($R_L = 0$) is

$$i_{2sc} = i_1 + g_m v_{BE}$$

And with the usually justified assumption $g_o \ll G_E$, the open circuit output voltage ($R_L = \infty$) is approximately

$$v_{2oc} \approx (i_1 + g_m v_{BE}) R_E$$

Hence, the effective *output resistance* is approximately

$$R_{out} = \frac{v_{2oc}}{i_{2sc}} \approx R_E \tag{2.48}$$

This is clearly the lowest value of all three stages (if we take R_E to be equivalent to R_C in the CB and CE case).

Equation (2.46) can be simplified even more because in most cases the base current i_1 is much smaller than the collector current $g_m v_{BE}$. Thus we write

$$v_2 \approx g_m v_{BE} / (G_E + G_L)$$

or

$$v_{BE} \approx (G_E + G_L) v_2 / g_m$$

The latter expression and Eq. (2.45) give then the *voltage gain*

$$A_v = \frac{v_2}{v_1} \approx g_m / (g_m + G_E + G_L) < 1 \tag{2.49}$$

In practical cases, the choice is to make $g_m \gg (G_E + G_L)$ so that

$$A_v \approx 1$$

The output current is

$$i_2 = v_2 G_L$$

With the usual case of $g_o \ll (G_E + G_L)$ and Eq. (2.46), the *current gain* is then

$$A_i = \frac{i_2}{i_1} \approx \frac{(1 + g_m r_\pi) G_L}{G_E + G_L} = \frac{(1 + \beta) G_L}{G_E + G_L} \tag{2.50}$$

2.4.3 Biasing Methods for Junction Transistors

In the preceding discussion of the three basic transistor stages, it was assumed that a dc current source I_B determined the quiescent level of the collector current (Fig. 2.18). This scheme is repeated in the left circuit of Fig. 2.19a. A simple way to simulate such a current source is shown on the right of Fig. 2.19a. Obviously, we can write

$$I_{0B} = \frac{V_{CC} - V_{BE}}{R_B} \approx \frac{V_{CC}}{R_B} = \text{constant}$$

because the supply voltage V_{CC} is usually kept constant at a level of several volts, whereas V_{BE}, the voltage drop across the forward-biased base–emitter diode, is only a fraction of a volt.

With a constant base current I_{0B}, the quiescent emitter current of a simple CE stage is then $I_E = (1 + \beta_0) I_{0B}$. (For the purpose of developing an improved biasing scheme we consider the emitter current instead of the nearly equal collector current $I_C = \beta_0 I_{0B}$.)

The disadvantage of base current bias is obvious from the graph of Fig. 2.19c, which illustrates what happens when a repair requires that a transistor (curve 1) be replaced by another transistor of the same type but different β (curve 2). Thus for

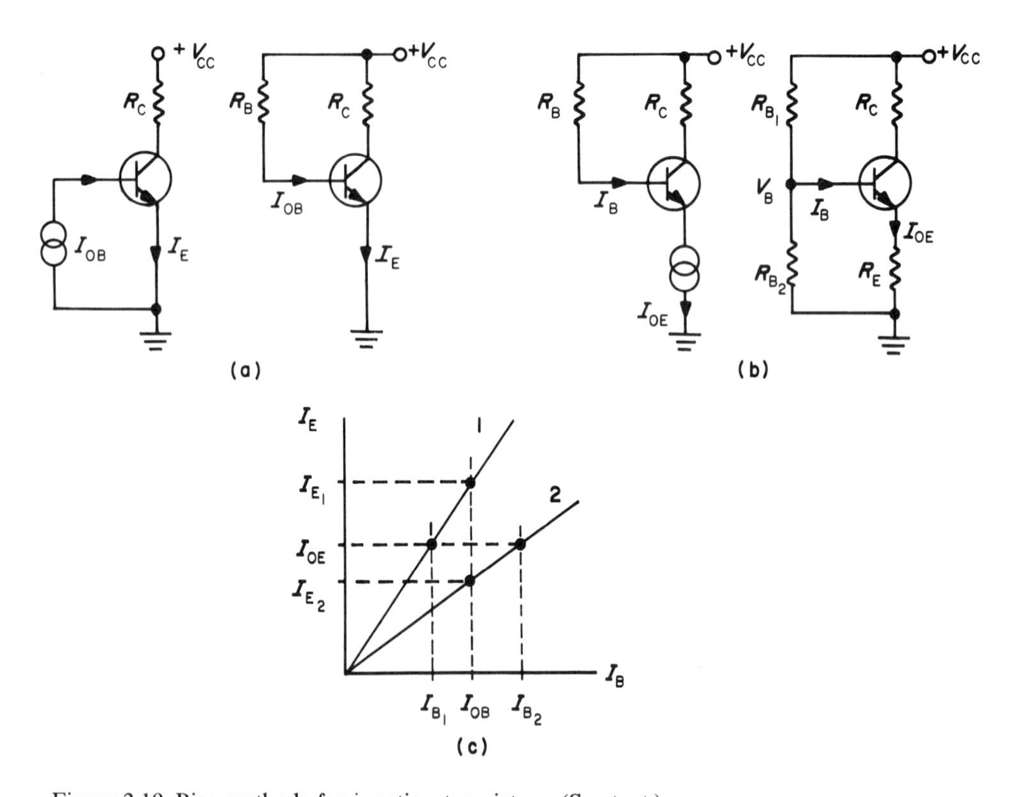

Figure 2.19 Bias methods for junction transistors. (See text.)

the same base current I_{0B}, two different emitter currents, I_{E1} and I_{E2}, and hence two different operating points will result.

The alternative is to fix the emitter current and hence the quiescent collector current ($I_{0C} \approx I_{0E}$). According to Fig. 2.19c, this approach leads to variations of the quiescent base current (I_{B1} and I_{B2}). But these are usually inconsequential as long as the external circuitry can accommodate them.

Methods for establishing emitter current bias are shown in Fig. 2.19b. A current source like that of Fig. 2.21a could be used. A more common device is the *self-bias* scheme with an emitter resistor R_E and a base voltage divider $R_{B_1} - R_{B_2}$ as shown on the left of Fig. 2.19b. With self-bias, the quiescent emitter current is

$$I_{0E} = (V_B - V_{BE})/R_E \approx V_B/R_E$$

The constraints on the design of the base voltage divider are conflicting. On one hand, R_{B_1} and R_{B_2} should be as large as possible so that the effective input impedance of the stage is kept high. On the other hand, the parallel combination of R_{B_1} and R_{B_2} should not be so large that the highly temperature-dependent base leakage current causes a voltage drop sufficient to change the bias. Fortunately, self-bias improves bias stability: any increase in the emitter current raises the voltage drop across R_E and thus reduces V_{BE}, the forward bias on the base–emitter junction.

The combination of self-bias and *RC* coupling in a standard ac amplifier stage is our next topic.

2.4.4 The RC-Coupled Common Emitter Amplifier Stage

By definition, the RC-coupled stage amplifies only ac signals. Hence, the emitter bias resistor R_E is bypassed by a large capacitor C_E (Fig. 2.20a). The simple ac equivalent circuit of the RC-coupled CE stage is shown in Fig. 2.20b. We shall use the simple hybrid-π model of the transistor of Fig. 2.17b to analyze the circuit.

First we define the equivalent input resistance to the right of coupling capacitor C_c:

$$R'_B = r_\pi R_B/(r_\pi + R_B)$$

Next we limit our analysis to signal frequencies for which

$$(R_1 + R'_B) \gg \frac{1}{2\pi f C_c}$$

The effective base–emitter voltage is then

$$v_{BE} = \frac{R'_B}{R_1 + R'_B} v_1$$

and the output voltage is

$$v_2 = -i_C R_C = -v_{BE} g_m R_C = \frac{-g_m R_C R'_B v_1}{R_1 + R'_B}$$

Thus the ac voltage gain is

$$A_v = \frac{v_2}{v_1} = -g_m R_C R'_B/(R_1 + R'_B) \tag{2.51}$$

Equation (2.51) is analogous to the voltage gain equation for a JFET stage [Fig. 2.13, Eq. (2.21)] developed in Section 2.3.7.

This concludes our rudimentary discussion of transistor amplifier stages. It should suffice because integrated circuits are increasingly displacing discrete transistor designs in instrumentation. The remainder of this chapter is devoted to applications other than the prototypical transistor amplifier stage.

Figure 2.20 The RC-coupled and self-biased common emitter stage.

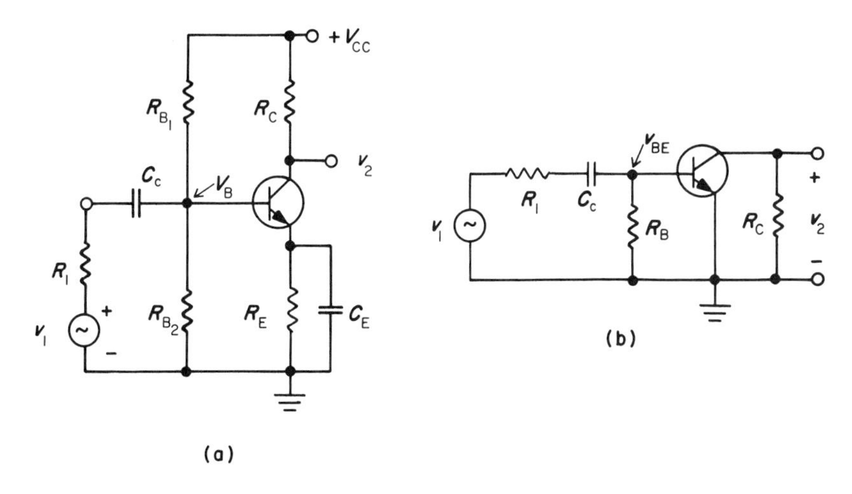

2.5 Special Transistor Circuits

In this section we shall consider a small sample of useful transistor circuits.

2.5.1 Current Sources

Aside from amplification, transistors should obviously be useful as constant current sources. Among the simplest of these are the FET circuits of Figs. 2.21a and 2.21b. In these, V_{GS} and, therefore, the drain current can be adjusted with the potentiometer R_S.

The circuits of Figs. 2.21c and 2.21d are so-called current mirrors. The point here is to set the reference current I_1 and make the other current, I_2, proportional (or equal) to the reference by adjusting R_2.

In the PNP example of Fig. 2.21c, transistor Q_1 is in the *diode connection* (the collector–base diode is short circuited). If matched transistors are used, the emitter–base voltages of the two transistors and, therefore, the emitter voltages will be quite similar. Thus we have

$$I_2 R_2 \approx I_1 R_1$$

Figure 2.21 Other useful transistor circuits. (See text.)

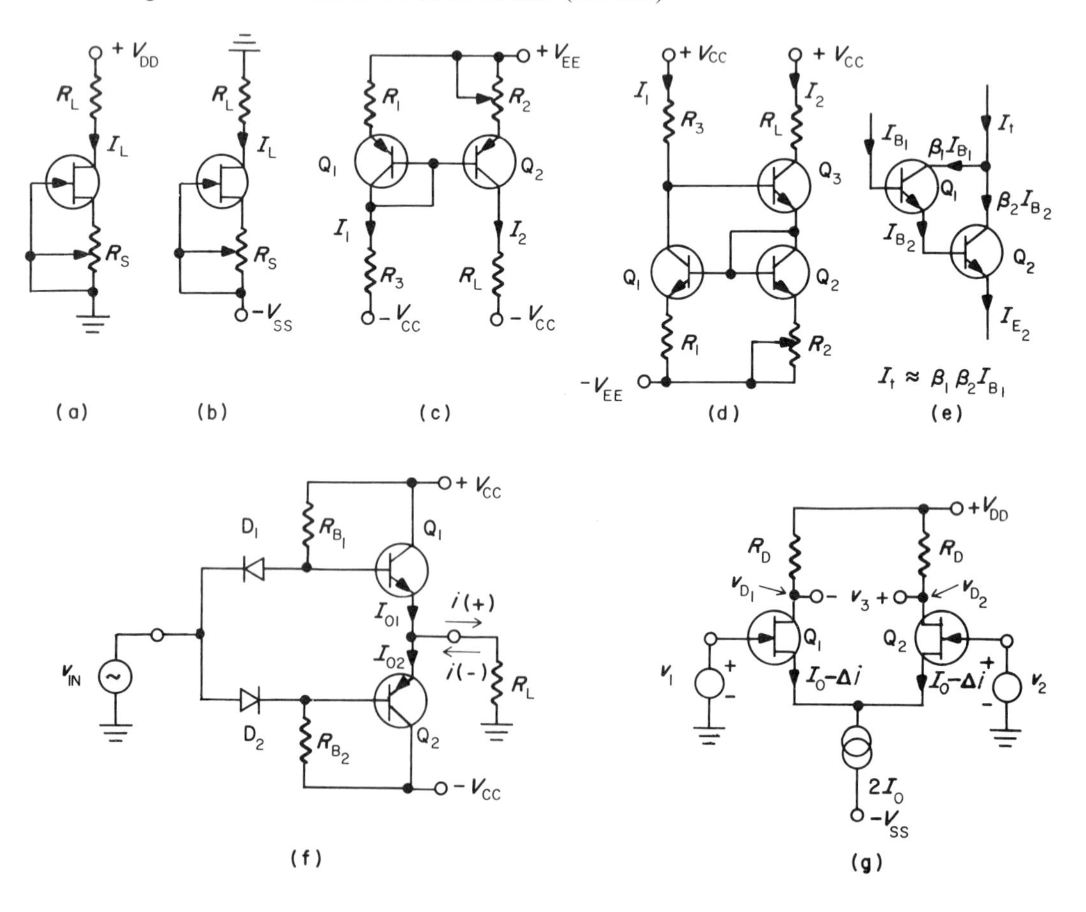

and

$$I_1 = \frac{V_{EE} - V_{CC} - V_{BE}}{R_1 + R_3} \approx \frac{V_{EE} - V_{CC}}{R_1 + R_3}$$

These two expressions define the current through the load R_L:

$$I_2 = \frac{R_1}{R_2} I_1 = \frac{R_1}{R_2} \left(\frac{V_{EE} - V_{CC}}{R_1 + R_3} \right)$$

Actually the current through R_1 is not I_1 but $I_1 - I_{B_2}$, and that through R_2 is $I_2 + I_{B_2}$ instead of I_2. The mismatch due to the base current can be reduced by adding a third transistor. An example of such an improved current source with NPN transistors is shown in Fig. 2.21d.

2.5.2 *The Darlington Transistor*

The Darlington connection, Fig. 2.21e, is a method of cascading two transistors so that they act as a single one but with a "super beta." A current balance shows that the effective beta of the cascade is approximately the product of the individual betas. Thus for the total "collector" current we have

$$I_t = \beta_1 I_{B_1} + \beta_2 I_{B_2} = \beta_1 I_{B_1} + (1 + \beta_1)\beta_2 I_{B_1} \approx \beta_1 \beta_2 I_{B_1}$$

an approximation justified for typical β values of 100 or more.

Complementary Darlington connections are also possible. For example, we may use an NPN transistor Q_1. Then we have

$$I_t = (1 + \beta_1) I_{B_1} + \beta_2 I_{B_2} = (1 + \beta_1) I_{B_1} + \beta_1 \beta_2 I_{B_1} \approx \beta_1 \beta_2 I_{B_1}$$

2.5.3 *The Push–Pull Stage*

The complementary "push–pull" emitter follower stage, (Fig. 2.21f) is often used as the output stage of a power amplifier or to boost the output current range of integrated circuits. The diodes D_1 and D_2 are forward biased with voltage drops similar to those of the base–emitter junctions of the transistors. They thus provide temperature compensation and the appropriate bias for the matched transistors so that Q_1 and Q_2 have the same quiescent current ($I_{01} = I_{02}$) for $v_{IN} = 0$. The two transistors act as emitter followers so that Q_1 supplies the current $i(+)$ for $v_{IN} > 0$, and Q_2 the current $i(-)$ for $v_{IN} < 0$.

2.5.4 *The Difference Amplifier*

The final example concerns the difference or differential amplifier of Fig. 2.21g. In its FET or bipolar junction transistor version, it is typically the input stage of instrumentation and operational amplifiers.

Assume that the two input voltages are

$$v_1 = V_0 + \frac{\Delta v}{2}$$

$$v_2 = V_0 - \frac{\Delta v}{2}$$

The difference is, therefore, $v_1 - v_2 = \Delta v$. Next we assume that the transistors are matched and that I_0 is chosen to equal I_{DSS} (i.e., $V_{GS1} = V_{GS2} = 0$). The drain currents are then

$$i_{D1} = I_0 + \Delta i = I_0 + \frac{\Delta v}{2} g_m$$

and

$$i_{D2} = I_0 - \Delta i = I_0 - \frac{\Delta v}{2} g_m$$

because the FET source terminal voltages simply rise by V_0 in order to maintain the quiescent bias $V_{GS} = 0$ and $I_{total} = 2I_0$. Thus the drain voltages are

$$v_{D1} = V_{DD} - i_{D1} R_D$$

and

$$v_{D2} = V_{DD} - i_{D2} R_D$$

We now define the output voltage as

$$v_3 = v_{D2} - v_{D1} = (i_{D1} - i_{D2}) R_D$$

With the expressions for the currents above, we obtain the output voltage

$$v_3 = \Delta v \, g_m R_D$$

Note that the *common mode* component of the input signals, V_0, does not appear in the expression for the output voltage: common mode "rejection" is an important application of differential amplifiers. (In practical amplifiers, common mode rejection is incomplete because of mismatch.)

2.5.5 Transistors as Voltage-Controlled Resistors and Switches

Below pinch-off, the FET is a voltage-controlled resistor. This suggests its use in a voltage-controlled voltage divider or "electronic" attenuator as shown in Fig. 2.22a. By varying the control voltage v_c between 0 and the pinch-off level V_p, the D–S channel resistance can be continuously varied from a minimum "on" value of about 100 Ω or less to a maximum "off" value of several megohms.

In Fig. 2.22b, the gate–source voltage is abruptly changed between 0 (on) and $-V_p$ (off). The FET now acts as a shunt switch and "chops" the input voltage v_{IN} into a series of pulses if the switching is periodically repeated.

Another application of the FET shunt switch is the quick discharge of a capacitor (Fig. 2.22c). The duration of the control pulse should be about five times the discharge time constant (capacitance × "on" resistance).

In the series switch arrangement ("analog gate"; Fig. 2.22d), the gate voltage must be properly referenced to the source voltage. Manufacturer data sheets on JFET and MOSFET switches usually give the appropriate information on drive (control) voltages, etc.

Bipolar junction transistors can also be used as switches. An example is given in Fig. 2.23a. With the control voltage at 0, the collector voltage is at $V_{CC} - R_C I_{CE0}$ or approximately V_{CC} (point A in Fig. 2.23b). A sudden transition of the control voltage to v_B generates a base current large enough to shift the transistor from the off state (point A on the load line, Fig. 2.23b) to the on state (point B). In the on or

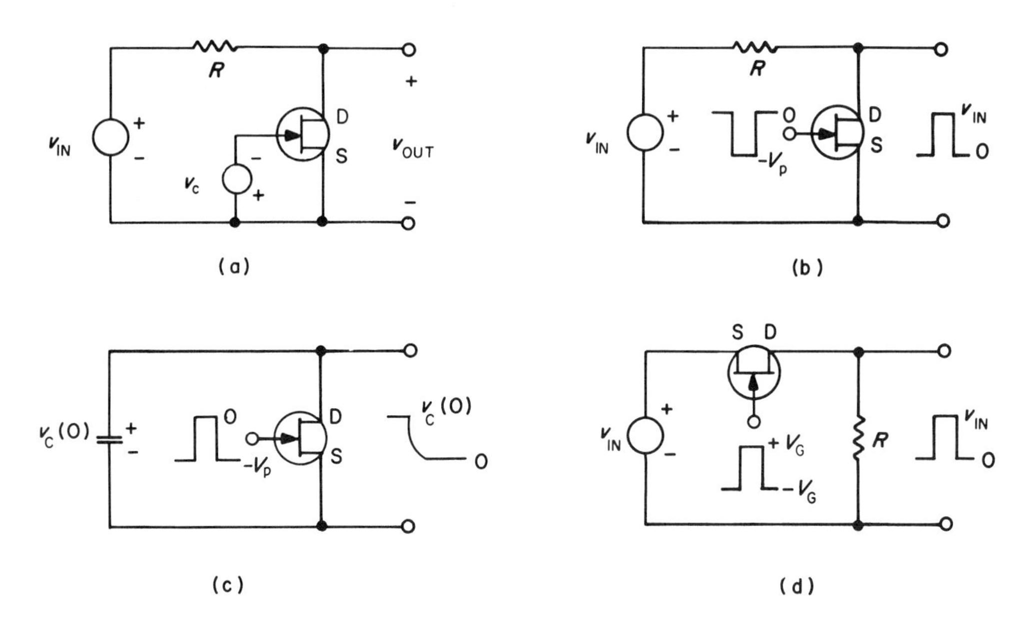

Figure 2.22 FETs as voltage-controlled resistors and switches. (See text.)

Figure 2.23 The junction transistors as a switch. (See text.)

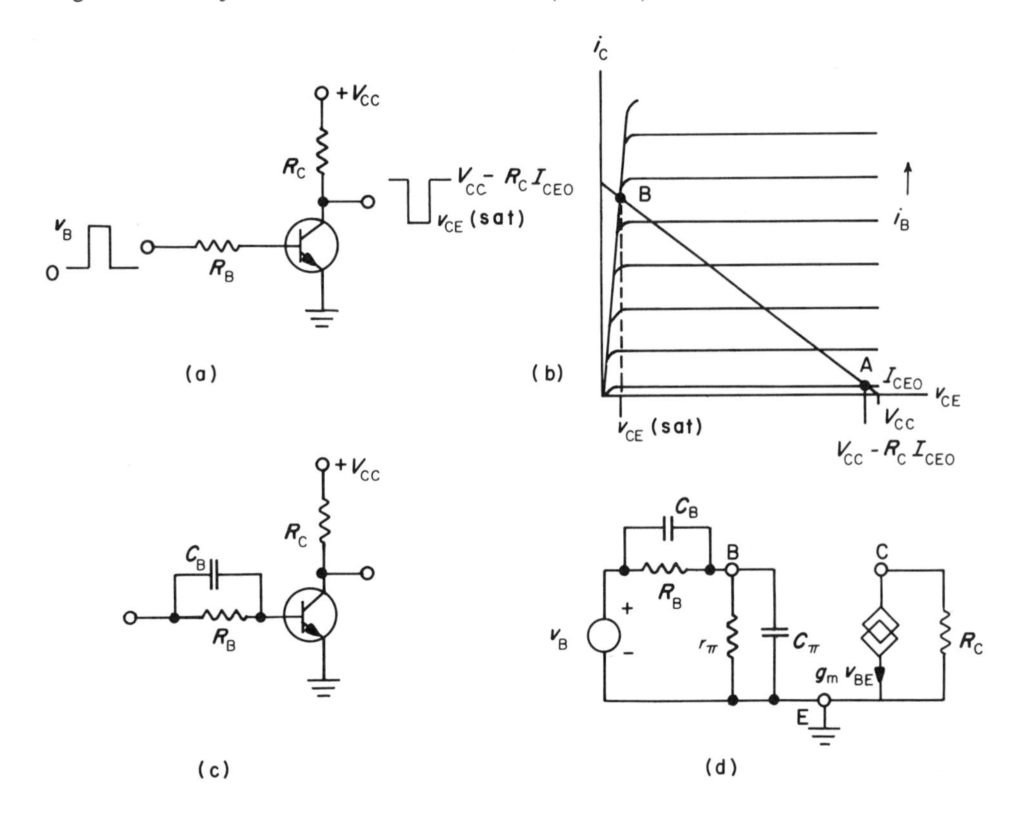

"saturated" state, the collector current is nearly V_{CC}/R_C, and the collector–emitter voltage is equal to v_{CE}(saturated), a mere fraction of a volt. In order to ensure saturation the base current should be about one tenth of the collector current; thus we use the approximation

$$I_B = \frac{v_B - V_{BE}}{R_B} \approx \frac{v_B}{R_B} \approx \frac{V_{CC}}{10R_C}$$

The base resistor R_B should be sufficiently large to protect the base–emitter junction.

During saturation switching, parasitic capacitance retards the switching process. One way to speed up the switching action is to bypass the base resistor with a capacitor C_B (Fig. 2.23c). A rather simple rationale for this method is based on the concept of the frequency-compensated voltage divider as shown in Fig. 2.23d. There it is assumed that the simple hybrid-π model is an adequate representation of the large-signal behavior of the transistor. Frequency compensation occurs if $R_B C_B = r_\pi C_\pi$. In practice, C_B is often selected by trial and error.

The switching action can also be speeded up if the base current is made smaller than $0.1 I_C$(saturated) but just large enough to reach point B on the load line (Fig. 2.23b).

This concludes Chapter 2. Further details on transistor design can be found in the references.

Problems

1. An N-channel JFET has the parameters $I_{DDS} = 2$ mA and $V_p = 2$ V. (a) Plot I_D versus V_{GS}. (*Hint*: Use Eq. (2.4) and make the plot as shown on the left of Fig. 2.7b.) (b) Plot the transconductance g_m defined as the absolute value of the derivative $dI_D/d|V_{GS}|$ of Eq. (2.4). (c) Design a self-biasing circuit according to Fig. 2.12e with a quiescent operating current $I_D = 1$ mA, an equivalent external gate resistance $R_G = R_a \| R_b \approx 1$ MΩ, a supply voltage $V_{DD} = +15$ V, and an equivalent gate voltage $V_G \approx 2.1$ V. (d) Assume that this design is used in the *RC*-coupled amplifier stage of Fig. 2.13a. Compute the voltage gain A_v for $R_D = 5$ kΩ and $R_1 = 100$ kΩ for a frequency for which C_c and C_S act as short circuits.

2. Derive the conversion formulas, Eqs. (2.32) and (2.33), for the current gains of junction transistors.

3. For the bias circuit on the right of Fig. 2.19a assume $V_{CC} = 15$ V, $V_{BE} \approx 0.7$ V, $\beta_0 = 100$, quiescent collector current $I_C = 1$ mA, $V_{CE} = 10$ V. (a) Determine R_B and R_C. (b) Assume that the transistor is replaced by another one with $\beta_0 = 50$. With R_B and R_C as determined previously, what will the new quiescent values I_C and V_{CE} be? (c) Design an alternative biasing scheme according to Fig. 2.19b. How does a change in $\beta_0 = 50$ affect the quiescent I_C and V_{CE} in this case?

4. Replace the drain resistors R_D of the differential amplifier of Fig. 2.21g with nonideal current sources which can be modeled by the parallel combination of an ideal current source $I_L = I_0$ with a resistor $R_L = 1000 R_D$. Determine the change in differential gain $v_3/(v_1 - v_2)$.

References

Angelo, E. J., Jr., *Electronics: BJTs, FETs and Microcircuits*, McGraw-Hill, New York, 1964. (Reprinted by R. E. Krieger, Huntington, NY, 1979.)

Casasent, D., *Electronic Circuits*, Quantum Publishers, New York, 1973.

Department of the Army, Technical Manual TM11-690: *Basic Theory and Application of Transistors*, Washington DC, 1959 (or later edition).

Gärtner, W. W., *Transistors—Principles, Design, and Applications*. Van Nostrand, Princeton, NJ, 1960.

Lowry, H. R., et al., *General Electric Transistor Manual*, General Electric Co., Liverpool, NY, 5th ed., 1960 (or later editions).

Siliconix, Inc., *Designing with Field-Effect Transistors*, McGraw-Hill, New York, 1981.

3

Introduction to Signals and Systems

Gunter N. Franz

This chapter has three major goals: (1) the efficient mathematical treatment of periodic and aperiodic (nonperiodic) signals; (2) the introduction of transformation methods that simplify the mathematical handling and analysis of signals and network and system equations; (3) an overview of strategies for the analysis of linear systems. More detailed treatments can be found in specialized texts such as Aseltine (1958), Desoer and Kuh (1969), Doetsch (1971), Bendat and Piersol (1966), D'Azzo and Houpis (1966), Papoulis (1962), Bracewell (1965), Van Valkenburg (1974), Wylie (1960), and Truxal (1955).

3.1 Periodic Signals: Fourier Series

Periodic signals repeat themselves in time and/or space. We concern ourselves in this chapter only with lumped systems; we shall ignore spatial periodicities. An example of a periodic signal is given in Fig. 3.1. From this example we can deduce the mathematical definition of a periodic signal:

$$f(t) \equiv f(t+T) \equiv f(t+kT) \tag{3.1}$$

where $k=0, \pm 1, \pm 2, \pm 3, \ldots$ for the time interval $-\infty < t < \infty$.

According to Eq. (3.1), a particular value of the function, e.g., $f(t_1)$, is repeated every T seconds for both positive and negative times. This repetition interval is defined as the

period or *periodic* time: T

We also define these parameters:

fundamental frequency: $\nu = \dfrac{1}{T}$

fundamental radian frequency: $\omega = 2\pi\nu = 2\pi/T$

The frequencies defined here refer to the *repetition* frequency of the waveform defined in the interval $0 < t < T$ (or any other interval of length T).

3.1.1 The Fourier Series Expansion

Mathematical analysis provides us with a theorem stating that periodic functions (meeting certain conditions) can be represented by an infinite series of sinusoids, the so-called *Fourier series*. The decomposition of periodic signals into well-behaved and

Figure 3.1 Example of a periodic function: T=periodic time; $f=1/T$= fundamental frequency; $\omega = 2\pi/T = 2\pi f$=fundamental radian frequency.

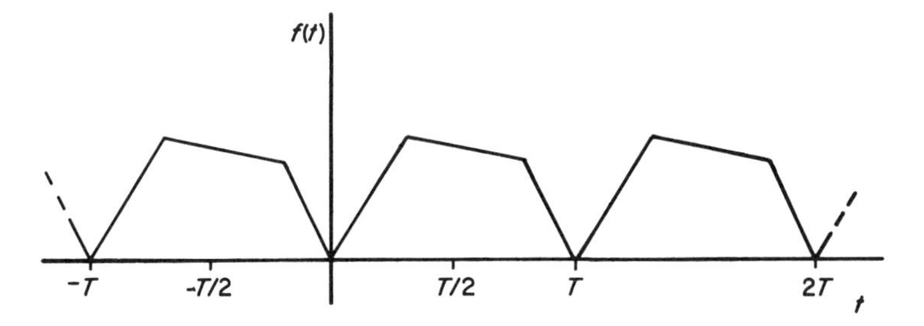

well-understood sinusoids makes the results of *steady state* sinusoidal frequency response analysis directly applicable to periodic signals of arbitrary shape. This is particularly true in the case of linear systems, where the superposition theorem allows us to synthesize the system response to any periodic signal as the additive superposition of the responses to the individual sinusoidal components of the signal.

There are several versions of the Fourier series. We give the ones most suitable for our purposes.

Fourier Series of a Periodic Signal

1.
$$f(t)=a_0+\sum_{n=1}^{\infty} a_n\cos(n\omega t)+b_n\sin(n\omega t) \tag{3.2}$$

with the Fourier Coefficients

$$a_0=\frac{1}{T}\int_0^T f(t)\,dt=\frac{1}{T}\int_{-T/2}^{T/2} f(t)\,dt \tag{3.3}$$

$$a_n=\frac{2}{T}\int_0^T f(t)\cos(n\omega t)\,dt=\frac{2}{T}\int_{-T/2}^{T/2} f(t)\cos(n\omega t)\,dt \tag{3.4}$$

$$b_n=\frac{2}{T}\int_0^T f(t)\sin(n\omega t)\,dt=\frac{2}{T}\int_{-T/2}^{T/2} f(t)\sin(n\omega t)\,dt \tag{3.5}$$

2.
$$f(t)=a_0+\sum_{n=1}^{\infty} r_n\cos(n\omega t-\phi_n) \tag{3.6}$$

where a_0 is defined by Eq. (3.3) and the new parameters are given by

$$r_n=\sqrt{a_n^2+b_n^2} \tag{3.7}$$

$$\phi_n=\tan^{-1}\left(\frac{b_n}{a_n}\right) \tag{3.8}$$

3.
$$f(t)=a_0+\sum_{n=1}^{\infty} r_n\sin(n\omega t+\psi_n) \tag{3.9}$$

where a_0 and r_n are defined by Eqs. (3.3) and (3.7) and the new parameter is given by

$$\psi_n=\tan^{-1}\left(\frac{a_n}{b_n}\right) \tag{3.10}$$

4.
$$f(t)=\sum_{n=-\infty}^{\infty} c_n e^{jn\omega t} \tag{3.11}$$

where

$$c_n=\frac{1}{T}\int_0^T f(t)e^{-jn\omega t}\,dt=\frac{1}{T}\int_{-T/2}^{T/2} f(t)e^{-jn\omega t}\,dt \tag{3.12}$$

and

$$c_0 = a_0 \tag{3.13}$$

For $n \neq 0$ ($n = \pm 1, \pm 2, \ldots$), the c_n coefficients are, generally, complex numbers:

$$c_n = \frac{a_n - jb_n}{2} \tag{3.14}$$

$$c_n = |c_n| e^{j\phi_n} = \frac{r_n}{2} e^{j\phi_n} \qquad n = \pm 1, \pm 2, \ldots \tag{3.15}$$

where a_n, b_n, r_n, and ϕ_n are given by Eqs. (3.4), (3.5), (3.7), and (3.8).

A number of observations should be made here.

1. The a_0 or c_0 terms, Eqs. (3.3) and (3.13), represent the equivalent dc level or average over the period T of the signal $f(t)$.
2. The sinusoidal components with radian frequency $\omega(n = 1)$, having the same repetition frequency as $f(t)$, are called the *fundamental components* or *fundamentals* of the series expansion.
3. The sinusoids aside from the fundamentals have radian frequencies which are *integer multiples* of the fundamental radian frequency ($\omega_n = n\omega$). These components are called the *nth-order harmonics* of the periodic signal. Note that not all harmonic frequencies need be present in the Fourier expansion of a periodic signal.
4. The series expansion can be in the form of cosines and sines, Eq. (3.2), with Fourier *amplitude coefficients* (a_n, b_n) specified by Eqs. (3.3)–(3.5). Not all coefficients need be present in a particular case. Specifically, we can state
   a. for signals without dc component, $a_0 = 0$;
   b. for even functions, i.e., $f(t) = f(-t)$, all $b_n = 0$;
   c. for odd functions, i.e., $f(t) = -f(-t)$, all $a_n = 0$.
5. The alternative expansions, Eqs. (3.6), (3.9), and (3.11), require the specification of a *magnitude* coefficient, Eqs. (3.7) and (3.15), and a *phase*, Eqs. (3.8) and (3.10). These parameters are particularly suitable for applying the results of sinusoidal steady state (phasor) analysis to periodic functions represented by Fourier expansions.
6. Just as the periodic function $f(t)$ itself is assumed to exist over $-\infty < t < \infty$, the sinusoidal Fourier components, too, are considered "switched on" at $t = -\infty$; i.e., we deal with a *sinusoidal steady state*.
7. The phase specified for the fundamental and each harmonic is measured relative to a corresponding unshifted *sinusoid* of the *same* frequency. The original periodic waveform, being a sum of harmonically related sinusoids, has no single phase parameter as such.
8. The expansion given by Eq. (3.11) is especially advantageous for us because it provides the means of representing a periodic voltage or current signal as a sum of phasors.

Example 3.1.

Fourier series for square waves without and with a dc level (or bias) (Fig. 3.2). From Eq. (3.3) we have with Fig. 3.2a

$$a_0=\frac{1}{T}\left[\int_0^{T/2}E\,dt-\int_{T/2}^{T}E\,dt\right]=0$$

The cosine amplitude coefficients a_n ($n\neq 0$) all vanish because the waveform is odd. The reader may verify this from Eq. (3.4).

The sine amplitude coefficients can be determined from Eq. (3.5). We write according to Fig. 3.2a

$$b_n=\frac{2}{T}\int_{-T/2}^{T/2}f_1(t)\sin(n\omega t)\,dt$$

or

$$b_n=\frac{-2E}{T}\int_{-T/2}^{0}\sin(n\omega t)\,dt+\frac{2E}{T}\int_0^{T/2}\sin(n\omega t)\,dt$$

After integration we obtain

$$b_n=\frac{-2E}{n\omega T}\left[\cos(n\omega t)|_0^{T/2}+\cos(n\omega t)|_0^{T/2}\right]$$

and with $n\omega T/2=n\pi$

$$b_n=\begin{cases}4E/n\pi & \text{for } n=1,3,5,7,\ldots\\ 0 & \text{for } n=2,4,6,8,\ldots\end{cases} \tag{3.16}$$

Figure 3.2 **(a)** Example of an odd function with zero average value. **(b)** Example of an even function with average value E. **(c)** Approximation of the square wave in **(a)** by the first 4 terms of its Fourier series.

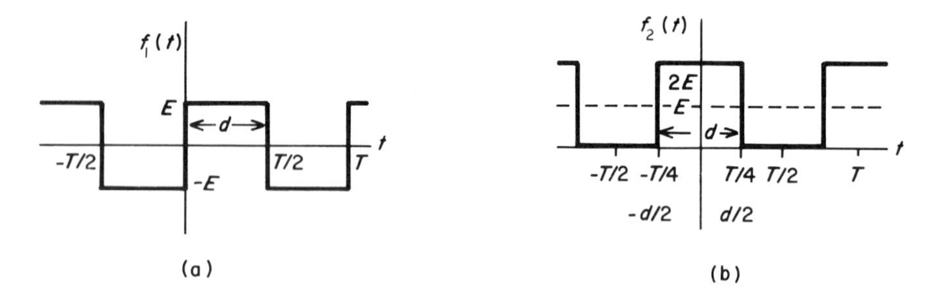

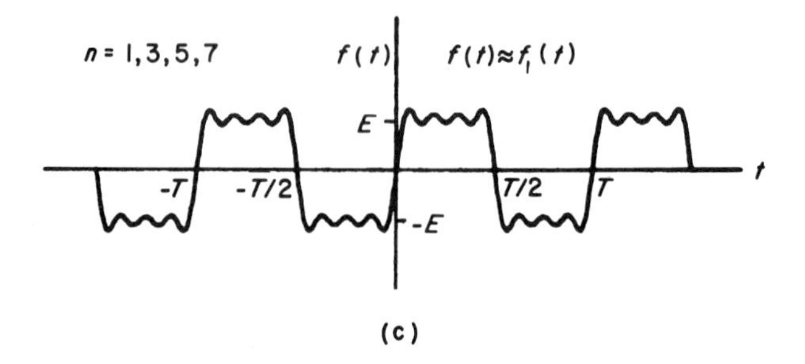

This can be written in more compact form as

$$b_k = \frac{4E}{(2k+1)\pi} \qquad \text{for} \quad k = 0, 1, 2, \ldots \tag{3.17}$$

The Fourier series is now completely specified; it is

$$f_1(t) = \frac{4E}{\pi} \sum_{k=0}^{\infty} \frac{1}{2k+1} \sin[(2k+1)\omega t] \tag{3.18}$$

or

$$f_1(t) = \frac{4E}{\pi}\left[\sin(\omega t) + \tfrac{1}{3}\sin(3\omega t) + \tfrac{1}{5}\sin(5\omega t) + \cdots\right]$$

The expansion, Eq. (3.18), is already suitable for the form given in Eq. (3.9), with $a_0 = 0$, $r_n = b_n$, and $\psi_n = 0$. Note that in this particular case, the Fourier series contains only *odd-numbered harmonics*. Furthermore, the amplitudes of the harmonics decline in inverse proportion to the harmonic multiple. In practice, this means that we can truncate this infinite series after a reasonable number of terms. In Fig. 3.2c we see an approximation of the square wave by the sum of the first four Fourier terms ($n = 1, 3, 5, 7$ or $k = 0, 1, 2, 3$).

Next we shall compute the Fourier parameters of the *complex Fourier series*, Eq. (3.11), for the square wave in Fig. 3.2b.

We use Eq. (3.12) and the functional definitions of Fig. 3.2b to obtain

$$c_n = \frac{1}{T}\int_{-T/2}^{T/2} f_2(t) e^{-jn\omega t}\, dt = \frac{2E}{T}\int_{-d/2}^{d/2} e^{-jn\omega t}\, dt$$

which upon integration yields

$$c_n = \frac{2Ed}{T} \frac{\sin(n\omega d/2)}{n\omega d/2} \tag{3.19}$$

For $n = 0$, this expression gives us

$$c_0 = a_0 = \frac{2Ed}{T} \tag{3.20}$$

because by a series expansion we can prove that the function $(\sin x)/x = 1$ for $x = 0$. The time average or equivalent dc level of this square wave is then proportional to the pulse duration/period ratio d/T. In this case we have $d = T/2$; hence from Eq. (3.20)

$$c_0 = a_0 = E$$

which is the level indicated by the broken line in Fig. 3.2b.

For $n \neq 0$, Eq. (3.19) yields zero for even n, and nonzero values of alternating sign for odd n.

With the aid of Eqs. (3.14) and (3.19) we can write

$$a_k = \frac{(-1)^k 4E}{(2k+1)\pi} \qquad \text{for} \quad k = 0, 1, 2, \ldots \tag{3.21}$$

The corresponding Fourier expansion for $f_2(t)$ in Fig. 3.2b is

$$f_2(t) = E + \frac{4E}{\pi} \sum_{k=0}^{\infty} \frac{(-1)^k}{2k+1} \cos[(2k+1)\omega t] \tag{3.22}$$

or

$$f_2(t) = E + \frac{4E}{\pi}\left[\cos(\omega t) - \tfrac{1}{3}\cos(3\omega t) + \tfrac{1}{5}\cos(5\omega t) - \cdots\right]$$

3.1.2 Discrete Fourier Spectra of Periodic Functions

The characterization of a periodic function by its Fourier series always amounts to the specification of two sets of numbers which uniquely determine the Fourier expansion. According to the form of Fourier expansion chosen, the two sets of numbers are either amplitude coefficients (a_n, b_n) or magnitudes and phases $(r_n, |c_n|; \phi_n, \psi_n)$. Whatever our choice, the different parameter sets specify the *same* Fourier series because of the unique one-to-one correspondence between the original periodic function and its Fourier expansion.

The definition and graphic display of the Fourier parameters as a function of frequency $(n\omega)$ or harmonic multiple (n) is called the *spectrum*, *Fourier spectrum*, or *frequency spectrum* of the periodic function. The spectrum, representing the parameters of the Fourier series, specifies the signal in the *frequency domain*. It is an alternative to defining the signal in the time domain.

Figures 3.3a and 3.3b show the spectra for the periodic signals defined in the time domain according to Figs. 3.2a and 3.2b, respectively. We note that except for the dc component $(n=0)$, the *absolute* magnitude of the Fourier components is the same for the two square waves: the two signals have the same *frequency "content"* in terms of magnitude, but they differ in the phase terms (not shown) as a comparison between Eqs. (3.18) and (3.22) shows.

The spectra in Fig. 3.3 are *discrete* and *one-sided*. They are discrete, because Fourier components can occur only at integer multiples of the fundamental frequency. They are one-sided because the summation in Eqs. (3.2), (3.6), and (3.9) extends only

Figure 3.3 Discrete Fourier (coefficient) spectra for the square waves in Fig. 3.2: **(a)** spectrum for the odd function; **(b)** spectrum for the even function with $a_0 = E$.

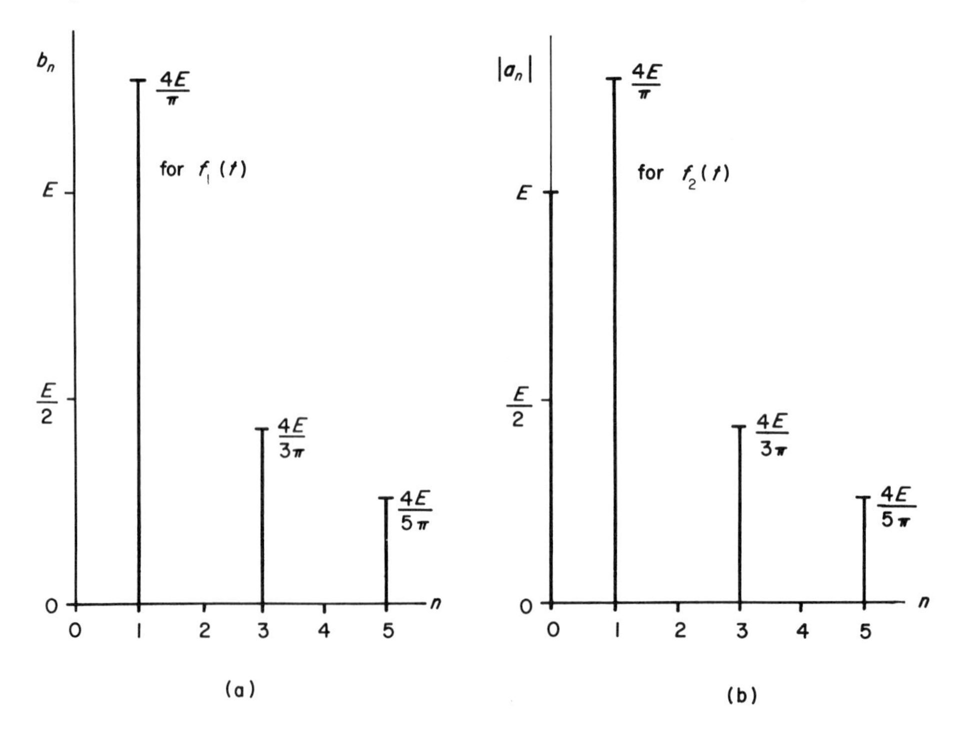

over positive n. The complex Fourier series, Eq. (3.11), requires a two-sided spectrum because the summation extends over both positive and negative n. The two-sided spectrum for $|c_n|$ is symmetric to the ordinate and at half the magnitude of the corresponding one-sided spectrum, Eq. (3.15). An example is given in Fig. 3.4a.

The theoretical significance of the spectrum concept rests partly on the fact that we can deduce the response of a linear system to an arbitrarily shaped periodic signal from the frequency response function of the system in two steps. First, we use the frequency response function to compute the response to each sinusoidal (or phasor) Fourier component. Then we synthesize the complete response by adding the individual component responses according to the superposition theorem.

In practice, however, this is rarely done. It is much more typical to use the frequency spectrum of a signal to specify the required frequency response characteristics of a system. For example, should we require faithful signal transmission, then the transmission system should have a frequency response function which shows neither significant attenuation nor phase shift for major spectral components of the signal. On the other hand, we might only be interested in the fundamental component of the signal. In that case we require a system with the characteristics of a bandpass filter excluding all harmonics and the dc component.

From the foregoing discussion it is obvious that the spectrum concept is more important to us than the actual mechanics of Fourier series expansions. The extension of the frequency spectrum concept to nonperiodic signals is the subject of the next section.

Figure 3.4 Transition from a periodic signal (a) to an aperiodic one (b).

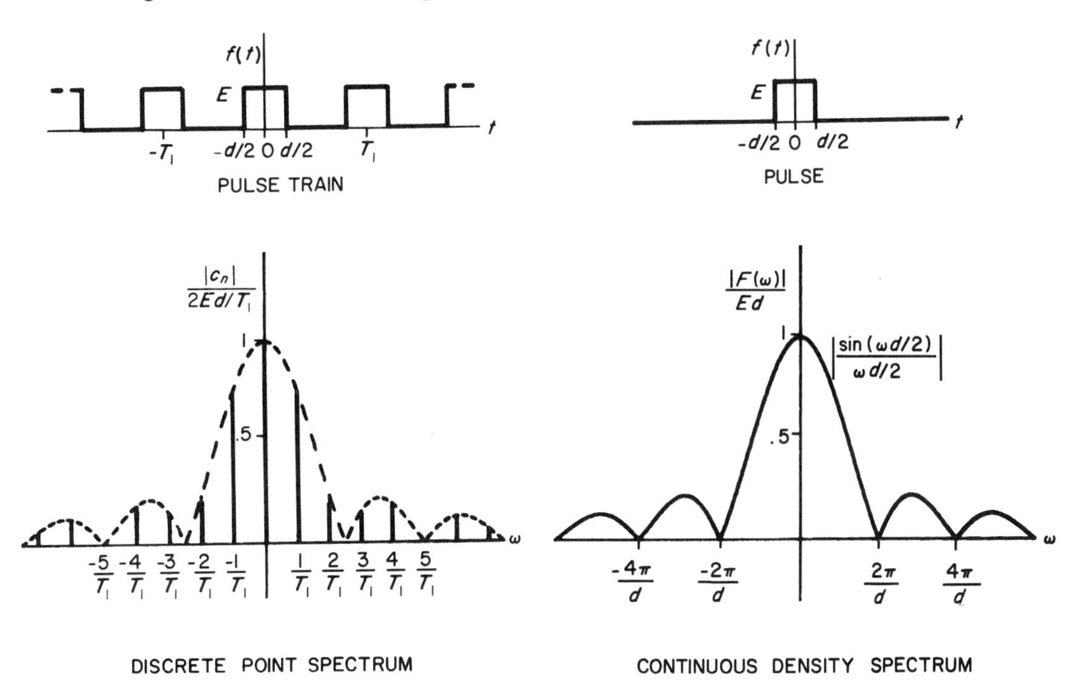

3.2 Aperiodic Signals: The Fourier Transformation

The spectrum concept and the decomposition of signals into steady state sinusoids can be adapted to the case of *aperiodic* or *nonperiodic* signals. The necessary adaptations involve limit processes such that (1) the finite period is transformed into an infinite one spanning $-\infty<t<\infty$, (2) the summation of the Fourier series [e.g., Eq. (3.11)] becomes an integration, and (3) the discrete point spectra change into the continuous spectral densities of magnitude and phase. We forego the detailed mathematical development and proceed directly into the definition of the resulting *Fourier integral* or *Fourier transformation* (or *transform*) and the discussion of its properties and application.

3.2.1 The Fourier Transformation

For functions $f(t)$ meeting the convergence condition

$$\int_{-\infty}^{\infty} |f(t)|\, dt<\infty \tag{3.23}$$

we define with the symbol $\mathcal{F}\{f(t)\}=F(j\omega)$ the

1. Fourier transformation, Fourier transform, Fourier integral:

$$\mathcal{F}\{f(t)\}=F(j\omega)=\int_{-\infty}^{\infty} f(t)e^{-j\omega t}\,dt \tag{3.24}$$

2. Inverse Fourier transformation:

a.

$$\mathcal{F}^{-1}\{F(j\omega)\}=f(t)=\frac{1}{2\pi}\int_{-\infty}^{\infty} F(j\omega)e^{j\omega t}\,d\omega \tag{3.25}$$

b.

$$\mathcal{F}^{-1}\{F(j\nu)\}=f(t)=\int_{-\infty}^{\infty} F(j\nu)e^{j2\pi\nu t}\,d\nu \tag{3.26}$$

with $\omega=\pi 2\nu$ and $F(j\nu)=F(j\omega)|_{\omega=2\pi\nu}$

3.

$$F(j\omega)=|F(j\omega)|e^{j\Phi(\omega)} \tag{3.27}$$

4. $|F(j\omega)|$: amplitude density (spectrum)
5. $\Phi(\omega)$: phase density (spectrum)

Equations (3.24) and (3.25) are the counterparts to Eqs. (3.12) and (3.11) for the Fourier series.

It is important to stress that $f(t)$ is uniquely determined by its Fourier transformation. Together, $f(t)$ and $F(j\omega)$ form a *transform pair* representing two equivalent formulations of the same signal.

The Fourier transformation $F(j\omega)$ is generally a complex function of (radian) frequency, Eq. (3.27). Its magnitude, $|F(j\omega)|$, is the *amplitude* or *magnitude density* (*spectrum*) defined over the entire range $-\infty<\omega<\infty$, with the quantity $|F(j\omega)|\,d\omega$ representing the contribution of oscillations in the infinitesimal frequency band $d\omega$. For a unique description of a signal we must include the *phase density* (*spectrum*) $\Phi(\omega)$. Many practical methods of spectral analysis deliver only a magnitude spectrum; these are sufficient if we are only interested in the relative contributions of the component frequencies and do not require unique signal identification.

Example 3.2. Density spectrum of a pulse.

This example illustrates the transition from a discrete point spectrum to a continuous density spectrum as we reduce a periodic function to an aperiodic one. Figure 3.4a shows a pulse train whose Fourier coefficients we have already computed with Eq. (3.19). The corresponding discrete magnitude spectrum is plotted in normalized form in the same figure. The equidistant spacing of the discrete spectral points, in terms of integer multiples of the fundamental radian frequency, $\omega_1=2\pi/T_1$, becomes denser as the period T_1 is lengthened. In the limit the pulse train is reduced to a single pulse (Fig. 3.4b). This produces a continuous density spectrum equivalent to the envelope of the discrete spectrum because from Eq. (3.24) we have for Fig. 3.4b.

$$F(j\omega)=\int_{-\infty}^{\infty} f(t)e^{-j\omega t}\,dt=E\int_{-d/2}^{d/2} e^{-j\omega t}\,dt=\frac{-E}{j\omega}\left(e^{-j\omega d/2}-e^{j\omega d/2}\right)$$

or, with the Euler formulas,

$$F(j\omega)=\frac{2E}{\omega}\sin(\omega d/2)=Ed\frac{\sin(\omega d/2)}{(\omega d/2)} \tag{3.28}$$

We observe that $F(j\omega)$ has its first zero crossings at the radian frequencies $\pm\omega_d=\pm 2\pi/d$. From Fig. 3.4b it is evident that the dominant portion of the spectrum is in the bilateral interval $(-\omega_d,+\omega_d)$. Note that this frequency interval increases as we make the pulse width d smaller and smaller. Obviously, faithful transmission of very short aperiodic signal requires a concomitantly large bandwidth of the system.

3.2.2 Properties of the Fourier Transformation

A summary of some important properties of the Fourier transformation is given here without proofs.

1. *The Fourier transformation is a linear operation*; *it obeys the superposition theorem*

$$\mathfrak{F}\{af_1(t)+bf_2(t)\}=aF_1(j\omega)+bF_2(j\omega) \tag{3.29}$$

2. *The amplitude modulation of sinusoids by a signal $f(t)$ corresponds to a shift of the signal's Fourier spectrum.*

a. $$\mathfrak{F}\{f(t)e^{j\omega_0 t}\}=F[j(\omega-\omega_0)] \tag{3.30}$$

b. $$\mathfrak{F}\{f(t)e^{-j\omega_0 t}\}=F[j(\omega+\omega_0)] \tag{3.31}$$

c. $$\mathfrak{F}\{f(t)\cos\omega_0 t\}=\tfrac{1}{2}F[j(\omega-\omega_0)]+\tfrac{1}{2}F[j(\omega+\omega_0)] \tag{3.32}$$

3. *Convolution in the time domain is equivalent to multiplication in the frequency domain*. (The significance of the convolution integral will be made evident later in the chapter.)

$$\mathcal{F}\left\{\int_{-\infty}^{\infty} f_1(\tau)f(t-\tau)\,d\tau\right\} = \mathcal{F}\left\{\int_{-\infty}^{\infty} f_1(t-\tau)f_2(\tau)\,d\tau\right\}$$

$$= F_1(j\omega)F_2(j\omega) \tag{3.33}$$

The next two properties are extremely important for the solution of differential and integrodifferential equations and for network analysis.

4. *Differentiation in the time domain is equivalent to multiplication with $j\omega$ in the frequency domain*. Like the ac phasor method, *the Fourier transformation reduces linear, constant coefficient differential equations into linear algebraic equations*.

$$\mathcal{F}\left\{\frac{d^n}{dt^n}f(t)\right\} = (j\omega)^n F(j\omega) \tag{3.34}$$

5. *Integration in the time domain corresponds to division by $j\omega$ in the frequency domain*.

$$\mathcal{F}\left\{\underbrace{\int_{-\infty}^{t}\cdots\int_{-\infty}^{t}}_{n\text{ times}} f(\tau)(d\tau)^n\right\} = \frac{1}{(j\omega)^n}F(j\omega) \tag{3.35}$$

3.2.3 Applications of the Fourier Transformation

3.2.3.1 Signal Spectra

The computation of spectral densities was the lead-off point for the introduction to the Fourier integral. We need not belabor the point any further.

3.2.3.2 Extension of the Phasor Method

The inverse Fourier transformation, Eq. (3.25), shows that aperiodic signals can be represented by the superposition of a continuum of phasors each having the complex magnitude $F(j\omega)\,d\omega$. If we write the equations for a linear, constant parameter network in the sinusoidal steady state in terms of ac phasors and ac immittances, the equations will also be applicable to Fourier transformable aperiodic signals by the superposition theorem. We simply replace the voltage and current phasors by Fourier transforms and leave the immittance terms intact.

3.2.3.3 Immittances

The Fourier transformation of the element laws of passive one-ports generalizes the ac immittance concept to *aperiodic* voltages and currents. For an unspecified one-port we introduce the immittances as follows.

The ac impedance $Z(j\omega)$ of a passive one-port is defined by

$$V(j\omega)=Z(j\omega)I(j\omega) \tag{3.36}$$

where $V(j\omega)$ and $I(j\omega)$ are the Fourier transformations of the aperiodic or periodic voltage–current pair at the terminals.
For aperiodic functions *only*, we can alternatively state

$$Z(j\omega)=\frac{V(j\omega)}{I(j\omega)} \tag{3.37}$$

The ac admittance $Y(j\omega)$ of a passive one-port is defined by

$$I(j\omega)=Y(j\omega)V(j\omega) \tag{3.38}$$

for both aperiodic and periodic current–voltage pairs at the terminals.
For aperiodic function *only*, we may alternatively use

$$Y(j\omega)=\frac{I(j\omega)}{V(j\omega)} \tag{3.39}$$

Ironically, Eqs. (3.37) and (3.39) are improper immittance definitions in the case of sinusoids or other periodic functions. In a later section we shall demonstrate that periodic signals are characterized by impulse spectra, and we may not divide by impulses.

Let us apply these definitions to the linear passive storage elements. For the capacitor we have the element law

$$v(t)=\frac{1}{C}\int_{-\infty}^{t} i(\tau)\,d\tau$$

with the sensible initial condition $v(-\infty)=0$. According to Eq. (3.35) we have the Fourier transformation

$$V(j\omega)=\frac{1}{j\omega C}I(j\omega),$$

and from Eqs. (3.36) or (3.37) follows the expression for the impedance

$$Z_C(j\omega)=\frac{V(j\omega)}{I(j\omega)}=\frac{1}{j\omega C}$$

The admittance is, of course,

$$Y_C(j\omega)=\frac{1}{Z_C(j\omega)}=j\omega C$$

These expressions are exactly the same as those for the ac immittances derived by the ac phasor method in Chapter 1.

For the inductor we write

$$v(t)=L\frac{di(t)}{dt}$$

and after Fourier transformation with Eq. (3.34)

$$V(j\omega)=j\omega LI(j\omega)$$

The impedance, according to Eqs. (3.36) or (3.37), is then

$$Z_L(j\omega)=\frac{V(j\omega)}{I(j\omega)}=j\omega L$$

The admittance follows from Eqs. (3.38) or (3.39):

$$Y_L(j\omega)=\frac{I(j\omega)}{V(j\omega)}=\frac{1}{j\omega L}$$

3.2.3.4 Applications to Linear Differential and Network Equations

As the Fourier transformation amounts to a spectral decomposition into *steady state* oscillations, we refer all initial conditions to $t=-\infty$ and set them to zero. It is now a simple matter to transform a linear, constant coefficient differential equation into an algebraic one:

$$a_n\frac{d^n}{dt^n}y(t)+\cdots+a_2\frac{d^2}{dt^2}y(t)+a_1\frac{d}{dt}y(t)+a_0y(t)=x(t)$$

converts upon Fourier transformation [Eq. (3.34)] to

$$Y(j\omega)\left[a_n(j\omega)^n+\cdots+a_2(j\omega)^2+a_1(j\omega)+a_0\right]=X(j\omega) \tag{3.40}$$

provided the Fourier transformation for the forcing function $x(t)$ exists.

Equation (3.40) can be solved for the transformed dependent variable. This gives us the

Fourier Transform Solution of the nth-Order Linear Differential Equation

$$Y(j\omega)=\frac{X(j\omega)}{a_n(j\omega)^n+\cdots+a_2(j\omega)^2+a_1j\omega+a_0} \tag{3.41}$$

The dependent variable in the time domain, $y(t)$, is now obtained from Eq. (3.41) with the aid of the inverse Fourier transformation, Eq. (3.25), or more simply from a table of transformation pairs.

The Fourier transformation of the forcing term $x(t)$ (input) is related to the transformation of the dependent variable $y(t)$ (output) by the *frequency response function* $G(j\omega)$ according to this expression

$$Y(j\omega)=G(j\omega)X(j\omega) \tag{3.42}$$

The frequency response function may be an immittance or represent a dimensionless but frequency-dependent transfer ratio, depending on the nature of $x(t)$ and $y(t)$.

From Eqs. (3.41) and (3.42) we deduce that the frequency response function of a linear constant coefficient differential equation is given by

$$G(j\omega)=\frac{1}{a_n(j\omega)^n+\cdots+a_2(j\omega)^2+a_1j\omega+a_0} \tag{3.43}$$

We can determine $G(j\omega)$ *experimentally* by simple *sinusoidal analysis*. In a linear system, a sinusoidal input creates a sinusoidal output. At frequency ω_1 the ratio of the output amplitude over the input amplitude gives us the magnitude $|G(\omega_1)|$, and the phase difference between output and input sinusoid equals the phase of $G(j\omega_1)$. It is common to plot the results in a Bode plot (see Chapter 1).

It can be proved that $G(j\omega)$ in Eq. (3.43) is also a Fourier transformation. By inverse Fourier transformation we obtain

$$\mathcal{F}^{-1}\{G(j\omega)\} = g(t) \tag{3.44}$$

the so-called *weighting function*.

With Eqs. (3.33) and (3.44) we can restate Eq. (3.42) in the time domain by convolution integrals:

$$\begin{aligned} y(t) &= \mathcal{F}^{-1}\{Y(j\omega)\} = \mathcal{F}^{-1}\{G(j\omega)X(j\omega)\} \\ &= \int_{-\infty}^{\infty} g(\tau)x(t-\tau)\,d\tau = \int_{-\infty}^{\infty} g(t-\tau)x(\tau)\,d\tau \end{aligned} \tag{3.45}$$

Equations (3.42) and (3.45) are fundamental equations in linear system theory. Their significance will be discussed further in Section 3.5. Here we are content to state that both equations express one and the same input–output relation: once in the frequency domain [Eq. (3.42)] and once in the time domain [Eq. (3.45)]. However, depending on the circumstances, one is easier to apply than the other. Whenever $G(j\omega)$ and $X(j\omega)$ are given in closed mathematical form by relatively simple functions, it is usually advantageous to determine $Y(j\omega)$ from Eq. (3.42) and (in most cases) look up the inverse $y(t)$ in tables of Fourier transformations. In cases where $g(t)$ and $x(t)$ or $G(j\omega)$ and $X(j\omega)$ are given in numerical form, computer methods must be used, and the choice between the two equations will depend on the availability of the appropriate programs.

3.2.4 The Discrete Fourier Transformation (DFT) and the Fast Fourier Transformation (FFT)

The use of the Fourier transformation for the spectral analysis of practical signals has increased in recent years because digital computers make short shrift of the tedious calculations. Digital computers, however, require that continuously recorded signals in the time domain be approximated by a *discrete* sequence of points. These points or *samples* represent the value of the signal every T_s seconds, where T_s is the sampling interval. (Sampling is discussed further in Section 3.4.)

With the transition to a discrete signal comes the conversion of the continuous integration process into a summation by the digital computer. In addition, the summation is restricted to a *finite* time interval.

If N is the total number of signal samples equidistant in time, T_s the sampling interval, and $T = NT_s = 1/\Delta f$ the total time interval ("epoch" or window time) over which the signal has been sampled (Δf is then the smallest frequency interval to be resolved), then we can define, by analogy with Eqs. (3.24) and (3.25), the *discrete Fourier transformation* (DFT) and its inverse:

$$X(n) = T_s \sum_{k=0}^{N-1} x(kT_s)e^{-j2\pi nk/N}, \qquad n = 0, 1, \ldots, N-1$$

and

$$x(kT_s)=\Delta f \sum_{n=0}^{N-1} X(n)e^{j2\pi nk/N}, \qquad k=0,1,\ldots,N-1$$

where $X(n)$ is the spectral component at frequency $n(\Delta f)$, and $x(kT_s)$ is the signal sample at time kT_s.

The *fast Fourier Transformation* (FFT) represents a particularly efficient algorithm for computing the DFT on digital computers. Many computer centers have programs for the FFT. But before such programs are used, the reader should attempt to understand the peculiarities and pitfalls of the DFT/FFT. A start can be made with these references: Brigham and Morrow (1967), Bergland (1969), Ramirez (1974), and Harris (1978).

3.3 The Laplace Transformation

One of the limitations of the Fourier transformation is the requirement for signals to meet the convergence condition, Eq. (3.23). This excludes many important signals such as step functions and sinusoids (though the latter can be handled by special means as discussed in Section 3.4). In this section we introduce a variation of the Fourier transformation which overcomes this difficulty. The new transformation is the *Laplace transformation*, whose properties are discussed in a manner most suitable for the "practitioner" by Doetsch (1971).

The application of the Laplace transform is directed not at the determination of the frequency spectra of signals but at the manipulation of linear integrodifferential equations and convolution integral equations. The point of the transformation is to perform a *mapping* which translates signal functions in the time domain into analytic functions in the so-called s domain and reduces the transcendental operations of differentiation and integration into the elementary operations of multiplication and division. Another advantage of the Laplace transformation is the systematic inclusion of initial conditions in the formulation of network equations. Finally, the Laplace transformation provides a sound justification for the use of generalized impedances $Z(s)$ and generalized admittances $Y(s)$.

3.3.1 The Laplace Transformation

Consider a signal $f(t)$ which is not Fourier transformable because it violates the convergence condition, Eq. (3.23). We now modify the signal function by multiplying it with the *convergence factor*

$$e^{-\sigma t} \qquad \sigma>0$$

With a suitable σ, many signal functions $f(t)$ meet a modified convergence criterion at least for the interval $0<t<\infty$:

$$\int_0^{\infty} |f(t)e^{-\sigma t}|\, dt<\infty \tag{3.46}$$

Because of Eq. (3.46) *we limit the Laplace formation to signals $f(t)$ that vanish for $t<0$.*

The Fourier transformation of $f(t)$ modified by the convergence factor is for $t>0$

$$\mathcal{F}\{f(t)e^{-\sigma t}\}=\int_0^\infty f(t)e^{-(\sigma+j\omega)t}\,dt$$

By setting $s=\sigma+j\omega$ we define the

1. One-sided Laplace transformation or Laplace transform:

$$\mathcal{L}\{f(t)\}=F(s)=\int_{0+}^\infty f(t)e^{-st}\,dt \tag{3.47}$$

2. Inverse Laplace transformation:

$$f(t)=\mathcal{L}^{-1}\{F(s)\}=\frac{1}{2\pi j}\int_{\sigma_0-j\omega}^{\sigma_0+j\omega}F(s)e^{st}\,ds \tag{3.48}$$

REMARK: We always assume $f(t)=0$ for $t<0$. The value of $f(t)$ at the origin is the right-hand limit $f(0+)$. For example, if we write $f(t)=\cos\omega t$ we actually mean

$$f(t)=\begin{cases}\cos\omega t & \text{for} \quad t>0\\ \cos\omega(0+)=1 & \text{for} \quad t=0+\\ 0 & \text{for} \quad t<0\end{cases}$$

Another way of expressing this is to say that all signals are "switched on" at $t=0$.

Equation (3.48) for the inversion of the Laplace transformation is rarely used except when numerical methods have to be used. The usual procedure is to look up the inverse in tables. A good table is given in Doetsch (1971). A short summary of transform pairs is given in Table 3.1.

An important class of signals are switching or step functions and rectangular pulses. These signals are handled especially well by the Laplace transformation as we can see in the following section.

3.3.2 The Laplace Transformation of Step and Pulse Functions

The *unit step function* or unit step is defined as

$$u_{-1}(t)=\begin{cases}1 & \text{for} \quad t>0\\ \frac{1}{2} & \text{for} \quad t=0\\ 0 & \text{for} \quad t<0\end{cases} \tag{3.49}$$

Its right-hand limit at the origin is then $u_{-1}(0+)=1$ and the left-hand limit is $u_{-1}(0-)=0$, the value $u_{-1}(0)=\frac{1}{2}$ being the average of the two by convention. Figure 3.5a gives a graph of the unit step.

The *delayed unit step*, $u_{-1}(t-t_0)$, is shown in Fig. 3.5b. Note that $u_{-1}(t-t_0)=0$ for $t<t_0$.

By subtracting a delayed unit step from a unit step we can generate a *pulse of unit height* (Fig. 3.5c). A pulse of unit height and duration d we designate $p_d(t)$ (Fig.

Table 3.1 Laplace Transform Pairs

$f(t)$	$F(s)$	$f(t)$	$F(s)$
1. $u_0(t)=\delta(t)$ (unit impulse, delta function)	1	7. $(e^{-at}-e^{-bt})$	$\frac{b-a}{(s+a)(s+b)}$
2. $u_{-1}(t)$ (unit step)	$\frac{1}{s}=s^{-1}$	7a. $(e^{-t/T_1}-e^{-t/T_2})$	$\frac{T_1-T_2}{(1+sT_1)(1+sT_2)}$
2a. $K=$constant	$\frac{K}{s}$	8. $\frac{(c-a)e^{-at}-(c-b)e^{-bt}}{b-a}$	$\frac{s+c}{(s+a)(s+b)}$
3. $u_{-2}(t)=t$ (unit ramp)	$\frac{1}{s^2}=s^{-2}$	9. $\sin\omega t$	$\frac{\omega}{s^2+\omega^2}$
4. e^{-at}	$\frac{1}{s+a}$	10. $e^{-at}\sin\omega t$	$\frac{\omega}{(s+a)^2+\omega^2}$
4a. $e^{-t/T}$	$\frac{T}{1+sT}$	11. $\cos\omega t$	$\frac{s}{s^2+\omega^2}$
5. te^{-at}	$\frac{1}{(s+a)^2}$	12. $e^{-at}\cos\omega t$	$\frac{s+a}{(s+a)^2+\omega^2}$
6. $(1-e^{-at})$	$\frac{a}{s(s+a)}$	13. $\frac{e^{-\zeta\omega}\sin\left(\omega_n\sqrt{1-\zeta^2}\,t\right)}{\omega_n\sqrt{1-\zeta^2}}$	$\frac{1}{s^2+2\zeta\omega_n s+\omega_n^2}$
6a. $(1-e^{-t/T})$	$\frac{1}{s(1+sT)}$	$(0\leqslant\zeta<1;\ \omega_n>0)$	

3.5d). Hence

$$p_d(t)=u_{-1}(t)-u_{-1}(t-d) \tag{3.50}$$

Pulses are also normalized in terms of *area* rather than height. A pulse of unit area and duration d (Fig. 3.5e) is then described by

$$\frac{1}{d}p_d(t) \tag{3.51}$$

A *pulse train* consisting of unit-height pulses of duration d repeated every T seconds (Fig. 3.5f) corresponds to the periodic function (for $t>0$)

$$p_T(t)=\sum_{n=0}^{\infty} p_d(t-nT) \tag{3.52}$$

Now we shall determine the Laplace transforms for these signals:

1. $u_{-1}(t)$—unit step:

$$\mathcal{L}\{u_{-1}(t)\}=\int_{0+}^{\infty}e^{-st}\,dt=-\frac{1}{s}e^{-st}\Big|_{0+}^{\infty}=\frac{1}{s} \tag{3.53}$$

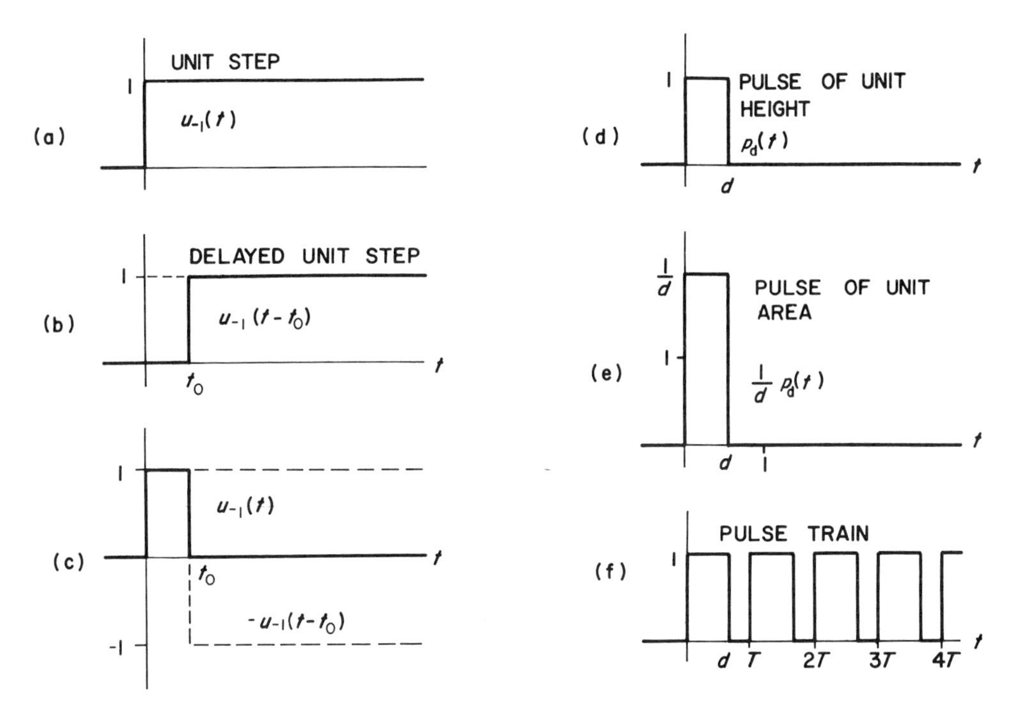

Figure 3.5 Step and pulse signals.

2. $u_{-1}(t-t_0)$—delayed unit step:

$$\mathcal{L}\{u_{-1}(t-t_0)\}=\int_{t_0}^{\infty} e^{-st}\,dt=-\frac{1}{s}e^{-st}\Big|_{t_0}^{\infty}=\frac{e^{-st_0}}{s} \tag{3.54}$$

We note that this transform equals the preceding transform multiplied by the "delay factor" e^{-st_0} or

$$\mathcal{L}\{u_{-1}(t-t_0)\}=e^{-st_0}\mathcal{L}\{u_{-1}(t)\}$$

This *delay rule* applies to other functions as well:

$$\mathcal{L}\{f(t-t_0)\}=e^{-st_0}F(s) \tag{3.55}$$

where $F(s)$ is the transform of $f(t)$.

3. $p_d(t)$—pulse of unit *height* and duration d:

$$p_d(t)=u_{-1}(t)-u_{-1}(t-d)$$

The Laplace transformation is a linear operation. We can, therefore, combine the individual transforms, Eqs. (3.53) and (3.54), according to the superposition principle:

$$\mathcal{L}\{p_d(t)\}=\mathcal{L}\{u_{-1}(t)\}-\mathcal{L}\{u_{-1}(t-d)\}$$

or

$$p_d(t)=\frac{1}{s}-\frac{e^{-sd}}{s}=\frac{1}{s}(1-e^{-sd})=P_d(s) \tag{3.56}$$

4. $(1/d)p_d(t)$—pulse of unit *area* and duration d:

$$\mathcal{L}\left\{\frac{1}{d}p_d(t)\right\}=\frac{1}{d}P_d(s)=\frac{1}{sd}(1-e^{-sd})$$

which follows from Eq. (3.56) because of the linearity of the Laplace transformation.

The above transform can be expanded into a series after insertion of the Taylor series for the exponential:

$$\mathcal{L}\left\{\frac{1}{d}p_d(t)\right\}=\frac{1}{sd}\left[1-\left(1-sd+\frac{s^2d^2}{2}-\frac{s^3d^3}{6}+\cdots\right)\right]$$

$$=1-\frac{sd}{2}+\frac{s^2d^2}{6}-\cdots \tag{3.57}$$

A pulse of special interest is generated when we let the duration become infinitesimally small while keeping the area under the pulse at unity. Such a "pulse" is called a *unit impulse* and is designated $u_0(t)$. (We devote Section 3.4 to a more detailed presentation of the properties of impulses.) We write accordingly

$$\lim_{d\to 0}\frac{1}{d}p_d(t)=u_0(t) \tag{3.58}$$

and for its Laplace transform from Eq. (3.57)

$$\mathcal{L}\{u_0(t)\}=\lim_{d\to 0}\mathcal{L}\left\{\frac{1}{d}p_d(t)\right\}=1 \tag{3.59}$$

The preceding "derivation" of Eqs. (3.58) and (3.59) skirts some mathematical difficulties. For a rigorous and correct treatment of the problem the reader should refer to Doetsch (1971) or Zemanian (1965).

The important result for us here is that an impulse "function" in the time domain is mapped into a constant in the s domain

$$\mathcal{L}\{ku_0(t)\}=k$$

5. $p_T(t)$—pulse train:

$$\mathcal{L}\{p_T(t)\}=\sum_{n=0}^{\infty}\mathcal{L}\{p_d(t-nT)\}=\sum_{n=0}^{\infty}e^{-snT}P_d(s)$$

$$=\frac{1}{s}(1-e^{-sd})\sum_{n=0}^{\infty}e^{-snT}$$

using Eqs. (3.52), (3.55), and (3.56) to synthesize the transformation without recourse to the integral.

The summation term is a geometric series which in the limit sums to

$$\frac{1}{1-e^{-sT}}$$

The transform of the pulse train is, therefore,

$$\mathcal{L}\{p_T(t)\}=\frac{1-e^{-sd}}{s(1-e^{-sT})}$$

From the preceding examples, the reader should have developed sufficient understanding of the Laplace transformation so that a listing of transform properties and a table of common transforms should be of value.

3.3.3 Properties of the Laplace Transformation. Table of Transforms

1. *The Laplace transformation is a linear operation*:

$$\mathcal{L}\{af_1(t)+bf_2(t)\}=aF_1(s)+bF_2(s) \tag{3.60}$$

2. *Change in time scale*:

$$\mathcal{L}\{f(kt)\}=\frac{1}{k}F\left(\frac{s}{k}\right), \qquad k>0 \tag{3.61}$$

3. *Time delay* ($t_0>0$):

$$\mathcal{L}\{f(t-t_0)\}=e^{-st_0}F(s) \tag{3.62}$$

with

$$f(t-t_0)=0, \qquad \text{for } t<t_0$$

4. *Exponential damping*:

$$\mathcal{L}\{e^{-\alpha t}f(t)\}=F(s+\alpha) \tag{3.63}$$

5. *Convolution in the time domain*:

$$\mathcal{L}\left\{\int_0^t f_1(\tau)f_2(t-\tau)\,d\tau\right\}=\mathcal{L}\left\{\int_0^t f_1(t-\tau)f_2(\tau)\,d\tau\right\}=F_1(s)F_2(s) \tag{3.64}$$

This is a most important result for the analysis of linear systems.

6. *Differentiation in the time domain*:

$$\mathcal{L}\left\{\frac{d}{dt}f(t)\right\}=sF(s)-f(0+) \tag{3.65}$$

and

$$\mathcal{L}\left\{\frac{d^nf(t)}{dt^n}\right\}=s^nF(s)-\sum_{k=0}^{n-1}f^{(k)}(0+)s^{n-k-1} \tag{3.66}$$

where $f^{(k)}(0+)$ is the limit value of the kth derivative for $t\rightarrow 0+$. It is assumed that $f(t)$ is differentiable for $t>0$.

This is a crucial transform property for the solution of linear, constant coefficient differential equations. With Eqs. (3.65) or (3.66) we *transform differential equations into algebraic equations which include the initial conditions explicitly*!

7. *Generalized derivatives in the time domain*:

$$\mathcal{L}\{\bar{D}^nf(t)\}=s^nF(s) \tag{3.67}$$

These derivatives make it possible to handle functions which do not have derivatives in the conventional (d/dt) sense. They are also called "derivatives in the distribution sense" or "derivations." The reader is referred to Doetsch (1971) and Zemanian (1965). A brief, nonrigorous introduction is given in Section 3.4.

8. *Integration in the time domain*:

$$\mathcal{L}\left\{\int_0^t f(\tau)\,d\tau\right\}=\frac{1}{s}F(s) \tag{3.68}$$

This property is useful for the transformation of integrodifferential equations.

A list of transform pairs is given in Table 3.1.

3.3.4 Applications of the Laplace Transformation

3.3.4.1 Immittances

In Chapter 1 we defined generalized immittances for the case of excitation with generalized phasors $\mathbf{A}e^{st}$. The inverse Laplace transform, Eq. (3.48), can be interpreted as the superposition of such phasors with "amplitude" $F(s)\,ds$. Therefore, by taking the Laplace transforms of the equation governing a passive one-port we can define generalized immittances in a more rigorous way.

For example, the element law of an inductor is

$$v=L\frac{di}{dt}$$

After Laplace transformation we have

$$V(s)=LI(s)-i(0+)$$

For the zero state, i.e., $i(0+)=0$, we define the impedance as

$$Z_{\mathrm{L}}(s)=\frac{V(s)}{I(s)}$$

and the admittance as

$$Y_{\mathrm{L}}(s)=\frac{I(s)}{V(s)}$$

The method also applies to one-ports containing several passive elements. An example is the parallel RLC circuit. Its governing equation for the total current (KCL) is

$$i(t)=C\frac{d}{dt}v(t)+\frac{v(t)}{R}+\frac{1}{L}\int_0^t v(\tau)\,d\tau+i_L(0+)$$

The Laplace transform, again assuming an initially relaxed system (zero state), is

$$I(s)=\left(sC+\frac{1}{R}+\frac{1}{sL}\right)V(s)$$

The admittance is then

$$Y_p(s)=sC+\frac{1}{R}+\frac{1}{sL}$$

The point here is that the Laplace transformation extends the immittance concept from signals of the type e^{st} to any Laplace transformable signal. Hence the following definition:

The generalized impedance $Z(s)$ of a passive one-port in the zero-state is defined by

$$Z(s)=\frac{V(s)}{I(s)} \tag{3.69}$$

where $V(s)$ and $I(s)$ are the Laplace transforms of the voltage–current pair at the terminals of the one-port.
The generalized admittance $Y(s)$ of a passive one-port in the zero-state is defined by

$$Y(s)=\frac{I(s)}{V(s)}=\frac{1}{Z(s)} \tag{3.70}$$

Nonzero initial conditions can be handled either by the method of "transform" networks (Section 3.6) or by the direct application of the Laplace transformation to the integrodifferential equations describing the network. The latter approach is our next topic.

3.3.4.2 Applications to Linear Differential and Network Equations

We consider the first-order differential equation

$$\frac{d}{dt}y(t)+a_0y(t)=x(t)$$

Transforming this equation with the aid of Eq. (3.65) gives

$$sY(s)-y(0+)+a_0Y(s)=X(s)$$

By a rearrangement of this *algebraic* equation we obtain

$$Y(s)(s+a_0)=X(s)+y(0+)$$

or

$$Y(s)=\frac{X(s)}{s+a_0}+\frac{y(0+)}{s+a_0} \tag{3.71}$$

This is the solution for $y(t)$ in the *s domain.*

Note that the solution has two parts. The first, due to the forcing term or input $x(t)$, is called the *zero-state response* in the s domain because it is the sole response for vanishing initial conditions [$y(0+)=0$], i.e., the zero state. The second represents the relaxation from the initial conditions; it is called the *zero-input response* in the s domain because it describes the unforced system [$x(t)=0$].

The generalized phasor method introduced in Chapter 1 gives us only the equivalent of the zero-state response because it ignores initial conditions.

We state the general case for the nth-order differential equations next.

The linear constant coefficient differential equation

$$a_n \frac{d^n}{dt^n} y(t) + \cdots + a_2 \frac{d^2}{dt^2} y(t) + a_1 \frac{d}{dt} y(t) + a_0 y(t) = x(t)$$

transforms into the *algebraic* equation

$$Y(s)\left(a_n s^n + \cdots + a_2 s^2 + a_1 s + a_0\right) = X(s) + R(s)$$

where $R(s)$ is the polynomial arising from all the initial condition terms according to Eq. (3.66).

Solving for $Y(s)$ gives

1. Laplace transform solution of the nth-order linear differential equation:

$$a_n \frac{d}{dt^n} y(t) + \cdots + a_2 \frac{d^2}{dt^2} y(t) + a_1 \frac{d}{dt} y(t) + a_0 y(t) = x(t)$$

$$Y(s) = \frac{X(s)}{a_n s^n + \cdots + a_2 s^2 + a_1 s + a_0} + \frac{R(s)}{a_n s^n + \cdots + a_2 s^2 + a_1 s + a_0} \tag{3.72}$$

2.

$$\mathcal{L}\{\text{zero-state response}\} = \frac{X(s)}{a_n s^n + \cdots + a_2 s^2 + a_1 s + a_0} = G(s)X(s) \tag{3.73}$$

3.

$$\mathcal{L}\{\text{zero-input response}\} = \frac{R(s)}{a_n s^n + \cdots + a_2 s^2 + a_1 s + a_0} = Q(s) \tag{3.74}$$

4.

$$Y(s) = G(s)X(s) + Q(s) \tag{3.75}$$

5. System function or transfer function:

$$G(s) = \frac{\mathcal{L}\{\text{zero-state response}\}}{\mathcal{L}\{\text{input}\}} = \frac{1}{a_n s^n + \cdots + a_2 s^2 + a_1 s + a_0} \tag{3.76}$$

or

$$\mathcal{L}\{\text{zero-state response}\} = \begin{bmatrix} \text{system or} \\ \text{transfer function } G(s) \end{bmatrix} \times \mathcal{L}\{\text{input}\} \tag{3.77}$$

6. Weighting or Green's function:

$$g(t) = \mathcal{L}^{-1}\{G(s)\} \tag{3.78}$$

7. Solution in terms of the weighting function:

$$y(t)=\int_0^t g(\tau)x(t-\tau)\,d\tau+\mathcal{L}^{-1}\{Q(s)\} \tag{3.79}$$

$$=\int_0^t g(t-\tau)x(\tau)\,d\tau+\mathcal{L}^{-1}\{Q(s)\} \tag{3.80}$$

REMARKS:

1. Equations (3.72) and (3.75) represent the *complete* transform solution of the *inhomogeneous* differential equation. This solution consists of two parts: the transform of the zero-state response, Eqs. (3.73) and (3.77), and the transform of the zero-input response, Eq. (3.74). Note that the zero-state response equals the particular integral of the differential equation, whereas the zero-input response is the solution of the homogeneous equation $[x(t)=0]$ and, therefore, represents the complementary function of the differential equation.
2. The system or transfer function $G(s)$, Eq. (3.76), relates the input transform to the transform of the zero-state response. It contains all the important information about the inherent response characteristics of the system described by the differential equation. The polynomial in the denominator of Eq. (3.76) is called the *characteristic polynomial* of the differential equation. The zeros of this polynomial, which are the *poles* of the transfer function $G(s)$, represent the *reciprocal time constants* and the *natural frequencies* of the system. We may prove this by factoring the characteristic polynomial, followed by partial fraction expansion of $G(s)$ (see Doetsch, 1971, or Aseltine, 1958) and look-up of the inverse transforms in Table 3.1.
3. For $s=j\omega$, the system function equals the frequency response function $G(j\omega)$, Eq. (3.42).
4. The system function may relate the Laplace transforms of a voltage–current pair at the same port, and thus be a driving point immittance. Or it could be a transfer immittance if voltage and current are associated with different ports. The system function is a dimensionless transfer function if like variables at different ports are related, e.g., the input and output voltages of an amplifier.
5. The immittances and transfer ratios derived by the generalized phasor method are identical to those derived by the Laplace transformation because the system function is limited to the zero-state response.
6. The solution $y(t)$ in the time domain can be determined *uniquely* by inverting Eq. (3.72) directly from tables after partial fraction expansion. An alternative is to determine first the inverse of the system function $g(t)$ according to Eq. (3.78) and then use one of the convolution integrals relating input $x(t)$, and output $y(t)$, as shown in Eqs. (3.79) and (3.80). The latter procedure would be advantageous whenever the *weighting or Green's function* $g(t)$ is available in numerical form so that a computer solution of Eqs. (3.79) or (3.80) is indicated. The response formulation in terms of convolution integrals has other conceptual advantages which will be considered in Section 3.5.

Example 3.3. The RL low-pass filter

The circuit is given in Fig. 1.33. The KVL equation is

$$L\frac{di(t)}{dt}+Ri(t)=v_1(t)$$

After Laplace transformation we have

$$sLI(s) - Li(0+) + RI(s) = V_1(s)$$

where $i(0+)$ is the initial current in the inductor. Defining the time constant $T = L/R$ and rearranging gives the transform solution

$$I(s) = \frac{V_1(s)/R}{1+sT} + \frac{Ti(0+)}{1+sT} \tag{3.81}$$

where the first term on the right is the Laplace transformation of the zero-state response and the second one that of the zero-input response.

The system function according to Eq. (3.76) is

$$G_{iv}(s) = \frac{I(s)}{V_1(s)} = \frac{1}{R(1+sT)} = \frac{1}{R+sL}$$

which is, of course, the total mesh impedance.

For the output voltage we have

$$v_2(t) = Ri(t)$$

or

$$V_2(s) = RI(s).$$

By Eq. (3.81) this gives us

$$V_2(s) = \frac{V_1(s)}{1+sT} + \frac{RTi(0+)}{1+sT} \tag{3.82}$$

For the zero-state response $[i(0+)=0]$, the system or transfer function is then

$$G(s) = \frac{V_2(s)}{V_1(s)} = \frac{1}{1+sT}$$

If the input source provides a step excitation to the circuit, we can write

$$v_1(t) = V_0 u_{-1}(t)$$

which has the Laplace transform

$$V_1(s) = \frac{V_0}{s}$$

We insert this expression into Eq. (3.82) and obtain

$$V_2(s) = \frac{V_0}{s(1+sT)} + \frac{RTi(0+)}{1+sT} \tag{3.83}$$

By means of entries 4a and 6a in Table 3.1 we invert Eq. (3.83) and express $v_2(t)$ as

$$v_2(t) = V_0(1 - e^{-t/T}) + Ri(0+)e^{-t/T} \tag{3.84}$$

3.4 Impulse Functions, Generalized Derivatives, Sampled Functions

Impulse functions and generalized derivatives were briefly mentioned in Sections 3.3.2 and 3.3.3, respectively. In this section we consider these and other related topics in more detail because of their importance to signal and system analysis. For mathematically rigorous introductions to these concepts, the reader may consult Doetsch (1971) or Zemanian (1965).

3.4.1 The Impulse Function

For the purpose of a nonrigorous, intuitive derivation of the impulse concept, we begin with a consideration of the unit ramp step $r(t)$ and its derivative. The unit ramp step is defined according to Fig. 3.6a by the function

$$r(t)=\begin{cases} 0 & \text{for} \quad t<0 \\ t/d & \text{for} \quad 0<t<d \\ 1 & \text{for} \quad t>d \end{cases}$$

Except for the break points at $t=0$ and $t=d$, the derivative is defined as

$$\frac{dr(t)}{dt}=\begin{cases} 0 & \text{for} \quad t<0 \\ 1/d & \text{for} \quad 0<t<d \\ 0 & \text{for} \quad t>d \end{cases}$$

Obviously, the derivative is a pulse of unit area and amplitude $1/d$ (Fig. 3.6b).

As we let the interval d approach zero, with the condition that the area of the derivative pulse stay at unity, the unit ramp step becomes the unit step $u_{-1}(t)$ (Fig. 3.6c) and the unit-area pulse becomes the unit impulse $u_0(t)$. The unit impulse is symbolized by an arrow in Fig. 3.6d; the number 1 next to the arrow indicates that the impulse has a *weight* or *strength* of unity.

The point here is to introduce the unit impulse as the "derivative" of the unit step even though the unit step is discontinuous at the origin and, therefore, has no derivative in the *conventional* sense. With the theory of distributions or generalized

Figure 3.6 Derivation of the unit impulse as the generalized derivative of the unit step.

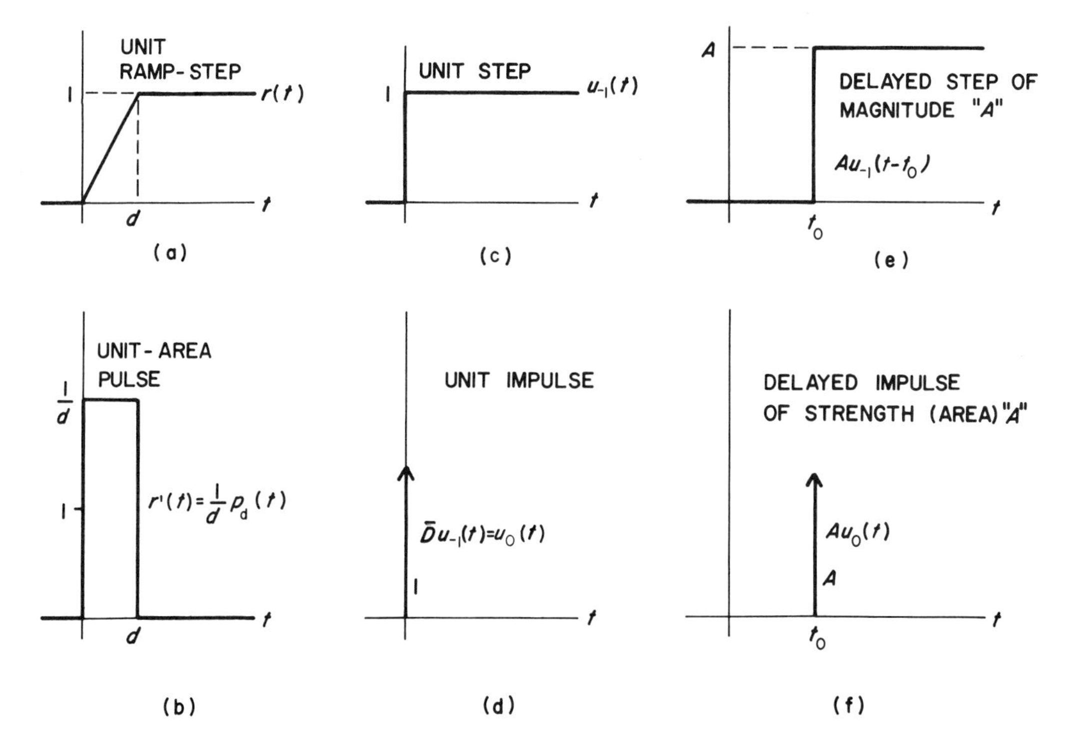

functions (Doetsch, 1971; Zemanian, 1965) it is possible to overcome the difficulties which conventional derivatives pose in such a case and define a *generalized derivative* (or *derivative in the distribution sense*) for which we use the operator symbol $\bar{D}$. Hence we write for the generalized derivative of the unit step function

$$\bar{D}u_{-1}(t)=u_0(t)$$

And for a step of delay t_0 and magnitude A (Fig. 3.6e) we have

$$\bar{D}Au_{-1}(t-t_0)=Au_0(t-t_0)$$

which represents a delayed impulse of weight or strength A (Fig. 3.6f). (Note that A is not the "height" of the impulse; see below.)

We now summarize the

Properties of the Unit Impulse Function

(1) $u_0(t)=\delta(t)=0 \quad \text{for} \quad t\neq 0$ (3.85)

(2) $u_0(t-t_0)=\delta(t-t_0)=0 \quad \text{for} \quad t\neq t_0$ (3.86)

(3) $\int_{-\infty}^{\infty} u_0(t)\,dt=1$ (3.87)

(4) $\int_{-\infty}^{\infty} Au_0(t-t_0)\,dt=A$ (3.88)

(5) $\int_{-\infty}^{\infty} Au_0(\tau-t_0)\,d\tau=\begin{cases}0 & \text{for } t<t_0\\ Au_{-1}(t-t_0) & \text{for } t>t_0\end{cases}$ (3.89)

(6) $\bar{D}u_{-1}(t)=u_0(t)$ (3.90)

(7) $\bar{D}[Au_{-1}(t-t_0)]=Au_0(t-t_0)$ (3.91)

(8) $\int_{-\infty}^{\infty} f(t)u_0(t-t_0)\,dt=f(t_0)$ (3.92)

(sifting or sampling property)

(9) $\int_{-\infty}^{\infty} f(t)u_0(t+t_0)\,dt=f(-t_0)$ (3.93)

(10) $\mathfrak{F}\{u_0(t)\}=1$ (3.94)

(11) $\mathfrak{L}\{u_0(t)\}=1$ (3.95)

(12) $\int_0^t f(\tau)u_0(t-\tau)\,d\tau=\int_0^t f(t-\tau)u_0(\tau)\,d\tau=f(t)$ (3.96)

(convolution with $f(t)$; $f(t)=0$ for $t<0$)

REMARKS: Equations (3.85) and (3.86) contain another common notation for the impulse function, namely, $\delta(t)$ and $\delta(t-t_0)$. In fact, the unit impulse function is often called the *δ-function* (delta function), Dirac δ-function, or Dirac impulse (after P.A.M. Dirac, who popularized its use in quantum mechanics).

Equations (3.87) and (3.88) emphasize the fact that the definite integral or "area" of an impulse function equals its weight or strength.

The definition of the unit impulse as the generalized derivative of the unit step is given by Eq. (3.90), while Eqs. (3.89) and (3.91) are a pair of complementary definitions in terms of definite integral and generalized derivative for impulses and steps of arbitrary magnitude and delay.

The integral in Eq. (3.92) and its symmetric companion, Eq. (3.93), define the sifting or sampling property of the unit impulse function. This property is important for the treatment of sampled signals, as we shall see in Section 3.4.4.

Both the Fourier and Laplace transforms of the unit impulse equal unity, Eqs. (3.94) and (3.95). We can insert Eq. (3.94) in Eq. (3.64), take the inverse transform, and thus prove Eq. (3.96) for the convolution. Equations (3.94) to (3.96) are extremely important for methods of system analysis (outline in Section 3.5).

3.4.2 Generalized Derivatives

We use the result of Eq. (3.91) in order to find the generalized derivative of arbitrary functions.

We restrict ourselves to generalized derivatives of the first order for functions with at most stepwise discontinuities. For the treatment of higher-order derivatives or discontinuities represented by impulses and their generalized derivatives, the reader should consult texts such as Doetsch (1971) or Zemanian (1965).

Consider the discontinuous function in Fig. 3.7a. This function can be decomposed into a sum consisting of a smooth, continuous function (Fig. 3.7b) and delayed step functions for each of the stepwise discontinuities (Fig. 3.7c). According to Fig. 3.7 we can write

$$\begin{aligned} f(t) = {} & f_1(t) + [f(t_1+) - f(t_1-)]u_{-1}(t-t_1) \\ & + [f(t_2+) - f(t_2-)]u_{-1}(t-t_2) \end{aligned} \tag{3.97}$$

where the first term, $f_1(t)$, is the function in Fig. 3.7b for which the conventional derivative exists. Defining the *conventional* derivative of $f(t)$ as

$$\frac{df(t)}{dt} = \frac{df_1(t)}{dt}$$

which by definition excludes the discontinuities, gives the *generalized* derivative of $f(t)$ by adding the generalized derivatives of the step function terms in Eq. (3.97) to the conventional derivative. Thus we have

$$\begin{aligned} \bar{D}f(t) = {} & \frac{d}{dt}f(t) + [f(t_1+) - f(t_1-)]u_0(t-t_1) \\ & + [f(t_2+) - f(t_2-)]u_0(t-t_2) \end{aligned}$$

The general rule for obtaining the generalized first derivative of a function with stepwise discontinuities is then to *take the conventional derivative for the continuous portion and add an impulse weighted by the value of the discontinuity for each stepwise discontinuity*.

Example 3.4.

Generalized derivative of a cosine "switched on" at time $t=0$. We define

$$u_{-1}(t)\cos\omega t = \begin{cases} \cos\omega t & \text{for } t>0 \\ 0 & \text{for } t<0 \end{cases}$$

For the discontinuity at the origin we set $\cos(0-)=0$ and $\cos(0+)=1$. Hence we have for $t \geqslant 0+$.

$$\bar{D}[u_{-1}(t)\cos\omega t] = \frac{d}{dt}\cos\omega t + u_0(t) = -\omega\sin\omega t + u_0(t)$$

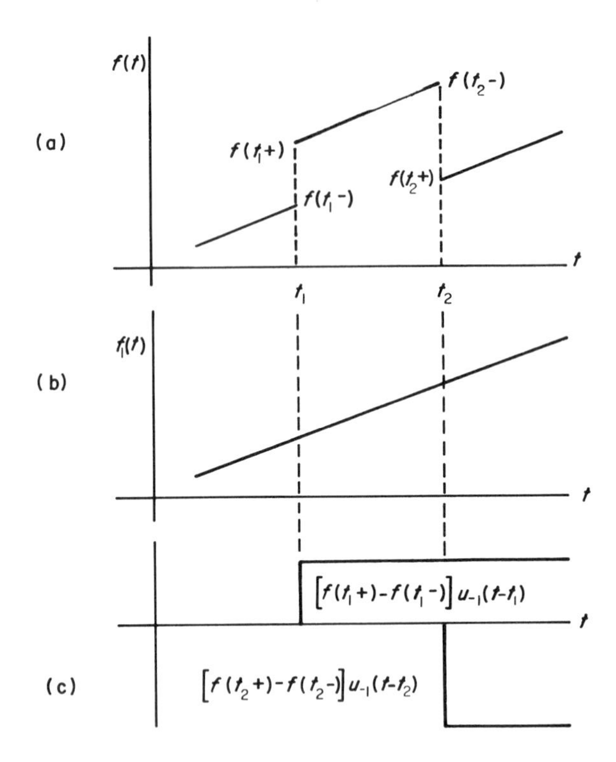

Figure 3.7 Decomposition of a discontinuous function into a continuous component and step functions.

If we Laplace transform this result, we obtain from Table 3.1

$$\mathfrak{L}\{-\omega \sin \omega t + u_0(t)\} = \frac{-\omega^2}{s^2+\omega^2} + 1 = \frac{s^2}{s^2+\omega^2}$$

which agrees with the Laplace transform according to Eq. (3.67):

$$\mathfrak{L}\{\bar{D} \cos \omega t\} = s\mathfrak{L}\{\cos \omega t\} = \frac{s^2}{s^2+\omega^2}$$

(It is not necessary to include the "switching function" $u_{-1}(t)$ because application of the one-sided Laplace transform implies that the switching process takes place.)

In the preceding example the discontinuity occurred at $t=0$; thus the generalized derivative includes an impulse at the origin. In the strict application of the Laplace integral this can lead to difficulties because its lower integration limit is set at $0+$. In order to avoid an inadvertant exclusion of an impulse at the origin we may assume that the impulse is actually defined for $t=0+$. A rigorous treatment though should make use of the theory of generalized functions (Doetsch, 1971).

3.4.3 Frequency Spectra of Periodic Signals (Impulse Spectra)

The classical Fourier transformation excludes such important functions as $\cos \omega t$ or $\sin \omega t$ because they violate the convergence condition, Eq. (3.23). However, the impulse concept allows us to generalize the Fourier transformation in order to include all periodic functions.

We introduce the generalization by postulating the existence of *impulse spectra*, i.e., spectra which are mathematically represented by impulse functions in the *frequency domain*. Specifically, let us assume that the following Fourier transforms exist:

$$F_1(j\omega)=2\pi u_0(\omega-\omega_0) \tag{3.98}$$

$$F_2(j\omega)=2\pi u_0(\omega+\omega_0) \tag{3.99}$$

We now ask, What are the inverse transforms of Eqs. (3.98) and (3.99) in the time domain? We insert these equations in the inversion formula, Eq. (3.25), and apply the sifting property of impulse functions, Eqs. (3.92) and (3.93). This procedure gives us

$$f_1(t)=\frac{1}{2\pi}\int_{-\infty}^{\infty} 2\pi u_0(\omega-\omega_0)e^{j\omega t}\,d\omega=e^{j\omega_0 t} \tag{3.100}$$

and

$$f_2(t)=\frac{1}{2\pi}\int_{-\infty}^{\infty} 2\pi u_0(\omega+\omega_0)e^{j\omega t}\,d\omega=e^{-j\omega_0 t} \tag{3.101}$$

The transforms of Eqs. (3.100) and (3.101) are, therefore, the Fourier transforms of the phasor signals $\exp(j\omega_0 t)$ and $\exp(-j\omega_0 t)$, respectively (Figs. 3.8a and 3.8b).

With Eqs. (3.98)–(3.101) we can now generate the Fourier transforms of the following four important classes of periodic signals.

1.
$$f(t)=\cos\omega_0 t=\tfrac{1}{2}(e^{j\omega_0 t}+e^{-j\omega_0 t})$$

Figure 3.8 Examples of impulse spectra for complex exponentials, **(a)** and **(b)**; sinusoids, **(c)** and **(d)**; periodic functions, **(e)**; and the sampling function, **(f)**. (See text.)

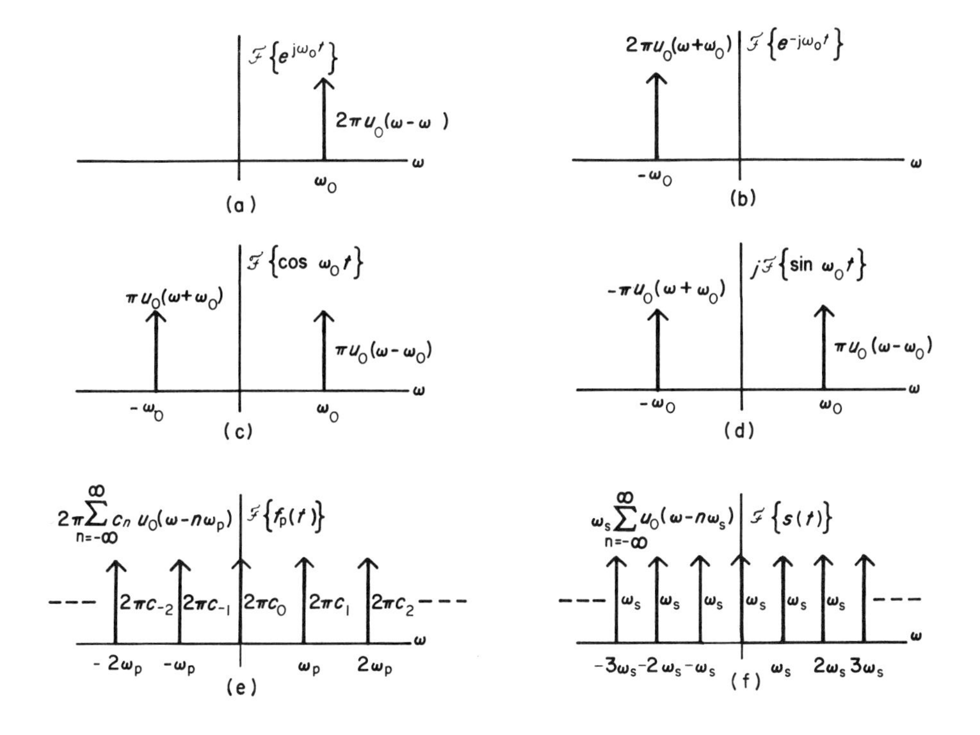

and

$$\mathcal{F}\{\cos\omega_0 t\} = \pi u_0(\omega - \omega_0) + \pi u_0(\omega + \omega_0)$$

The corresponding impulse spectrum is shown in Fig. 3.8c.

2.

$$f(t) = \sin\omega_0 t = \frac{1}{2j}(e^{j\omega_0 t} - e^{-j\omega_0 t})$$

and

$$\mathcal{F}\{\sin\omega_0 t\} = \frac{\pi}{j}u_0(\omega - \omega_0) - \frac{\pi}{j}u_0(\omega + \omega_0)$$
$$- j\pi u_0(\omega - \omega_0) + j\pi u_0(\omega + \omega_0)$$

The impulse spectrum of this transform is shown in Fig. 3.8d.

3. $f(t) = f_p(t)$, an arbitrary periodic function. First we write the complex Fourier series of the periodic function $f_p(t)$:

$$f_p(t) = \sum_{n=-\infty}^{\infty} c_n e^{jn\omega_p t}$$

The Fourier transformation of the exponential in the series is, according to Eqs. (3.98) and (3.100), $2\pi u_0(\omega - n\omega_p)$. The Fourier transformation of the whole Fourier series is, therefore,

$$\mathcal{F}\{f_p(t)\} = \sum_{n=-\infty}^{\infty} 2\pi c_n u_0(\omega - n\omega_p) \tag{3.102}$$

The corresponding Fourier *density* spectrum (Fig. 3.8e) differs from the conventional Fourier *coefficient* spectrum in that the weights of the impulse functions in the density spectrum are $2\pi c_n$, whereas in the discrete coefficient spectrum we have only c_n. (For simplicity's sake we assumed here that the c_n are real. Of course, they may be complex. In that case we must consider two spectra: the magnitude and phase spectra, or spectra for the real and imaginary parts of the c_n.) Note that the spectral impulses have equidistant spacing in multiples of the fundamental frequency ω_p.

4.

$$f(t) = s(t) = \sum_{n=-\infty}^{\infty} u_0(t - nT_s) \tag{3.103}$$

This is the so-called *sampling function* $s(t)$. The interval between two succeeding unit impulses, T_s, is the *sampling interval*; its reciprocal, $f_s = 1/T_s$, is the *sampling frequency*. We shall use the sampling function in the next section. For now we are only interested in the Fourier transformation of this function.

The sampling function is a periodic impulse train. We can therefore represent it by a Fourier *series*:

$$s(t) = \sum_{n=-\infty}^{\infty} c_n e^{jn\omega_s t} \qquad (\omega_s = 2\pi f_s)$$

We compute the Fourier series coefficients c_n according to Eqs. (3.13) and (3.92):

$$c_n=\frac{1}{T_s}\int_{-T_s/2}^{T_s/2} u_0(t)e^{-jn\omega_s t}\,dt=\frac{e^{-0}}{T_s}=\frac{1}{T_s}=f_s$$

We combine the preceding two equations to obtain the Fourier *series*:

$$s(t)=f_s\sum_{n=-\infty}^{\infty} e^{jn\omega_s t} \tag{3.104}$$

The Fourier transformation of Eq. (3.104) follows from Eq. (3.102):

$$S(j\omega)=\omega_s\sum_{n=-\infty}^{\infty} u_0(\omega-n\omega_s) \tag{3.105}$$

where $\omega_s=2\pi/T_s=2\pi f_s$.

The Fourier density spectrum of the sampling function is obviously also an impulse "train" in the frequency domain. The weight of each impulse equals the radian sampling frequency (Fig. 3.8).

3.4.4 Sampled Signals and Their Spectra

Digital signal analysis requires that continuous signals be transformed into discrete number sequences. The usual procedures is to *sample* the signal $x(t)$ every T_s seconds and generate the sequence of samples $x(T_s)$, $x(2T_s)$, $x(3T_s)$, etc. By choosing a sampling interval T_s as small as possible we can make the sample sequence an arbitrarily close approximation of the continuous signal.

The actual choice of the sampling interval T_s or its reciprocal, the sampling frequency f_s, is governed by two considerations. The first is based on the limitations of the electronic device doing the actual sampling (and on the memory space available for sample storage). The second concerns a theorem which states that the *sampling frequency f_s must be more than twice the highest frequency present in the continuous signal.* If that highest frequency is f_1, then the minimal sampling frequency should just exceed $2f_1$. Furthermore, the theorem implies that *the complete, continuous signal can be recovered from the sample sequence provided the sample* frequency exceeds twice the bandwidth ($f_s>2f_1$).

Consider the continuous function $x(t)$ (Fig. 3.9) which we wish to convert into a sampled function. Our basic assumption is that $x(t)$ is *band limited* (e.g., $x(t)$ is the result of low-pass filtering). For the mathematical description of the sampling process we exploit the sifting or sampling property of the impulse function, Eq. (3.92). Using the sampling function $s(t)$ in Eq. (3.103) (Fig. 3.9), we generate the desired sample sequence (Fig. 3.9) by

$$\int_{-\infty}^{\infty} x(t)s(t)\,dt=\int_{-\infty}^{\infty} x(t)\sum_{n=-\infty}^{\infty} u_0(t-nT_s)\,dt=\sum_{n=-\infty}^{\infty} x(nT_s) \tag{3.106}$$

An alternative to Eq. (3.106) is to describe the sequence of samples by a sequence of impulses weighted by the sample values, namely,

$$x_s(t)=\sum_{n=-\infty}^{\infty} x(nT_s)u_0(t-nT_s) \tag{3.107}$$

where $x_s(t)$ symbolizes the sampled function (Fig. 3.9).

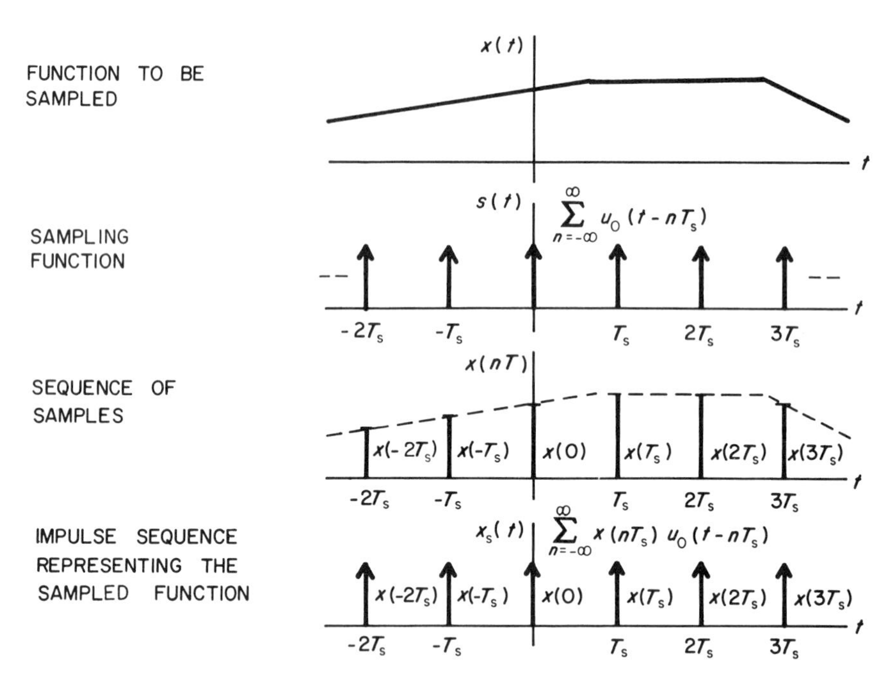

Figure 3.9 Sampling of a continuous signal. (See text).

As the impulse functions are nonzero only at $t=\pm T_s$, $\pm 2T_s$, we can nonrigorously write in place of Eq. (3.107)

$$x_s(t)=x(t)s(t) \tag{3.108}$$

We obtain the Fourier transform of the product in Eq. (3.108) by the complex convolution integral (see, e.g., p. 27 in Papoulis, 1962):

$$X_s(j\omega)=\frac{1}{2\pi}\int_{-\infty}^{\infty} X[j(\omega-\Omega)]S(j\Omega)\,d\Omega \tag{3.109}$$

Inserting the transform of the sampling function, Eq. (3.105), in the convolution integral of Eq. (3.109) gives

$$X_s(j\omega)=\frac{1}{2\pi}\int_{-\infty}^{\infty} X[j(\omega-\Omega)]2\pi f_s \sum_{n=-\infty}^{\infty} u_0(\Omega-n\omega_s)\,d\Omega$$

The last expression, finally, can be simplified by the sifting property, Eq. (3.92), to yield the Fourier transform of the sampled signal:

$$X_s(j\omega)=f_s\sum_{n=-\infty}^{\infty} X[j(\omega-n\omega_s)] \tag{3.110}$$

From Eq. (3.110) we conclude that *sampling the signal in the time domain produces an infinite number of discretely spaced "side bands" which are scaled and shifted replicas of the original spectrum $X(j\omega)$.*

The graphic interpretation of Eq. (3.110) in Fig. 3.10 explains at once the significance of the sampling frequency and why it is possible to recover the original signal (except for a scale factor) from a sample sequence obtained at more than twice the highest signal frequency.

Figure 3.10a shows the band-limited spectrum of the unsampled continuous signal. Sampling at *less* than twice the bandwidth ($\omega_s<2\omega_1$) produces a sequence of

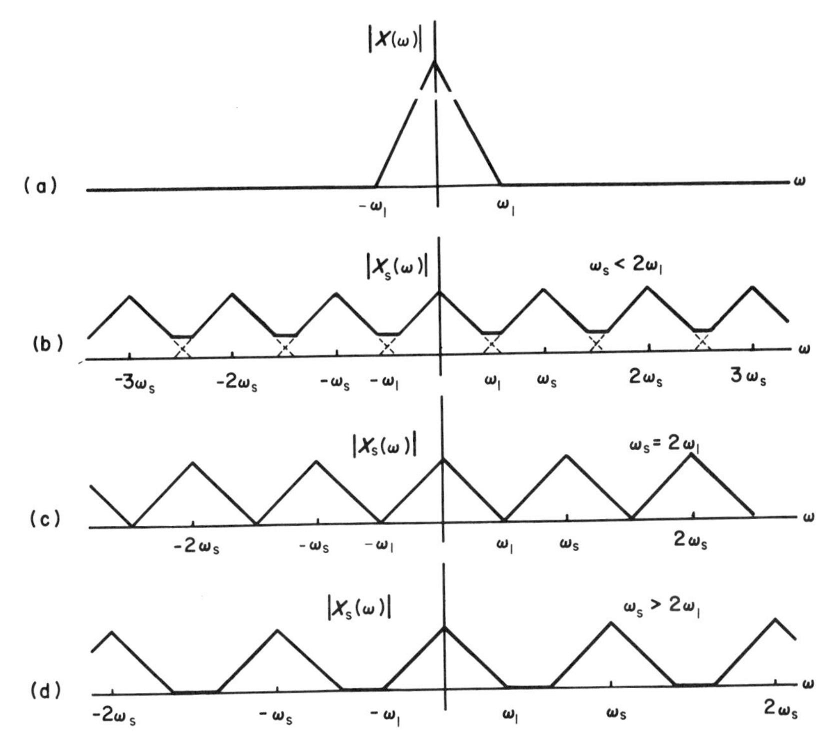

Figure 3.10 Spectral effects of sampling: **(a)** spectrum of unsampled, band-limited signal; **(b)**, **(c)**, and **(d)** spectra of the signal sampled at various sampling rates.

overlapping side bands (Fig. 3.10b). If we were to attempt to recover the "middle" spectrum ($n=0$) with an ideal low-pass filter covering the range $-\omega_1 < \omega < \omega_1$, we would obtain a *distorted* version of the original spectrum because portions of the neighboring side bands have been folded into the primary frequency interval. The generation of unwanted spectral components because of too low a sampling frequency is called *aliasing*.

Sampling at exactly twice the highest signal frequency ($\omega_s = 2\omega_1$) brings about a complete separation of main and side bands (Fig. 3.10c), but problems still arise in the case of sinusoidal components at exactly half the sampling frequency. Prudence and the properties of actually realizable low-pass filters dictate that the sampling frequency be made quite a bit larger than twice the highest signal frequency ($\omega_s > 2\omega_1$). Then the component spectra of the sampled signal are well separated (Fig. 3.10d), and the original signal can be reconstructed through amplification and low-pass filtering over the frequency band from 0 to ω_1.

It is good practice to low-pass filter the signal *before* sampling in order to avoid aliasing from unwanted high-frequency components present in the original signal.

3.5 Linear Network and System Analysis

In Chapter 1 we discussed various methods of deriving sets of equations describing the behavior of lumped network models. In earlier sections of this chapter we considered the application of the Fourier and Laplace transform methods to equations typical for linear, constant parameter networks. We also recognized that the transform methods are powerful generalizations of the phasor methods used in Chapter 1. In this section we shall extend the application of transform methods and introduce some general strategies for the description of linear, constant parameter systems.

3.5.1 Linear, Constant Coefficient Networks: Transform Networks

A common way to apply transform methods to the solution of network equations is to derive the integrodifferential equations first and then to transform these equations into algebraic ones with the Fourier or Laplace transformation. Here we describe a method which bypasses the integrodifferential equations and yields the transformed algebraic equation directly. The key to this more efficient method is the concept of the *transform network*. Its counterpart in Chapter 1 was the phasor network.

3.5.1.1 Fourier Transform Networks

This procedure is best suited for networks in the zero state, i.e., all passive storage elements have zero initial conditions.

Let us start with the simple circuit in Fig. 3.11a. *We obtain the Fourier transform equivalent of the network by labeling all passive elements with their appropriate ac impedances* [Eqs. (3.36) and (3.37)] *or ac admittances* [Eqs. (3.38) and (3.39)] *and replacing all current and voltage variables with their Fourier transforms*. In the case of our example, the resultant transform network is shown in Fig. 3.11b.

We may now write KCL and KVL equations, or mesh or node equations, etc., based on the transform network. Whether the passive elements are treated as impedances or admittances depends on the choice of equations. For example, impedances are appropriate for a set of mesh equations. Thus for Fig. 3.11b we can write these mesh equations by inspection:

$$I_1(j\omega)(R_1 + j\omega L) - I_2(j\omega)j\omega L = V_0(j\omega)$$

$$-I_1(j\omega)j\omega L + I_2(j\omega)\left(R_2 + j\omega L + \frac{1}{j\omega C}\right) = 0$$

We can solve this set of simultaneous equations for the transformed mesh currents and obtain the solution in the time domain by the inverse Fourier transformation. In most instances, however, it will be sufficient to derive a frequency response function (Section 3.5.5) from the transform equations without making the inversion to the time domain.

Figure 3.11 A network **(a)** and its Fourier transform version **(b)**.

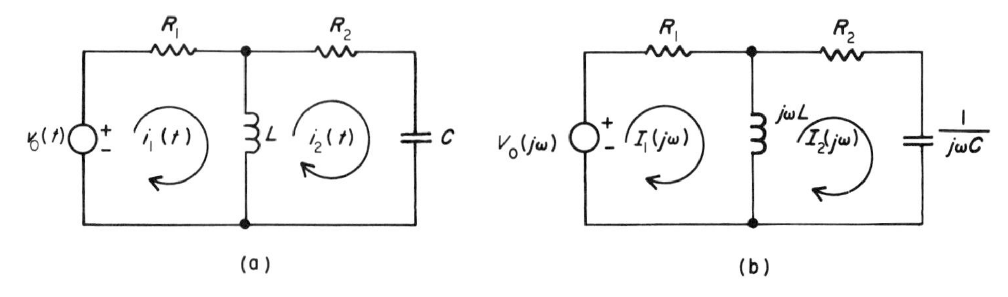

3.5.1.2 Laplace Transform Networks

The major advantage of this method over the previous one is that initial conditions are introduced with the aid of explicit model elements. *In the Laplace transform equivalent of a network* (1) *all passive elements are labeled with their appropriate generalized immittances* [Eqs. (3.69) and (3.70)], (2) *appropriate source elements are added to reflect the initial conditions of storage elements* (Fig. 3.12), *and* (3) *all voltage and current variables are replaced by their Laplace transforms.*

The addition of source elements which represent the initial conditions of the storage elements follows naturally from the element laws. For example, for the inductor (Fig. 3.12a) we have the element law

$$v_L = L\frac{di_L}{dt}$$

which upon Laplace transformation changes to

$$V_L(s) = L[sI_L(s) - i_L(0+)] = sLI_L(s) - Li_L(0+)$$

The network equivalent of this element law is given in Fig. 3.12b. The Norton equivalent of this series combination is equally suitable (Fig. 3.12c).

For a capacitor (Fig. 3.12d) we can write the element law

$$v_C = \frac{1}{C}\int_{0+}^{t} i_C\,dt + v_C(0+)$$

Laplace transformation of this equation yields the expression

$$V_C(s) = \frac{1}{sC}I_C(s) + \frac{v_C(0+)}{s}$$

as the basis for the series combination in Fig. 3.12e and the corresponding Norton equivalent of Fig. 3.12f.

Figure 3.12 Storage elements and their Laplace transform equivalents.

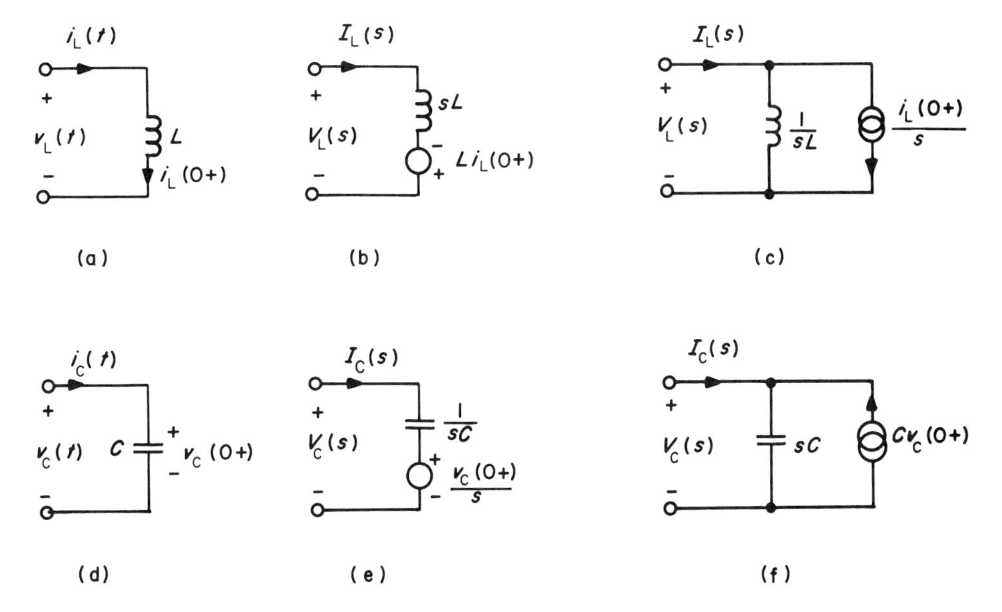

Let us return now to the network of Fig. 3.11a. The corresponding Laplace transform equivalent suitable for mesh analysis is the transform network in Fig. 3.13a. An equivalent network suitable for nodal analysis (Fig. 3.13b) makes use of the Norton equivalents in Fig. 3.12. As an example, we write the node equations for the transform network in Fig. 3.13b by inspection:

$$V_1(s)\left(G_1+G_2+\frac{1}{sL}\right)-V_2(s)G_2=G_1V_0(s)-\frac{i_L(0+)}{s}$$

$$-V_1(s)G_2+V_2(s)(G_2+sC)=Cv_C(0+)$$

These equations may be solved for one or both of the transformed node voltages by any method suitable for simultaneous linear equations (see, e.g., Wylie, 1960).

In the next four sections we direct our attention to methods based on input–output analysis of linear networks or, more generally, systems. The main strategy in every case is to predict the response (output) to an arbitrary input from the response to a *specific test* input. The response to the specific test input is either derived from an explicit set of network equations (such as the ones just written for Fig. 3.12a on the basis of Figs. 3.12b or 3.13b) or determined by experimental observation. From a practical and/or theoretical point of view, the signal waveforms most suitable as test inputs are steps, impulses (and pulses), and sinusoids. (For the use of white noise signals see Truxal, 1955.) We consider steps first.

3.5.2 The Step Response

We make the following assumptions about the system (or network) in question: (1) The system is linear and time invariant; i.e., the superposition principle holds and the system parameters are constant. (2) The system is in the zero state, i.e., all initial conditions vanish, or the system is at least resting at a stable operating point. In the latter case, a step input will cause the output variable to deviate from its operating point (steady state resting value for zero input). We then consider this deviation as the (incremental) "output." (3) The system has one externally accessible point (or

Figure 3.13 Laplace transform network for the circuit in Fig. 3.11a.

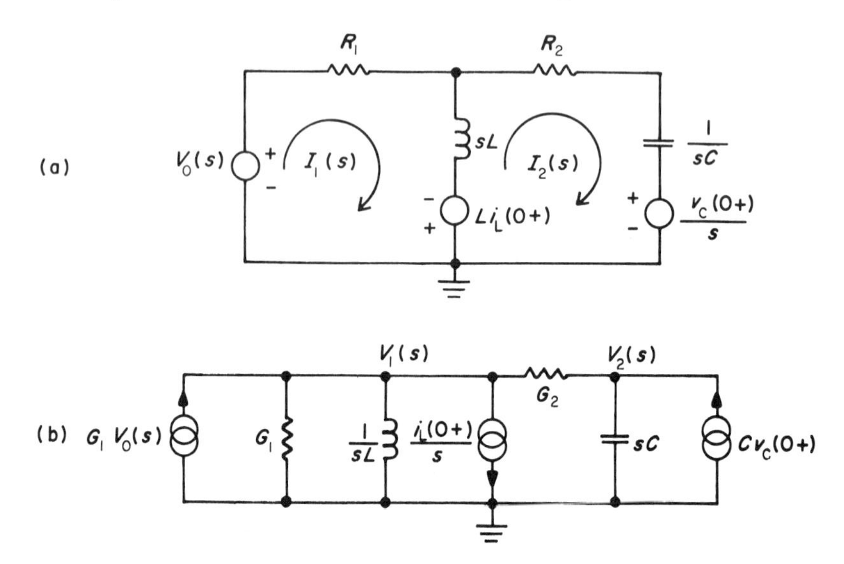

port) at which we can apply the input $x(t)$, and a (usually separate) point (or port) at which we can measure the output $y(t)$.

3.5.2.1 Unit Step Response

If the input to the system is a unit step function, i.e.,

$$x(t)=u_{-1}(t)$$

then we record the unit step response $h(t)$, i.e.,

$$y(t)=h(t)$$

An example of such an input–output pair is shown in Fig. 3.14a. For a step of arbitrary magnitude A and delay t_1 we obtain a delayed and proportionately scaled step response. Thus we write

$$x(t)=Au_{-1}(t-t_1)\overset{\text{system}}{\rightarrow}y(t)=Ah(t-t_1) \tag{3.111}$$

An important reason for the popularity of step response analysis is the fact that modern signal generators supply excellent approximations to the ideal step waveform. Of additional interest to us, in the present context, is the possibility of deriving the system response to an input of *arbitrary* waveform from the step response, as we see next.

3.5.2.2 General System Response from the Step Response

We begin by approximating the waveform of an arbitrary input signal $x(t)$ with a "staircase" generated by a sum of weighted step functions. For example, for the interval $t_2<t<t_3$ we can approximate the waveform on the left in Fig. 3.14b by the

Figure 3.14 **(a)** The step response of a system. **(b)** The approximate response to an arbitrary input derived from the step response. (See text.)

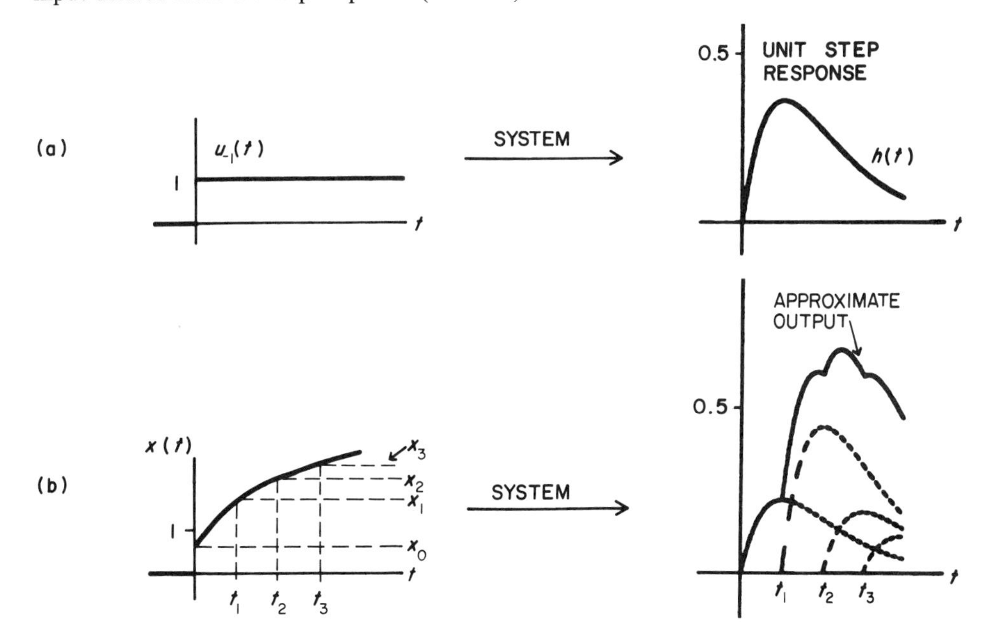

sum

$$x(t) \approx x_0 u_{-1}(t) + (x_1 - x_0) u_{-1}(t - t_1) + (x_2 - x_1) u_{-1}(t - t_2)$$

Each of the step functions in this approximation generates a step response, Eq. (3.111), so that by the superposition theorem we can write for the approximate output

$$y(t) \approx x_0 h(t) + (x_1 - x_0) h(t - t_1) + (x_2 - x_1) h(t - t_2)$$

for the interval $t_2 < t < t_3$. The total approximate response is shown as the solid curve on the right in Fig. 3.14b; the individual component responses, shown as broken traces in Fig. 3.14b, are weighted and shifted replicas of the unit step response in Fig. 3.14a.

We now generalize this result as follows. Assuming constant delay increments of duration T, we define the input increment at time $t = nT$ as

$$\Delta x(nT) = x(nT) - x[(n-1)T]$$

The corresponding staircase approximation of the input signal is then

$$x(t) \approx x(0+) u_{-1}(t) + \sum_{n=1}^{\infty} \Delta x(nT) u_{-1}(t - nT) \tag{3.112}$$

With the superposition theorem and Eqs. (3.111) and (3.112), we can, therefore, write the approximate output as

$$y(t) \approx x(0+) h(t) + \sum_{n=1}^{\infty} \Delta x(nT) h(t - nT)$$

We now introduce the difference quotient $\Delta x(nT)/T$ and rewrite the expression as

$$y(t) \approx x(0+) h(t) + \sum_{n=1}^{\infty} \frac{\Delta x(nT)}{T} h(t - nT) T$$

Proceeding from this equation, we obtain the exact equation for the output by a limit process whereby the delay interval T becomes the differential dt, the difference quotient a differential quotient, and the sum a convolution integral. As a result we can state

The zero-state response $y(t)$ to an arbitrary (but smooth for $t > 0$) input $x(t)$ is given by

$$y(t) = x(0+) h(t) + \int_0^t \frac{dx(\tau)}{d\tau} h(t - \tau)\, d\tau \tag{3.113}$$

or

$$y(t) = x(0+) h(t) + \frac{dx(t)}{dt} * h(t) \tag{3.114}$$

or

$$y(t) = \frac{d}{dt} x(t) * h(t) \tag{3.115}$$

or

$$y(t) = \mathcal{L}^{-1}\{Y(s)\} = \mathcal{L}^{-1}\{sX(s)H(s)\} \tag{3.116}$$

Equations (3.113)–(3.116) are various forms of *Duhamel's formula*. The symbolic "star product" in Eqs. (3.114) and (3.115) is the widely used notation for the convolution integral, namely,

$$f_1(t)*f_2(t)=f_2(t)*f_1(t)=\int_0^t f_1(\tau)f_2(t-\tau)\,d\tau \tag{3.117}$$

Equation (3.115) is equivalent to Eqs. (3.113) and (3.114) under assumptions which will be met in most practical cases (Doetsch, 1971, pp. 119–120).

The point of Eqs. (3.113)–(3.115) is that *an experimentally or analytically generated step response $h(t)$ can be used to compute the response to an arbitrary input by convolution*. These convolution equations characterize the system behavior as well as any other system equation. Systems may even be treated as "black boxes" of undetermined interior structure as long as they are linear, time invariant, and in the zero state (or at a stable operating point), and as long as a step response can be recorded.

In the case of an experimentally obtained step response, the convolution can be performed by a digital computer without the need to fit the step response data to an explicit mathematical expression. If a closed form of the step response is available or if it has been obtained by network analysis, it is often advantageous to use the Laplace transformation and Eq. (3.116). From Eq. (3.116) we have the transform equation

$$Y(s)=sX(s)H(s) \tag{3.118}$$

which rivals Eq. (3.73) in simplicity.

With Eqs. (3.64) and (3.65) we can rewrite Eq. (3.118) and obtain

$$Y(s)=sH(s)X(s)=\left\{\mathcal{L}\left[\frac{dh(t)}{dt}\right]+h(0+)\right\}X(s)$$

or

$$Y(s)=\mathcal{L}\left[\frac{dh(t)}{dt}\right]X(s)+h(0+)X(s)$$

or

$$y(t)=h(0+)x(t)+\int_0^t\frac{dh(\tau)}{d\tau}x(t-\tau)\,d\tau \tag{3.119}$$

Equation (3.119) is a form of Duhamel's formula which is complementary to Eq. (3.113). Again, we assume that certain conditions are met by $h(t)$ and $x(t)$, expecially that $h(t)$ is smooth for $t>0$ (see Doetsch, 1971, pp. 119–120). In the next section, we shall use this equation to derive one of the most important equations in linear system analysis.

3.5.3 The Impulse Response

We make the same assumptions about the properties of a system as stated at the beginning of Section 3.5.2. Although it may be practically more difficult to generate a good approximation of an impulse signal as opposed to a unit step, the impulse response is of greater conceptual importance than the step response. However, impulse and step response are closely related, as we shall discover next.

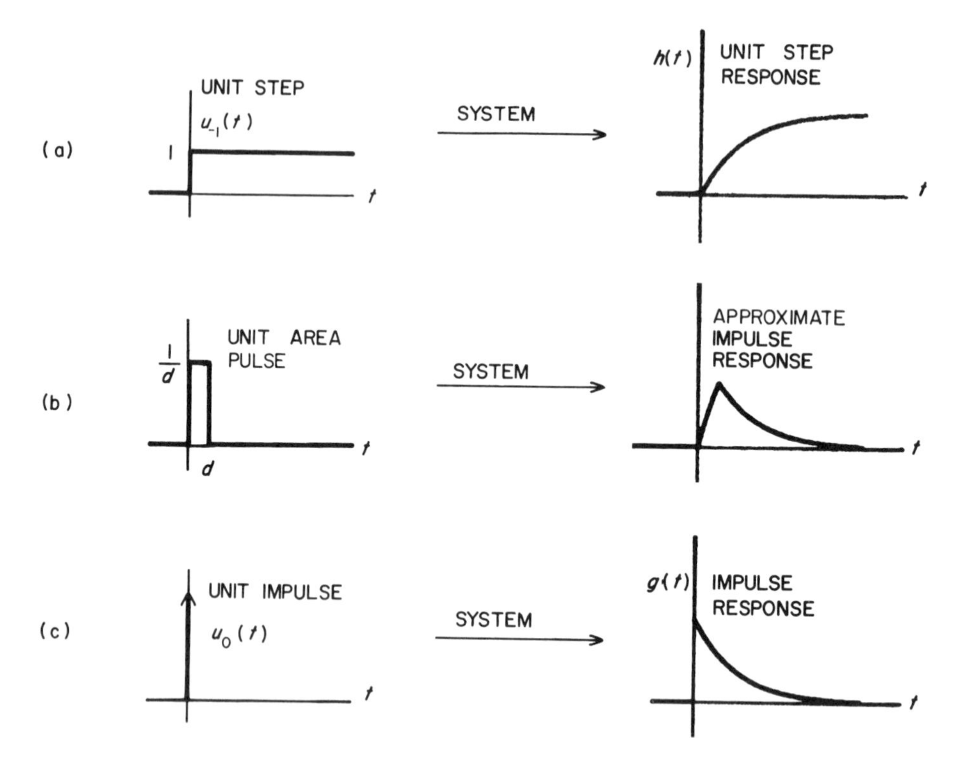

Figure 3.15 The derivation of the impulse response from the step response. (See text).

3.5.3.1 Unit Impulse Response

The unit impulse response, or *impulse response*, $g(t)$ is easily derived from the step response.

Consider Fig. 3.15. We assume that the step response of a system $h(t)$ is given according to Fig. 3.15a. As an approximation to a unit impulse input we take the *unit area pulse*, which, according to Eqs. (3.50) and (3.51), can be defined by the superposition of two step functions, i.e.,

$$\text{unit area pulse (of duration } d) = \frac{u_{-1}(t) - u_{-1}(t-d)}{d}$$

Obviously, then, the system response to this pulse input is the scaled superposition of two step responses. As the pulse represents the approximation to a unit impulse, we consider the pulse response as the *approximation of the impulse response* and write the difference quotient

$$g(t) \approx \frac{h(t) - h(t-d)}{d}$$

This approximation is shown in Fig. 3.15b. (Note that the difference of the shifted step responses must be divided by the pulse duration d in order to make this an approximation of a *unit* impulse response.)

Making d infinitesimally small converts the difference quotient representing the unit area pulse into the generalized derivative representing the unit impulse; i.e.,

$$u_0(t) = \bar{D}u_{-1}(t)$$

From this follows that *the exact impulse response is the generalized derivative of the step response* because the differentiation process is linear. In general, for a step response $h(t)$, with stepwise discontinuities $h(0+)$, $\Delta h(t_1)$, $\Delta h(t_2)$, etc., we have (see Sections 3.4.1 and 3.4.2)

$$g(t)=\bar{D}h(t)=\frac{dh(t)}{dt}+h(0+)u_0(t)$$
$$+\Delta h(t_1)u_0(t-t_1)+\Delta h(t_2)u_0(t-t_2)+\cdots \tag{3.120}$$

In the special (but frequent) case where $h(t)$ is smooth for $t>0$, and $h(0+)=0$, we can simplify Eq. (3.120) to

$$g(t)=\frac{dh(t)}{dt} \tag{3.121}$$

This is the situation in Fig. 3.15c, where the impulse response is the ordinary derivative of the step response (for $t>0$).

An equally common case is a step response with a discontinuity at the origin, $h(0+)\neq 0$, but otherwise smooth appearance for $t>0$. We then have

$$g(t)=\bar{D}h(t)=\frac{dh(t)}{dt}+h(0+)u_0(t) \tag{3.122}$$

Obviously, if the impulse response is the (generalized) derivative of the step response, it is conversely true that *the step response is the definite integral of the impulse response*. For example, integrating Eq. (3.122) gives

$$\int_0^t g(\tau)\,d\tau=\int_0^t dh(\tau)+h(0+)\int_0^t u_0(\tau)\,d\tau$$

With the assumption that the impulse function is "properly" placed at the origin so that Eq. (3.87) holds, the integration yields

$$\int_0^t g(\tau)\,d\tau=h(t)-h(0+)+h(0+)$$

or

$$h(t)=\int_0^t g(\tau)\,d\tau \tag{3.123}$$

Furthermore, Laplace transformation of Eq. (3.123) gives us the simple formulas

$$H(s)=\frac{1}{s}G(s) \tag{3.124}$$

and

$$G(s)=sH(s) \tag{3.125}$$

Equation (3.125) is, of course, the Laplace transform of the general formula, Eq. (3.120), if we apply the rules stated as Eqs. (3.66) or (3.67).

Equations (3.120) and (3.123)–(3.125) express the fundamental relations between the impulse response and the step response of a linear, time invariant system. In many instances, the less general equations (3.121) and (3.122) will suffice. In the case of experimentally obtained responses, we may use numerical differentiation or integration to obtain the corresponding impulse responses. Another common method is to fit experimental curves with sums of exponentials which have simple Laplace transforms. Equations (3.124) and (3.125) are then very handy. Unfortunately, exponential curve fits tend to be rather inaccurate.

Example 3.5. Step and impulse responses of a first-order low-pass filter.

The RC version of such a filter and its step response are given in Fig. 1.34. For a unit step we set $V_0 = 1$ V so that the unit step response is

$$h_{LP}(t) = 1 - e^{-t/T} \qquad (T = RC) \tag{3.126}$$

This is actually the response shown in Fig. 3.15a. The function is obviously smooth for $t > 0$, and we have $h_{LP}(0+) = 0$. We, therefore, determine the impulse response by applying Eq. (3.121) to Eq. (3.126):

$$g_{LP}(t) = \frac{d}{dt} h_{LP}(t) = \frac{1}{T} e^{-t/T} \tag{3.127}$$

This is the impulse response plotted in Fig. 3.15c.

Laplace transformation of Eqs. (3.126) and (3.127) yields

$$H_{LP}(s) = \frac{1}{s} - \frac{T}{1+sT} = \frac{1}{s(1+sT)} \tag{3.128}$$

and

$$G_{LP}(s) = \frac{1}{1+sT} = sH_{LP}(s) \tag{3.129}$$

Eqs. (3.128) and (3.129) obviously agree with Eqs. (3.124) and (3.125).

3.5.3.2 General System Response from the Impulse Response

Because of the definite relation between impulse response and step response, it stands to reason that the impulse response, too, can be used to predict the system output to an input of arbitrary waveform. For example, we may take Eq. (3.122) and combine it with Duhamel's formula in the form of Eq. (3.119) in order to obtain the output $y(t)$ for the arbitrary input $x(t)$:

$$y(t) = \int_0^t g(\tau) x(t-\tau)\, d\tau$$

This result can be generalized so that we present the following relations between impulse response and general response.

> The zero-state response $y(t)$ to an arbitrary input $x(t)$ is given by
>
> $$y(t) = \int_0^t g(\tau) x(t-\tau)\, d\tau = \int_0^t g(t-\tau) x(\tau)\, d\tau = g(t) * x(t) \tag{3.130}$$
>
> or
>
> $$Y(s) = G(s) X(s) \tag{3.131}$$
>
> or
>
> $$Y(j\omega) = G(j\omega) X(j\omega) \tag{3.132}$$
>
> where
>
> $g(t)$ is the *unit-impulse response*,
> $G(s) = \mathfrak{L}\{g(t)\}$ is the *system* or *transfer function*,
> $G(j\omega) = \mathfrak{F}\{g(t)\}$ is the *frequency response function*.

Equations (3.130)–(3.132) are basic relations in linear system analysis. The convolution integral, Eq. (3.130), is especially suited for numerical integration with digital computers.

It is a common practice to equate the impulse response with the weighting (or Green's) function, although that is strictly not correct (Doetsch, 1971, pp. 52–54) because of problems at the origin. Conceptually, the impulse response is the *zero-state* response of the *forced* system subjected to an impulse input, whereas the weighting function can be thought of as the *zero-input* response of the *unforced system* with zero initial conditions *except* for the highest-order initial condition being unity. [Use Eqs. (3.66), (3.73), and (3.74).]

The Laplace and Fourier transforms, Eqs. (3.131) and (3.132), of the convolution integral are especially simple; they form the basis for the material in the next two sections. (See also Eqs. (3.76) and (3.42), respectively.)

3.5.4 The System or Transfer Function

Once more we consider a linear system (Fig. 3.16a) with the properties postulated at the beginning of Section 3.5.2. We take $y(t)$ as the *zero-state response* to an input $x(t)$. By Eq. (3.131) we can write

$$Y(s)=G(s)X(s)$$

Figure 3.16 **(a)** Symbolic representation of a linear zero-state system. **(b)** Description by the system or transfer function. **(c)** Description by the frequency response function. (See text.)

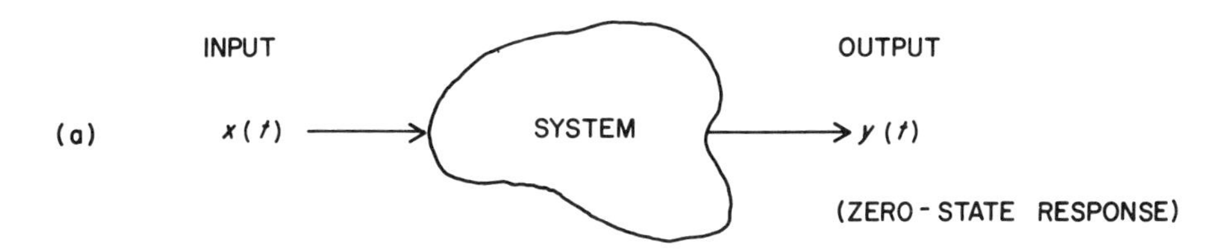

(b) $X(s) \rightarrow$ [$G(s)$] $\rightarrow Y(s)$

$Y(s) = G(s)X(s)$

(c) $X(j\omega) \rightarrow$ [$G(j\omega)$] $\rightarrow Y(j\omega)$

$Y(j\omega) = G(j\omega)X(j\omega)$

From this we define

> The System Function or Transfer Function
>
> $$G(s)=\frac{Y(s)}{X(s)}=\frac{\mathfrak{L}\{\text{zero-state response}\}}{\mathfrak{L}\{\text{input}\}} \quad (3.133)$$
>
> $$G(s)=\mathfrak{L}\{\text{impulse response}\} \quad (3.134)$$
>
> $$G(s)=\mathfrak{L}\{\text{weighting function}\} \quad (3.135)$$

It is important to keep in mind that the system or transfer function is the ratio of the *transformed* output and input variables. In contrast, the relation between output and input variables in the *un*transformed state is given by convolution integrals such as Eqs. (3.113), (3.119), or (3.130). The simple product relation of the transformed variables according to Eqs. (3.131) and (3.133) is symbolically expressed in Fig. 3.16b: The transformed input variable $X(s)$ "entering" the system "box" is multiplied with the system transfer function $G(s)$ in order to yield the transformed output variable, $Y(s)$ "leaving" the box. In the *block diagram* formulation (see Section 3.5.6) this product is implied by generally accepted convention; it is, therefore, not necessary to state it explicitly, as is done in Fig. 3.16b.

The definitions of Eqs. (3.134) and (3.135) follow naturally from the discussion in Section 3.5.3.

Example 3.6. Transfer function of the ideal differentiator

We consider an idealized device which can generate the generalized derivative of a signal. According to Eq. (3.67) its transfer function should be

$$G_D(s)=s \quad (3.136)$$

Let us now apply two signals to this device.

1. $x_1(t)=u_{-1}(t)$, the unit step. With $X_1(s)=1/s$ and Eqs. (3.131) and (3.136) we have for the transformed output

   $$Y_1(s)=G_D(s)X_1(s)=\frac{s1}{s}=1$$

   and

   $$y_1(t)=u_0(t)$$

   according to Table 3.1.
2. The switched-on cosine function.

   $$x_2(t)=\begin{cases}\cos\omega t & \text{for } t>0\\ 0 & \text{for } t<0\end{cases}$$

   $$x_2(0+)=1$$

   Laplace transformation of this signal (Table 3.1) yields

   $$X_2(s)=\frac{s}{s^2+\omega^2}$$

   The transformed output is now

   $$Y_2(s)=G_D(s)X_2(s)=\frac{s^2}{s^2+\omega^2}$$

Long division converts this transform into a proper fraction. Thus

$$Y_2(s) = 1 - \frac{\omega^2}{s^2 + \omega^2}$$

which inverts according to Table 3.1 to

$$y_2(t) = u_0(t) - \omega \sin \omega t$$

The impulse term on the right arises, of course, from the stepwise discontinuity of the switched cosine at the origin, whereas the second term represents the classical derivative of the cosine function.

We shall make extensive use of the transfer function concept in Sections 3.5.6 and 3.5.7. Next, we introduce a related function, namely, the frequency response function.

3.5.5 *The Frequency Response Function*

The *frequency response function* or *frequency response* can be defined and interpreted in many ways. As a starting point we may take Eqs. (3.42) or (3.132), which relate the Fourier transform of the system output $Y(j\omega)$ to the transformed input $X(j\omega)$. In any event we can make the following generalizations.

Definitions of the Frequency Response Function $G(j\omega)$

1.

$$G(j\omega) = |G(j\omega)|\, e^{j\Phi(\omega)} = \mathcal{F}\left\{\begin{matrix}\text{impulse response or} \\ \text{weighting function} \\ g(t)\end{matrix}\right\} \tag{3.137}$$

2. For the input $x(t) = e^{j\omega t}$ and the *steady state sinusoidal* output $y_{ss}(t)$:

$$G(j\omega) = \frac{y_{ss}(t)}{e^{j\omega t}} = \frac{\text{steady state output}}{e^{j\omega t}} \tag{3.138}$$

3. For the input $x(t) = \cos \omega_0 t$ and the *steady state sinusoidal* output $y_{ss}(t)$:

$$y_{ss}(t) = |G(j\omega_0)| \cos[\omega_0 t + \Phi(\omega_0)] \tag{3.139}$$

$$|G(j\omega_0)| = |G(\omega_0)| = \frac{\text{output amplitude}}{\text{input amplitude}} \tag{3.140}$$

(amplitude ratio, amplitude characteristic, magnitude)

$$\Phi(\omega_0) = \text{output phase} - \text{input phase} \tag{3.141}$$

(Phase, phase characteristic)

4. For a Fourier transformed input $X(j\omega)$ that is free of impulse functions:

$$G(j\omega) = \frac{\mathcal{F}\{\text{output}\}}{\mathcal{F}\{\text{input}\}} = \frac{Y(j\omega)}{X(j\omega)} \tag{3.142}$$

---

1. The frequency response function $G(j\omega)$, as defined by Eq. (3.137), is the Fourier transform analog to the transfer function $G(s)$. In fact, $G(j\omega)$ can be derived from $G(s)$ by setting $s = j\omega$ in the expression for $G(s)$, provided the real parts of the poles of $G(s)$ are negative.
2. It is very instructive to derive Eq. (3.138). In the time domain, the input and output are related by the convolution integral according to Eq. (3.130). With $x(t) = e^{j\omega t}$ we have

$$y(t) = \int_{-\infty}^{t} g(\tau) e^{j\omega(t-\tau)}\, d\tau = e^{j\omega t} \int_{-\infty}^{t} g(\tau) e^{-j\omega\tau}\, d\tau$$

(The lower limit is set at $t = -\infty$ because of the steady state input.) With the assumption that $g(t)$ is Fourier transformable, we replace the definite integral by two integrals of appropriate limits and obtain

$$y(t) = e^{j\omega t}\left[\int_{-\infty}^{\infty} g(\tau) e^{-j\omega\tau}\, d\tau - \int_{t}^{\infty} g(\tau) e^{-j\omega\tau}\, d\tau\right]$$

The first integral is the Fourier transform of the impulse response, i.e., the frequency response function $G(j\omega)$, whereas the second integral vanishes as $t \to \infty$. Therefore, if we wait long enough for the second integral to become negligibly small we shall be left with the *sinusoidal steady state* response $y_{ss}(t)$, for which we can write

$$y_{ss}(t) = G(j\omega) e^{j\omega t} \tag{3.143}$$

Note that Eq. (3.143) is, for this special case, a simple *product* rather than a complicated convolution integral. The ratio definition of the frequency response function in terms of steady state time domain variables, Eq. (3.138), follows immediately from Eq. (3.143).

Equation (3.143) is, of course, nothing but a restatement of the ac phasor method presented in Chapter 1. In this context, the application is much broader because the frequency response method can be extended to periodic nonsinusoidal and even aperiodic signals.

Equation (3.139) provides the rationale for the oldest and most important experimental method to determine the frequency response (function) of a linear system. The idea is (a) to apply a sinusoidal input [Eq. (3.139) is still valid if we replace the cosine with the sine], (b) to measure the *steady state* output amplitude and divide it by the input amplitude in order to obtain the magnitude $|G(\omega)|$, (c) to measure the phase lead ($\Phi > 0$) or lag ($\Phi < 0$) of the output sinusoid with reference to the input, (d) to repeat the whole process for all frequencies of interest, and (e) to plot the magnitude and phase data in a *Bode plot* (Chapter 1).

The whole procedure can be simplified by using an input sinusoid of unit amplitude because then the output amplitude is already equal to $|G(\omega)|$. Furthermore, it is possible to use swept frequency oscillators which automatically supply a sinusoidal signal of continuously varying frequency for the automatic recording of frequency response functions. (Obviously, the frequency change must be sufficiently slow for the system to reach a sinusoidal steady state.)

The definition given by Eq. (3.142) is the Fourier transform equivalent of the definition of the transfer function according to Eq. (3.133). The ratio expression follows naturally from Eq. (3.132) but is actually not suitable for periodic input signals that are represented by impulses in the frequency domain (Section 3.4.3). For these reasons it is safer to stay with the implicit definition of the frequency response function as stated by Eq. (3.132). This product relation is symbolically represented by the block diagram notation of Fig. 3.16c.

We now illustrate the frequency response method with a simple example.

Example 3.7. The frequency response function of a first-order low-pass filter

As a prototype we take the RC circuit of Fig. 1.33. The impulse response for this filter is, according to Eq. (3.127), with $T = RC$,

$$g(t) = \begin{cases} (1/T)e^{-t/T} & \text{for } t > 0 \\ 0 & \text{for } t < 0 \end{cases}$$

We insert this expression in Eq. (3.137) and obtain the frequency response function

$$G(j\omega) = \mathfrak{F}\{g(t)\} = \frac{1/T}{(1/T) + j\omega} = \frac{1}{1 + j\omega T}$$

An alternative is to use the ac impedances of the filter elements and compute the frequency response function as a voltage divider ratio:

$$G(j\omega) = \frac{V_2(j\omega)}{V_1(j\omega)} = \frac{1/j\omega C}{R + (1/j\omega C)} = \frac{1}{1 + j\omega RC} = \frac{1}{1 + j\omega T}$$

The result is the same as before. This second method represents the explicit application of Eq. (3.142) and the implicit use of Eq. (3.138) (phasor method).

3.5.6 Block Diagrams

In Fig. 3.16 we introduced the *block diagram* symbolism representing the multiplicative rule for the *transformed* output/input pair and the system transfer function (or frequency response function). This formalism can be extended for the description of complex systems which are made up of subsystems with specified transfer functions. *The subsystems are represented by blocks labeled with the corresponding transfer functions.*

In linear systems we can expect two basic connection patterns for the subsystems: *cascading* (the output of one block is the input of the next block) and additive *superposition*, i.e., the algebraic summation of transformed variables.

The cascade pattern of interconnection can be explained with Figs. 3.17a and 3.17b. If the first block with transfer function $G_1(s)$ in Fig. 3.17a produces the (transformed) output $X_1(s) = G_1(s)X(s)$, then the second block produces the output $X_2(s) = G_2(s)X_1(s) = G_1(s)G_2(s)X(s)$, etc. After the nth block is included we have

$$Y(s) = G_1(s)G_2(s)\cdots G_n(s)X(s)$$

The product of these n transfer functions is then the equivalent of the *single* transfer function

$$G(s) = G_1(s)G_2(s)\cdots G_n(s)$$

This equivalence is symbolically represented by the single block in Fig. 3.17b.

The underlying assumption for this consolidation of the n subsystems into a single equivalent system (representing the product of the individual transfer functions) is that *the physical connection of the subsystems does not change their transfer function*; in other words, the subsystems are not load sensitive. This might be so because of inherent properties of the subsystems or because we have inserted isolation amplifiers between the output terminals of one system and the input terminals of the subsequent system.

The unweighted linear combination (superposition) of transformed system variables is graphically indicated either by the symbol in Fig. 3.17c or by that of Fig.

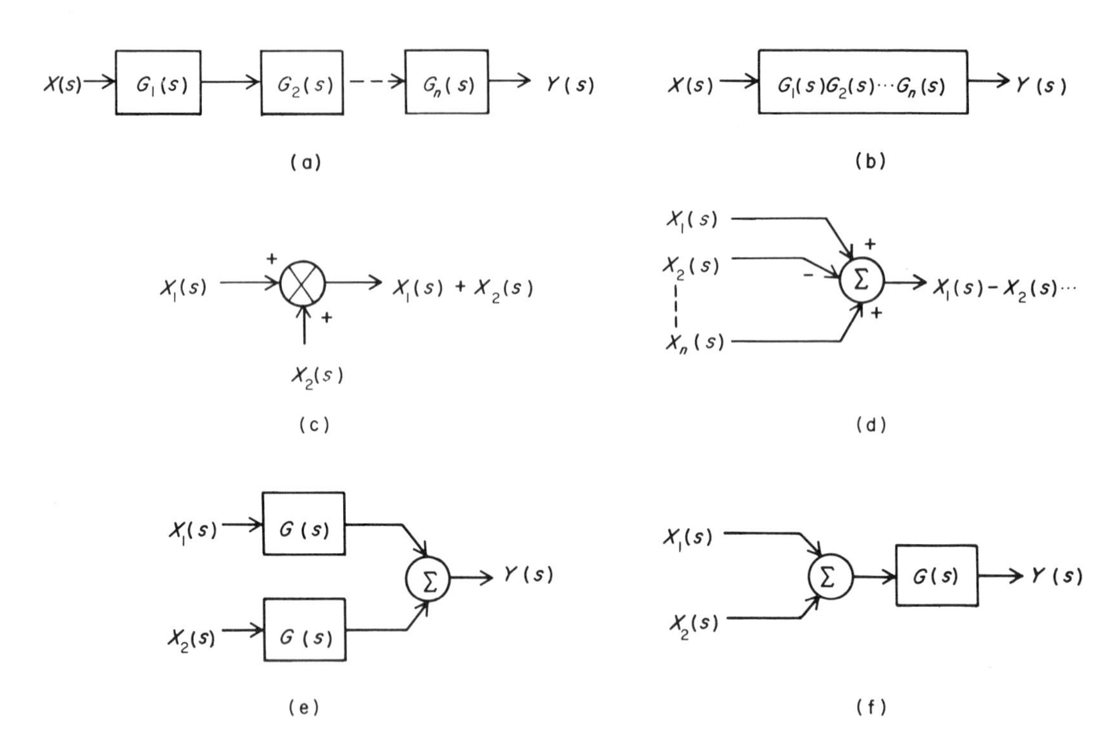

Figure 3.17 Block diagram representations. (See text.)

3.17d. If there is any question whether the superposition is additive or subtractive, appropriate signs (+ or −) are added to the summation symbols. The notation in Fig. 3.17c means $X_1(s)+X_2(s)$, and that in Fig. 3.17d means $X_1(s)-X_2(s)\cdots+X_n(s)$.

A transfer function block can often be shifted "through" a summation symbol (or "summing point"), as a comparison between Figs. 3.17e and 3.17f shows. In the first instance we have

$$Y(s)=G(s)X_1(s)+G(s)X_2(s)$$

and in the second

$$Y(s)=[X_1(s)+X_2(s)]G(s)$$

which is, of course, mathematically equivalent to the first.

Example 3.8.

In the simplified circuit diagram of Fig. 3.18a, two voltage sources are connected to two identical twin-T filters (represented here as two-ports labeled TT). The broken lines separate the circuit into three cascaded "gain" stages with the transfer functions $G_1(s)=G_{TT}(s)$, i.e., the transfer function of the filters, $G_2(s)=\mu$, i.e., the gain of the ideal difference amplifier, and $G_3=R_2/(R_1+R_2)$, i.e., the voltage divider ratio.

The functional interrelation of the five port voltages is "translated" from the complicated circuit diagram of Fig. 3.18a to the simple block diagram of Fig. 3.18b. The summing point and each block are the symbolic representation of one of the equations needed to describe the circuit.

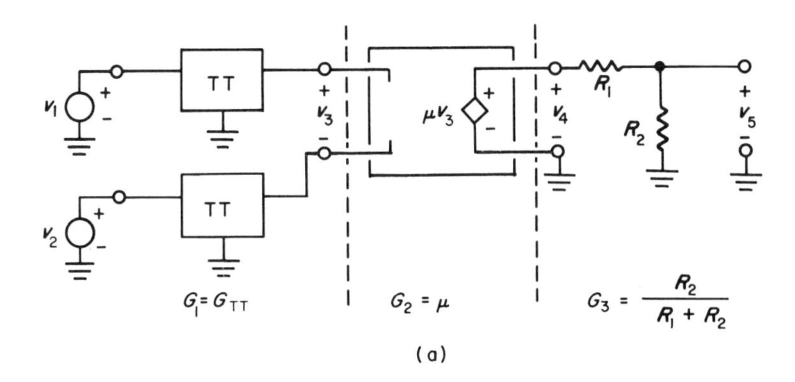

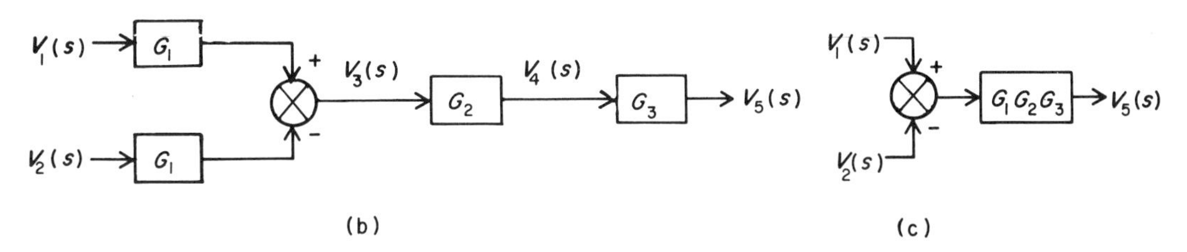

Figure 3.18 Example for the transformation of a network diagram into block diagrams.

In order to derive the dependence of the (transformed) output $V_5(s)$ on the (transformed) inputs $V_1(s)$ and $V_2(s)$, we start at the end of the block diagram and write by inspection

$$V_5(s) = G_3(s)V_4(s) = G_2(s)G_3(s)V_3(s) = G_1(s)G_2(s)G_3(s)[V_1(s) - V_2(s)]$$

The equivalent, consolidated block diagram is suggested by the last expression, as shown in Fig. 3.18c.

3.5.7 Feedback

In the circuit example of Fig. 3.18, the input signals progressed from stage to stage (or block to block). The topology of the block diagram of Fig. 3.18b, particularly, makes it clear that there are no circular or return pathways for the signal flow. Such a system is called a *feed forward* system. There are, however, arrangements where the output of a system or subsystem is "fed back" directly or in modified form to the input of an earlier stage within the system. This means that part of the signal flow must follow a circular pattern along a so-called *feedback loop*.

The presence of feedback can have a profound effect on the behavior of a system. Designers use feedback in order to (1) *stabilize system parameters* (e.g., a gain) and make them largely immune to component variations, temperature, etc., (2) *improve system parameters* (e.g., rise time, bandwidth), (3) *generate novel circuits*, (4) *introduce or suppress oscillations*, and (5) *perform analog computations* with the aid of feedback-connected operational amplifiers (Chapter 4). There are, of course, other uses of feedback.

We shall now consider the simple circuit of Fig. 3.19a in order to develop some basic properties of feedback systems. The output v_2 is fed back through a twin-T

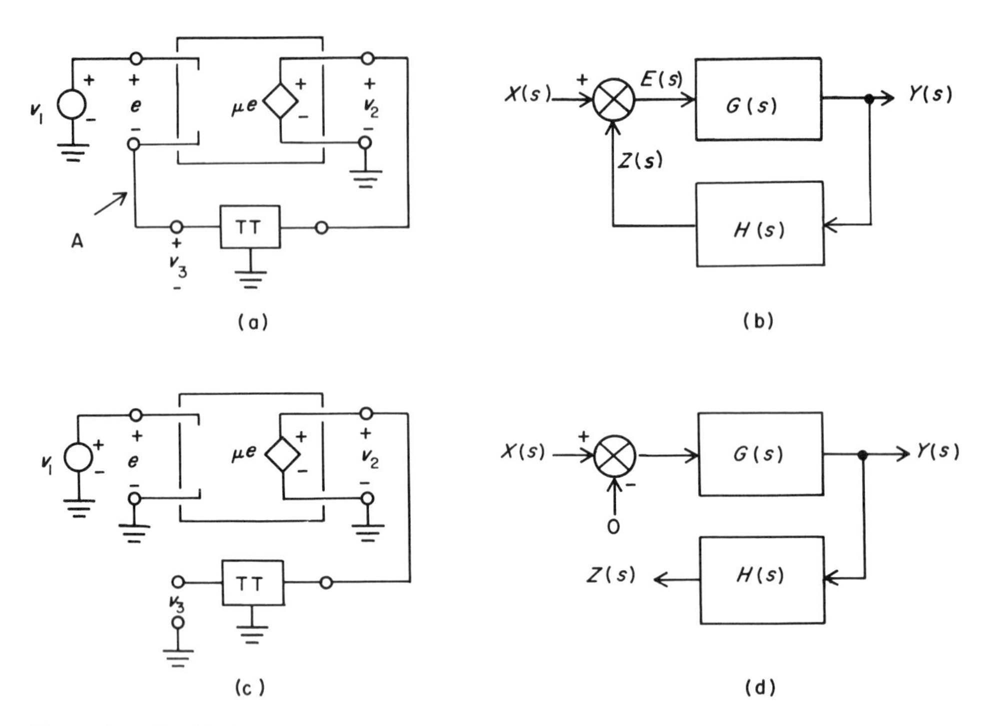

Figure 3.19 Feedback: **(a)** example of a feedback network; **(b)** canonical block diagram of a linear feedback system; **(c)** the network example with the feedback loop opened for the determination of the open loop frequency response function; **(d)** general scheme for determination of the open loop transfer function. (See text.)

notch filter to the *negative* input terminal of an idealized difference amplifier, which is represented by a controlled voltage source. The *net* voltage at the input terminals of the amplifier is the difference between the input source voltage v_1 and the feedback voltage v_3. Obviously, the feedback signal v_3 *opposes* the input signal v_1 because it is being subtracted (connection to the negative terminal). We, therefore, classify this as a case of *negative feedback*.

In order to reduce the system equations to algebraic form we use Laplace transforms. In the forward path of the circuit in Fig. 3.19a we have only the difference amplifier with the simple transfer function $G(s)=\mu$. The associated equation is

$$V_2(s)=G(s)E(s)=\mu E(s) \tag{3.144}$$

For the input of the difference amplifier we write

$$E(s)=V_1(s)-V_3(s) \tag{3.145}$$

The feedback signal is given by

$$V_3(s)=G_{TT}(s)V_2(s) \tag{3.146}$$

where $G_{TT}(s)$ is the transfer function of the twin-T filter. Next, we combine Eqs. (3.145) and (3.146) and obtain

$$E(s)=V_1(s)-G_{TT}(s)V_2(s) \tag{3.147}$$

We substitute Eq. (3.147) in Eq. (3.144) and solve for the output:

$$V_2(s)=\frac{G(s)}{1+G(s)G_{\mathrm{TT}}(s)}V_1(s) \tag{3.148}$$

This is the fundamental equation for a linear negative feedback system.

3.5.7.1 Canonical Form of a Feedback System

We now make some changes in nomenclature in order to cast the preceding equations into a more general form. Let us adopt the following designations: the (transformed) *output*, $Y(s)$; the (transformed) *input*, $X(s)$; the (transformed) *feedback signal*, $Z(s)$; the (transformed) *error*, $E(s)$; the *forward transfer function* (or forward gain), $G(s)$; the *feedback transfer function* (or feedback gain), $H(s)$; the *loop transfer function* or *open loop transfer function*, $G_{\mathrm{OL}}(s)=G(s)H(s)$; the *closed loop transfer function*, $G_{\mathrm{CL}}(s)=G(s)/[1+G(s)H(s)]$. (Note that, in this context, $H(s)$ is *not* the Laplace transform of a step response as defined in Section 3.5.2)

With the new nomenclature, Eqs. (3.144)–(3.148) are now written

$$Y(s)=G(s)E(s) \tag{3.149}$$

$$E(s)=X(s)-Z(s) \tag{3.150}$$

$$Z(s)=H(s)Y(s) \tag{3.151}$$

$$E(s)=X(s)-H(s)Y(s) \tag{3.152}$$

$$Y(s)=\frac{G(s)}{1+G(s)H(s)}X(s)=\frac{G(s)}{1+G_{\mathrm{OL}}(s)}X(s)=G_{\mathrm{CL}}(s)X(s) \tag{3.153}$$

Equations (3.149)–(3.153) completely describe the *linear*, *negative feedback* (*control*) *system*, which is shown in its *canonical form* in the block diagram of Fig. 3.19b. We could have written the first three equations of this set by inspection of Fig. 3.19b. Equation (3.149) represents the *forward path*, Eq. (3.151) the *feedback path*, Eq. (3.150) the *summing junction* (subtraction of the feedback signal from the input), and Eq. (3.153) the *overall output–input relation*.

The summing junction in Fig. 3.19b is the functional equivalent of a *comparator*: The feedback signal $Z(s)$ is compared to the input $X(s)$; whenever the two disagree, an error $E(s)=X(s)-Z(s)$ is generated. By combining Eqs. (3.149) and (3.153) we show the error to be a function of the open loop transfer function, namely,

$$E(s)=\frac{1}{1+G(s)H(s)}X(s)=\frac{1}{1+G_{\mathrm{OL}}(s)}X(s) \tag{3.154}$$

For finite $X(s)$, the error becomes very small for $|G_{\mathrm{OL}}(s)|\gg 1$.

The open loop transfer function can be recorded experimentally by breaking the feedback loop before it joins the summing junction. In the actual circuit of Fig. 3.19a this is equivalent to cutting the wire between the filter output and the negative input terminal of the amplifier (point A). With this terminal grounded, we apply the input v_1 and measure the filter output v_3 (Fig. 3.19c). The corresponding block diagram (with the general nomenclature) is shown in Fig. 3.19d, where $Z(s)$ is to be recorded as a function of $X(s)$ because the open loop transfer function is

$$G_{\mathrm{OL}}(s)=G(s)H(s)=\frac{Z(s)}{X(s)}$$

The properties of the open loop transfer function are especially important for the stability of a feedback system (Section 3.5.7.2).

The effect of feedback on a simple circuit is illustrated in the next example.

Example 3.9. Unity feedback systems and buffers.

In Fig. 3.20a, the voltage-controlled voltage source is the equivalent circuit of an idealized amplifier with gain A, input resistance R_2, and a direct feedback connection from the output to the input. Let us distinguish two cases.

Case 1: $R_1 \neq 0$, finite; $R_2 = \infty$. Under these conditions, the input current i is zero and we can write the next two equations by inspection:

$$e = v_1 - v_2 \tag{3.155}$$

$$v_2 = Ae \tag{3.156}$$

By substituting for the error voltage e in Eq. (3.156) the expression of Eq. (3.155) we obtain

$$v_2 = \frac{A}{1+A} v_1 \tag{3.157}$$

and

$$e = \frac{1}{1+A} v_1 \tag{3.158}$$

We can generalize Eqs. (3.157) and (3.158) by setting $G(s) = A$, $H(s) = 1$, $Y(s) = V_2(s)$, and $X(s) = V_1(s)$. The new equations are

$$Y(s) = \frac{G(s)}{1+G(s)} X(s) = G_{CL}(s) X(s) \tag{3.159}$$

and

$$E(s) = \frac{1}{1+G(s)} X(s) \tag{3.160}$$

The corresponding block diagram (Fig. 3.20b) is the canonical form of a *unity feedback system*. Equations (3.159) and (3.160) are obviously specific cases of Eqs. (3.153) and (3.154) for $H(s) = 1$.

Figure 3.20 Unity gain feedback: **(a)** a network example; **(b)** the canonical block diagram of a unity gain feedback system.

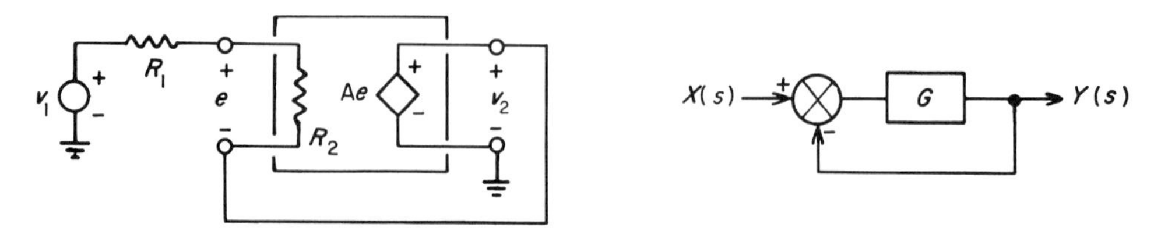

The circuit of Fig. 3.20a is most commonly used with a very large gain factor A. From Eqs. (3.157) and (3.158) we obtain the limits

$$\lim_{A \to \infty} v_2 = v_1 \tag{3.161}$$

$$\lim_{A \to \infty} e = 0 \tag{3.162}$$

Thus, we see that a very large loop gain A makes the output v_2 "follow" the input v_1 [Eq. (3.161)], *reduces the error e* to practically zero [Eq. (3.162)], and *stabilizes* the closed loop gain $G_{CL} = A/(1+A)$ at unity. Accordingly, such a feedback system with very large loop gain is called a *unity gain follower*.

Case 2: $R_1 = 0$; $R_2 \neq 0$, finite. Equations (3.155)–(3.162) and the block diagram of Fig. 3.20b are still valid. We might be tempted to assume that the input resistance loading the source v_1 is simply R_2. This is incorrect. The input current is

$$i = \frac{e}{R_2}$$

The input resistance is, therefore,

$$R_{\text{in}} = \frac{v_1}{i} = \frac{v_1 R_2}{e}$$

which with Eq. (3.158) becomes

$$R_{\text{in}} = (1+A) R_2 \tag{3.163}$$

Obviously, we can make R_{in} very large (and the input current very small) by increasing the gain A. Therefore, even with finite R_2, this circuit acts as a *buffer* and unity gain follower.

3.5.7.2 Stability

The whole point of negative feedback is to provide stabilization of a system. Imprudent application of the negative feedback technique may, however, have the opposite effect. The design criteria and strategies for stable negative feedback systems are extensively treated in the control system literature (e.g., D'Azzo and Houpis, 1966; Truxal, 1955). Here we shall only consider how instability might arise in the first place and then discuss a simple way of determining stability from frequency response data.

The corrective action of negative feedback entails subtraction of the feedback signal at the summing junction (Fig. 3.19b). This subtraction is equivalent to imposing a uniform frequency-*independent* phase shift of $-180°$ on all spectral components of the feedback signal. In general, however, this is not the only phase shift, as we see next.

The circular structure of a feedback system requires that the signal emerging from the summing junction is modified by the transfer functions in the forward path $G(s)$ and feedback path $H(s)$ before it is returned to the summing junction. The spectral components of the signal, therefore, undergo additional frequency-*dependent* phase shifts as specified by the loop frequency response function $G_{OL}(j\omega) = G(j\omega)H(j\omega)$. For a particular frequency, this additional phase shift might be $-180°$, so that the total phase shift upon circulation through the loop and summing junction is $-360°$ for this spectral component.

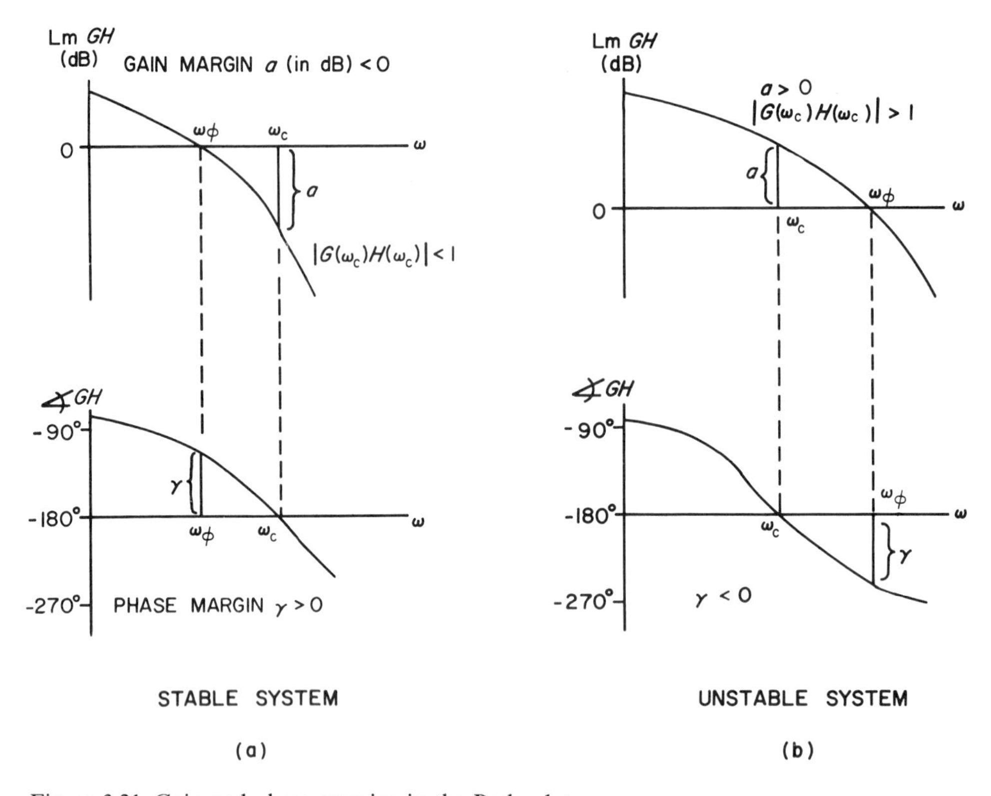

Figure 3.21 Gain and phase margins in the Bode plots.

Of course, a sinusoid shifted by $-360°$ is again in phase with its spectral counterpart in the input signal entering the summing junction. This spectral component is therefore enhanced by *positive* feedback rather than attenuated by negative feedback. If the corresponding loop gain $|GH|$ is less than unity, the net effect is an amplification of that spectral component. But for $|GH|>1$, the vicious circle of positive feedback quickly escalates the magnitude of that spectral component toward infinity. The practical result is either saturation or even destruction of the system.

In this overly simple approach to stability, two questions about the open loop frequency response function must be answered: (1) Does $G(j\omega)H(j\omega)$ have a phase lag of $-180°$ for any frequency? (2) For that frequency, is the loop gain unity or greater, i.e. $|G(j\omega)H(j\omega)|\geqslant 1$ or $\mathrm{Lm}\,GH>0$ dB?

The best way to answer these questions is to inspect the Bode plot (Chapter 1) of the open loop frequency response function. An example is given in Fig. 3.21a. In this case there is indeed a phase lag of $-180°$ at the radian frequency ω_c. But the associated gain magnitude $|G(\omega_c)H(\omega_c)|$ is much smaller than unity; in fact, the log-magnitude of the gain is $-|a|$ dB below the zero-dB (i.e., unity gain) line. In terms of stability this represents a safety margin. We, therefore, call the difference between actual gain (at ω_c) and unity gain the *gain margin* of the system. A negative gain margin ($a<0$, $|GH|<1$) indicates stability.

Conversely, we could have asked whether the phase lag amounted to $-180°$ or more at the radian frequency corresponding to unity gain. In our example of Fig. 3.21a, unity gain occurs at ω_ϕ. The corresponding phase lag is $\phi=(-180°)+\gamma$ with

$\gamma > 0$. Again, we have stability because the phase lag is less than $-180°$ for frequencies for which the gain equals or exceeds unity. By analogy with the gain margin, we can define the *phase margin* of the system as the difference between the actual phase lag at the unity gain frequency (ω_ϕ) and $-180°$. A positive phase margin ($\gamma > 0$) indicates stability.

The counterexample of an unstable system is represented by the Bode plot in Fig. 3.21b. In this case the gain margin is positive and the phase margin negative.

The primary virtue of this simple approach is that we may judge the stability of the closed loop system from the frequency response of the *open loop* system. The latter is often known beforehand or can be determined experimentally before closure of the loop according to Fig. 3.19d. Should the open loop frequency response indicate instability, then the Bode plot must be "reshaped" through the addition of compensation networks of appropriate frequency response characteristics. Various approaches to this design problem are discussed in the references mentioned at the beginning of this section.

A more rigorous approach to the problem of stability would entail a study of the poles and zeros of the closed loop transfer function $G_{CL}(s)$. The inverse Laplace transform of this function is the impulse response of the system. An *unbounded* impulse response would indicate an *unstable* system. Such would be the case if $G_{CL}(s)$ had poles in the right half of the s plane (poles with positive real parts). An impulse response in the form of a sustained sinusoidal oscillation, while bounded, can also be considered a case of instability. Such a response would arise from poles with zero real parts.

The crucial expression for the stability of a linear feedback system is the denominator term of the closed loop transfer function. This *characteristic equation* [see Eq. (3.153)] is

$$1 + G_{OL}(s) = 1 + G(s)H(s) \tag{3.164}$$

or, in terms of frequency response functions,

$$1 + G_{OL}(j\omega) = 1 + G(j\omega)H(j\omega) \tag{3.165}$$

Common methods to investigate stability are (1) the stability criterion of Routh and the root-locus method of Evans for Eq. (3.164), and (2) the Nyquist criterion for Eq. (3.165). The references cited at the beginning of this section contain detailed expositions of these methods.

Problems

1. Determine the Fourier series of the integral of the waveform in Fig. 3.2a.

2. Compute the Fourier transform of (a) $f_1(t) = e^{-\alpha t}$, (b) $f_2(t) = \alpha t e^{-\alpha t}$.

3. Prove through integration by parts that

$$\mathfrak{F}\left\{\frac{d}{dt}f(t)\right\} = j\omega F(j\omega)$$

4. Determine the Laplace transform of $f(t) = e^{-\alpha t}$ according to Eqs. (3.47) and (3.63).

5. $\mathfrak{L}\{t\} = ?$

6. Determine the Laplace transform of $f(t) = \alpha t e^{-\alpha t}$ according to Eqs. (3.47) and (3.63).

7. Determine the step response of the *RC* high-pass filter of Fig. 1.33 according to the method of Example 3.3.

8. Determine the sampling rate if at least the 9th harmonic of the signal in Fig. 3.2a needs to be recovered. Will there be a problem with aliasing?

9. In Example 3.4, determine the step response [Eq. (3.126)] from the impulse response [Eq. (3.127)] using the convolution integral of Eq. (3.130).

10. Use the transfer function method to determine the response of the *RC* low pass of Example 3.5 to

$$x(t) = u_{-1}(t)\cos \omega t$$

11. For the same filter determine the response to

$$x(t) = e^{-\alpha t}, \qquad \alpha \neq \frac{1}{T}$$

12. Repeat Problem 11 but use the frequency response function method.

References

Aseltine, J. A., *Transform Method in Linear System Analysis*, McGraw-Hill, New York, 1958. (Clear and concise)

Bendat, J. S., and A. G. Piersol, *Measurement and Analysis of Random Data*, John Wiley, New York, 1966.

Bergland, G. D., "A Guided Tour of the Fast Fourier Transform," *IEEE Spectrum*, *6*(7):41–52, July 1969. (This article has an excellent set of references)

Bracewell, R., *The Fourier Transform and Its Applications*, McGraw-Hill, New York, 1965.

Brigham, E. O., and R. E. Morrow, "The Fast Fourier Transform," *IEEE Spectrum*, *4*(12):63–70, December 1967.

D'Azzo, J. J., and C. H. Houpis, *Feedback Control System Analysis and Synthesis*, 2nd ed., McGraw-Hill, New York, 1966. (Quite suitable for beginners)

Desoer, C. A., and E. S. Kuh, *Basic Circuit Theory*, McGraw-Hill, New York 1969. (Does not discuss Fourier methods, but is excellent otherwise)

Doetsch, G. *Guide to the Applications of the Laplace and Z-Transforms*, 2nd ed., Van Nostrand Reinhold, New York, 1971. (One of the most effective presentations of the Laplace transformation by a master in the field; it also has a good set of tables)

Harris, F. J., "On the Use of Windows for Harmonic Analysis with the Discrete Fourier Transform," *Proc. IEEE*, *66*(1):51–83, January 1978.

Papoulis, A., *The Fourier Integral and Its Application*, McGraw-Hill, New York, 1962.

Ramirez, R. W., "The Fast Fourier Transform's Errors Are Predictable, Therefore Manageable," *Electronics*, *47*(12):96–102, 1974. (A simple but effective discussion of the problem)

Truxal, J. G., *Automatic Feedback Control System Synthesis*, McGraw-Hill, New York, 1955.

Van Valkenburg, M. E., *Network Analysis*, 3rd, ed., Prentice-Hall, Englewood Cliffs, NJ, 1974. (Clear and straightforward)

Wylie, C. R., Jr., *Advanced Engineering Mathematics*, McGraw-Hill, New York, 1960. (A good reference text)

Zemanian, H., *Distribution Theory and Transform Analysis*, McGraw-Hill, New York, 1965.

4

Analog Design Using Integrated Circuits

Paul B. Brown

In this chapter we shall discuss the use of the most recently introduced electronics component, the integrated circuit (IC), in analog design. In the next chapter we shall discuss digital design. The distinction between analog and digital is a clear-cut one which separates two very different kinds of design problems and should be clarified briefly at this point.

Analog signals are those signals which are intended to be continuously variable, possessing an infinite number of stable states over a finite voltage range. Examples of such signals include audio signals such as those derived from microphones and applied to loudspeakers, radio signals, and the outputs of such transducers as thermocouples used in electronic temperature measurement. *Digital signals*, on the other hand, are intended to be discretely variable; that is, they possess only a finite number of stable states, although in the real world they must pass through all intermediate values as they switch from one state to another. Such signals include the signals derived from thermostats (temperature-controlled switches) and those used to drive relays; they include the signals to the different light-emitting diodes in an LED display on a digital wristwatch; and they include the on–off signals in the circuitry within digital wristwatches, digital calculators, and digital computers. Almost all digital signals used today are *binary*: they possess only two stable states.

Analog and digital circuits are those which process analog and digital signals. They differ fundamentally in their design. Analog circuits must be capable of producing continuously variable (but not necessarily linear) outputs, often as a function of continuously variable inputs. Some analog devices, such as some types of oscillators, have no inputs. Digital circuits are specifically designed to produce only a finite number of stable outputs (two in the case of binary devices), regardless of their inputs. A third class of circuits is used to couple analog and digital circuits. We shall discuss these in Chapter 7.

4.1 Introduction to Integrated Circuits

The invention and subsequent use of the IC has already revolutionized electronics design as much as did the invention of the transistor. ICs, each consisting of as many as a thousand or more transistors, diodes, and resistors, have resulted in a miniaturization of circuit size by factors of as much as thousands relative to similar circuits built with separate components, such as transistors, diodes, and resistors, which are referred to in this context as *discrete devices*. The use of ICs has directly resulted in a number of other advantages which will become apparent as we describe them in more detail. Perhaps the most important one from the standpoint of the electronics designer, and therefore for readers of this text, is the following:

> Integrated circuits perform easily understood functions, and have simple characteristics, allowing the designer to treat them as "black boxes" that can be interconnected to perform a wide variety of useful functions without any understanding of their internal mechanisms.

That is, the designer has available to him devices which will amplify, perform analog and digital computations, or any of a number of other functions, and which possess very simple rules for their use. These devices can be treated as "black boxes" by the designer: all the design complexity inside them can be ignored, and he needs only to apply inputs and use their outputs. In computer terms, he has at his disposal a "higher-level language" for design than he would using discrete components. Thus, for example, he can buy, in a black box, a differential amplifier with gain adjustable from 1 to 1000, of the desired bandwidth: he need not concern himself with designing it to be linear, to have a flat frequency response, or to be compensated for temperature. All that is taken care of in the design of the IC.

Integrated circuits are manufactured by a series of steps which result in the simultaneous fabrication of transistors, diodes, resistors, conductors, and insulators, all on the same substrate. Although somewhat more complicated than transistor manufacture, the principles are essentially the same. Thousands of components composing a circuit can fit on an area of silicon a fraction of a centimeter square, and a number of such circuits, sometimes hundreds, can be manufactured simultaneously on a single silicon wafer. Then the wafer is cut into "chips," each containing one circuit. These chips are bonded to a carrier with connecting pins on it, and connections are made between the pins and the circuit. The package is sealed, and a completed IC results. The procedure of manufacturing an IC is only a few times more expensive than that for a transistor, yet it would take many transistors to make one: hence, *an IC is slightly more expensive than a transistor, but usually much less expensive than the total cost of the discrete components which would be needed to build a similar circuit*. Different techniques are used to manufacture MOS and bipolar ICs, as is the case with the corresponding transistor types. Although it is possible to manufacture them on the same chip, most ICs which use both types of technology use "hybrids," for which the MOS components and bipolar components are on separate chips, both mounted in the same package and wired to each other as well as to the connector pins.

The resulting package, in any case, is mechanically durable and hermetically sealed. The most commonly used packages in this volume are the circular metal can and the *dual inline package* (DIP).

We have already mentioned several advantages of integrated circuits: low cost, durability, dependability, and the possibility of black box design. Another set of advantages accrues from the physical nature of these devices. Their small size, of course, permits miniaturization from tens to thousands of times greater than with transistors. Modularity of construction is enhanced; servicing is easier, both in terms of diagnosis of faults and their repair. Circuit reliability is enhanced by virtue of a substantial reduction of the number of connections made in construction of an IC-based device: although an IC has more connectors than a transistor, it contains many transistors, all of which would have to be mechanically and electrically connected with discrete design (design using discrete transistors and diodes). The interconnections within an IC are among the most reliable in existence, and are mechanically protected from the environment by the IC case.

One complexity in design realization does arise from this miniaturization, however. The increased density of signals per unit area of circuits results in an increased density of interconnections, requiring more advanced interconnection schemes. As a consequence, printed circuit and connector technology have had to keep pace, resulting in such developments as multi-layered printed circuit boards and improved connector designs. Ironically, the resulting complexity of interconnection has resulted in printed circuit boards and connectors which are often substantially more expensive than the integrated circuits. Whereas in the past the active components (transistors) often constituted the major construction expense for a device, now the active components (ICs) are usually the least expensive. Nevertheless, most circuits can be built for a very small fraction of the cost which would be incurred using discrete components.

One last advantage of these devices should be mentioned. By using them appropriately, variations in their properties can be made to have negligible effect on circuit performance. Thus, variations in gain, dc offset, and bandwidth of an IC as a function of temperature or age, or variations from one IC to another, can be rendered unimportant in the performance of the circuit as a whole. How this is possible will be explained in subsequent sections.

4.2 Introduction to Operational Amplifiers and the Negative Feedback Configuration

By far the most versatile and widely used analog integrated circuit is the operational amplifier. It was the first widely used analog IC introduced commercially, and now several varieties are available which cover the majority of applications for less than a dollar. In an age of inflation, electronics prices have held steady or even dropped largely because of a progressive decrease in the cost of integrated circuits such as the operational amplifier.

We can best define an operational amplifier by specifying the characteristics of an ideal device. Such an operational amplifier (op amp) is a differential amplifier with single-ended output which has infinite gain, infinite common mode rejection, infinite bandwidth, zero delay from input to output, infinite input impedance, zero output impedance, and no noise. Of course, no such device can be realized, but even an inexpensive op amp such as the 741, the one most commonly referred to in this book, has specifications which are usually an adequate approximation of these characteristics.

The frequency response function, a particularly important specification for op amps, is illustrated for the 741 in Fig. 4.1. Note that the 741 has a frequency response which is flat from dc to a corner frequency f_c of 10 Hz, at almost 100 dB (100,000) with a 6-dB/octave roll-off. The phase lags in such a way as to ensure stability. The gain function falls to unity at about 1 MHz.

Operational amplifiers are so named because, in conjunction with feedthrough and feedback elements, they are used to operate on an input signal: e.g., their modification of the signal can be specified by an operator, which is determined by the nature of the feedthrough and feedback circuits and is relatively insensitive to variation among op amps. The operator may be a linear one such as amplification, filtering, or analog computation, or it may be a nonlinear one such as clipping or rectification. We shall present the most commonly used operators and their realizations in this chapter.

The configuration most commonly used with op amps is the negative feedback configuration, introduced in Chapter 3. The op amp compares the input and output signals, as modified by interposed feedthrough and feedback circuits. We can describe the operation at the intuitive level by saying the op amp is connected in such a way as to minimize the difference between the feedthrough and feedback signals. By modifying the relations between these signals with appropriate

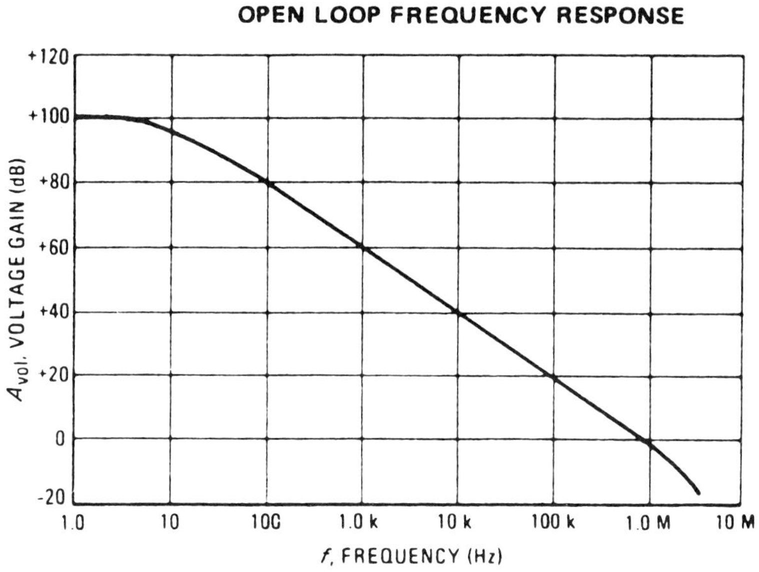

OPEN LOOP FREQUENCY RESPONSE
+120
+100
+80
+60
+40
+20
0
-20
A vol. VOLTAGE GAIN (dB)
1.0
10
100
1.0 k
10 k
100 k
1.0 M
10 M
f, FREQUENCY (Hz)

Figure 4.1 Frequency response of a 741 op amp.

feedthrough and feedback circuits, we can thus control the relation between input and output signals.

Figure 4.2 illustrates this point. The op amp is symbolized by the triangle with positive and negative inputs e^- and e^+ and output e_2. The boxes represent arbitrary circuits, linear or nonlinear. The input signal e_1 is transformed by the feedthrough circuit into some signal e_1', and the output e_2 is transformed by the feedback circuit into some signal e_2'.

The amplifier itself amplifies the difference between positive and negative inputs by its gain m, which we will assume for now to be infinite over an infinite bandwidth (no filtering of the signal), with zero delay. Therefore,

$$e_2 = m(e_1' - e_2') \tag{4.1}$$

Substituting e^+ and e^- for e_1' and e_2', respectively,

$$e^+ - e^- = \frac{e_2}{m} \tag{4.2}$$

Figure 4.2 An op amp with positive feedthrough and negative feedback.

e1
feedthrough
e1'
e +
e -
e2
feedback
e2'

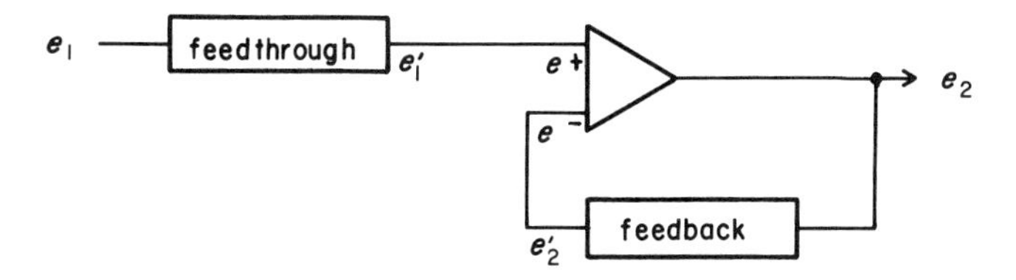

This means that, if m is very large, or approaches infinity,

$$\lim_{m \to \infty} e^{+} = \frac{e^{2}}{m} + e^{-} = e^{-} \tag{4.3}$$

This is a very useful result: for very large m, the two inputs are virtually identical. The virtual identity will hold as long as e_2 does not exceed the output limits of the op amp. The important consequences of this virtual identity will become clearer as we proceed, but we can say now that it accounts for the most important aspect of op amp behavior, namely:

> An operational amplifier in the negative feedback configuration can be used to perform operations on signals that are predominantly determined by the virtual identity of its inputs and by feedthrough and feedback elements. Temperature- and age-caused variations in gain of the operational amplifier are rendered insignificant in the performance of such configurations.

The output is related to the input by

$$m' = \frac{e_2}{e_1} \tag{4.4}$$

This is the *closed loop gain*. The op amp gain m is called the *open loop gain*.

The requirement that m be very large is easily met. Even relatively inexpensive op amps have gains of 10^4–10^6. Only when m' itself must be high will any significant error develop; this error is readily calculated. It is called the *closed loop gain error*, and is the ratio of closed loop gain to open loop gain, m'/m. Its reciprocal, m/m', is called the *gain margin*.

The following, a derivation of the above results for the frequency domain, is for the more advanced reader.

Figure 4.3 illustrates the general case of Fig. 4.2, expressed in terms of frequency response. Here the input signal is $E_1(j\omega)$, the feedthrough frequency response function is $G_1(j\omega)$, the output is $E_2(j\omega)$, and the feedback frequency response function is $G_2(j\omega)$. The two inputs to the op amp, $E^+(j\omega)$ and $E^-(j\omega)$, must be

$$E^{+}(j\omega) = E_1(j\omega) \cdot G_1(j\omega) \tag{4.5}$$

Figure 4.3 Block diagram for analysis of frequency response of positive feedthrough, negative feedback configuration.

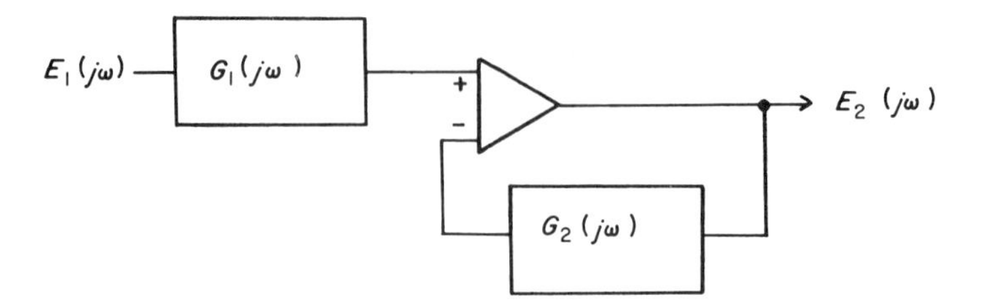

and

$$E^{-}(j\omega)=E_2(j\omega)\cdot G_2(j\omega) \tag{4.6}$$

If we specify the frequency response of the amplifier as $M(j\omega)$, we can deduce the relation of $E_2(j\omega)$ to $E_1(j\omega)$. First, we describe $E_2(j\omega)$ in terms of $E^{-}(j\omega)$, $E^{+}(j\omega)$, and $M(j\omega)$:

$$E_2(j\omega)=M(j\omega)\cdot\left[E^{+}(j\omega)-E^{-}(j\omega)\right] \tag{4.7}$$

That is, the output is equal to the difference between the inputs times the gain, for all frequency components. Substituting for $E^{+}(j\omega)$ and $E^{-}(j\omega)$ from (4.5) and (4.6) in (4.7) gives

$$E_2(j\omega)=M(j\omega)\cdot\{[E_1(j\omega)\cdot G_1(j\omega)]-[E_2(j\omega)\cdot G_2(j\omega)]\} \tag{4.8}$$

Solving for $E_2(j\omega)$ gives

$$E_2(j\omega)\cdot\{1+[M(j\omega)\cdot G_2(j\omega)]\}=M(j\omega)\cdot E_1(j\omega)\cdot G_1(j\omega)$$

$$E_2(j\omega)=\frac{G_1(j\omega)}{[1/M(j\omega)]+G_2(j\omega)}E_1(j\omega) \tag{4.9}$$

As $M(j\omega)$ becomes very large,

$$\lim_{M(j\omega)\to\infty} E_2(j\omega)=\lim_{M(j\omega)\to\infty}\frac{G_1(j\omega)E_1(j\omega)}{[1/M(j\omega)]+G_2(j\omega)}=\frac{G_1(j\omega)}{G_2(j\omega)}E_1(j\omega) \tag{4.10}$$

That is, the output frequency composition is related to the input frequency composition as the ratio of feedthrough to feedback frequency responses, as long as

$$\frac{1}{M(j\omega)}\ll M'(j\omega) \tag{4.11}$$

This is usually easily achieved, since $10^4<M<10^6$ for the majority of op amps over a bandwidth from dc to 10^4 Hz. A number of op amps with even higher gains and bandwidths are available, but are only rarely needed if good design practices are adhered to.

The ratio

$$\frac{E_2(j\omega)}{E_1(j\omega)}=\frac{G_1(j\omega)}{G_2(j\omega)}=M'(j\omega) \tag{4.12}$$

defines the *closed loop frequency response* of the circuit. The op amp gain $M(j\omega)$ is referred to as the *open loop frequency response*. Note that, as long as the nonidealities of the op amp are negligible, the closed loop frequency response is determined entirely by the feedthrough and feedback elements. In the majority of circuits these are composed of passive elements (R, L, C), which are inherently quite linear and show only minor variations with temperature and aging. Thus the designer can virtually eliminate the problems of temperature and component aging that plague discrete design.

The circuit of Fig. 4.4 shows an important variation on the general circuit of Figs. 4.2 and 4.3, wherein $e^{+}=e_1$ and $e^{-}=e_2$.

From Eq. (4.3)

$$e^{+}=e^{-}, \qquad \text{so that} \qquad e_2=e_1, \tag{4.13}$$

that is,

$$m'=\frac{e_2}{e_1}=1 \tag{4.14}$$

This circuit is called a *unity gain follower*, or *follower* for short. In terms of frequency

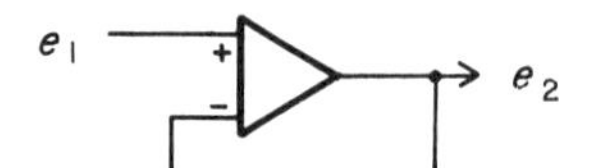

Figure 4.4 Direct connection of input signal to positive input, output signal to negative input.

domain analysis, $G_1(j\omega)=G_2(j\omega)=1$, so from Eq. (4.9)

$$\frac{e_2}{e_1}=\frac{G_1(j\omega)}{[1/M(j\omega)]+G_2(j\omega)}=\frac{1}{1+[1/M(j\omega)]}=\frac{M(j\omega)}{M(j\omega)+1} \tag{4.15}$$

For frequencies at which $m \gg 1$, gain is essentially unity. But for frequencies at which $M(j\omega)$ is not much greater than 1, the gain error becomes significant. The closed loop gain error is equal to the difference between the theoretical gain and the actual closed loop gain, approximately equal to the reciprocal of the open loop gain function $M(j\omega)$. Typically this is 10^{-3}–10^{-6} up to at least 10 kHz. Op amps with high bandwidths can have such a gain error up to several megahertz.

The frequency response of the unity gain follower configuration is illustrated in Fig. 4.5, for ideal and real (741) op amps, along with open loop frequency responses for both types of op amps.

The unity gain follower has a high input impedance, typically greater than 10^5 Ω, and in special FET input op amps, 10^{12} Ω or greater; the output impedance is low, typically less than 100 Ω, and in power op amps it can be reduced to a fraction of an ohm. Thus, the unity gain follower is an *impedance transformer*. Such a configuration is used to lower the output impedance of a circuit with high output impedance or to raise the input impedance of a circuit with low input impedance.

Figure 4.6 illustrates the potentiometric feedback configuration. A voltage divider is inserted into the feedback circuit, so that only a fraction of e_2 is fed back as e^-. Such a circuit has a net gain greater than unity:

$$e^+=e^-=e_1=e_2\frac{R_2}{R_1+R_2}$$

Figure 4.5 Frequency response of open loop op amp and for the unity gain follower of Fig. 4.4.

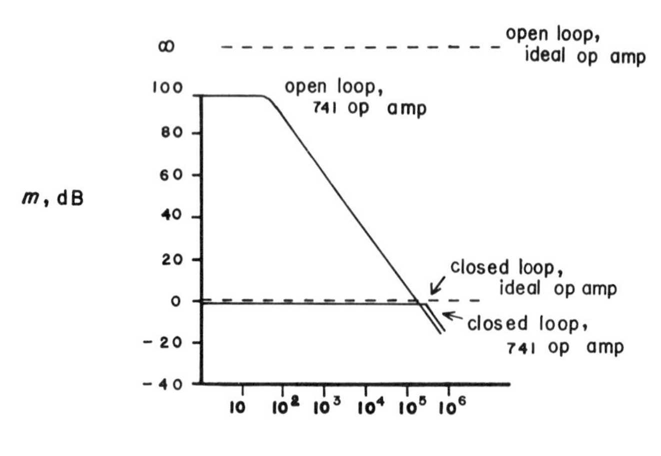

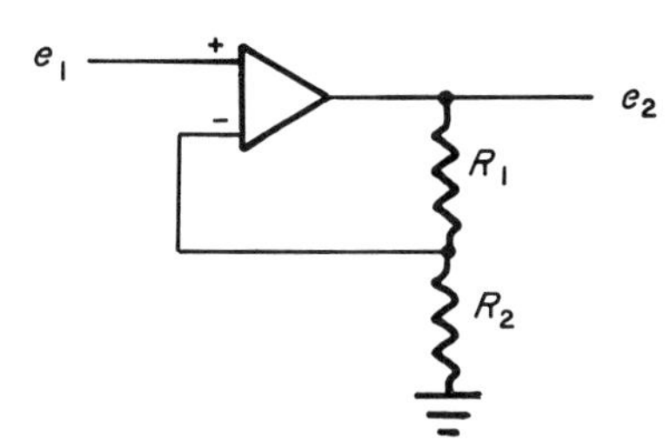

Figure 4.6 Potentiometric feedback.

Solving for m'

$$m' = \frac{e_2}{e_1} = \frac{R_1 + R_2}{R_2} \tag{4.16}$$

which is necessarily greater than or equal to unity for all values of R_1. This circuit preserves the high input and low output impedance properties of the unity gain follower and provides greater than unity gain. The voltage divider may be a potentiometer, in which gain m' can be varied from 1 to m, the open loop gain of the first op amp.

In Fig. 4.7, the potentiometric feedback circuit is used as the feedback element in another circuit. Here,

$$e_3 = e_2 \frac{R_1 + R_2}{R_2} \tag{4.17}$$

We know that the inputs to the first op amp are virtually identical, i.e., $e_1 = e_3$. Substituting e_1 for e_3 in Eq. (4.17),

$$e_1 = e_2 \frac{R_1 + R_2}{R_2}$$

Solving for e_2 gives

$$e_2 = e_1 \frac{R_2}{R_1 + R_2} \tag{4.18}$$

Thus, the circuit gain m' is *less* than unity. It retains the impedance transforming properties of the unity gain follower since the input impedance is still that of the first op amp's positive input, and the output impedance is still that of the first op amp's output. This circuit is rarely used because the same function can be achieved

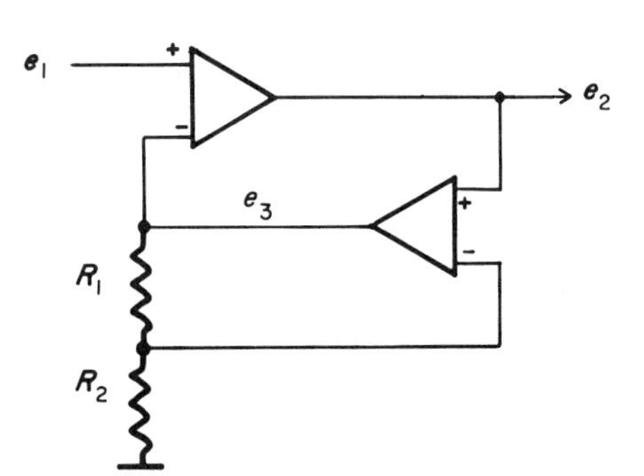

Figure 4.7 Gain element in feedback circuit.

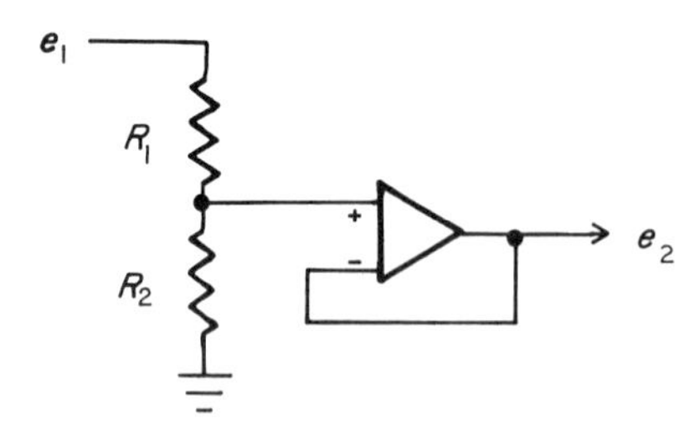

Figure 4.8 Potentiometric feedthrough to unity gain follower.

by the potentiometric feedthrough circuit of Fig. 4.8, where

$$e_2 = e^+ = e_1 \frac{R_2}{R_1 + R_2} \tag{4.19}$$

which is identical to the circuit of Fig. 4.7. Some of the impedance-reducing properties are lost, however; the resistance R_2 must be substantially lower than the op amp input impedance, yet $R_1 + R_2$ must be substantially higher than the output impedance of the source of signal e_1. This is generally not a serious problem, because the R_1–R_2 voltage divider can be trimmed with a potentiometer to achieve the desired gain to any practical degree of precision, compensating for decreased voltage transfer. Alternatively, an input impedance transformer can be used to buffer the input signal to the divider.

The other major class of negative feedback op amp circuits is illustrated in Fig. 4.9. Here the feedthrough is to the negative input of the op amp, as is the feedback. We can calculate the output e_2 as follows: First, e_2 is the difference of the op amp inputs times the gain:

$$e_2 = m(e^+ - e^-)$$

e^+ is grounded:

$$e^+ = 0$$

Substituting gives

$$e_2 = -me^-$$

or

$$\frac{e_2}{m} = -e^-$$

For m very large,

$$\lim_{m \to \infty} e^- = \frac{-e_2}{m} = 0 \tag{4.20}$$

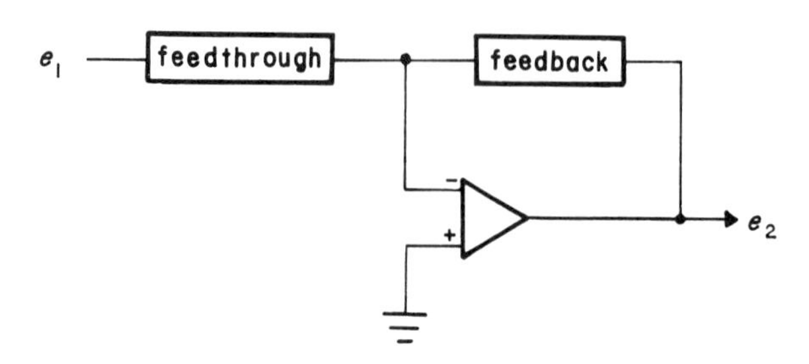

Figure 4.9 Feedthrough to the negative input, feedback to the negative input, positive input grounded.

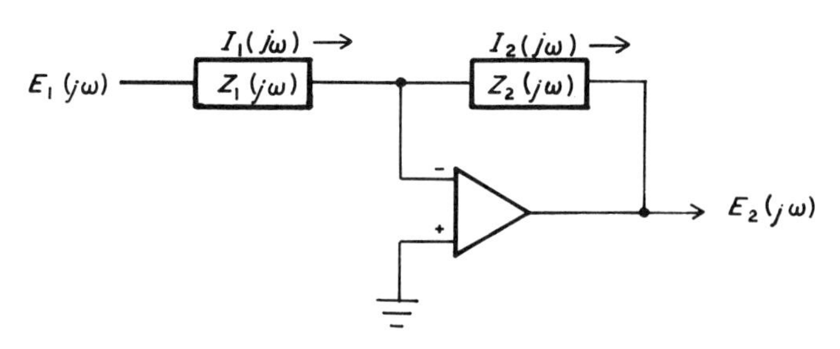

Figure 4.10 Negative feedthrough and feedback expressed in terms of frequency response.

That is, e^- is very close to ground, as long as the operating limits (voltage, gain, and frequency) of the op amp are not approached. Thus, the negative input is once again virtually identical to the positive input and is said to be a *virtual ground*. The ease of design with the assumption of a virtual ground will be illustrated in a few specific cases which are of great general utility. First, however, we shall derive the frequency response function for the advanced reader. Others may skip the following section without loss of continuity.

The circuit of Fig. 4.10 illustrates the general case for negative feedthrough and feedback. The positive and negative inputs are both zero since the positive input is grounded and the negative input is virtually identical (virtual ground):

$$E^+(j\omega) = E^-(j\omega) = 0$$

The feedthrough and feedback circuits therefore constitute impedances connecting $E_1(j\omega)$ and $E_2(j\omega)$ through a node whose voltage is always zero. Since the amplifier draws no current,

$$I_1(j\omega) = [E_1(j\omega) - E^-(j\omega)]/Z_1(j\omega) = E_1(j\omega)/Z_1(j\omega)$$

$$I_2(j\omega) = [E_2(j\omega) - E^-(j\omega)]/Z_2(j\omega) = E_2(j\omega)/Z_2(j\omega)$$

and

$$I_1(j\omega) + I_2(j\omega) = 0$$

Substituting and solving for $E_2(j\omega)$ give

$$E_2(j\omega) = -E_1(j\omega)\frac{Z_2(j\omega)}{Z_1(j\omega)} \tag{4.21}$$

Figure 4.11 illustrates the resistive negative feedback configuration. Feedthrough and feedback elements are both resistors. Assuming op amp input current to be zero (infinite input impedance) and applying Kirchhoff's junction rule give

$$i_{R_1} + i_{R_2} = 0 \tag{4.22}$$

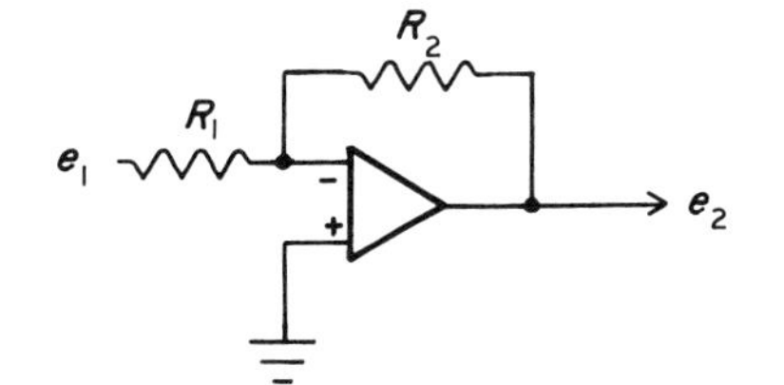

Figure 4.11 Resistive feedthrough and feedback in the circuits of Figs. 4.9 and 4.10.

That is, the net current into or out of the node at the negative input must be zero. We know the two currents, because we know the voltage drop across each resistor (assuming $e^- = 0$):

$$i_{R_1} = (e_1 - e^-)/R_1 = e_1/R_1$$

$$i_{R_2} = (e_2 - e^-)/R_1 = e_2/R_1$$

Substituting for i_{R_1} and i_{R_2} in Eq. (4.22) gives

$$e_1/R_1 + e_2/R_2 = 0$$

Solving for e_2 gives

$$e_2 = -e\frac{R_2}{R_1} \tag{4.23}$$

The closed loop gain is therefore

$$m' = \frac{e_2}{e_1} = -\frac{R_2}{R_1} \tag{4.24}$$

Note the simplicity of derivation given virtual identity of inputs, especially in the case of a virtual ground. In only a few steps, it is possible to solve for the output given feedthrough and feedback. Consider the following example.

Example 4.1. Negative resistive feedback with a gain of $-10\times$.

From Eq. (4.24),

$$-\frac{R_2}{R_1} = m' = -10$$

so that $R_2 = 10R_1$. Assuming an input impedance of 100 kΩ and an output impedance of 100 Ω, it should be safe to use $R_1 = 500$ Ω and $R_2 = 5$ kΩ, for an error of less than 10%.

We shall use a number of other circuits employing virtual ground at the negative input in this volume.

4.3 Introduction to Analog Computation

4.3.1 Addition

An adder is defined as a device which performs the sum

$$e_2 = e_{1a} + e_{2a} + e_{3a} + \cdots + e_{na}$$

where e_2 is the output and $e_{1a}, \ldots,$ are inputs.

A simple adder circuit is diagrammed in Fig. 4.12. Each of n inputs is fed through an identical resistor R. From Kirchhoff's junction rule,

$$i_{1a} + i_{2a} + i_{3a} \cdots + i_{na} = 0$$

Also,

$$e^+ = e^- = e_2$$

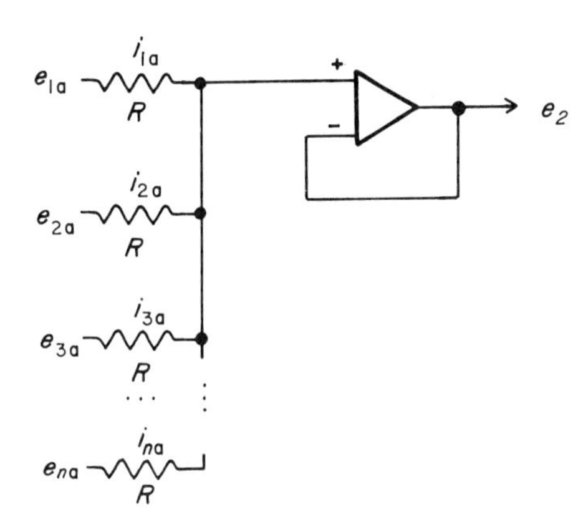

Figure 4.12 An adder (summing amplifier).

Substituting iR drops gives

$$(e_{1a}-e_2)/R+(e_{2a}-e_2)/R+(e_{3a}-e_2)/R+\cdots+(e_{na}-e_2)/R=0$$

Solving for e_2 gives

$$e_2=\frac{1}{n}\sum_{i=1}^{n} e_{ia} \tag{4.25}$$

That is, the output is the sum of the inputs $\Sigma_{i=1}^{n} e_{ia}$ divided by the number of inputs n.

If potentiometric feedback is used, we can multiply the output by m', as in Fig. 4.13, and e_2 becomes

$$e_2=\frac{m'}{n}\sum_{i=1}^{n} e_{ia}$$

where m' is the gain of the potentiometric feedback circuit. If the resistor values in the feedback circuit are $(n-1)R_1$ and R_1, as illustrated in Fig. 4.13, then using Eq. (4.16), we can calculate the gain:

$$m'=\frac{[(n-1)(R_1)]+R_1}{R_1}=n$$

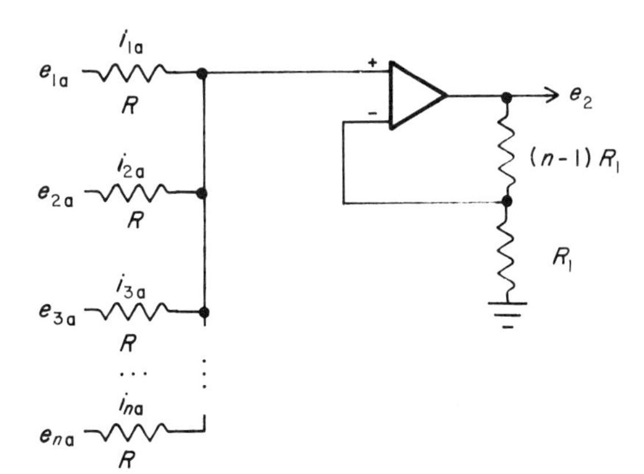

Figure 4.13 Use of potentiometric feedback to remove scaling factor n, due to n resistive inputs.

so that

$$e_2 = \sum_{i=1}^{n} e_{ia} \tag{4.26}$$

and we have eliminated the $1/n$ scaling factor from Eq. (4.25). In fact, by adjusting m', we can realize any scaling factor we like. Note that adding or removing resistive inputs changes the scaling factor, unless the resistances in the potentiometric feedback are changed. This is less of a disadvantage than it may seem, because in real situations, the designer incorporates the maximum number of inputs he anticipates ever using, and grounds all inputs not currently in use, effectively adding zero at each unused input.

4.3.2 Subtraction

Figure 4.14 illustrates a subtractor, where n inputs are all subtracted from zero. Here, we know $e^- = 0$, since it is a virtual ground. Therefore, the current through each resistor, including the feedback resistor, can be expressed directly, and all the currents into the negative input sum to zero (Kirchhoff's junction rule):

$$0 = i_{1s} + i_{2s} + i_{3s} + \cdots + i_{ns} + i_f$$
$$= e_{1s}/R + e_{2s}/R + e_{3s}/R + \cdots + e_{ns}/R + e_2/R_f$$

Solving for e_2 gives

$$e_2 = -\frac{R_f}{R} \sum_{i=1}^{m} e_{is} \tag{4.27}$$

The scaling factor R_f/R can be adjusted to unity by setting $R_f = R$.

4.3.3 Adder–Subtractor

A general case which subsumes all adders and subtractors, and all combinations of both, is illustrated in Fig. 4.15. Note that each resistor is uniquely assigned; there is no constraint such as was used above, forcing all input resistances to the same value. There are n adder inputs $e_{1a}, \ldots, e_{na}$, and m subtractor inputs $e_{1s}, \ldots, e_{ms}$. The feedback resistor is R_f.

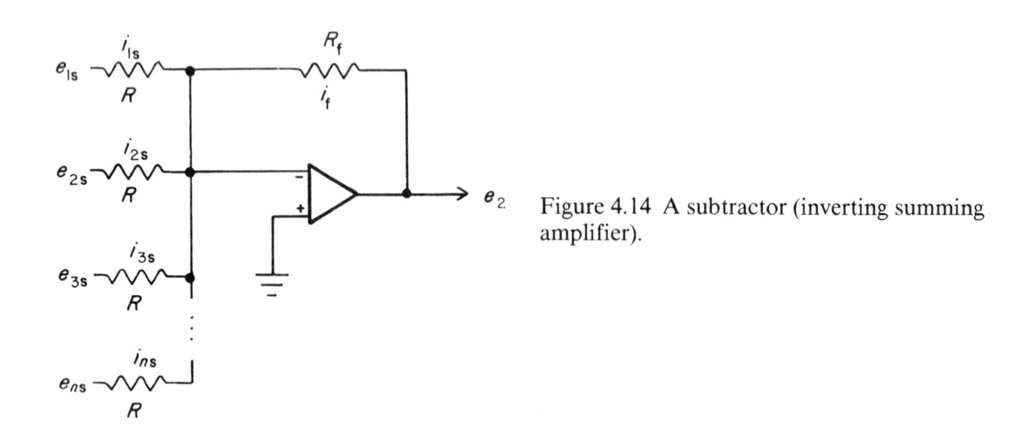

Figure 4.14 A subtractor (inverting summing amplifier).

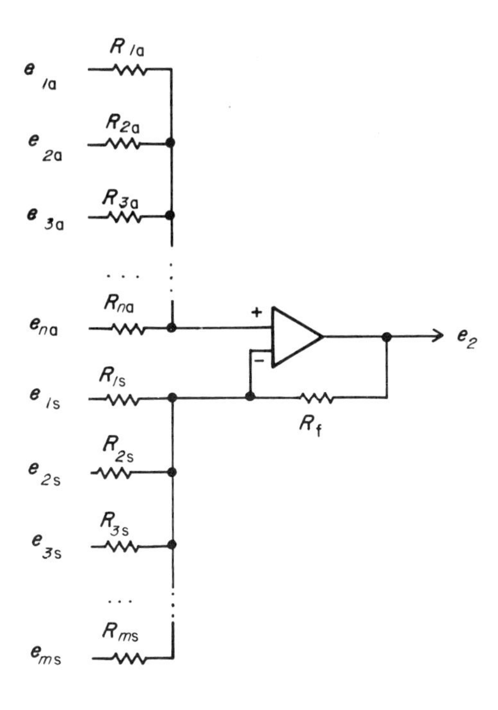

Figure 4.15 Generalized adder–subtractor (summing amplifier with inverting and noninverting inputs).

The current at the positive input is determined only by the adder input, as developed for the adder circuit:

$$(e_{1a}-e^{+})/R_{1a}+(e_{2a}-e^{+})/R_{2a}+\cdots+(e_{na}-e^{+})/R_{na}=0 \tag{4.28}$$

Similarly, at the negative input,

$$(e_{1s}-e^{-})/R_{1s}+(e_{2s}-e^{-})/R_{2s}+\cdots+(e_{ms}-e^{-})/R_{ms}+(e_2-e^{-})/R_f=0 \tag{4.29}$$

The problem in reducing such equations further lies in combining fractions with different denominators. This problem can be eliminated by adding one constraint:

$$\sum_{i=1}^{n}\frac{1}{R_{ia}}=\frac{1}{R_f}+\sum_{j=1}^{m}\frac{1}{R_{js}} \tag{4.30}$$

That is, the sum of all the conductances on the adder side of the circuit is equal to the sum of all the conductances (including the feedback resistor) on the subtractor side.

Recall that $e^{+}=e^{-}$ because of virtual identity of the two inputs. Substituting e^{+} for e^{-} and rearranging shows that Eqs. (4.27) and (4.28) can be expressed as

$$e_{1a}/R_{1a}+e_{2a}/R_{2a}+\cdots+e_{na}/R_{na}=e^{+}\sum_{i=1}^{n}\frac{1}{R_{ia}} \tag{4.31}$$

and

$$e_{1s}/R_{1s}+e_{2s}/R_{2s}+\cdots+e_{ms}/R_{ms}+e_2/R_f=e^{-}\left(\frac{1}{R_f}+\sum_{j=1}^{m}\frac{1}{R_{js}}\right) \tag{4.32}$$

Since we have adopted the constraint of Eq. (4.30), the right-hand terms of Eqs.

(4.31) and (4.32) are equal. Therefore, the left-hand terms are equal:

$$e_{1a}/R_{1a}+e_{2a}/R_{2a}+\cdots+e_{na}/R_{na}=e_{1s}/R_{1s}+e_{2s}/R_{2s}+\cdots+e_{ms}/R_{ms}+e_2/R_f$$

Solving for e_2 gives

$$e_2=R_f\left(\sum_{i=1}^{n}\frac{e_{ia}}{R_{ia}}+\sum_{j=1}^{m}\frac{e_{ja}}{R_{ja}}\right)$$

That is, e_2 is the weighted sum of all adder inputs minus the weighted sum of all subtractor inputs. The weight (gain) for each term is $m'_{ia}=R_f/R_{ia}$ for each adder term or $m'_{js}=R_f/R_{js}$ for each subtractor term. If R_f is known, each R_{ia} or R_{js} can be selected on the basis of the desired coefficient m'_{ia} or m'_{js}. Thus, to realize the equation

$$e=\sum_{i=1}^{n}m'_{ia}e_{ia}-\sum_{j=1}^{m}m'_{js}e_{js}$$

$$m'_{ia}=\frac{R_f}{R_{ia}} \quad \text{and} \quad m'_{ja}=\frac{R_f}{R_{ja}}$$

so that

$$R_{ia}=\frac{R_s}{m'_{ia}} \quad \text{and} \quad R_{ja}=\frac{R_f}{m'_{js}} \tag{4.33}$$

How do we meet the constraint of Eq. (4.30)? This is done quite easily. First, determine the values of all R_{ia} and R_{js} according to their relative weights. Thus, for example, if we wish to realize the computation,

$$4e_1+3e_2+0.4e_3-5e_4-7e_5-0.9e_6=e_2$$

then we could set $R_f=10\ \text{k}\Omega$ and solve for each R according to the desired weighting factor:

$$R_{ia}=\frac{R_f}{m'_{ia}}$$

or

$$R_{ja}=\frac{R_f}{m'_{js}}$$

or for the expression in Eq. (4.33),

$$R_i=\frac{R_f}{|m'_i|}$$

The m'_i coefficients and their corresponding resistances are calculated to obtain proper weights for all adder and subtractor inputs. Then find the difference between conductances on the two sides of Eq. (4.30):

$$\frac{1}{R_x}=\sum_{i=1}^{n}\frac{1}{R_{ia}}-\left(\frac{1}{R_f}+\sum_{j=1}^{m}\frac{1}{R_{js}}\right) \tag{4.34}$$

If this difference is positive, that is,

$$\sum_{i=1}^{n} \frac{1}{R_{ia}} > \left(\frac{1}{R_f} + \sum_{j=1}^{m} \frac{1}{R_{js}} \right)$$

$1/R_x$ should be added to the subtractor circuit as a resistor to ground. Then

$$\sum_{i=1}^{n} \frac{1}{R_{ia}} = \frac{1}{R_f} + \frac{1}{R_x} + \sum_{j=1}^{m} \frac{1}{R_{js}}$$

and Eq. (4.30) is satisfied.

If the difference in Eq. (4.34) is negative, that is,

$$\sum_{i=1}^{n} \frac{1}{R_{ia}} < \left(\frac{1}{R_f} + \sum_{j=1}^{m} \frac{1}{R_{js}} \right)$$

$1/R_x$ should be added to the adder circuit as a resistor to ground. Then

$$\frac{1}{R_x} + \sum_{i=1}^{n} \frac{1}{R_{ia}} = \frac{1}{R_f} + \sum_{j=1}^{m} \frac{1}{R_{js}}$$

and Eq. (4.30) is satisfied. Let us run through a sample design problem.

Solved problem. We wish to realize the following analog computation:

$$e_2 = e_1 + 2e_2 + 3e_3 - 0.5e_4 - 0.7e_5 - 1.1e_6 \tag{4.35}$$

Let us set R_f as 10 kΩ, a convenient value. We can now solve for all m'_{ia} and m'_{js} terms: see Table 4.1. To meet the constraint of Eq. (4.30) we can sum the reciprocals of the adder terms and subtract the reciprocals of feedback and subtractor terms [Eq. 4.34]:

$$\frac{1}{R_x} = \underbrace{\frac{1}{10\text{ k}\Omega} + \frac{1}{5\text{ k}\Omega} + \frac{1}{3.3\text{ k}\Omega}}_{\text{adders}} - \underbrace{\frac{1}{10\text{ k}\Omega}}_{\text{feedback}} - \underbrace{\frac{1}{20\text{ k}\Omega} - \frac{1}{14.3\text{ k}\Omega} - \frac{1}{9.1\text{ k}\Omega}}_{\text{subtractors}}$$

$$= -\underbrace{\frac{1}{5.5\text{ k}\Omega}}_{R_x}$$

Therefore, $R_x = 5.5$ kΩ and is an adder term. The circuit is illustrated in Fig. 4.16.

Table 4.1

Right-hand term	m_{ia} or $\lvert m_{js} \rvert$	R_{ia} or R_{js}
1	1	$R_f/m'_{ia} = 10$ kΩ$/1 = 10$-kΩ adder
2	2	$R_f/m'_{ia} = 10$ kΩ$/2 = 5$-kΩ adder
3	3	$R_f/m'_{ia} = 10$ kΩ$/3 = 3.3$-kΩ adder
4	0.5	$R_f/m'_{js} = 10$ kΩ$/0.5 = 20$-kΩ subtractor
5	0.7	$R_f/m'_{js} = 10$ kΩ$/0.7 = 14.3$-kΩ subtractor
6	1.1	$R_f/m'_{js} = 10$ kΩ$/1.1 = 9.1$-kΩ subtractor

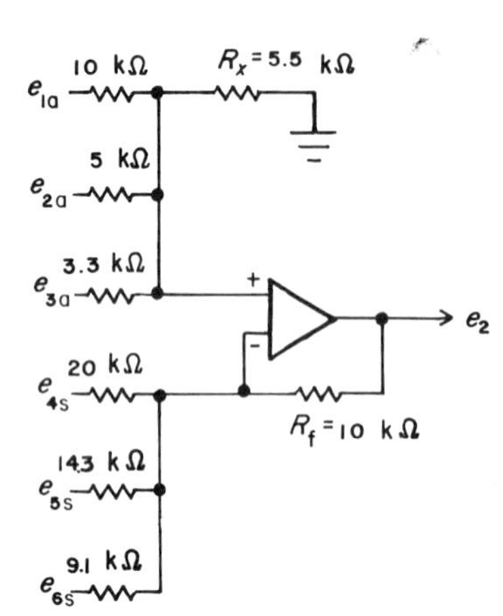

Figure 4.16 A sample design problem. The circuit is a realization of the equation $e_2 = e_{1a} + 2e_{2a} + 3e_{3a} - 0.5e_{4s} - 0.7e_{5s} - 1.1e_{6s}$.

4.3.4 Differentiation and Integration

The curious reader may wonder why we do not discuss the other basic arithmetic computations, multiplication and division, before the apparently more esoteric operations of differentiation and integration. The reason is simple: the implementation of an integrator or differentiator is much easier than implementation of multiplication or division. This is because the component laws governing passive components include the operations of differentiation and integration, but not those of multiplication and division. By differentiation and integration, we mean *with respect to time*. By multiplication and division, we mean the multiplication or division of one signal by another.

Recall from Chapter 1 that

$$i = C\frac{dE}{dt}$$

Consider the circuit of Fig. 4.17. Here, a capacitor is the feedthrough component and a resistor is the feedback component. We know that

$$i_1 + i_2 = 0$$

$$i_1 = C\frac{de_1}{dt}$$

and

$$i_2 = e_2/R$$

Substituting for i_1 and i_2 gives

$$C\frac{de_1}{dt} + \frac{e_2}{R} = 0$$

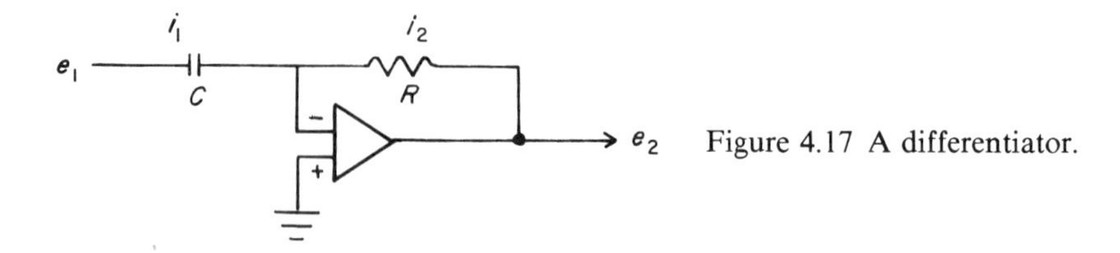

Figure 4.17 A differentiator.

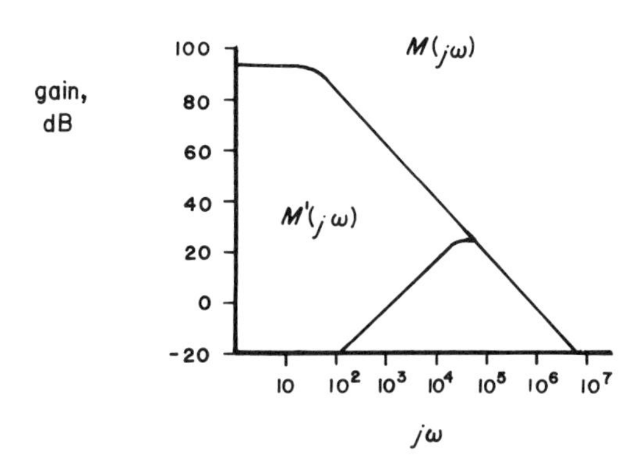

Figure 4.18 Frequency response of differentiator in Fig. 4.17, compared with frequency response of open loop 741 op amp: $R=100\ \text{k}\Omega$, $C=0.1\ \mu\text{F}$ ($RC=10^{-2}$ sec).

Solving for e_2 gives

$$e_2 = -RC\frac{de_1}{dt} \tag{4.36}$$

That is, the output voltage is proportional to the derivative of the input voltage. In terms of frequency response, recall from Eq. (4.21) that

$$\frac{E_2(j\omega)}{E_1(j\omega)} = -\frac{Z_2(j\omega)}{Z_1(j\omega)}$$

For the differentiator of Fig. (4.18),

$$M'(j\omega) = \frac{E_2(j\omega)}{E_1(j\omega)} = -\frac{R}{1/j\omega C} = -RCj\omega \tag{4.37}$$

The frequency response of the differentiator is indicated in Fig. 4.18. Within the frequency limits of the op amp, the device is a first-order high-pass filter. We will describe in more detail the filter characteristics of this and a number of other filters in Chapter 6. For use as a practical differentiator, an intelligent choice of R and C must be made: a rule of thumb is to select them according to the frequency composition of the input signal, so that unity gain is at or near the highest frequency component: this will prevent excessive amplification of the highest frequencies, which would result in truncation of peaks ("clipping") whenever they exceed the operating limits of the amplifier. From Eq. (4.37), $M'(j\omega)$ is unity when $|-RCj\omega|=1$, that is, when $RC=1/j\omega$ for the selected value of ω. Of course, this rule of thumb will not apply in some applications.

In some cases, it is useful to modify the differentiator as shown in Fig. 4.19a. Here, a series input resistor R has been added. The frequency response is

$$M'(j\omega) = \frac{Z_2(j\omega)}{Z_1(j\omega)} = -\frac{R_2}{R_1+(1/j\omega C)} = -\frac{R_2Cj\omega}{R_1Cj\omega+1}$$

This frequency response is illustrated in Fig. 4.19b. The modified differentiator is useful in applications where a regular differentiator would produce excessive high-frequency gain, for example, when processing signals with significant amounts of high-frequency noise. For frequencies above $j\omega = R_1C_1$ the frequency response is nearly flat, with $M'(j\omega) \cong R_2/R_1$.

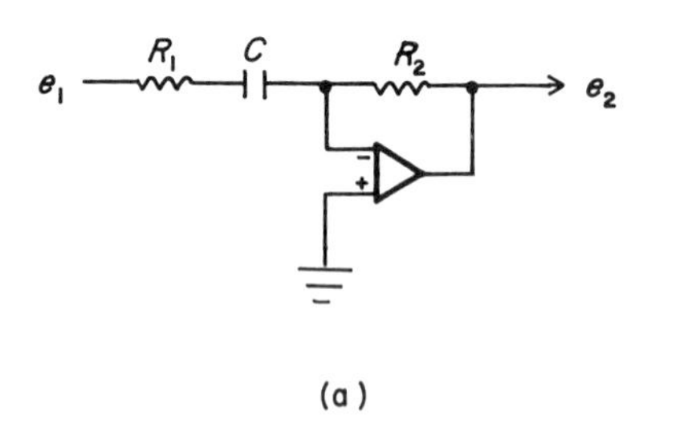

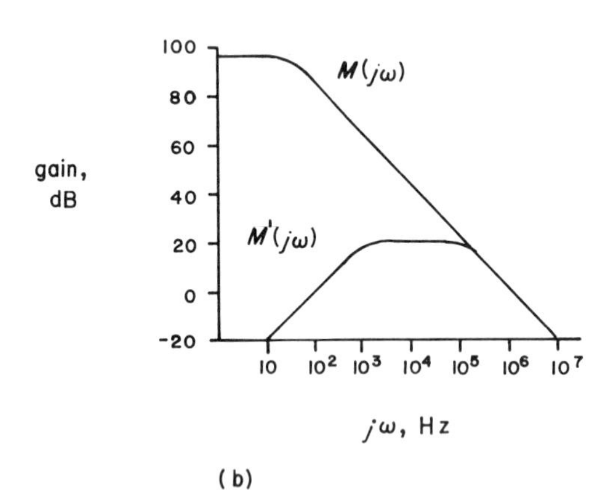

Figure 4.19 Modified differentiator. **(a)** Use of a series resistor R_1 to diminish high-frequency response peaking. **(b)** Frequency response of modified differentiator, with $R_1 = 1$ kΩ, $R_2 = 100$ kΩ, $C = 0.1$ μF.

Figure 4.20a illustrates an integrator. Here, the capacitor is the feedback element and the resistor is the feedthrough element. As before, $i_1 + i_2 = 0$, and $e^- = 0$, but now

$$i_1 = \frac{e_1}{R}$$

and

$$i_2 = C\frac{de_2}{dt}$$

Therefore,

$$\frac{e_1}{R} = -C\frac{de_2}{dt}$$

or

$$de_2 = \frac{-e_1\,dt}{RC}$$

Integrating gives

$$e_2 = -\frac{1}{RC}\int e_1\,dt + K \tag{4.38}$$

where K is an integration constant, equal to the initial value of e_2.

Applying Eq. (4.21) again gives

$$M(j\omega) = \frac{E_2(j\omega)}{E_1(j\omega)} = -\frac{Z_2(j\omega)}{Z_1(j\omega)} = \frac{-1/j\omega C}{R} = \frac{1}{-RCj\omega} \tag{4.39}$$

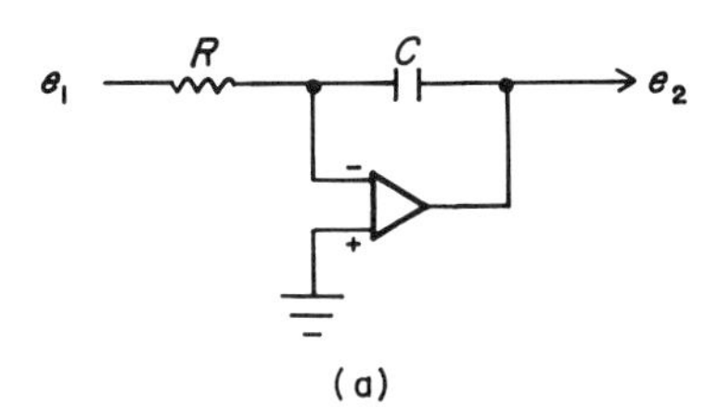

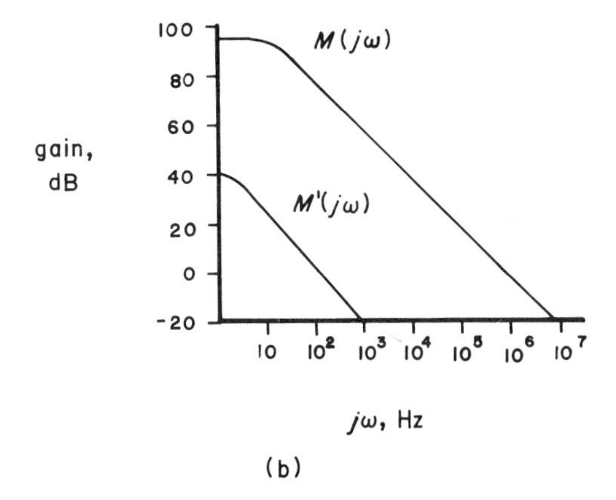

Figure 4.20 **(a)** An integrator. **(b)** Frequency response of the integrator, with $R = 100\ \text{k}\Omega$ and $C = 0.1\ \mu\text{F}$.

The frequency response of an integrator is illustrated in Fig. 4.20b. Examination of Eq. (4.39) reveals that the device has *infinite gain for dc*, and very high gain for low frequencies. For this reason, the circuit of Fig. 4.20 is almost never used as shown. This is because even very small offset voltages on the op amp input will result in a ramplike drift of the output, until the output saturates at the upper or lower voltage limit of the op amp. Instead, one of the modifications of Fig. 4.21a or 4.21b is used. Choice of modification depends on the application. The switch of Fig. 4.21 is used to convert the op amp into a zero gain device; the output is reset to a value which is virtually zero:

$$M'(j\omega) = \frac{-Z_2(j\omega)}{Z_1(j\omega)} = \frac{0}{R} = 0$$

This would be used in applications where a true integral is desired over a relatively brief time, with resetting to zero just prior to the integration period. The signal must be large relative to dc errors in the op amp. Several applications using this modification will be demonstrated in Chapter 6.

The modification of Fig. 4.21b is a "leaky integrator": the capacitor is slowly discharged by R_2 to prevent drift with dc offset.

The frequency response is

$$M'(j\omega) = \frac{-Z_2(j\omega)}{Z_1(j\omega)} = -\frac{1/(1/R_2 + j\omega C)}{R_1} = \frac{R_2}{R_1 + R_1 R_2 C j\omega} \qquad (4.40)$$

The frequency response is illustrated in Fig. 4.21c. As $j\omega$ approaches zero, the gain ap-

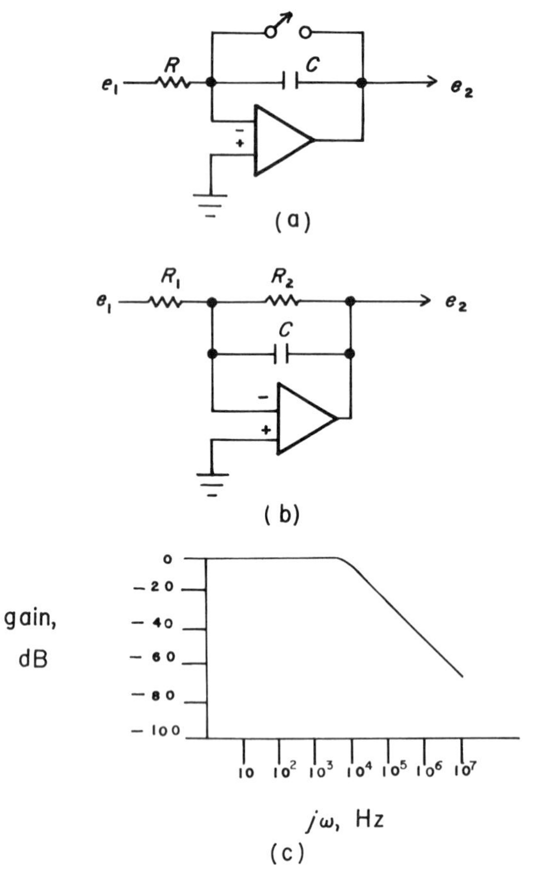

Figure 4.21 **(a)** A resetting integrator. **(b)** A leaky integrator. **(c)** Frequency response of the leaky integrator with $R_1 = 100\ \text{k}\Omega$, $R_2 = 100\ \text{k}\Omega$, $C = 1000$ pF.

proaches R_2/R_1. The corner frequency is determined by

$$M'(j\omega) = \frac{R_2}{R_1 + R_1 R_2 j\omega_c} = 0.5$$

or, using values of $R_1 = R_2 = 100\ \text{k}\Omega$ and $C = 1000$ pF,

$$0.5 = \frac{10^5}{10^5 + 10^5 10^5 10^{-9} j\omega_c}$$

Solving for $j\omega_c$ gives

$$j\omega_c = 10^4\ \text{Hz},$$

and $f_c = j\omega_c/2\pi \simeq 1.59 \times 10^3$ Hz. Thus the leaky integrator has unity gain to frequencies around 1–2 kHz, with the device acting as an integrator for higher frequencies.

For $f < 10$ kHz,

$$e_2 \cong \frac{R_2 e_1}{R_1} \cong e_1$$

and for $f > 100$ kHz,

$$e_2 \cong \frac{1}{R_1 C} \int e_1 \, dt \cong 10^{-4} \int e_1 \, dt$$

The leaky integrator thus avoids the problem of drift caused by dc offset and slow signals. It is useful as a short-term integrator, an integrator with "short memory."

Integrators and leaky integrators are low-pass filters, as can be seen from the frequency responses of Figs. 4.20 and 4.21c. Their filter properties will be considered in detail in Chapter 6.

Under some circumstances, it is desirable to differentiate low frequencies and integrate high frequencies. A combined differentiator–integrator and its frequency response are illustrated in Fig. 4.22. The feedthrough and feedback impedances are

$$Z_1(j\omega) = R_1 + 1/j\omega C_1$$

$$Z_2(j\omega) = R_2/(1 + R_2 C_2 j\omega)$$

so that

$$M'(j\omega) = \frac{Z_2(j\omega)}{Z_1(j\omega)}$$

$$= \frac{R_2/(1 + R_2 C_2 j\omega)}{R_1 + 1/j\omega C_1}$$

$$= \frac{R_2 C_1 j\omega}{(1 + R_1 C_1 j\omega)(1 + R_2 C_2 j\omega)} \tag{4.41}$$

Figure 4.22 **(a)** Differentiator–integrator. **(b)** Frequency response of differentiator–integrator.

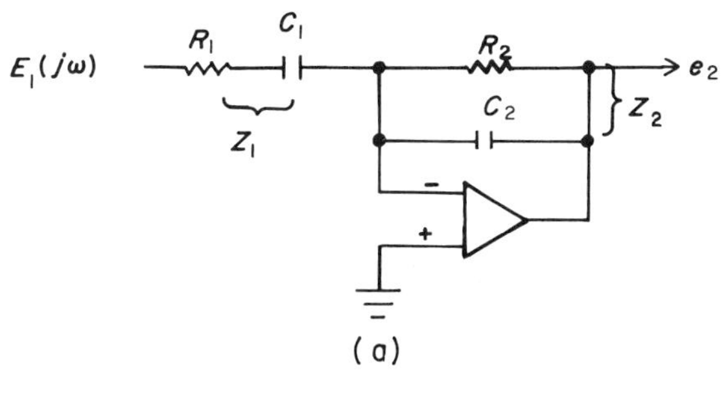

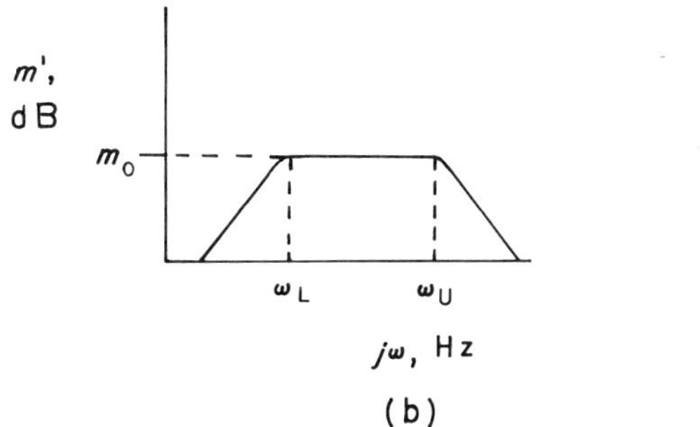

In the graph of Fig. 4.22b, ω_L and ω_U are the corner frequencies for the low-frequency roll-off and the high-frequency roll-off, respectively. They are determined as follows:

$$\omega_L = \frac{1}{R_1 C_1}$$

$$\omega_U = \frac{1}{R_2 C_2}$$

If $\omega_L < \omega_U$, $m_0 = R_2/R_1$.

For frequencies below ω_L, the device is a differentiator; for frequencies above ω_U, the device is an integrator; for frequencies between ω_L and ω_U, the device is an amplifier with gain $m' = m_0 = R_2/R_1$. From the frequency response it can be seen that the device is a bandpass filter.

4.3.5 Logarithms

The most common method of finding logarithms relies on the use of a property of bipolar transistors: the emitter–base voltage of a forward biased silicon transistor closely approximates the logarithm of its collector current over as many as nine decades of collector current. This is the basis for the logarithm circuit of Fig. 4.23, where

$$e_2 = \log e_1$$

The first amplifier drives the transistor with a collector current equal to $-e_1/10\ \text{k}\Omega$, producing a voltage between emitter and base of $\log e_1/10\ \text{k}\Omega$, for positive e_1. For negative e_1, the transistor is reverse biased, so a diode is used to protect the transistor from breakdown. This means the circuit only works for positive voltage. The second amplifier takes the difference between base and emitter voltages, corrects the sign, and amplifies by a unity gain.

In practice, the designer should not build his own log devices because temperature compensation requirements due to the use of an active feedback element introduce undesirable design complexities. Instead, logarithmic amplifier modules are available commercially which have all the design problems taken care of by the manufacturer. Their costs are competitive with any similar devices which could be built with op amps. These modules use a similar design to that discussed here.

Antilogarithms are computed by either using a transistor in the feedthrough network or by placing the log module in the negative feedback circuit of another op amp.

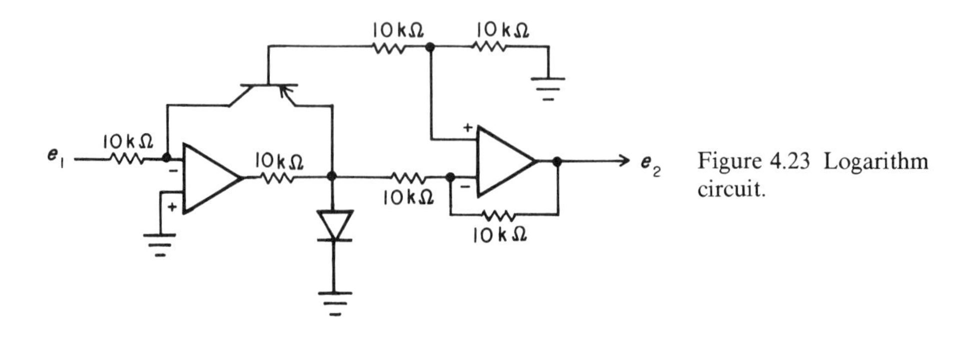

Figure 4.23 Logarithm circuit.

4.4 Analog Computation Symbols and Diagrams

The astute reader will have anticipated that black box design can be hierarchical. That is, just as an IC can be treated as a black box without regard for the actual circuit design of the IC, circuits built with ICs can themselves be treated as black boxes: in other words, more complex functions than addition, subtraction, differentiation, and integration can be performed by the appropriate combination of adders, subtractors, differentiators, and integrators. To enable the designer to synthesize circuits at this higher level of abstraction, we recommend the use of standard symbols for the representation of analog computing elements. These symbols, while they are not of uniformly high value as mnemonics for their functions, are nevertheless a great time-saver and allow the designer to synthesize complex functions without regard to the design of each element. Diagramming methodology is that used by scientists and engineers to set up analog computer programs.

The more common standard symbols used in analog computing are given in Fig. 4.24. In general, coefficients such as those in Eq. (4.33) and Fig. 4.16 are adjusted with potentiometers, symbolized by the notation of Figs 4.24a and 4.24b; however, fixed coefficients are represented as in the notation of Figs. 4.24e, 4.24f, 4.24g, 4.24i, and 4.24k. The divider, Fig. 4.24b, would be used in circuits where a voltage is set between two extremes with a potentiometer, in potentiometric feedback, and in other applications. The high-gain amplifier, Fig. 4.24c, is used primarily in the diagramming of negative feedback op amp configurations, as the op amp symbol. The inverter symbol is poorly chosen because there is no indication of change of polarity, and it could be mistaken for an amplifier or operational amplifier symbol. However, it is the accepted notation for inversion of a signal. Note, incidentally, that more archaic notation assumes all components with op amps in them (e.g., Figs. 4.24c–4.24g) do invert the signal because most of them were formed using negative feedback. We will never assume this, except for the inverter. When reading other literature, check the sign conventions!

The summing amplifier, Fig. 4.24e, can be incorporated implicitly into all other devices by simply attaching multiple inputs, as in Figs. 4.24f and 4.24g. Subtractor inputs have negative coefficients. Note that the integrator (Fig. 4.24f) has an additional input for specification of its output value at $t=0$. The time $t=0$ is defined as the time at which the integrator is reset to its initial value. The function generator, Fig. 4.24h, is a generalized black box used to represent any function of any number of variables except time. An integrator with constant input can generate a ramp voltage, however, which can be used as an electric analog representation of time, if necessary. Also, signal generators, which produce time-varying signals without external inputs, can be represented as function generators without inputs, to produce time-varying electric signals. We adopt this convention to prevent confusion between functions of voltages, which may be time-varying voltages, and functions of time.

The multiplier depicted in Fig. 4.24i has a nonunity coefficient a for one input. It could have coefficients specified for either or both inputs. Except for high-gain amplifiers, unspecified coefficients are interpreted as unity (one).

The servo amplifier of Fig. 4.24j is used to represent a negative feedback amplifier used to force a transducer output (linear or angular position, velocity, or acceleration; temperature; light intensity; etc.) to vary proportionally with the input voltage.

Component	Symbol	Mathematical relation
(a) coefficient-setting potentiometer	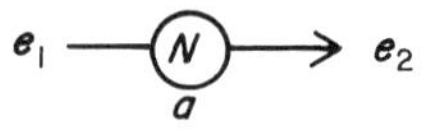 e_1 N e_2 a	$e_2 = ae_1$
(b) voltage divider	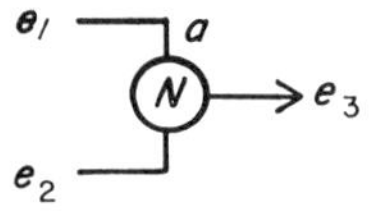 e_1 a N e_3 e_2	$e_3 = ae_1 + (1-a)e_2$
(c) high gain dc amplifier	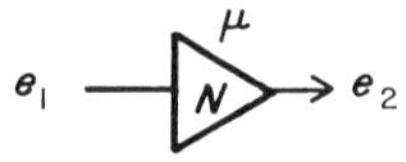 μ e_1 N e_2	$e_2 = \mu e_1$ (in some notations, $e_2 = -\mu e_1$)
(d) inverter	 e_1 N e_2	$e_2 = -e_1$
(e) summing amplifier (adder)	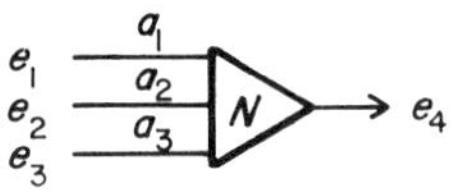 a_1 a_2 a_3 e_1 e_2 e_3 N e_4	$e_4 = a_1e_1 + a_2e_2 + a_3e_3$ (in some notations, $e_4 = -(a_1e_1 + a_2e_2 + a_3e_3)$)
(f) integrator	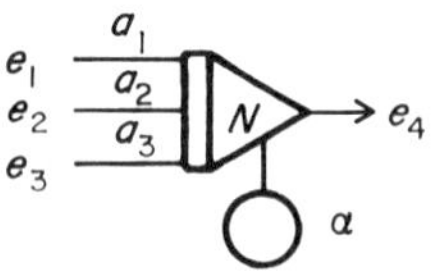 a_1 a_2 a_3 e_1 e_2 e_3 N e_4 α	$e_4 = \int_0^t (a_1e_1 + a_2e_2 + a_3e_3)dt + \alpha$, where α = initial condition. (some notations reverse signs)
(g) differentiator	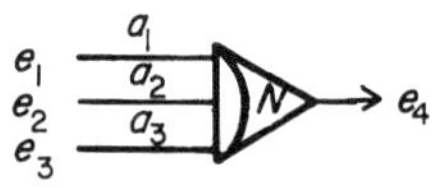 a_1 a_2 a_3 e_1 e_2 e_3 N e_4	$e_4 = \frac{d}{dt}(a_1e_1 + a_2e_2 + a_3e_3)$ (some notations reverse sign)
(h) function generator	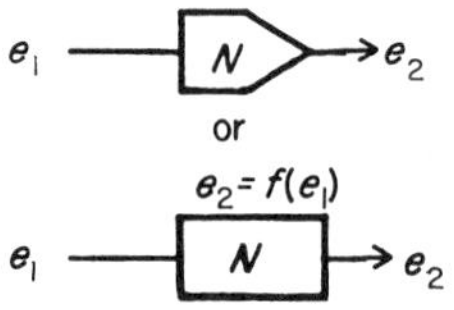 e_1 N e_2 or $e_2 = f(e_1)$ e_1 N e_2	$e_2 = f(e_1)$
(i) multiplier	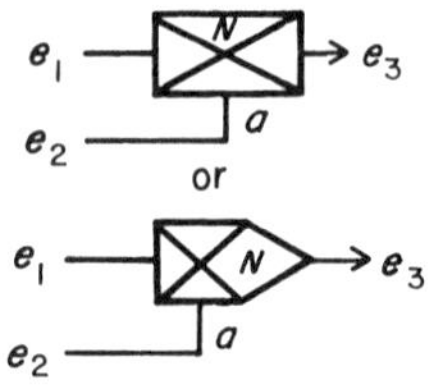 e_1 N e_3 e_2 a or e_1 N e_3 e_2 a	$e_3 = ae_1e_2$

Figure 4.24 Analog computation symbols. The N in each element is an identifying number; in a circuit, each would have a different number.

Component	Symbol	Mathematical relation
(j) servo amplifier and transducer(s)	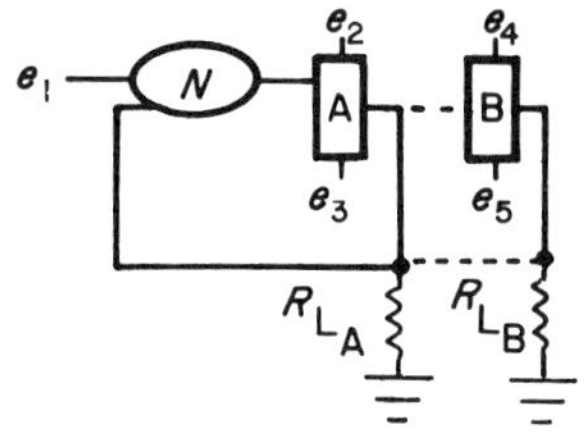 e_1, N, A, B, e_2, e_3, e_4, e_5, R_{L_A}, R_{L_B}	—
(k) limiting amplifier, a_1 = upper limit and a_2 = lower limit	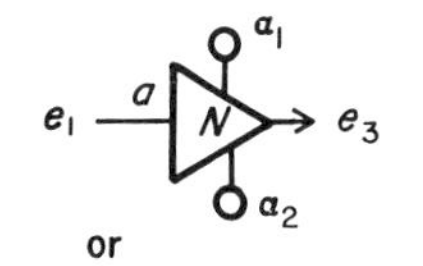 e_1, a, N, e_3, a_1, a_2 or e_1, a, N, e_3, a_1, a_2 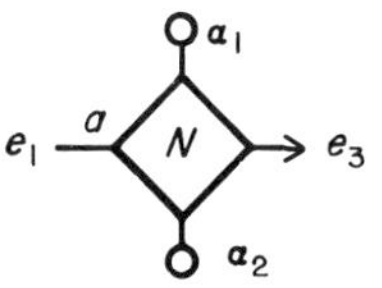	$e_3 = a e_1$ for $a_2 < a e_1 < a_1$ $e_3 = a_2$ for $a e_1 \leq a_2$ $e_3 = a_1$ for $a e_1 \geq a_1$ $(a_1 > a_2)$
(l) comparator	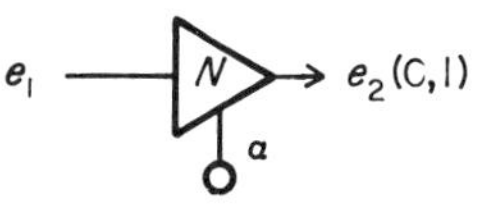 e_1, N, $e_2(C,1)$, a	$e_2 = 0$ if $e_1 < a$ $e_2 = 1$ if $e_1 > a$ e_2 = same as most recent value if $e_1 = a$

Figure 4.24 (*continued*)

The output is transduced back into a proportional electrical signal and is used to provide feedback for error correction.

The limiting amplifier, Fig. 4.24k, "clips" or truncates the output voltage whenever it reaches specified upper and lower limits. This is a nonideality of all devices driven with finite voltage power supplies, and clipping is sometimes used to protect low-voltage devices from being overdriven by excessive inputs.

The comparator, Fig. 4.24l, compares an input voltage with a reference voltage, which may itself be variable, and has a binary output: it is on if the input exceeds the reference, and off if it is less than the reference. If equal to the reference, it retains the state it held during the most recent inequality.

Example 4.2. Equation of motion of a falling body

The position of a falling body is described by

$$\frac{d^2x}{dt^2} = -g$$

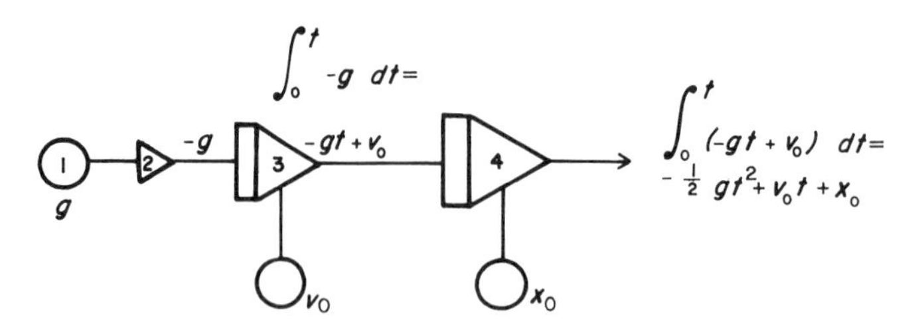

Figure 4.25 Simulation of the position of a falling body.

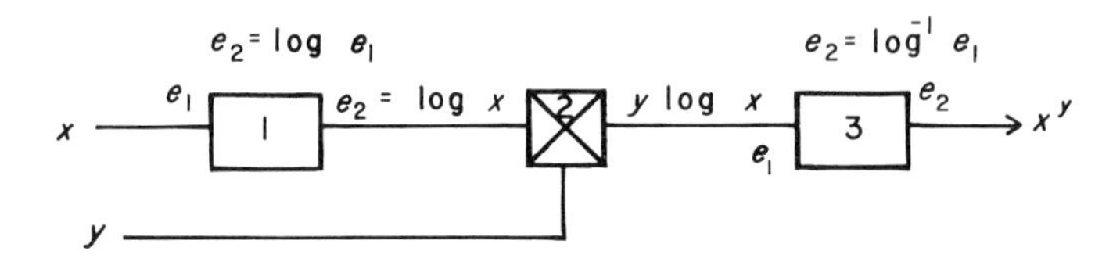

Figure 4.26 Raising a voltage to a power.

where x is distance from the ground and g is gravitational acceleration. Integrating gives

$$v=\frac{dx}{dt}=-gt+v_0$$

where v is velocity and v_0 is a constant of integration, the velocity at $t_0=0$. Integrating again gives

$$x=-\tfrac{1}{2}gt^2+v_0t+x_0$$

where x_0 is the initial position, another constant of integration. This equation can be simulated with the circuit of Fig. 4.25. The output voltage is equal to the position of the falling body. The gravitational acceleration is represented by the constant voltage input to the first integrator. At time $t_0=0$, both integrators are reset to x_0 and v_0. Subsequently, as integration proceeds, the output of the first integrator is equal to the body's velocity and the output of the second integrator is equal to its position.

Example 4.3. Exponentiation

We can raise a voltage to any power, including negative and fractional powers, with the circuit of Fig. 4.26. The logarithm of the input voltage is derived with the first module, this is multiplied by the desired power, and the antilog is obtained. Since

$$\log x^y = y \log x$$

the output is equal to x^y. Thus, e.g., if $y=2$, the circuit is a squaring circuit, and if $y=1/2$, the circuit is a square root circuit. Note that negative inputs are not allowed as values of x. Most log modules produce a zero output with negative input.

4.5 Some Nonlinear Functions

One circuit we have already discussed utilized the nonlinear properties of some transistors to produce one type of nonlinear module: the logarithmic module. The diode, which also has nonlinear characteristics, can be used to generate nonlinear functions. One of them is the absolute value: $e_2=|e_1|$.

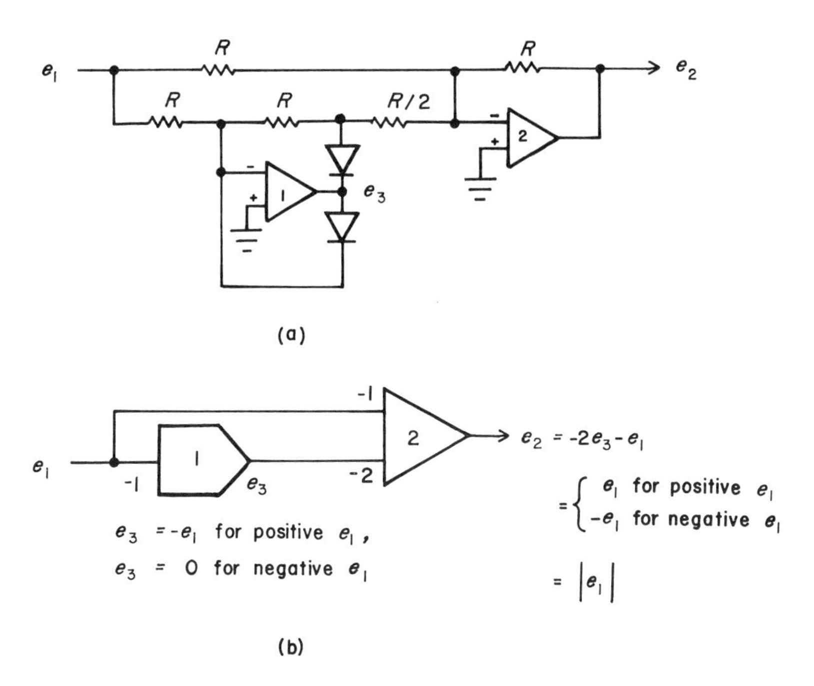

Figure 4.27 An absolute value circuit **(a)**, schematized in analog computation symbology in **(b)**.

Absolute value circuits can be used to convert signal voltages to equal-valued positive voltages. Absolute value circuits are useful as input to squaring circuits which use logarithmic modules. Since the output of a squaring operation is always positive, the results are correct, and the circuits can then operate on negative inputs as well as positive ones. The circuit of Fig. 4.27a is an absolute value circuit. Consider the behavior of the circuit for positive and negative voltages at the input. Note that the first op amp has two negative feedback loops: one is a direct feedback through the lower diode, which will conduct to the virtual ground negative input for any positive voltage, with an effective impedance of zero. Thus, for any negative input, the output will be forced to virtual zero by the shunting diode to the virtual ground. For positive input, the lower diode becomes an effectively infinite resistance for the positive output and hence does not participate in the negative feedback. Instead, the upper diode begins to conduct. Since the net feedback resistance is unaffected by the forward biased diode, it must be equal to R, the same as the input resistance. Hence the output of the first op amp is

$$e_3 = -e_1$$

for positive inputs, and

$$e_3 = 0$$

for negative inputs. Then e_3 and e_1 are fed to the second amplifier, which is a subtractor (inverting summing amplifier). The gain of this device is $R/(R/2)=2$ for the input from the first op amp, and $R/R=1$ for the input voltage. Therefore the output is

$$e_2 = -2e_3 - e_1$$
$$= \begin{cases} -2(-e_1) - e_1 = e_1 & \text{for positive } e_1 \\ -2(0) - e_1 = -e_1 & \text{for negative } e_1 \end{cases}$$
$$= |e_1|$$

Diodes can also be used to produce dead zones in the input voltage range over which the op amp acts as an amplifier. Such a device is illustrated in Fig. 4.28a. For values of e_1 for which the output of the upper divider is positive, the diode bridge shunts that output to ground. For negative outputs of the upper potentiometer, the bridge conducts current to the op amp negative input rather than shunting it to ground. We know this to be true because the negative input is a virtual ground.

Similar reasoning applies to the lower potentiometer output. If the output is negative, it is shunted to ground by the bridge; if positive, current goes to the op amp negative input. Clearly, the lower potentiometer output is always more negative

Figure 4.28 **(a)** A dead zone amplifier. **(b)** The input–output curve for a dead zone amplifier. **(c)** Sample input and output waveforms.

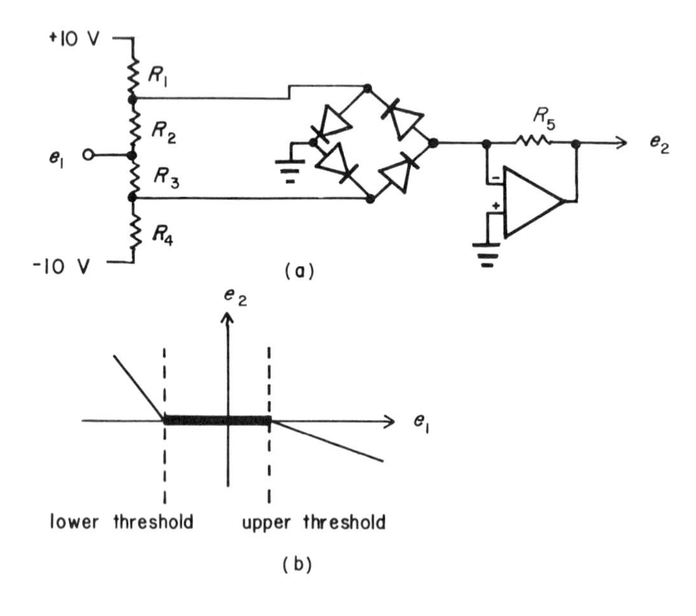

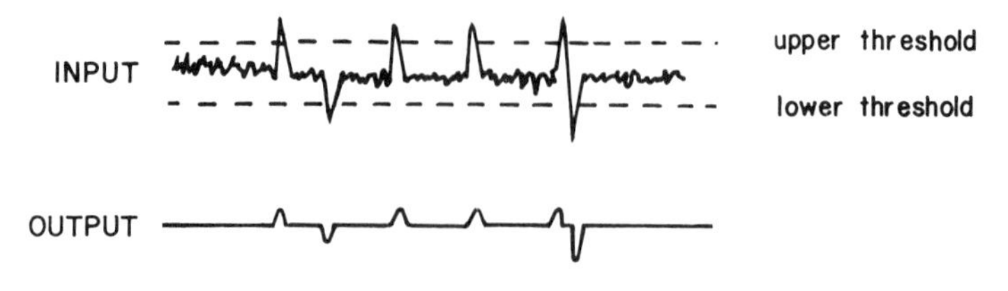

(c)

than the upper one; therefore, the upper potentiometer output can never conduct to the op amp at the same time as the lower one. However, there can be a range of input voltages over which both potentiometer outputs are shunted to ground, and the op amp output is zero. This is illustrated in Fig. 4.28b. Above and below this range, the op amp will see an output current directly proportional to e_1 plus or minus one or the other threshold:

$$-i_{R_5} = i_{R_1} + i_{R_2} = \frac{e_1}{R_2} + \frac{+10\text{ V}}{R_1}$$

for

$$\left(\frac{e_1}{R_2} + \frac{+10}{R_1}\right) \leqslant 0$$

i.e.,

$$e_2 = i_{R_5} R_5 = -R_5\left(\frac{e_1 R_1 + 10 R_2}{R_2 R_1}\right)$$

for

$$e_1 \leqslant -10\text{ V}\frac{R_5}{R_1}$$

Similarly,

$$-i_{R_5} = i_{R_3} + i_{R_4} = \frac{e_1}{R_3} + \frac{-10\text{ V}}{R_4}$$

for

$$\left(\frac{e_1}{R_2} + \frac{-10\text{ V}}{R_4}\right) \geqslant 0$$

i.e.,

$$e_2 = i_{R_5} R_5 = -R_5\left(\frac{e_1 R_3 - 10 R_4}{R_3 R_4}\right)$$

for

$$e_1 \geqslant 10\text{ V}\frac{R_5}{R_3}$$

Note that gains (slopes of input–output function for positive and negative output voltages) are separately determined for negative and positive voltages; generally, for equal slopes, R_3 and R_1 are equal in value, and R_2 and R_4 are varied to adjust dead zone limits. Note that by using a pure subtractor whose negative input is virtual ground, these resistances can be varied without changing the gain of the circuit.

Sample input and output waveforms are given in Fig. 4.28c.

Just as it is sometimes convenient to eliminate baseline noise, it is also occasionally necessary to truncate large excursions from baseline, either to protect circuits from overvoltages, or to expand the baseline. Different methods of accomplishing this are illustrated in Fig. 4.29. In Fig. 4.29a the input is buffered by a follower, and passed through a 100-kΩ resistor to a diode pair. These diodes are nonconducting for voltages within the clipping limits because they are reverse biased. Under these

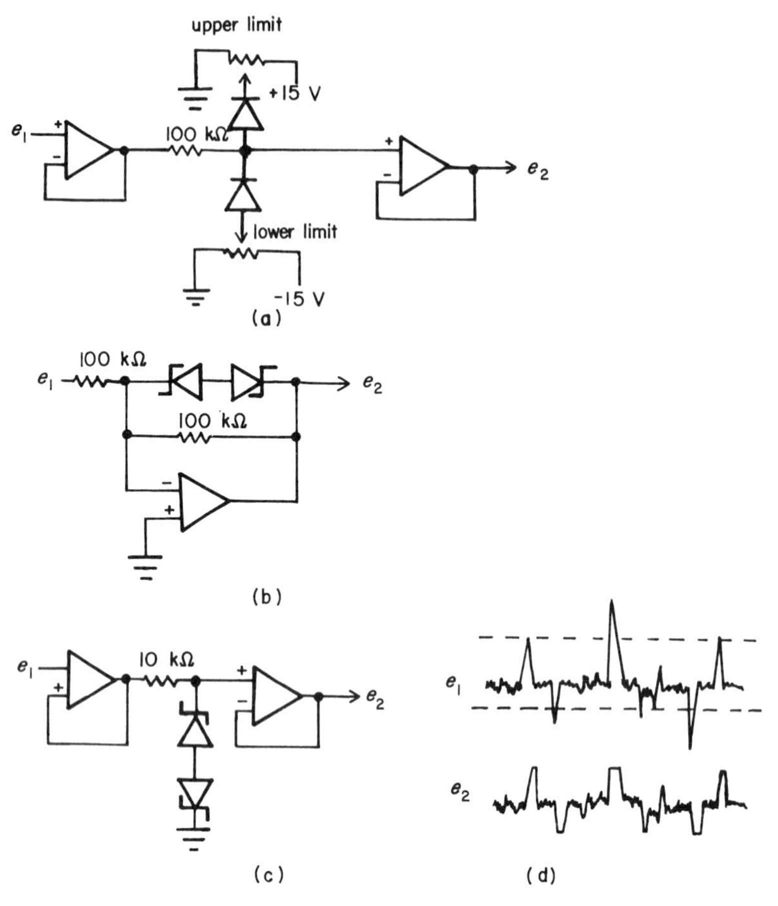

Figure 4.29 Methods of limiting output voltage: **(a–c)** three different circuit configurations; **(d)** sample input and output waveforms.

conditions the signal is passed unchanged to the input of the second follower. However, if the signal becomes more positive than the upper limit or more negative than the lower limit, the corresponding diode will conduct, since it is now forward biased. This holds the voltage at the upper or lower limit, as long as the diode is forward biased. Note, however, that since the potentiometer has a 1-kΩ resistance, there will be some current drop across the potentiometer resistance, and the peaks may not be clipped absolutely flat. Buffering the two potentiometers with unity gain followers would present 50–100-Ω impedances to ground to the diodes, and 100-kΩ potentiometers could be used, diminishing current drain and improving the clipping. Both input and output followers are optional, depending on the signal source impedance and the load impedance.

The circuit of Fig. 4.29b uses a pair of zener diodes in the feedback circuit of an inverting amplifier. If the output voltage dictated by the unity gain of the resistive network is within the zener limits, neither zener diode will conduct. However, above the breakdown voltage of either diode it will conduct; at that voltage the other diode is forward biased and will conduct also. The diodes thus constitute a short circuit, bypassing the feedback resistor and passing any current necessary to maintain a virtual ground at the negative input and hold the output at the zener voltage.

The circuit of Fig. 4.29c uses zener diode limiting at the output. The zeners replace the two diodes and threshold-setting potentiometers of Fig. 4.29a, and therefore the two followers are once again optional, depending on source and load impedances. Although the zeners are more convenient, they are not adjustable.

By using a number of diodes in the feedthrough circuit, a piecewise linear approximation to any arbitrary nonlinear gain function can be obtained. This principle is illustrated in Fig. 4.30. The diodes only begin to conduct to virtual ground when their forward bias exceeds the diode potential. The input signal e_1 is connected to n voltage dividers, each labeled R_{A_i}, R_{B_i}, i from 1 to n. The other end of each divider goes to a negative reference voltage E^-. Each divider output is connected through a diode with positive current polarity to the virtual ground negative input of the op amp. When a voltage divider's output is less than the diode potential, the diode connected to that divider does not conduct, and therefore does not contribute to the virtual ground current. Note that no diode can conduct unless the input voltage e_1 exceeds the positive diode potential. This means the device will only work for sufficiently positive voltages, but this limitation can be overcome by adding an offset potential to the input voltage.

Each diode has a threshold, determined by the resistors in the voltage divider. Also, each will contribute an amount of current to the summing junction of the subtractor which is determined by the voltage divider potential and the series input resistor:

$$i_{D_j} = \frac{\left[e_1 - \left(R_{A_j}/R_{B_j}\right)E^-\right]}{R_{A_j}} = \left(\frac{e_1}{R_{A_j}} - \frac{E^-}{R_{B_j}}\right)$$

Thus, depending on the resistances chosen, each diode will contribute to the total input current with a specifiable threshold and weighting factor. The net gain of the circuit for a given input voltage is a function of the number of diodes conducting and the current contribution from each diode. Note that the gain slope can only decrease with increased voltage; this can be modified by cascading nonlinear circuits with appropriate offset and polarity adjustments.

The designer should generally not try to synthesize his own nonlinear transconductors using this method. The technique requires temperature compensation and other sophisticated modifications to be practical. A number of manufacturers supply such modules for various nonlinear functions such as logarithms, exponentiation, and transcendental functions, which are well engineered and cost less than a custom-designed one.

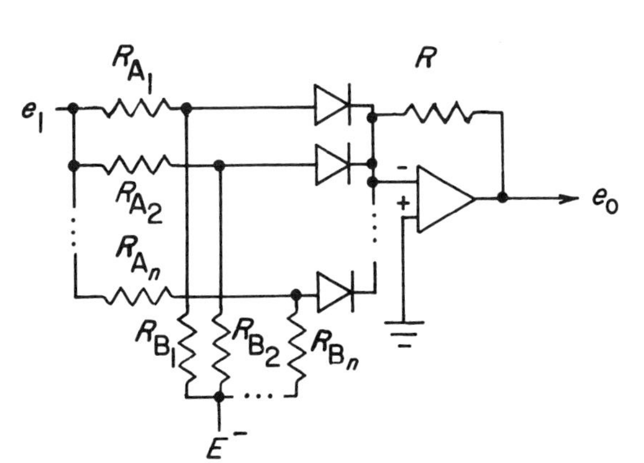

Figure 4.30 Nonlinear transconductor using a set of diodes with different breakpoints to produce a piecewise linear approximation to an arbitrary nonlinear voltage gain. (After Fig. 2.23 in *Application Manual for Operational Amplifiers*, 1968, by Philbrick/Nexus Research.)

4.6 Multiplication and Division

Once the capability for logarithms and absolute values is established, multiplication is easily accomplished. For both inputs positive, the multiplier of Fig. 4.31a is adequate. This is called a one-quadrant multiplier because if the two inputs are plotted on x, y coordinates, all legal input values fall in the first quadrant. The principle is straightforward: the logarithm of one signal is added to the other, and the antilogarithm computed. Since

$$\log xy = \log x + \log y$$

it follows that

$$\log^{-1}(\log x + \log y) = xy$$

The device only works in the first quadrant because logarithm modules only operate on positive inputs.

To operate in all four quadrants, a more complicated circuit is required, as illustrated in Fig. 4.31b. Sum and difference voltages $(x+y)$ and $(x-y)$ are computed, and each of these is squared, to obtain $(x^2+2xy+y^2)$ and $(x^2-2xy+y^2)$. The difference between these two voltages is taken, weighting each by a factor of $1/4$:

$$\tfrac{1}{4}(x^2+2xy+y^2) - \tfrac{1}{4}(x^2-2xy+y^2) = xy$$

The squaring circuits are broken down in Fig. 4.31c, where it can be seen that they consist of an absolute value circuit, log circuit, multiplication by 2, and anti-

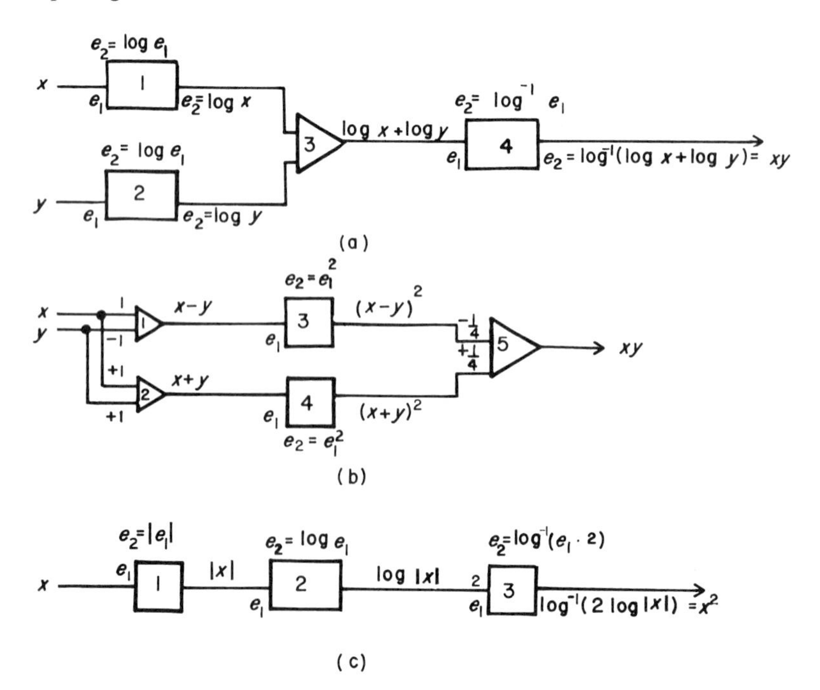

Figure 4.31 **(a)** A one-quadrant multiplier. **(b)** A four-quadrant multiplier. **(c)** One type of squaring circuit.

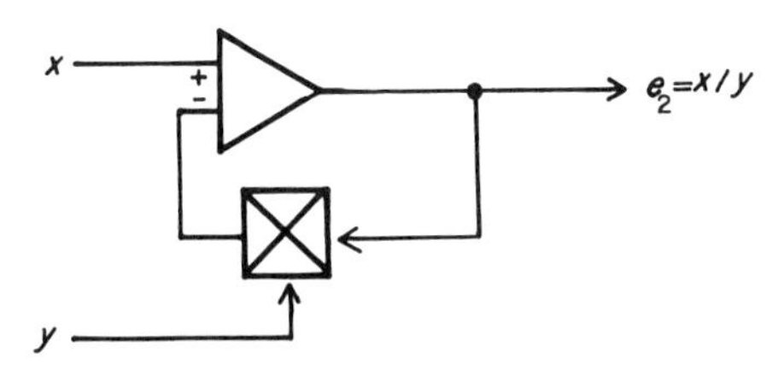

Figure 4.32 A divider.

logarithm:

$$x^2=\log^{-1}(s\log|x|)$$

The designer should not bother building his own multipliers. Multiplier modules are available which offer higher performance and lower cost than the average designer could attain with op amps or log modules.

Since division is the inverse of multiplication, a multiplier module can be inserted in the negative feedback loop of an op amp to make a divider, as is illustrated in Fig. 4.32. The two inputs to the op amp are virtually identical:

$$e^+=x$$

$$e^-=e_2y$$

Therefore, since $e^+=e^-$, we can substitute for e^+ and e^-:

$$x=e_2y$$

Solving for e_2 gives

$$e_2=\frac{x}{y}$$

Some multipliers come with divider inputs. The designer can either use these or use the method of Fig. 4.32. The former choice is preferable whenever feasible because of added simplicity and because occasionally stability problems arise in the circuit of Fig. 4.32 requiring degradation of frequency response to restore stability.

4.7 Positive Feedback

Figure 4.33a illustrates an op amp with resistive positive feedback. We can immediately make some observations about signals in this circuit:

$$i_{R_1}+i_{R_2}=0$$

from Kirchhoff's rules. Since the negative input is grounded,

$$e^-=0$$

The op amp will still try to obey the equation

$$e_2=m(e^+-e^-) \tag{4.42}$$

but as we shall see, it will be unable to do so because we must acknowledge the existence of maximum outputs of V^+_{max} and V^-_{max} imposed by the power supply levels and the characteristics of the op amp. That is,

$$e_2=m(e^+-e^-)$$

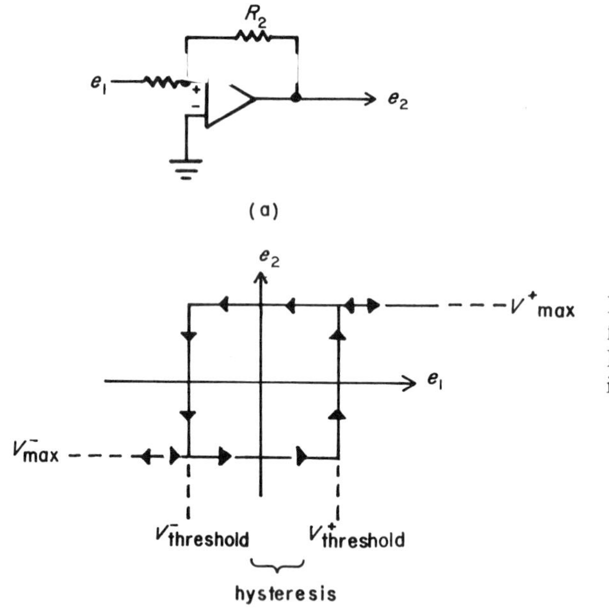

Figure 4.33 **(a)** Resistive positive feedback configuration. **(b)** Input–output curve of circuit in **(a)**.

only if

$$V^-_{max} \leqslant m(e^+ - e^-) \leqslant V^+_{max} \tag{4.43}$$

Since $e^- = 0$, the constraint becomes

$$V^-_{max} \leqslant me^+ \leqslant V^+_{max} \tag{4.44}$$

For an op amp with gain of 100,000 and $V^{\pm}_{max}$ of ± 10 V, e^+ would have to be within the range of ± 100 μV, a very small range indeed, less in fact that the dc offset of most op amps.

Consider, however, the operation of an ideal op amp with no output voltage limits, given different values of e_1. If $e_1 = 0$, and e_2 is also initially zero, e^+ must be zero since no current flows through the voltage divider $R_1 - R_2$ from e_1 to e_2. With $e^+ = 0$,

$$e_2 = m(e^+ - e^-) = m(0 - 0) = 0 \tag{4.45}$$

The output will remain zero as long as the input is precisely zero.

Consider e_1 now to become nonzero by an arbitrarily small amount δ. If $MR_1 > R_1 + R_2$, the output will rapidly swing to V^+_{max} if δ is positive, and V^-_{max} if δ is negative.

Therefore, the state $e_1 = e_2 = 0$ is *unstable* if $MR_1 > R_1 + R_2$. R_2 would have to be greater than MR_1, or 10^4–10^6 times larger than R_1, for the typical op amp for this condition *not* to occur.

There are, however, two stable states for this configuration: V^+_{max} and V^-_{max}. Given $e_2 = V^+_{max}$,

$$e^+ = \frac{(e_1 - V^+_{max})R_2}{R_1 + R_2} + V^+_{max} \tag{4.46}$$

Since $e_2 = me^+$, e^+ must become negative before the op amp can depart from $V_{\max}^+$. If e^+ does become negative, e_2 will move *regeneratively* to a new stable state, $V_{\max}^-$. For e^+ to be negative,

$$e^+ \leqslant \frac{(e_1 - V_{\max}^+)R_2}{R_1 + R_2} + V_{\max}^+ < 0 \tag{4.47}$$

Solving for e_1,

$$\frac{e_1 R_2 - V_{\max}^+ R_2 + V_{\max}^+ R_1 + V_{\max}^+ R_2}{R_1 + R_2} < 0$$

$$e_1 R_2 + V_{\max}^+ R_1 < 0$$

$$e_1 < -V_{\max}^+ \frac{R_1}{R_2}$$

That is, e_1 must be less than $-V_{\max}^+(R_1/R_2)$, which we shall call $V_{\text{threshold}}^-$:

$$V_{\text{threshold}}^- = -V_{\max}^+ \frac{R_1}{R_2} \tag{4.48}$$

This is the least negative value of e_1 necessary to force e_2 from $V_{\max}^+$ to $V_{\max}^-$. Similarly, we can define $V_{\text{threshold}}^+$ as the least positive value necessary to force e_2 from $V_{\max}^-$ to $V_{\max}^+$:

$$V_{\text{threshold}}^+ = -V_{\max}^- \frac{R_1}{R_2} \tag{4.49}$$

Thus, if R_1/R_2 is finite, $V_{\text{threshold}}^+$ is greater than zero and $V_{\text{threshold}}^-$ is less than zero. This means that the output will follow a different trajectory as the input e_1 goes from less than $V_{\text{threshold}}^-$ past $V_{\text{threshold}}^+$, then in the opposite direction from more than $V_{\text{threshold}}^+$ to less than $V_{\text{threshold}}^-$. This phenomenon is *hysteresis*, illustrated in the input–output relation of Fig. 4.33b.

The uses of such a device will become clear when an analogous configuration is demonstrated to serve as a binary memory (flipflop) in Chapter 5, and as a threshold detector in Chapter 7.

4.8 Performance Specifications of Analog Building Blocks

We have alluded several times in this chapter to the fact that real op amps and other analog devices are not the ideal devices upon which we have based our design theory. We have also indicated that for most applications it is possible to select devices whose specifications are adequate approximations to ideality. By "adequate" we mean that performance of the circuit as a whole is within the margin of error required by the user. Obviously deviations from ideality must be more rigorously examined for analog devices being incorporated into precision test and measuring equipment than for such imprecise applications as public address systems, dc power supplies to drive incandescent light bulbs, or crude audio monitors used for signal tracing when searching for defects in a malfunctioning circuit. In this section we shall define the most commonly used performance specifications.

4.8.1 Input Characteristics

1. *Input impedance*. As explained in Chapters 1 and 2, input impedance is the effective impedance between two terminals when subjected to an input signal. That is, if we apply a test signal voltage and measure the current drawn by the two terminals, we can compute the equivalent impedance between the terminals. This impedance and the output impedance of the signal source constitute a voltage divider whose transfer function can be computed to determine extent of signal loss, in terms of voltage, current, or power.

   In performance specifications provided by manufacturers of analog devices, input impedances are usually expressed as input resistance and capacitance (inductance is usually negligible). For devices with differential inputs, such as op amps, input impedance characteristics are usually specified both differentially and for either input with reference to ground. An ideal device has infinite input impedance, i.e., infinite resistance, zero capacitance, and zero inductance.
2. *Input bias current*. This is the average value of the two input bias currents of an IC differential input stage. This is a function of the large signal current gain of the input stage. An ideal device has zero input bias current.
3. *Input offset current*. This is the *difference* between input bias currents into the two input terminals of a differential input IC. It therefore represents an imbalance in voltage drops across the input resistances and is a measure of the effect of source impedance on common mode rejection. An ideal device has zero input offset current.
4. *Input offset voltage*. This is the voltage that must be applied across the terminals to bring the output voltage to zero. It is a measure of the matching tolerance in the differential amplifier stages. It is typically 1 mV or less for an IC op amp, and can often be compensated by injecting current through offset adjust inputs to the devices. Generally a potentiometer is used to zero the offset. The input offset is amplified by the gain of the device at its output: hence high-gain devices must be trimmed to lower input offset voltages than low-gain devices, to achieve the same output dc error. For an ideal device, this offset is zero.
5. *Input offset drift*. This is the average dc offset drift over time or temperature. It is zero for an ideal device.
6. *Input voltage range*. This is the voltage range at the input beyond which the device will not meet operating specifications. Usually exceeding this range will result in clipping or other nonlinearities. An ideal op amp has no input voltage limit.
7. *Input noise voltage*. With the inputs shorted to ground, a finite noise voltage will appear on the output of any real device. The input noise voltage is the input noise which would produce the observed output noise in an open loop configuration. It is zero for an ideal device.

4.8.2 Transfer Characteristics

1. *Frequency response*. This is the open-loop gain and phase plot (Bode plot) of the device.
2. *Large-signal voltage gain*. This is the maximum output voltage swing divided by the differential input voltage change required to produce the maximum output voltage swing. It is infinite in an ideal op amp.
3. *Common mode rejection ratio* (*CMRR*). The common mode signal is the average of the two input signals. Common mode rejection is the cancellation of the common mode signal by a differential amplifier or input stage. The CMRR is the ratio, often expressed in dB, of the output signal to a common mode input signal used to produce it. It is thus a measure of the ability of the device to cancel a common mode signal. The ideal device has infinite CMRR.
4. *Supply voltage rejection ratio* (*SVRR*). This is the ratio of change in output voltage produced by a change in supply voltage to the supply voltage change, usually expressed

as mV/V or μV/V. Thus, a 0.5-mV/V SVRR means that a 1-V supply voltage change will produce a 0.5-mV output voltage change. SVRR=0 in an ideal device.

5. *Rise time*. This is the length of time the output takes to go from 10% of its final value to 90% of its final value given a zero rise time step input. It is zero in an ideal device.
6. *Overshoot*. This is the degree to which the output transiently exceeds its final output value in response to a step input, usually expressed as percent of final output value. It is zero in an ideal device.
7. *Settling time*. This is the time after which the output will not deviate by more than a specified amount from its final output state. It is zero for an ideal device.
8. *Linearity*. This is the percent deviation of the output voltage from the predicted value based on a constant gain for all input voltages, usually obtained by comparing the observed output for all input voltages and finding the greatest difference from the best-fitting straight line, under open loop conditions. It is zero for an ideal device.

4.8.3 Output Characteristics

1. *Output impedance*. Usually output resistance is the only significant component of output impedance. It is the resistance seen by a signal source which tries to change the output voltage, defined as the voltage change divided by the amount of current required to produce it. An ideal device has zero output impedance ($R=0$, $C=0$, $L=0$).
2. *Maximum output voltage swing*. This is the maximum change in output voltage over which the op amp will perform to specification. Although the device may be capable of larger voltage swings, it will begin to show nonlinearities. This specification is dependent on supply voltage. An ideal device has infinite maximum output voltage swing.
3. *Maximum output current*. This is the maximum current which the output circuit can deliver. It is infinite in an ideal device.

4.9 Modification of Input and Output Characteristics

4.9.1 Changes of Input Circuitry

We have already presented some circuits which can be used to modify input characteristics. The unity gain follower is usually used as an impedance transformer, and if FET input op amps are used, with input impedances in excess of 10^{12} Ω and output impedances of less than 100 Ω, this can result in an impedance reduction of greater than 10^{10}, enough for almost all applications.

When using such high input impedances, it is wise to reduce all stray capacitances. Minimize all input cable lengths, using low-capacitance cables or omitting shields altogether. Unfortunately, it is precisely the high-impedance circuits which are most prone to electrostatic pickup, and hence they sometimes must be shielded.

To minimize the effect of shield capacitance and stray capacitance, two strategies are commonly used: *guard shielding* and *negative capacitance feedback*. In guard shielding, illustrated in Fig. 4.34a, the impedance-reduced signal out of the amplifier is applied to the shield rather than grounding the shield. Then the shield capacitance sees the same voltage on each side, and the voltage drop across the shield is always zero. This results in an absence of changes of voltage across the shield capacitance, effectively removing any capacitive load to ground. Nevertheless, interference signals see a low impedance to ground because of the low output impedance of the op amp.

Negative capacity feedback actually restores high frequencies at the input by feeding them back through a capacitor. The equivalent input circuit, R_1C_1 of Fig. 4.34b, depicts the shunting effect of C_1 for high frequencies:

$$Z_1 = R_1 + \frac{1}{j\omega C_1} = \frac{R_1 C_1 j\omega + 1}{j\omega C_1}$$

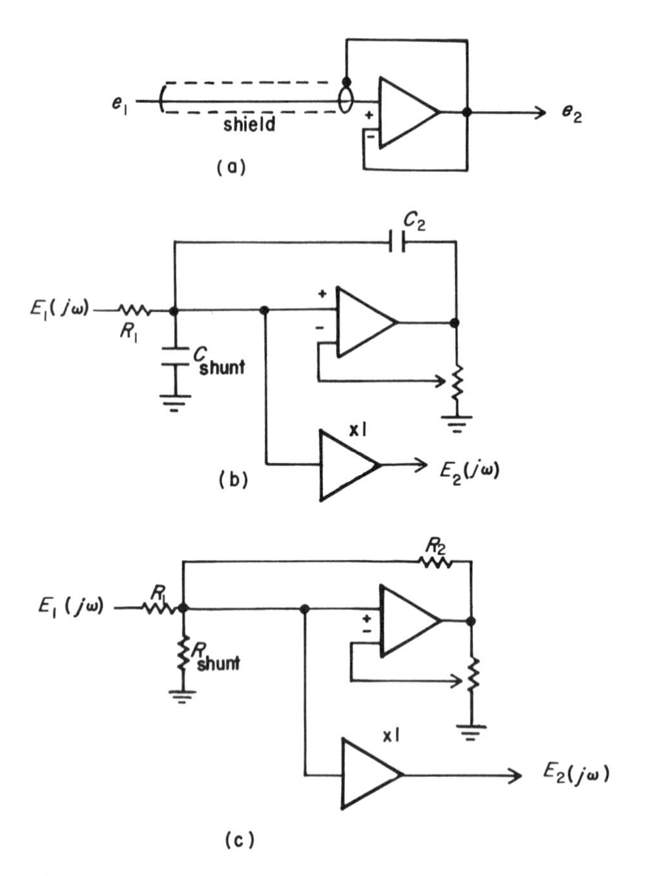

Figure 4.34 **(a)** Guard shield. **(b)** Negative capacity feedback. **(c)** Current pumping.

and the positive feedback element,

$$Z_2 = \frac{1}{j\omega C_2}$$

Thus, although the input circuit is a low-pass filter with maximum gain of unity for low frequencies, the positive feedback element is a high-pass filter with gain of greater than unity for high frequencies.

Since this is a positive feedback circuit with a phase lag introduced in the feedback loop, it can act as an oscillator (become unstable) if the gain is too high. Therefore the feedback gain must be adjusted to optimally compensate for input capacitance: if the gain is too low, input capacitance is inadequately compensated and high frequencies are unnecessarily attenuated; if too high, high frequencies are exaggerated and ringing or oscillation may occur.

Note also, that the capacitive positive feedback will tend to enhance high-frequency noise in the signal and introduced by the electronics.

Stray shunt resistances, which are unavoidable in some applications, can be compensated for by a method similar to negative capacity feedback, called *current pumping*. The circuit, illustrated in Fig. 4.34c, is identical to that of Fig. 4.34b, except that a resistor is used as a positive feedback element rather than a capacitor. Optimal operation occurs when a current is injected into the input node that is exactly equal to the current lost by the input shunting resistance.

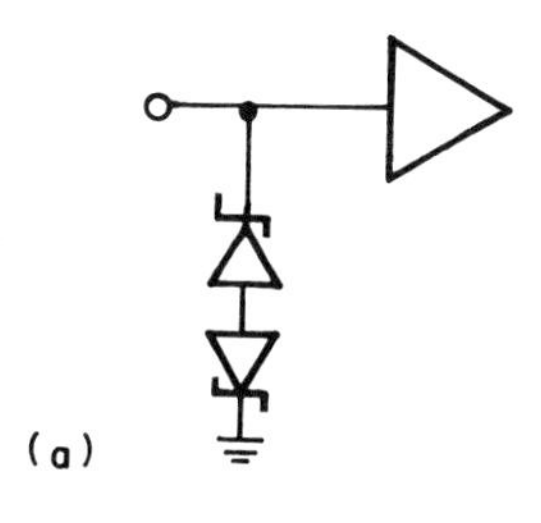

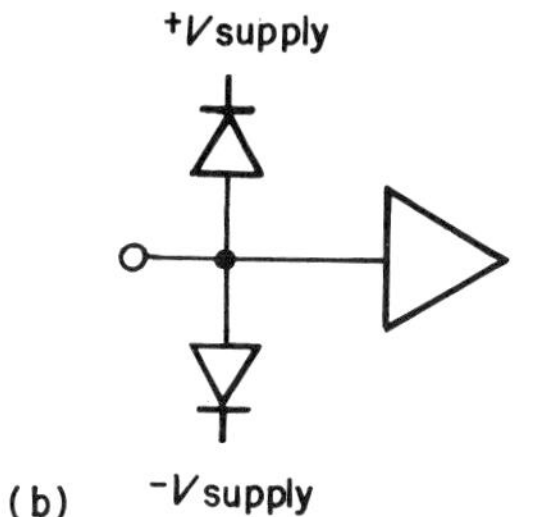

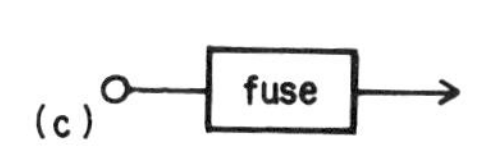

Figure 4.35 Input protection: **(a, b)** input overvoltage protection; **(c)** input current protection.

Methods of protecting the input of a device from overvoltages (voltages exceeding the maximum input voltage) are illustrated in Fig. 4.35. These circuits simply consist of diode configurations which shunt excessive voltages to ground or to the power supply lines. Shunting to ground is safer because there is no danger to the power supply. When currents are too great, however, the diodes will "blow," providing an open circuit, which will then fail to protect the input. For this reason, an input series element may be needed, consisting of an element which will open the input line before the shunt protection can be damaged. Such a fuse may be a pair of parallel low-current diodes pointed in opposite directions (do not forget that this will produce diode drops), a fast-blowing fuse (which must blow faster than the zeners), or even a cheap op amp which will blow out before the zeners do. The fusible element must not be soldered into the circuit; use a socket for easy replacement.

Occasionally the input to a circuit is from a current source, and a voltage signal is desired. The simplest conversion method for this situation is illustrated in Fig. 4.36. Since the input current must cancel the feedback current,

$$i_1 = -i_2$$

and the summing junction is a virtual ground:

$$e_2 = -i_1 R_2$$

R_2 can be chosen for desired proportionality. Any number of currents can be summed at the input junction and converted simultaneously.

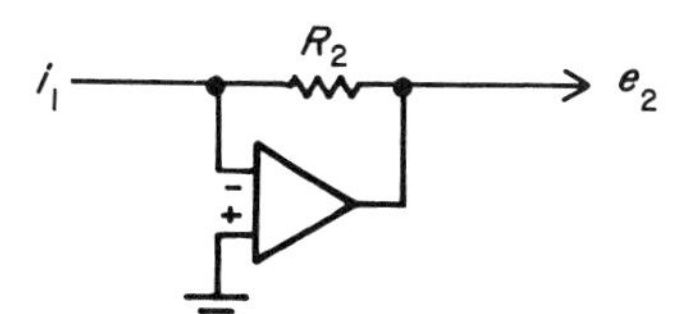

Figure 4.36 Current-to-voltage conversion.

4.9.2 Output Modifications

The most common requirements for output modifications are voltage or current range, voltage source to current source conversion, and protection of the output against overload. In addition, some op amps are sensitive to cable capacitance, and must be properly decoupled to prevent instability and oscillation. We shall discuss all of these.

To extend the current range of an analog circuit is the same as lowering its effective output impedance over its entire output voltage range. The simplest method is to add a current amplifier, often called a power amplifier or servo amplifier, to the output. These are available as modules, and as long as they have adequate specifications, they are usually reasonably priced and easy to use. Alternatively, power transistors can be used in emitter–follower configuration, as shown in Fig. 4.37. Since there is a zone of voltages at the op amp output in which neither transistor is adequately biased to conduct, the output is fed back to the op amp to force the output to follow the desired course. This will also compensate for the base–emitter diode drops of the transistors and for load changes.

The simplest way to boost the voltage output of a circuit is to buy a high-voltage op amp for the output stage. Such devices are cheaper and more reliable than the discrete designs which most amateur designers can come up with.

Figure 4.38 illustrates the use of a single op amp for voltage–current conversion, i.e., as a current source. The current i_2 through the load R_L must be equal to the input current i_1, which is

$$i_1 = e_1/R_1$$

This is true regardless of the load resistance, provided that the op amp limits are not exceeded. The circuit has the additional virtue of providing a regulated current in the event of a load whose characteristics are nonlinear or time varying.

Most op amps are now internally protected against overload. Some have protection against overload for short periods of time, and these can be protected with a series fuse of appropriate rating. Emitter–follower circuits such as the one of Fig. 4.37 can be used, with inexpensive low-power transistors in sockets; if the current limit is exceeded, the transistor will open-circuit, and can be replaced with a new one.

Output voltages can be limited by the circuits of Fig. 4.29; generally, this is safer than reducing the gain of the device to a sufficiently low level so that the maximum input times the gain is less than the desired voltage limit because transients may still occur that exceed the limit.

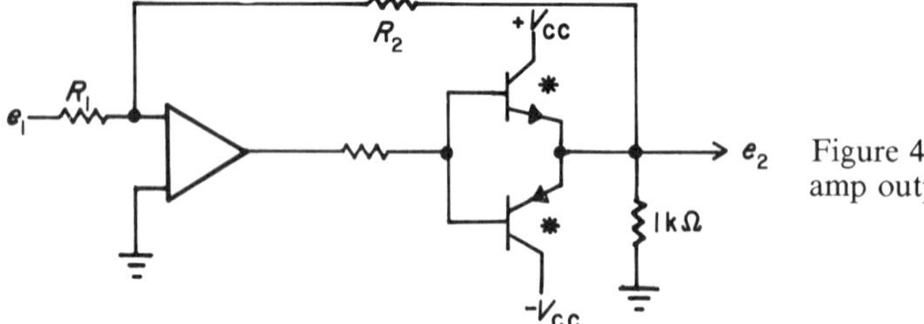

Figure 4.37 Current-boosting op amp output.

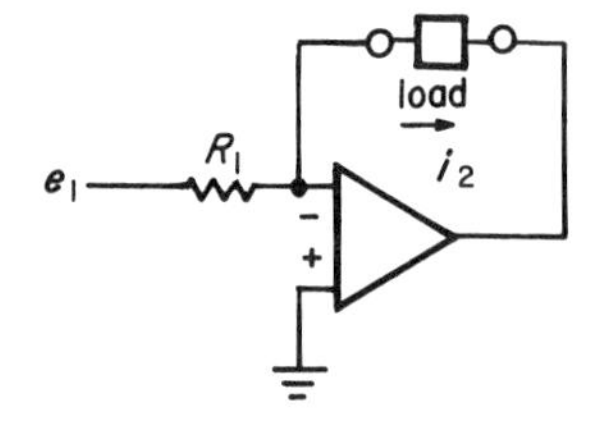

Figure 4.38 Current source.

4.10 Instrumentation Amplifiers

In order to adjust the gain of an operational amplifier used as a differential amplifier, it is usually necessary to vary the feedback resistor or one or more input resistors. This in turn causes a change in common mode rejection, which must be kept high (at least 60 dB, often 100 dB) in most differential amplifier applications. This then requires trimming a potentiometer to ground on the adder or subtractor circuit, which then affects the gain. Thus, one has to successively approximate the desired gain and CMRR by repeatedly tweaking these adjustments.

There are two possible solutions to this problem. One is to use a fixed gain differential amplifier as input stage, with a variable gain single-ended amplifier following it. This works well if the resulting variable dc offset of the output with changes in gain is acceptable.

Alternatively, one can use an instrumentation amplifier. These ICs and modules either have fixed gains or gains which can be adjusted by varying a single external resistor. The input stage characteristics are unaffected by gain changes, so that typically very high CMRR, very low dc offset changes, and often very low noise and high bandwidth can be maintained throughout the entire voltage range.

4.11 Higher-Level Building Blocks

Especially in the area of consumer electronics, quite complex analog functions are being realized in the form of single ICs or modules. Radio and television demodulation and demultiplexing, automatic gain and frequency control, chroma demodulators, and so forth are gradually replacing discrete components, lowering the cost of these devices and the cost of their servicing, and often improving the quality of their performance. Analog computational devices are appearing in home appliances and in automobiles as well.

In Chapter 6, we shall present methods of waveform generation using analog components. Many of the op amp-based methods are rapidly becoming obsolete as they become incorporated into single ICs. Devices exist for voltage-controlled frequency oscillators, frequency measurement, wave shaping, and other analog processes which are inexpensive and remarkably precise.

In Chapter 7, we shall present the rapidly growing field of hybrid analog/digital devices. These may represent the signal-processing world of the future because of certain unique advantages of digital versus analog computation. Once again, miniaturization and the incorporation of complex functions in ICs and modules is progressing at a rapid pace. Devices which occupied whole printed circuit boards are now available at lower costs in single ICs. This means that methods of digital signal processing which, although theoretically preferable, were impractical, are becoming quite practical. It also means that the modern scientist must become familiar with the world of digital signal processing, a subject we introduce in Chapter 7.

Problems

1. The following amplifier is used to provide switch variable gain changes which are tenfold (decade switch), with a vernier multiplier from 1.0 to 11. Insert the missing resistor values.

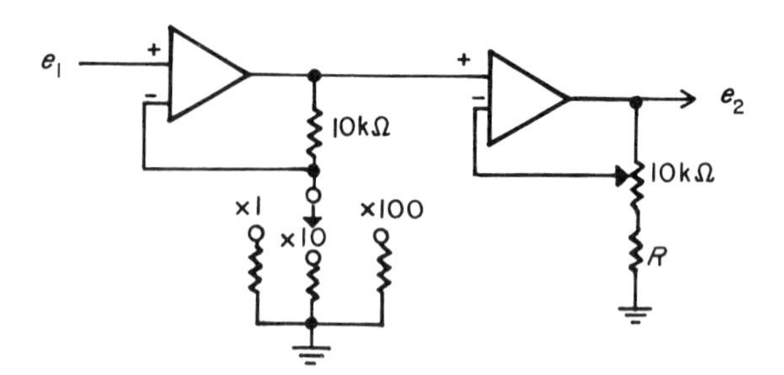

2. An op amp is built into a circuit such that the total circuit has a gain error of 0.01. How much does the gain error change if the op amp is replaced with another op amp whose gain is 10% lower than the original one?

3. Design an adder–subtractor to realize the following equations, using a 10-kΩ feedback resistor:

$$y = 16x_1 + 8x_2 + 0.3x_3 - 40x_4 - 0.6x_5$$

4. Block diagram a circuit to realize the following function, using analog computation symbols:

$$dy = a_1 + a_2x + a_3x^2\, dx$$

5. Design a bandpass filter which is an integrator for frequencies above 10 Hz, a differentiator for frequencies below 0.01 Hz, and has a gain of 6 dB for intermediate frequencies.

6. Using a 10-kΩ resistor as R, design an integrator for the bandwidth of 0–100 kHz. Design a differentiator with the same R, for the same bandwidth.

7. Design a decade-variable integrator with fixed R of 100 kΩ and switch selectable C_1 for $f_c = 1$ kHz, 10 kHz, and 100 kHz; design a differentiator with the same R, and for the same bandwidths.

8. Write the transfer function for the following circuit assuming an ideal op amp.

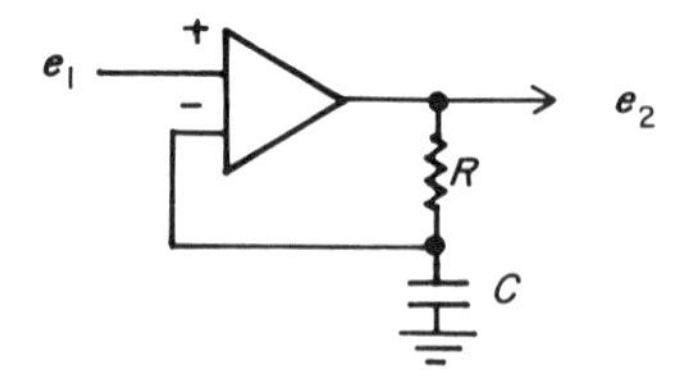

9. What is the transfer function of the following circuit assuming an ideal op amp?

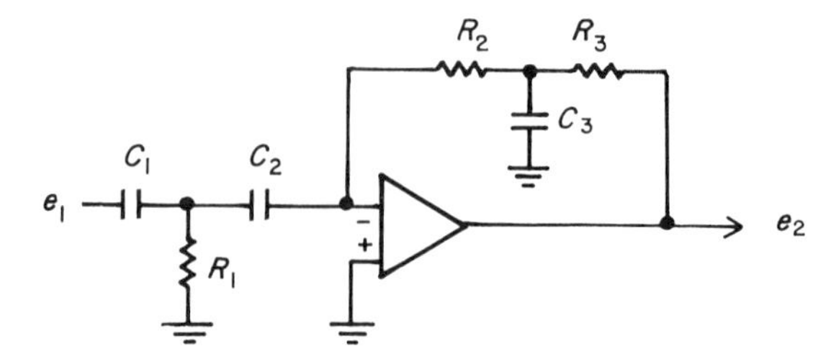

10. Diagram the following analog computations using analog computing symbology:

(a) $v_x = v\cos\theta$

(b) $v_y = v\sin\theta$

(c) $v = v_x^2 + v_y^2$

(d) $\theta = \tan^{-1} v_y / v_x$

(e) $a = \dfrac{d^2x}{dt^2}$

(f) $[NH_4^+][OH^-] \underset{k_1}{\overset{k_2}{\rightleftharpoons}} [NH_4OH]$

(g) $pH = pK + \log \dfrac{[HCO_3^-]}{a_1(PCO_2)}$

(h) $y = \begin{Bmatrix} 1 & \text{if } d^2s/dt^2 < 0 \\ 0 & \text{if } d^2x/dt^2 > 0 \end{Bmatrix}$, where $1 = V_{max}^+$, $0 = V_{max}^-$

11. Define the operation of the following circuit, in the form of an equation relating the output z to the inputs x and y.

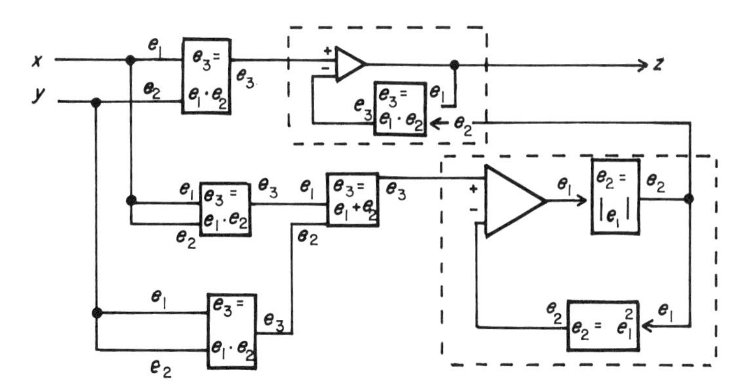

[*Hint*: define the inputs and outputs of each of the labeled analog computing elements in numerical order, keeping in mind that the large dashed computing elements have a function which is the inverse of the function of the feedback element: that is, the input of the op amp is equal to the output of the feedback element, and the output of the op amp is one or both of the inputs to the feedback element. Thus, after writing the equation for the feedback element, the output of the op amp can be solved for.]

Selected References

Eimbinder, J. *Designing with Linear Integrated Circuits*. John Wiley, New York, 1969.

Furlow, Bill. *Circuit Design Idea Handbook*. Cahners Books, Boston, 1972.

Korn, G. A., and T. M. Korn. *Electronic Analog and Hybrid Computers*, 2nd ed. McGraw-Hill, New York, 1972.

Lenk, John D. *Manual for Integrated Circuit Users*. Reston Publ. Co., Reston, VA, 1973.

5

Digital Devices

Paul B. Brown

5.1 Analog Versus Digital Signals

In Chapter 4, we introduced the distinction between digital and analog signals and proceeded to lay down the rudiments of analog signal processing and computation. In this chapter, we shall introduce digital design and on the basis of these two chapters, go on to more advanced applications in following chapters.

It will be recalled that we defined analog signals as those which can occupy a continuum of stable states, whereas digital signals can occupy only a finite number of stable states. This is analogous to the distinction between real numbers and integers: over a finite interval, there is an infinite number of real numbers and a finite number of integers. Although there are digital applications in which signals may occupy more than two stable states, the vast majority of current applications use electronics that have only two: these are called *binary* signals. The two states may be two different voltage levels, saturation or cutoff of a transistor, direction of current flow, or some other electronic variable. They may also correspond to the position of a two-position switch or the output of some circuit with an analog input, such as the discriminator introduced in Chapter 4 as an analog computing element, and the positive resistive feedback configuration.

Almost all digital devices are binary because the majority of physical parameters are most easily used for information carrying as on–off or plus–minus phenomena. Thus, a switch is either closed or open; the magnetic polarization of a particle on magnetic tape is polarized in one direction or another; current or voltage is either positive or negative, or on or off; a certain position on a computer card or computer tape has a hole in it or it does not. Many of these parameters could be varied to have any finite number of different values: we could allow a voltage to have any of ten different stable values; punched holes could have any of three different sizes; and so forth. But it is easiest, and therefore most reliable, to use only two-valued parameters for information storage, transfer, and manipulation. Each physical element which has such an on–off two-valued domain is known as a binary digit, or bit, for short. Information theory tells us that the bit is the smallest possible unit of information.

The immense importance of binary circuits and signals lies in the fact that they can be used to encode information, including numbers, logical states (true or false), the letters of the alphabet, or in fact anything which can be attributed a value, be it nominal, ordinal, or interval value. Technology has evolved to the point where complex operations can be performed on large numbers of digital signals at very high speeds, with truly remarkable dependability. Today, entire calculators and even small computers can be incorporated in single ICs, and their high speeds permit computations in the digital domain which often outstrip analogous analog signal processing. Indeed, as will be seen in Chapter 7, technology has evolved to the point where it is more efficient in many applications to replace such analog processes as signal processing and signal recording with a conversion to digital representation of the signal, digital processing, and either reconversion to analog form or digital recording. Even before these advances, which have occurred only in the last decade, digital processing had a strong demand in digital computers, which have been in great use since the 1950s.

In the ensuing sections we will develop the fundamentals of binary logic and binary arithmetic (Sections 5.2 and 5.3). Later sections will proceed from design at the level of elementary logic operations using *small-scale integration* (SSI), through more complex devices using *medium-scale integration* (MSI), to the very complex devices such as microcomputers, using *large-scale integration* (LSI).

5.2 Principles of Binary Logic

The fundamentals of binary logic have been a subject for study, in one form or another, since the dawn of history. Rules for testing the truth or falsity of propositions are understood at the intuitive level to some extent by every normal adult. The basis for all logical operations lies in the use of *logical variables*, to which can be assigned binary values, true or false. A set of elementary logical operators can be easily defined by means of *truth tables*. All of binary logic, also called Boolean logic, can be based on as few as one of these operators. We shall define a number of basic operators with truth tables and enumerate the basic laws for concatenation of operations.

The simplest operator, NOT, or logical inversion, is illustrated in Table 5.1. A logical variable p, when operated on by the logical operator NOT, becomes NOT p. The values of NOT p are listed in Table 5.1 for all possible values of p. Since a logical variable can only have two values, true or false, NOT p can also have only two values: it can be seen that NOT p is true if p is false, and NOT p is false if p is true. This reversal of value, from true to false or false to true, is the reason why NOT is often referred to as *inversion*, or sometimes INV. Other notational systems exist, and these will be reviewed later, along with corresponding alternate notations for other logical operators. However, for brevity we shall immediately introduce the use of binary notation in place of true and false. A logical 0 (zero) is here defined as false, and a logical 1 (one) is defined as true.

The operators EQUALS (equivalence), AND, and OR (INCLUSIVE OR) are defined in truth tables in Table 5.2; both true–false and binary notation are used, to familiarize the reader with their interchangeability. Note that all these operators

Table 5.1 Definition of the Logical Operator NOT with a Truth Table

p	NOT p
True	False
False	True

Table 5.2 Truth Table Definitions of EQUALS (Equivalence), AND, and OR.

p[a]	q[a]	$p = q$[a]	p AND q[a]	p OR q[a]
true	true	true	true	true
true	false	false	false	true
false	true	false	false	true
false	false	true	false	false
p[b]	q[b]	$p = q$[b]	p AND q[b]	p OR q[b]
1	1	1	1	1
1	0	0	0	1
0	1	0	0	1
0	0	1	0	0

[a] Using true–false values. [b] Using binary notation.

Table 5.3 Truth Table Definitions of NAND, NOR, and EOR (EXCLUSIVE OR) Operators

p	q	p NAND q	p NOR q	p EOR q
0	0	1	1	0
0	1	1	0	1
1	0	1	0	1
1	1	0	0	0

operate on pairs of variables. These, and the operator NOT, can be concatenated to operate on larger numbers of variables, as we shall describe later.

Note that the operators have definitions closely agreeing with their English usage. Thus, p EQUALS q is true if p and q are both 1 or both 0, but not if one is 1 and the other is 0. Also, p AND q is true only if both p and q are true. Finally p OR q is true if p is true, q is true, or both are true.

Three other logical operators, NAND, NOR, and EXCLUSIVE OR (EOR), are defined using binary notation in Table 5.3. NAND (abbreviation for NOT AND) and NOR (abbreviation for NOT OR) are the inverse of AND and OR, respectively, and EOR is the inverse of EQUALS.

Logical expressions can be written by stringing together variables and operators. In order to make notation less cumbersome, the logical operators can be expressed by symbols, just as arithmetic operators PLUS, MINUS (NEGATE), EQUALS, TIMES, and DIVIDED BY are expressed as $+$, $-$, $=$, $\times$ or $\cdot$, and / or $\div$ in arithmetic expressions and algebraic equations. Some notations for logical operators are indicated in Table 5.4. We shall use the notation in the top line of entries. Note that EQUALS is logically equivalent to IMPLIES. The symbol $=$ is used for EQUALS, the symbols $\supset$ and $\rightarrow$ are used for IMPLIES.

The symbols at the foot of the table are used most often to represent electronic realization of logical operators. Generally, EQUALS is not used in electronic operations, but it is sometimes used to perform comparisons, to detect equality of logical variables. Since it is the inverse of EOR, it is represented by the EOR symbol with a circle at its output. In general, incorporation of logical inversion is accomplished by adding a circle to the output. Inputs to devices can also be inverted by adding circles at the inputs.

Table 5.4 Notational Symbols Used for Logical Operators

NOT, INV	AND	OR	EQUALS	NAND	NOR	EOR
$\bar{p}$	$p \wedge q$	$p \vee q$	$p = q$	$p \overline{\wedge} q$	$p \overline{\vee} q$	$p \oplus q$
$\dot{p}$	$p \cap q$	$p \cup q$	p IMPLIES q	$\overline{p\, q}$		
$-p$	$p \times q$	$p + q$	$p \supset q$	$\overline{(p \wedge q)}$	$\overline{(p \vee q)}$	
$\sim p$	pq		$p \rightarrow q$	$\overline{(p \times q)}$	$\overline{(p + q)}$	
p → $\bar{p}$	p, q → $p \wedge q$	p, q → $p \vee q$	p, q → $p = q$	p, q → $p \overline{\wedge} q$	p, q → $(p \overline{\vee} q)$	p, q → $p \oplus q$

Table 5.5 Truth Table Demonstrating Laws of Tautology[a]

A	$A \vee A$	$A \wedge A$
1	1	1
0	0	0
identical		

[a]All three columns are identical, and therefore logically equivalent.

5.2.1 The Laws of Logic

Parentheses may be used for grouping expressions: $(A \overline{\vee} B) \overline{\wedge} (C \wedge D)$ means "the quantity (A NOR B) NAND the quantity (C AND D)." Parentheses can also be nested, as in algebra. Logically equivalent terms can be interchanged: Given $A = B$, A can be substituted for B and B can be substituted for A in any expression.

The following laws of logic are basic to proper use of binary logic and should be committed to memory. The accompanying truth tables demonstrate the validity of these laws.

LAW OF IDENTITY: $A = A$ (5.1)

LAWS OF TAUTOLOGY (Table 5.5): $A \wedge A = A$ (5.1a)

$A \vee A = A$ (5.1b)

COMMUTATIVE LAWS (Table 5.6): $A \wedge B = B \wedge A$ (5.2a)

$A \vee B = B \vee A$ (5.2b)

$A \oplus B = B \oplus A$ (5.2c)

DISTRIBUTIVE LAWS (Table 5.7): $A \wedge (B \vee C) = (A \wedge B) \vee (A \wedge C)$ (5.3a)

$A \vee (B \wedge C) = (A \vee B) \wedge (A \vee C)$ (5.3b)

LAW OF DOUBLE NEGATION (Table 5.8): $-(-A) = A$ (5.4)

LAWS OF ABSORPTION (Table 5.9): $A \wedge (A \vee B) = A$ (5.5a)

$A \vee (A \wedge B) = A$ (5.5b)

ASSOCIATIVE LAWS (Table 5.10): $A \wedge (B \wedge C) = (A \wedge B) \wedge C$ (5.6a)

$A \vee (B \vee C) = (A \vee B) \vee C$ (5.6b)

MISCELLANEOUS USEFUL RELATIONS:

$A \wedge 1 = A$ $\quad A \vee 1 = 1$ (5.7a)

$A \wedge 0 = 0$ $\quad A \vee 0 = A$ (Table 5.11) (5.8a)

$A \wedge \bar{A} = 0$ $\quad A \vee \bar{A} = 1$ (5.9a, b)

$(A = -B) = (-B = A) = (-A = B) = (B = -A)$ (Table 5.12) (5.10)

$(A \wedge B) \vee (A \wedge \bar{B}) = A$ (Table 5.13) (5.11)

$(A \vee B) \wedge (A \vee \bar{B}) = A$ (5.12)

$-(A \wedge B) = \bar{A} \vee \bar{B}$ de Morgan's theorem (5.13)

$-(A \vee B) = \bar{A} \wedge \bar{B}$ (Table 5.14) (5.14)

Table 5.6 Proof of Commutative Laws.

A	B	$A \wedge B$	$B \wedge A$	$A \vee B$	$B \vee A$	$A \oplus B$	$B \oplus A$
0	0	0	0	0	0	0	0
0	1	0	0	1	1	1	1
1	0	0	0	1	1	1	1
1	1	1	1	1	1	0	0
		identical		identical		identical	

Table 5.7 Proof of Distributive Laws

A	B	C	$B \wedge C$	$A \vee (B \wedge C)$	$A \vee B$	$A \vee C$	$(A \vee B) \wedge (A \vee C)$	$B \vee C$	$A \wedge (B \vee C)$
0	0	0	0	0	0	0	0	0	0
0	0	1	0	0	0	1	0	1	0
0	1	0	0	0	1	0	0	1	0
0	1	1	1	1	1	1	1	1	0
1	0	0	0	1	1	1	1	0	0
1	0	1	0	1	1	1	1	1	1
1	1	0	0	1	1	0	1	1	1
1	1	1	1	1	1	1	1	1	1

$A \vee (B \wedge C)$ and $(A \vee B) \wedge (A \vee C)$: identical

continued:

$A \wedge B$	$A \wedge C$	$(A \wedge B) \vee (A \wedge C)$
0	0	0
0	0	0
0	0	0
0	0	0
0	0	0
0	0	0
0	1	1
1	0	1
1	1	1

$A \wedge (B \vee C)$ and $(A \wedge B) \vee (A \wedge C)$: identical

Table 5.8 Proof of Law of Double Negation

A	$-A$	$-(-A)$
0	1	0
1	0	1

A and $-(-A)$: identical

Table 5.9 Proof of Laws of Absorption

A	B	$A \vee B$	$A \wedge (A \vee B)$	$A \wedge B$	$A \vee (A \wedge B)$
0	0	0	0	0	0
0	1	1	0	0	0
1	0	1	1	0	1
1	1	1	1	1	1

(A and $A \wedge (A \vee B)$: identical; A and $A \vee (A \wedge B)$: identical)

Table 5.10 Proof of Associative Laws

A	B	C	$B \wedge C$	$A \wedge (B \wedge C)$	$A \wedge B$	$(A \wedge B) \wedge C$	$B \vee C$	$A \vee (B \vee C)$	$A \vee B$	$(A \vee B) \vee C$
0	0	0	0	0	0	0	0	0	0	0
0	0	1	0	0	0	0	1	1	0	1
0	1	0	0	0	0	0	1	1	1	1
0	1	1	1	0	0	0	1	1	1	1
1	0	0	0	0	0	0	0	1	1	1
1	0	1	0	0	0	0	1	1	1	1
1	1	0	0	0	1	0	1	1	1	1
1	1	1	1	1	1	1	1	1	1	1

($A \wedge (B \wedge C)$ and $(A \wedge B) \wedge C$: identical; $A \vee (B \vee C)$ and $(A \vee B) \vee C$: identical)

Table 5.11 Proof of Relations (5.7)–(5.9)

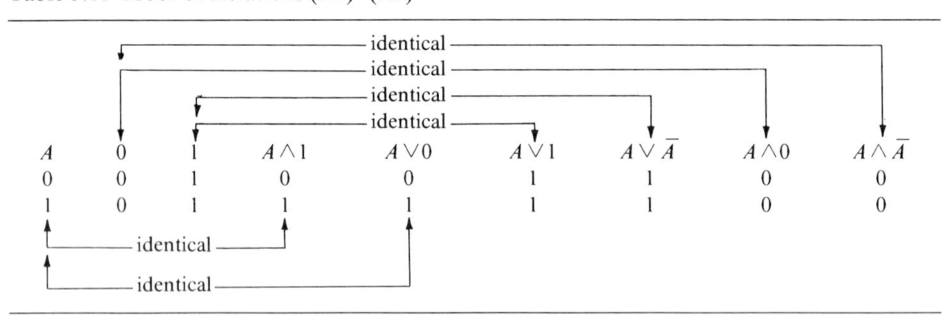

A	0	1	$A \wedge 1$	$A \vee 0$	$A \vee 1$	$A \vee \bar{A}$	$A \wedge 0$	$A \wedge \bar{A}$
0	0	1	0	0	1	1	0	0
1	0	1	1	1	1	1	0	0

(A and $A \wedge 1$: identical; A and $A \vee 0$: identical; 1 and $A \vee 1$: identical; 1 and $A \vee \bar{A}$: identical; 0 and $A \wedge 0$: identical; 0 and $A \wedge \bar{A}$: identical)

Table 5.12 Proof of Relation (5.10)

A	B	$-A$	$-B$	$A = -B$	$-B = A$	$-A = B$	$B = -A$
0	0	1	1	0	0	0	0
0	1	1	0	1	1	1	1
1	0	0	1	1	1	1	1
1	1	0	0	0	0	0	0

($A = -B$, $-B = A$, $-A = B$, $B = -A$: identical)

Table 5.13 Proof of Relations (5.11) and (5.12)

A	B	$\overline{B}$	$A \wedge B$	$A \wedge \overline{B}$	$(A \wedge B) \vee (A \wedge \overline{B})$	$A \vee B$	$A \vee \overline{B}$	$(A \vee B) \wedge (A \vee \overline{B})$
0	0	1	0	0	0	0	1	0
0	1	0	0	0	0	1	0	0
1	0	1	0	1	1	1	1	1
1	1	0	1	0	1	1	1	1

(A and $(A \wedge B) \vee (A \wedge \overline{B})$: identical; A and $(A \vee B) \wedge (A \vee \overline{B})$: identical)

Table 5.14 Proof of de Morgan's Theorem

A	B	$\overline{A}$	$\overline{B}$	$A \wedge B$	$-(A \wedge B)$	$\overline{A} \vee \overline{B}$	$A \vee B$	$-(A \vee B)$	$\overline{A} \wedge \overline{B}$
0	0	1	1	0	1	1	0	1	1
0	1	1	0	0	1	1	1	0	0
1	0	0	1	0	1	1	1	0	0
1	1	0	0	1	0	0	1	0	0

($-(A \wedge B)$ and $\overline{A} \vee \overline{B}$: identical; $-(A \vee B)$ and $\overline{A} \wedge \overline{B}$: identical)

De Morgan's theorem implies that given the operators INV and AND, we can synthesize OR (5.13); and given INV and OR, we can synthesize AND (5.14). Using NAND or NOR as primitives, we can synthesize all of the other operators:

$$\overline{A} = A \overline{\wedge} A = A \overline{\vee} A = A \overline{\wedge} 1 = A \overline{\vee} 0 \qquad \text{(Table 5.15)} \qquad (5.15)$$

$$A \wedge B = (A \overline{\vee} 0) \overline{\vee} (B \overline{\vee} 0) = (A \overline{\wedge} B) \overline{\wedge} 1 \qquad \text{(Table 5.16)} \qquad (5.16)$$

$$A \vee B = (A \overline{\wedge} 1) \overline{\wedge} (B \overline{\wedge} 1) = (A \overline{\vee} B) \overline{\vee} 0 \qquad \text{(Table 5.17)} \qquad (5.17)$$

$$A \overline{\wedge} B = ((A \overline{\vee} 0) \overline{\vee} (B \overline{\vee} 0)) \overline{\vee} 0 \qquad \text{(Table 5.18)} \qquad (5.18)$$

$$A \overline{\vee} B = ((A \overline{\wedge} 1) \overline{\wedge} (B \overline{\wedge} 1)) \overline{\wedge} 1 \qquad \text{(Table 5.19)} \qquad (5.19)$$

Table 5.15 Synthesis of INV Using Only NAND or NOR

A	$\overline{A}$	$A \wedge A$	$A \overline{\wedge} A$	$A \vee A$	$A \overline{\vee} A$	$A \overline{\wedge} 1$	$A \overline{\vee} 0$
0	1	0	1	0	1	1	1
1	0	1	0	1	0	0	0

($\overline{A}$, $A \overline{\wedge} A$, $A \overline{\vee} A$, $A \overline{\wedge} 1$, $A \overline{\vee} 0$: identical)

Table 5.16 Synthesis of AND Using NAND or NOR

A	B	$A \overline{\vee} 0$	$B \overline{\vee} 0$	$(A \overline{\vee} 0) \overline{\vee} (B \overline{\vee} 0)$	$A \overline{\wedge} B$	$(A \overline{\wedge} B) \overline{\wedge} 1$	$A \wedge B$
0	1	1	0	0	1	0	0
1	0	0	1	0	1	0	0
0	0	1	1	0	1	0	0
1	1	0	0	1	0	1	1
				↑	identical	↑ identical	↑

Table 5.17 Synthesis of OR Using Only NAND or NOR

A	B	$A \vee B$	$A \overline{\wedge} 1$	$B \overline{\wedge} 1$	$(A \overline{\wedge} 1) \overline{\wedge} (B \overline{\wedge} 1)$	$A \overline{\vee} B$	$(A \overline{\vee} B) \overline{\vee} 0$
0	0	0	1	1	0	1	0
0	1	1	1	0	1	0	1
1	0	1	0	1	1	0	1
1	1	1	0	0	1	0	1
		↑ identical			↑	identical	↑

Table 5.18 Synthesis of NAND Using Only NOR

A	B	$A \overline{\wedge} B$	$A \overline{\vee} 0$	$B \overline{\vee} 0$	$(A \overline{\vee} 0) \overline{\vee} (B \overline{\vee} 0)$	$((A \overline{\vee} 0) \overline{\vee} (B \overline{\vee} 0)) \overline{\vee} 0$
0	0	1	1	1	0	1
0	1	1	1	0	0	1
1	0	1	0	1	0	1
1	1	0	0	0	1	0
		↑		identical		↑

Table 5.19 Synthesis of NOR Using Only NAND

A	B	$A \overline{\vee} B$	$A \overline{\wedge} 1$	$B \overline{\wedge} 1$	$(A \overline{\wedge} 1) \overline{\wedge} (B \overline{\wedge} 1)$	$((A \overline{\wedge} 1) \overline{\wedge} (B \overline{\wedge} 1)) \overline{\wedge} 1$
0	0	1	1	1	0	1
0	1	0	1	0	1	0
1	0	0	0	1	1	0
1	1	0	0	0	1	0
		↑		identical		↑

5.2.2 Examples of the Use of Logical Operators

Example 5.1. The design of an EOR device

The choice of design for an Exclusive OR (EOR) device depends on the kinds of logical operators available. In words, EOR can be defined as "the output is 1 if and only if one input OR the other is 1, *AND* both are not high at the same time." A logical equation would be

$$A \oplus B = (A \vee B) \wedge (\overline{A \wedge B}) = (A \vee B) \wedge (A \mathbin{\overline{\wedge}} B) \tag{5.20}$$

The designer would thus need an OR, an AND, and a NAND device to implement the EOR function. The following are implementations using only (a) OR and INV, (b) AND and INV:

(a) $A \oplus B = (A \vee B) \wedge (\bar{A} \vee \bar{B})$ [from Eq. (5.13)]

$= ((\overline{A \vee B}) \vee (\overline{\bar{A} \vee \bar{B}}))$ [from Eq. (5.13)]

(b) $A \oplus B = (\overline{\bar{A} \wedge \bar{B}}) \wedge (\overline{A \wedge B})$ [from Eq. (5.14)]

Example 5.2. Control of a light from multiple switches

A pair of switches is used to control the lights on a stairway, so that flipping a switch at the top or the bottom of the stairs will turn the lights off if they are on, or on if they are off. If A = state of switch 1, and B = state of switch 2, use logical operators to relate A and B to determine the state Z (on or off), of the light. A truth table would be

A	B	Z
P	Q	0
$\bar{P}$	Q	1
P	$\bar{Q}$	0
$\bar{P}$	$\bar{Q}$	1

$P = 0$ or 1, $Q = 0$ or 1

That is, regardless of the value of A (P or $\bar{P}$), there is a value of B where $Z = 0$, and another value where $Z = 1$; and regardless of the value of B (Q or $\bar{Q}$), there is a value of A where $Z = 0$, and another value where $Z = 1$. Given all possible alternatives of P and Q we come up with four truth tables, all of which will be adequate:

A	B	Z
1	1	0
0	1	1
1	0	1
0	0	0

(a) $P = 1, Q = 1$

A	B	Z
0	1	0
1	1	1
0	0	1
1	0	0

(b) $P = 0, Q = 1$

A	B	Z
1	0	0
0	0	1
1	1	1
0	1	0

(c) $P = 1, Q = 0$

A	B	Z
0	0	0
1	0	1
0	1	1
1	1	0

(d) $P = 0, Q = 0$

Tables (a) and (d) are logically equivalent (different order of entries, but each A, B pair has same resultant C value), and are equivalent to the operation $A \oplus B = Z$ (see Table 5.13). Tables (b) and (c) are logically equivalent, and have Z equal to the inverse of Z in tables (a) and (b); hence $(A = B) = Z$ (meaning, "if '$A = B$' is true, $Z = 1$"). Thus, an EOR or EQUALS operation will suffice to control the stairway light.

Example 5.3. Generalize the design problem of the previous example to a stairway with n landings, each with a switch, so that flipping any switch changes the state of the stairway light(s).

Here, given the state of the light Z for two switches, the state Y of the light for three switches, adding switch C, can be expressed as

$$Y = C \oplus Z, \quad \text{or} \quad Y = (C = Z)$$

For a fourth switch D, the state X of the light is

$$X = D \oplus Y, \quad \text{or} \quad X = (D = Y)$$

Thus, the series can be continued indefinitely, taking the resultant of $n-1$ switches and interacting it with the nth switch with an EOR or EQUALS. In general, for switches numbered 1 to n, and $n-1$ states, we can express the iterative equation:

$$\text{For } i = 1 \text{ to } n, \text{ state } i = (\text{state } i-1) \oplus (\text{switch } i), \text{ or}$$

$$(\text{state } i-1) = (\text{switch } i)$$

The reader can verify that this will work for any number of switches, with a truth table.

Example 5.4. Control of a furnace and forced air ducts

A furnace and the forced air ducts to the rooms of a house are controlled as follows: A duct (one for each room) is open if the thermostat for that room is demanding heat. The furnace goes on if at least two rooms require heat. The logical equation for the furnace control signal, Z, is determined from the five room thermostat signals, A–E, by the following logical equation:

$$Z = (A \wedge (B \vee C \vee D \vee E)) \vee (B \wedge (C \vee D \vee E)) \vee (C \wedge (D \vee E)) \vee (D \wedge E)$$

This equation sets $Z = 1$ if any pair of thermostat inputs A–E is on. This simple logical system permits selective heating of rooms demanding heat, with additional fuel savings by virtue of the requirement that two rooms must demand heat before the furnance is turned on.

These fundamental logical operations are adequate for all logical manipulations. They are hence adequate for hardware implementation of devices such as computers, calculators, and other binary machines that utilize logical operations. Before describing such machines, it is necessary to provide a basic foundation in the number systems used by such machines, which is presented in the next section.

5.3 Binary, Octal, and Hexadecimal Arithmetic

We are all familiar with the everyday number system used by the entire modern world: decimal numbers. Using the integers 0–9 and a positional weighting notation, we can express all integers, all rational and irrational numbers, and all numbers with decimal points (which are themselves rational numbers, if they have a finite number of decimal places). Most readers will have forgotten, if they ever really understood, the actual basis for this notation system and for the simple arithmetic operations of addition, subtraction, multiplication and division. Since all of the arguments used to describe these aspects of the decimal system and its arithmetic are used for other number systems, we shall review the familiar decimal system first.

5.3.1 Decimal Arithmetic

An integer is a "round number," consisting of a string of digits, length 1 or more, each digit with a value 0, 1, 2, 3, 4, 5, 6, 7, 8, or 9. The following are integers: 0, 123, 42, 1, 628, and −49. The last example has a negative sign; numbers with no sign are assumed to be positive. In each integer, the value of the number is the sum of the weighted values of the digits which make it up. The rightmost digit has its face value (the value of the digit) times 10^0. Thus, the number 9 is equal to 9×10^0. The second digit from the right has a value equal to its face value times 10^1, the third is equal to its face value times 10^2, etc. The value of an n-bit integer is equal to the sum of the weighted values of its digits:

$$S=\sum_{i=0}^{n} 10^i m_i \tag{5.21a}$$

where S is the value of the integer, there are $n+1$ digits in the integer, and m_i is the ith digit. The zeroth digit is the rightmost one. For example,

$$\begin{aligned}
0&=0\times 10^0\\
123&=3\times 10^0+2\times 10^1+1\times 10^2\\
42&=2\times 10^0+4\times 10^1,\\
1{,}628&=8\times 10^0+2\times 10^1+6\times 10^2+1\times 10^3\\
-49&=-9\times 10^0-4\times 10^1
\end{aligned}$$

Note that for negative numbers the integer is equal to the sum of weighted *negative* digits.

Rational and irrational numbers are the ratio of two integers or irrational multiples of rational numbers, respectively.

Numbers with decimal points are simply an extension of integers, with digits to the right of the decimal point being the product of their face value and negative powers of ten. The first digit to the right of the decimal point is multiplied by 10^{-1}, the second by 10^{-2}, etc. Hence the value of a number with a decimal point is equal to

$$S=\sum_{i=n}^{p} 10^i m_i \tag{5.21b}$$

where n is a negative number representing the number of places to the right of the decimal point, and $p+1$ is the number of places to the left. Thus,

$$\begin{aligned}
11.3&=3\times 10^{-1}+1\times 10^0+1\times 10^1\\
0.0047&=7\times 10^{-4}+4\times 10^{-3}+0\times 10^{-2}+0\times 10^{-1}+0\times 10^0\\
-30.046&=-6\times 10^{-3}-4\times 10^{-2}-0\times 10^{-1}-0\times 10^0-3\times 10^1
\end{aligned}$$

The relative values of the ten one-digit integers are defined by counting. Starting with zero, counting produces the sequence $0,1,2,3,4,5,6,7,8,9$, and then $0+1\times10^1$, or 10. The last increment, or counting operation, resulted in adding one to the highest single-digit integer, 9, to produce 0, with a carry to the next digit to the left:

```
 09
+1
---
 10
 ||
 |└──── 9 goes to zero
 └───── a carry to the next position increments the 0 to 1
```

By means of carrying whenever a nine counts to ten, we can generate more and more digits to the left, resulting in a limitless sequence of numbers. This kind of representation system has the advantage that every integer is represented by only one numerical representation, every numerical representation corresponds to only one integer, and thus the set of all integers maps onto the set of all numbers in the numerical representation system.

Addition is a modified form of counting. To add n to m, we increment (count up) n times, starting at m. Thus,

$3+2=5$, where 5 is the result of two increments to 3

(this is the method used by people who add on their fingers);

$10+11=21$, where 21 is the result of eleven increments to ten.

At the tenth increment, 19 went to 20, because the nine went to zero with a carry incrementing the 1 in the second place to a two. In general, when an addition is expressed as $m+n=p$, or

$$\begin{array}{r} m \\ +n \\ \hline p \end{array}$$

m is the *addend*, or starting value, n is the *augend*, or number of increments and p is the *sum*, or final value.

Subtraction is the addition of a negative number. A negative number n is the representation of n decrements, or backward counts. Thus, -4 is the result of counting backwards from 0 four times: $0,-1,-2,-3,-4$. Adding a negative number n to another number m is performed by decrementing:

$3-2=1$, where 1 is the result of two decrements from 3.

When decrementing a digit through zero, the next decrement produces a nine and a *borrow* from the next digit to the left:

```
 20
-1
---
 19
 ||
 |└──── decrementing 0 produces 9
 └───── a borrow from the 2 produces 1
```

Decrementing zero produces -1: further decrements produce successively larger-magnitude negative numbers: $-2,-3,-4$, etc. These numbers follow the same sequence with successive decrements as do positive numbers for successive increments.

Subtracting n from p is the same as decrementing p n times. Thus, in $p-n=m$, or

$$\begin{array}{r} p \\ -n \\ \hline m \end{array}$$

p is the *subtrahend*, or initial value, n is the *minuend*, or number of decrements, and m is the *difference*, or final value. If decrementing takes the number through zero, the difference is of sign opposite to the sign of the subtrahend, and the magnitude is determined as follows: If $p-n=-m$, where $|m|$ is the magnitude of the negative difference, then $|m|=n-p$. Thus, since $47-63$ is a negative number, the magnitude (absolute value) of the difference is $|63-47|=16$. Therefore $47-63=-16$.

Finally, the addend or subtrahend may be negative. In such cases, i.e., $-m+n=\pm p$, addition or subtraction of n is still defined as the number of increments or decrements from the initial value $-m$.

If a negative number is incremented through 0, $-p+n=m$, the result is a positive number m, whose magnitude is $|m|=n-p$. Thus, $-47+63=63-47=16$. Decrementing a negative number can be expressed as the negation of the result of incrementing a corresponding positive number: $-121-36=-(121+36)=-(157)=-157$.

From these rules, addition and subtraction tables can be derived, for all combinations of single digits. Multidigit addition and subtraction are iterative processes:

carries	$+11$	borrows	-1	borrows	-1
addend	173	subtrahend	173	addend	-173
augend	$+38$	minuend	-38	augend	$+38$
sum	211	difference	135	sum	-135

carries	$+(-1)$	borrows	-1	borrows	-1
subtrahend	-173	addend	-38	subtrahend	38
minuend	-38	augend	$+173$	minuend	-173
difference	-211	sum	135	difference	-135

Chain addition or subtraction (addition or subtraction of more than two numbers) is the same as the sequential addition or subtraction of one number at a time. It differs only in that all the digits in a column are added or subtracted, cumulating borrows or carries, so that only one borrow or carry is performed per column, which, along with eliminating the need to write down individual sums or differences, saves considerable time.

Multiplication by m by n is the addition of m to zero n times. Thus, 3 times $3=3+3+3=9$. In the multiplication m times $n=p$, or

$$\begin{array}{rr} & m \\ \text{times} & n \\ \hline & p \end{array}$$

m is the *multiplicand*, or initial value, n is the *multiplier*, or number of times m is added to the starting value, and p is the *product*, or final value. Multiplication of multidigit numbers is accomplished by multiplying the multiplicand by the weighted value of each digit in the multiplier, adding carries as needed:

```
  128
×  32
-----
    1   carry
  246   individual product
   2    carry
 3640   individual product
-----
 4096   product=sum of carries and individual products
```

Usually this is written

```
  128
×  32
-----
  256
 384
-----
 4096
```

for simplicity of notation.

Division, expressed as

$$p/n=m, \quad p \div n=m, \quad \text{or} \quad n\overline{)p}^{\,m}$$

is accomplished by repeatedly subtracting n from p, until the difference is less than n. The number of times n has been subtracted from p is m. Here, p is the *dividend*, or initial value, n is the *divisor* or number to be subtracted, and m is the *quotient*, or number of subtractions. If the last difference, d, is nonzero, it is called the *remainder*. By adding zeros to the right of the decimal point, the precision of the quotient can be extended indefinitely, rendering the nonzero remainder arbitrarily small.

Multidigit division is accomplished by finding m digit by digit, left to right, filling in missing digits temporarily with zeros. The tentative value of m is multiplied by n each time and subtracted from p, which is the new temporary value of p for the determination of the next digit of m. Thus,

```
       0              00              003              003.0
31)123.4        31)123.4        31)123.4         31)030.4
   000.0    →      000.0    →      093.0    →       000.0
   123.4           123.4           030.4             30.4=remainder
```

5.3.2 *Binary Arithmetic*

We shall develop binary arithmetic along lines parallel to our development of decimal arithmetic. If the explanation becomes obscure at any point, the reader can review the corresponding section of decimal arithmetic.

Binary integers are strings of digits, just as are decimal ones. But in the binary system, digits may only have two values: 0 or 1. The following are integers: 0, 1010, and -1101. The value of an integer is the sum of the weighted values of its digits. Here, the successive weights, expressed in binary, of digits from right to left are 10_2^0, 10_2^1, 10_2^2, etc., just as they are in decimal (the subscript 2 for a number indicates binary notation):

$$S = \sum_{i=0}^{n} 10_2^i m_i \tag{5.22}$$

For example,

$$\begin{aligned} 0 &= 0 \times 10_2^0 \\ 1010 &= 0 \times 10_2^0 + 1 \times 10_2^1 + 0 \times 10_2^2 + 1 \times 10_2^3 \\ -1101 &= -1 \times 10_2^0 - 0 \times 10_2^1 - 1 \times 10_2^2 - 1 \times 10_2^3 \end{aligned}$$

Just as successive weights in the decimal system are powers of 10_{10}, they are powers of 10_2 in the binary system. How much is 10_2 in decimal? This is defined as the decimal equivalent of the largest legal integer in the new number system (binary in this case) plus 1. Therefore,

in decimal in binary

$$\begin{array}{rcr} 1_{10} & = & 1_2 \\ +1_{10} & = & +1_2 \\ \hline 2_{10} & = & 10_2 \end{array}$$

That is, in binary, a zero sum with a carry of one results, and in decimal a 2 results. Since the addends and augends were equal, the sums must be equal, so that

$$2_{10} = 10_2$$

We refer to 2_{10} as being the *base* or *radix* of the *binary* number system, as expressed in decimal. Successive digits of an integer have weighting factors which are successively higher powers of the base: 1, 2, 4, 8, 16, etc. for the binary system (base 2_{10}); 1, 8, 64, 512, etc., for the octal system (base 8_{10}); and 16, 256, 4096, etc., for the hexadecimal system (base 16_{10}). If the base is r, a number is expressed in its own number system as

$$S_r = \sum_{i_r=0}^{n_r} 10_r^{i_r} m_{i_r} \tag{5.23}$$

The value of a number with a point (decimal point, binary point, octal point, etc., depending on the number system) is

$$S_r = \sum_{i=n_r}^{p_r} 10_r^{i_r} m_{i_r} \tag{5.24}$$

where n is the number of places to the right of the point and $p+1$ is the number of places to the left.

Counting consists of sequencing through successive values of integers in the rightmost place, carrying to the next and further places to the left as the rightmost digit goes to zero. In binary, counting up from zero gives: 0, 1, 10, 11, 100, 101, 110, 111, etc. Step by step, $0+1=1$; $1+1=10$ (rightmost digit goes to zero, carry into next digit); $10+1=11$; $11+1=100$ (carry *ripples* two places: in decimal, the first two-digit ripple carry occurs at $99+1=100$); $100+1=101$; $101+1=110$ (carry); $110+1=111$. The next count, to 1000, would involve a three-place ripple carry, similar to $999+1=1000$ in decimal.

Addition in binary is exactly the same as in decimal, but with a different addition table:

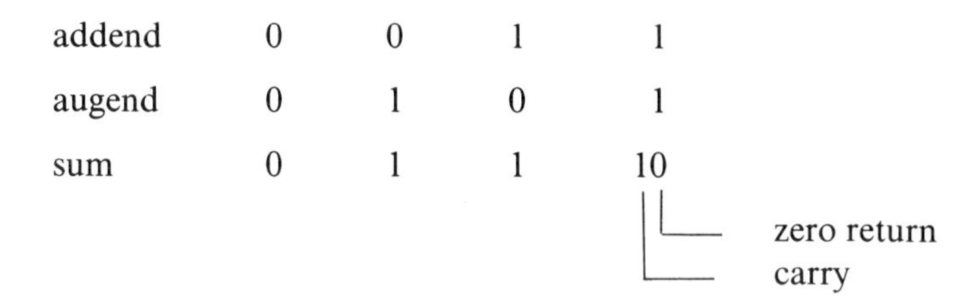

addend	0	0	1	1
augend	0	1	0	1
sum	0	1	1	10

A subtraction table is derived from the addition table:

subtrahend	10	1	1	0
minuend	1	0	1	0
difference	1	1	0	0

borrow clears 1 in 10_2 place

Examples are

carries	1	borrows	−1
addend	1011001	subtrahend	1101111
augend	10110	minuend	−10110
sum	1101111	difference	1011001

With a binary point:

```
 1011.001        1101.111
 +10.110         −10.110
 --------        --------
 1101.111        1011.001
```

The binary multiplication table is the same as for corresponding digits in decimal:

multiplicand	0	0	1	1
multiplier	0	1	0	1
product	0	0	0	1

Multidigit multiplication is accomplished exactly as in decimal, e.g.,

```
  110.1
   10.1
  -----
   1101
     0
 1101
--------
10000.01
```

The binary division table is

dividend	0	0	1	1	
divisor	0	1	0	1	* = undefined
quotient	*	0	*	1	
remainder	*	0	*	0	

Thus,

```
          1.000
      --------
1101 )1101.011
      1101
      ----
        00
        ---
         0
         01
         00
        ---
          11 = remainder
```

We shall discuss conversion from one number system to another after introducing the octal and hexadecimal number systems.

5.3.3 Octal Arithmetic

Although the binary number system is the one used by most digital arithmetic devices, it is a cumbersome one. For example, the number 1000 in decimal is written as 1111101000 in binary. Whereas the decimal number requires four digits, ten *binary digits*, or *bits* for short, are required. By grouping the bits in groups of three, and converting them to their decimal equivalents, we can reduce the number of digits required to represent a binary number. Thus,

$$1111101000_2 = 1|111|101|000_2 = 1750_8$$

This notation, using a radix of 8, is called *octal* notation. Conversion between octal and binary is very simple, as we will describe in Section 5.3.5, and most users of binary arithmetic actually use octal or hexadecimal (base 16) notation as a shorthand for representing binary numbers.

The octal addition and multiplication tables are given in Table 5.20. As an exercise, the reader is encouraged to derive subtraction and division tables. The

Table 5.20a Octal Addition

Addend		0	1	2	3	4	5	6	7
Augend	0	0	1	2	3	4	5	6	7
	1	1	2	3	4	5	6	7	10
	2	2	3	4	5	6	7	10	11
	3	3	4	5	6	7	10	11	12
	4	4	5	6	7	10	11	12	13
	5	5	6	7	10	11	12	13	14
	6	6	7	10	11	12	13	14	15
	7	7	10	11	12	13	14	15	16

Table 5.20b Octal Multiplication

Multiplicand		0	1	2	3	4	5	6	7
Multiplier	0	0	0	0	0	0	0	0	0
	1	0	1	2	3	4	5	6	7
	2	0	2	4	6	10	12	14	16
	3	0	3	6	11	14	12	22	25
	4	0	4	10	14	20	24	30	34
	5	0	5	12	17	24	31	36	43
	6	0	6	14	22	30	36	44	52
	7	0	7	16	25	34	43	52	61

reader can see that 0–7 are the legal digits, with 1+7=10. The following are examples of addition, subtraction, multiplication and division in octal:

```
  734.7       121.63        216.63            216.21
   13.2       −42.71          ×72       72)20073.32
  +50.1       ------       -------         164
 ------        56.72        435 46         ---
 1026.2                    17474 5         147
                           --------         72
                           20132.16         ---
                                            553
                                            534
                                            ---
                                             173
                                             164
                                             ---
                                               72
                                               72
                                               --
                                                0 =remainder
```

As an exercise, the reader should verify these calculations with Table 5.20.

Table 5.21a Hexadecimal Addition

Augend		0	1	2	3	4	5	6	7	8	9	A	B	C	D	E	F
Addend	0	0	1	2	3	4	5	6	7	8	9	A	B	C	D	E	F
	1	1	2	3	4	5	6	7	8	9	A	B	C	D	E	F	10
	2	2	3	4	5	6	7	8	9	A	B	C	D	E	F	10	11
	3	3	4	5	6	7	8	9	A	B	C	D	E	F	10	11	12
	4	4	5	6	7	8	9	A	B	C	D	E	F	10	11	12	13
	5	5	6	7	8	9	A	B	C	D	E	F	10	11	12	13	14
	6	6	7	8	9	A	B	C	D	E	F	10	11	12	13	14	15
	7	7	8	9	A	B	C	D	E	F	10	11	12	13	14	15	16
	8	8	9	A	B	C	D	E	F	10	11	12	13	14	15	16	17
	9	9	A	B	C	D	E	F	10	11	12	13	14	15	16	17	18
	A	A	B	C	D	E	F	10	11	12	13	14	15	16	17	18	19
	B	B	C	D	E	F	10	11	12	13	14	15	16	17	18	19	1A
	C	C	D	E	F	10	11	12	13	14	15	16	17	18	19	1A	1B
	D	D	E	F	10	11	12	13	14	15	16	17	18	19	1A	1B	1C
	E	E	F	10	11	12	13	14	15	16	17	18	19	1A	1B	1C	1D
	F	F	10	11	12	13	14	15	16	17	18	19	1A	1B	1C	1D	1E

Table 5.21b Hexadecimal Multiplication

Multiplicand		0	1	2	3	4	5	6	7	8	9	A	B	C	D	E	F
Multiplier	0	0	0	0	0	0	0	0	0	0	0	0	0	0	0	0	0
	1	0	1	2	3	4	5	6	7	8	9	A	B	C	D	E	F
	2	0	2	4	6	8	A	C	E	10	12	14	16	18	1A	1C	1E
	3	0	3	6	9	C	F	12	15	18	1B	1E	21	24	27	2A	2D
	4	0	4	8	C	10	14	18	1C	20	24	28	2C	30	34	38	3C
	5	0	5	A	F	14	19	1E	23	28	2D	32	37	3C	41	46	4B
	6	0	6	C	12	18	1E	24	2A	30	36	3C	42	48	4E	54	5A
	7	0	7	E	15	1C	23	2A	31	38	3F	46	4D	54	5B	62	69
	8	0	8	10	18	20	28	30	38	40	48	50	58	60	68	70	78
	9	0	9	12	1B	24	2D	36	3F	48	51	5A	63	6C	75	7E	87
	A	0	A	14	1E	28	32	3C	46	50	5A	64	6E	78	82	8C	76
	B	0	B	16	21	2C	37	42	4D	58	63	6E	79	84	8F	9A	A5
	C	0	C	18	24	30	3C	48	54	60	6C	78	84	90	9C	A8	B4
	D	0	D	1A	27	34	41	4E	5B	68	75	82	8F	9C	A9	B6	C3
	E	0	E	1C	2A	38	46	54	62	70	7E	8C	9A	A8	B6	C4	D2
	F	0	F	1E	2D	3C	4B	5A	69	78	87	76	A5	B4	C3	D2	E1

5.3.4 Hexadecimal Arithmetic

In some applications, for example in the case of sixteen-bit binary numbers, it is more convenient to group the bits by fours instead of by threes.

Each group of four bits can be represented as a single digit in the hexadecimal system, which has a base of 16. The first ten digits are 0–9, as in decimal, but six additional digits must be used to represent the decimal values 10–15. For these, the letters A–F are used. Thus, counting from 0 to 10, the sequence is 0, 1, 2, 3, 4, 5, 6, 7, 8, 9, A, B, C, D, E, F, 10.

Table 5.21 gives the addition and multiplication tables for the hexadecimal system. Sample calculations are given:

Addition	Subtraction	Multiplication	Division
D1.C	8A3	247E	3A0.4
+83.E	−B6	×8A3F	4A.C)10F0B.00
155.A	7ED	22362	E04
		6D7A	2ECB
		16CEC	2EB8
		123F0	1300
		143BAB2	1250
			050

5.3.5 Conversions among Number Systems

The general formula for computation of a number's value in these number systems is

$$S = \sum_{i=n}^{p} 10_r^i m$$

This formula provides the means for conversion from one number system to another. The rule for conversion is

To convert from one number system (which we will call the *old* system) to another (which we will call the *new* system) compute the value of the number from the following equation:

$$S = \sum_{i=n}^{p} r^i m_i \tag{5.25}$$

where

S = the computed value in the new number system,

i = the position of a digit in the *old* number, as expressed in the *new* number system,

n = the rightmost position in the *old* number, as expressed in the *new* number system,

p = the leftmost position in the *old* number, as expressed in the *new* number system,

r = the base of the *old* number's number system, expressed in the *new* number system,

m = the value of the ith digit of the *old* number, expressed in the *new* number system.

Thus, for example:

$$162_8 = 2\times 8^0 + 6\times 8^1 + 1\times 8^2 = 114_{10}$$
$$= 2\times 8^0 + 6\times 8^1 + 1\times 8^2 = 2_{16} + 30_{16} + 40_{16} = 72_{16}$$
$$= 10\times 1000^0 + 110\times 1000^1 + 1\times 1000^2 = 10_2 + 110000_2 + 1000000_2$$
$$= 1110010_2$$

$$38A_{16} = 10\times 16^0 + 8\times 16^1 + 3\times 16^2 = 10 + 128 + 768 = 906_{10}$$
$$= 12\times 20^0 + 8\times 20^1 + 3\times 20^2 = 12 + 200 + 1400 = 1612_8$$
$$= 1010\times 10000^0 + 1000\times 10000^1 + 11\times 10000^2 = 1110001010_2$$

$$1111110011 = 1\times 2^0 + 1\times 2^1 + 0\times 2^2 + 0\times 2^3 + 1\times 2^4$$
$$+ 1\times 2^5 + 1\times 2^6 + 1\times 2^7 + 1\times 2^8 + 1\times 2^9$$
$$= 1 + 2 + 0 + 0 + 16 + 32 + 128 + 256 + 512$$
$$= 1011_{10}$$
$$= 1\times 2^0 + 1\times 2^1 + 0\times 2^2 + 0\times 2^3 + 1\times 2^4 + 1\times 2^5$$
$$+ 1\times 2^6 + 1\times 2^7 + 1\times 2^{10} + 1\times 2^{11}$$
$$= 1 + 2 + 0 + 0 + 20 + 40 + 100 + 200 + 400 + 1000 = 1763$$
$$= 1\times 2^0 + 1\times 2^1 + 0\times 2^2 + 0\times 2^3 + 1\times 2^4 + 1\times 2^5$$
$$+ 1\times 2^6 + 1\times 2^7 + 1\times 2^8 + 1\times 2^9$$
$$= 1 + 2 + 0 + 0 + 10 + 20 + 40 + 80 + 100 + 200$$
$$= 3E3_{16}$$

$$324_{10} = 4\times A^0 + 2\times A^1 + 3\times A^2 = 4 + 14 + 12C = 144$$
$$= 4\times 12^0 + 2\times 12^1 + 3\times 12^2 = 4 + 24 + 454 = 504_0$$
$$= 100\times 1010^0 + 10\times 1010^1 + 11\times 1010^2 = 100 + 10100$$
$$+ 100101100 = 101000100$$

This method we might refer to as the *sum of products* method. There is another procedure, which relies on the same basic principle, which consists of iteratively dividing by powers of the base of the "new" number system, as expressed in the "old" number system, and summing the quotients and the final remainder. It is based on the fact that, if [Eq. (5.25)]

$$S = \sum_{i=n}^{p} r^i m_i$$

then

$$m_j = \text{REM}\, q_{j-1}/r, \qquad j = n,\ldots,p$$

where

n = position of rightmost digit of *new* number,

p = position of leftmost digit of *new* number,

j = position of digit of *new* number being computed,

r^j = the *new* number base raised to the jth power, expressed in the *new* number system,

REM q_{j-1}/r = remainder represented in new number system, after dividing q_{j-1} by r, both represented in the *old* number system,

q_{j-1} = quotient obtained in previous division, or in computing the first m, m_n, the original number, expressed in the *old* number system.

Therefore we can state the method of *iterative division* for conversion between number system:

> To convert a number from one number representation system (the *old* one) to another (the *new* one), iteratively perform the operation
>
> $$m_j = \text{REM}\, q_{j-1}/r,\ j = n, \ldots, p$$
>
> where q_{j-1} is the *old* number when finding m_n, or q_{j-1} is the quotient from the previous division when finding any subsequent m_j.

Thus,

$$\begin{array}{rl} & 0 \rightarrow 1 \\ 12 & \overline{)1} \rightarrow 1 \\ 12 & \overline{)13} \rightarrow 4 \\ 12_8 & \overline{)162_8} = \overline{114_{10}} \end{array}$$

$$\begin{array}{rl} & 0 \rightarrow 7 \\ 20 & \overline{)7} \rightarrow 2 \\ 20_8 & \overline{)162_8} = \overline{72_{16}} \end{array}$$

$$\begin{array}{rl} & 0 \rightarrow 1 \\ 2 & \overline{)1} \rightarrow 1 \\ 2 & \overline{)3} \rightarrow 1 \\ 2 & \overline{)7} \rightarrow 0 \\ 2 & \overline{)16} \rightarrow 0 \\ 2 & \overline{)34} \rightarrow 1 \\ 2 & \overline{)71} \rightarrow 0 \\ 2_8 & \overline{)162_8} = \overline{1110010_2} \end{array}$$

$$\begin{array}{rl} & 0 \rightarrow 9 \\ A & \overline{)9} \rightarrow 0 \\ A & \overline{)5A} \rightarrow 6 \\ A_{16} & \overline{)38A_{16}} = \overline{906_{10}} \end{array}$$

$$\begin{array}{rl} & 0 \rightarrow 1 \\ 8 & \overline{)1} \rightarrow 6 \\ 8 & \overline{)E} \rightarrow 1 \\ 8 & \overline{)71} \rightarrow 2 \\ 8_{16} & \overline{)38A_{16}} = \overline{1612_8} \end{array}$$

$$\begin{array}{rl} & 0 \rightarrow 1 \\ 2 & \overline{)1} \rightarrow 1 \\ 2 & \overline{)3} \rightarrow 1 \\ 2 & \overline{)7} \rightarrow 0 \\ 2 & \overline{)E} \rightarrow 0 \\ 2 & \overline{)1C} \rightarrow 0 \\ 2 & \overline{)38} \rightarrow 1 \\ 2 & \overline{)71} \rightarrow 0 \\ 2 & \overline{)E2} \rightarrow 1 \\ 2 & \overline{)1C5} \rightarrow 0 \\ 2 & \overline{)38A_{16}} = \overline{1110001010_2} \end{array}$$

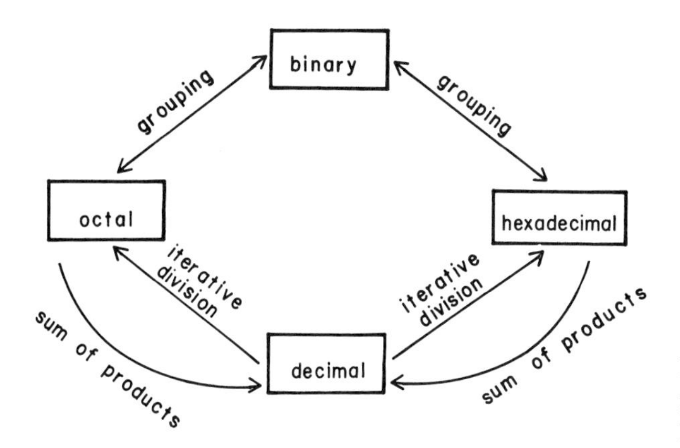

Figure 5.1 Recommended conversion methods for going from one number system to another. Follow the shortest path from the "old" number system to the "new" one; for equal path lengths (e.g., decimal to binary), pick according to individual preference.

$$
\begin{array}{l}
0 \rightarrow 1 \\
16\,\overline{)1} \rightarrow 4 \\
16\,\overline{)20} \rightarrow 4 \\
16_{10}\overline{)324_{10}} = 144_{16}
\end{array}
\qquad
\begin{array}{l}
0 \rightarrow 5 \\
8\,\overline{)5} \rightarrow 0 \\
8\,\overline{)40} \rightarrow 4 \\
8_{10}\overline{)324_{10}} = 504_{8}
\end{array}
\qquad
\begin{array}{l}
0 \rightarrow 1 \\
2\,\overline{)1} \rightarrow 0 \\
2\,\overline{)2} \rightarrow 1 \\
2\,\overline{)5} \rightarrow 0 \\
2\,\overline{)10} \rightarrow 0 \\
2\,\overline{)20} \rightarrow 0 \\
2\,\overline{)40} \rightarrow 1 \\
2\,\overline{)81} \rightarrow 0 \\
2\overline{)162} \rightarrow 0 \\
2\overline{)324_{10}} = 101000100_{2}
\end{array}
$$

The choice of a conversion method is largely a matter of taste. Clearly, conversion between binary and either octal or hexadecimal is trivial, being best accomplished by grouping bits into three or fours, respectively, and having memorized the binary equivalents of eight octal or 16 hexadecimal digits. Many people perform octal/hexadecimal or hexadecimal/octal conversion via an intermediate binary representation. In conversion of decimal to octal or hexadecimal, it is easier for most people to use the method of iterative division, since all the arithmetic is in decimal. When converting from octal or hexadecimal to decimal, many people prefer the sum of products method because all the arithmetic is in decimal. Conversion from decimal to binary is best accomplished via octal or hexadecimal as an intermediate, as in conversion from binary to decimal. Figure 5.1 summarizes these generalizations.

5.4 Other Number Representation Systems

The binary, octal, decimal, and hexadecimal systems all have one aspect in common: a number has a value equal to the sum of all the digits, each digit multiplied by a positional weighting factor. There are other ways to represent numbers, some of which are in sufficiently common use to merit discussion.

5.4.1 Ones and Twos Complement Representation of Negative Numbers

In general, machines which manipulate numbers do not have a separate "sign" character to represent the sign of a number. Since there are only two possible signs, the addition of sign to one unsigned number must require another bit. This is reasonable, since the negative number domain is the same size as the corresponding positive domain (neglecting zero). Thus, one mechanism would be to have a number composed of a sign bit plus magnitude bits. For example,

sign¬
0|011|100|101 = 345_8
1|011|100|101 = -345_3

if the leftmost bit is the sign bit and the rest of the number is the unsigned magnitude.

This conversion is rarely used because addition of an unsigned number to a signed number would give a result which is incorrect:

0	011	100	101	=	$(+345_8)$
+ 1	011	100	101	=	(-345_8)
1	100	001	010	≠	0_8

To overcome this problem, the *ones complement* and *twos complement* number systems were devised. In ones complement, all positive numbers are represented as in binary notation, and negative numbers are represented as *the bit-by-bit inversion of the equal magnitude positive numbers*. Thus,

−0	011	100	101	=	1	100	011	010
−1	100	011	010	=	0	011	100	101

If a number is added to its complement, the sum has all digits equal to 1:

0	011	100	101
+ 1	100	011	010
1	111	111	111

To make this sum come out correctly, we add one, and ignore the overflow:

1	111	111	111
0	000	000	001
0	000	000	000

In fact, any ones complement negative number can be added to any ones complement positive number plus one, to yield a correct sum, providing the largest number represented has a magnitude which does not include the leftmost bit.

Since a one must always be added for any ones complement negative number, the twos complement representation was devised, in which a negative number is equal to the ones complement of the equal magnitude positive number *plus one*. Hence the twos complement of 0 011 100 101 becomes 1 100 011 010 + 1 = 1 100 011 011. Now when the number is added to its twos complement,

$$
\begin{array}{rcccc}
 & 0 & 011 & 100 & 101 \\
+ & 1 & 100 & 011 & 011 \\
\hline
 & 0 & 000 & 000 & 000
\end{array}
$$

The sum is zero, as it should be. In fact, the sum of any number of twos complement numbers has the correct twos complement value. For example,

$$
\begin{array}{rccccr}
 & 0 & 110 & 100 & = & 64_8 \\
+ & 1 & 001 & 111 & = & +(-61_8) \\
\hline
 & 0 & 000 & 011 & & 3_8
\end{array}
$$

$$
\begin{array}{rccccr}
 & 1 & 010 & 110 & = & -52_8 \\
+ & 1 & 110 & 100 & = & +(-14_8) \\
\hline
 & 1 & 001 & 010 & & -66_8
\end{array}
$$

To subtract a number, the simplest method in terms of hardware is to twos complement it and add. We shall discuss hardware implementation later in this chapter.

5.4.2 Binary-Coded Decimal

In some applications, for example, many applications where a human operator enters decimal numbers on a keyboard and reads a decimal display, it is desirable to use a code which represents each decimal digit as a binary number, necessarily four bits long (the numbers eight and nine require four bits to represent in binary). In such a code, each group of four digits acts as a four-bit binary number in a counting operation, until nine is incremented. Then the number must go to zero, and a carry must be generated in the next higher-order digit. Such a code is called binary-coded decimal (BCD). Table 5.22 illustrates the BCD codes for $0\text{–}21_{10}$.

Table 5.22 BCD Equivalents of Decimal 0–21

Decimal	BCD	Decimal	BCD
0	0000 0000	11	0001 0001
1	0000 0001	12	0001 0010
2	0000 0010	13	0001 0011
3	0000 0011	14	0001 0100
4	0000 0100	15	0001 0101
5	0000 0101	16	0001 0110
6	0000 0110	17	0001 0111
7	0000 0111	18	0001 1000
8	0000 1000	19	0001 1001
9	0000 1001	20	0010 0000
10	0001 0000	21	0010 0001

Table 5.23 Gray Code for Decimal 0–21

Decimal	Gray code	Decimal	Gray Code
0	00000	11	01110
1	00001	12	01010
2	00011	13	01011
3	00010	14	01001
4	00110	15	01000
5	00111	16	11000
6	00101	17	11001
7	00100	18	11001
8	01100	19	11010
9	01101	20	11110
10	01111	21	11111

BCD decimal points are handled as in all number systems: the value of each BCD digit is equal to its decimal equivalent times the power of ten indicated by its position relative to the decimal point.

Generally arithmetic manipulations are accomplished by converting BCD to binary, performing the computations, and converting back to BCD. Some processors compute directly in BCD, but BCD computation is beyond the scope of this introductory text.

5.4.3 Optimizing Codes

In some applications, codes are used which are not variants on the sum of products principle common to all the number systems we have examined so far. That is, a given position in a digit string is not always the same weight. These are generally used for modified binary representations, as opposed to number systems with radix other than 2. For example, the Gray code is used to represent successive states in a sequence where it is desired that only one bit change from one state (number) to the next. The Gray code is illustrated for the numbers $0–21_{10}$ in Table 5.23.

5.5 Hardware Logic Implementation

The most common implementation of elementary logic functions is an IC technology using bipolar (junction) transistors, called *transistor-transistor logic* (TTL, T^2L). All such devices require a +5 V power supply and a ground lead, and have input and output characteristics which fall within certain specifications.

5.5.1 Gates

Regardless of the technology, the basic devices of logical design are realizations of AND, OR, NAND, NOR, and INV. The pictorial representations of such devices are given in Fig. 5.2. The basic AND and OR operations are indicated in Figs. 5.2a and 5.2b; inversion of the value of a logic signal is symbolized by small circles at inputs or outputs of the devices. The incorporation of one or more inversions into such devices can be useful as a means of cutting down on number of devices in a

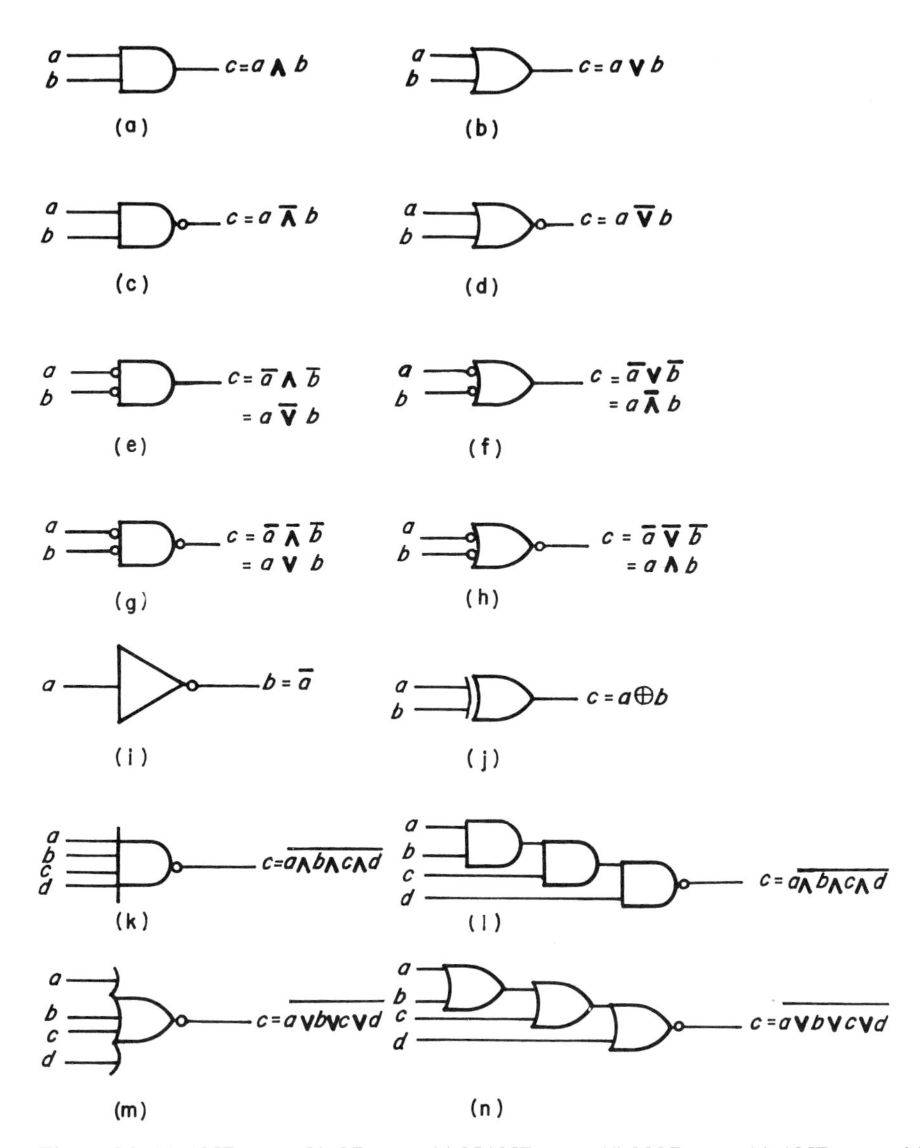

Figure 5.2 **(a)** AND gate. **(b)** OR gate. **(c)** NAND gate. **(d)** NOR gate. **(e)** AND gate with inverting inputs. **(f)** OR gate, inverting inputs. **(g)** NAND gate, inverting inputs. **(h)** NOR gate, inverting inputs. **(i)** Inverter. **(j)** Exclusive OR (EOR) gate. **(k,l)** 4-input NAND. **(m,n)** 4-input NOR.

circuit. The additional circuitry required is negligible compared to the amount of circuitry already in the IC, and no additional input or output leads are required. Each device, with or without inversions, costs less than ten cents. The devices of Figs. 5.2a and 5.2b are called *gates* because the logic level at one input can be used to control the ability of the output to change with changes of logic level at the other input.

The one-input device of Fig. 5.2i is called an *inverter*, since it directly performs the logical inversion operation. The devices of Figs. 5.2k–5.2n are a four-input

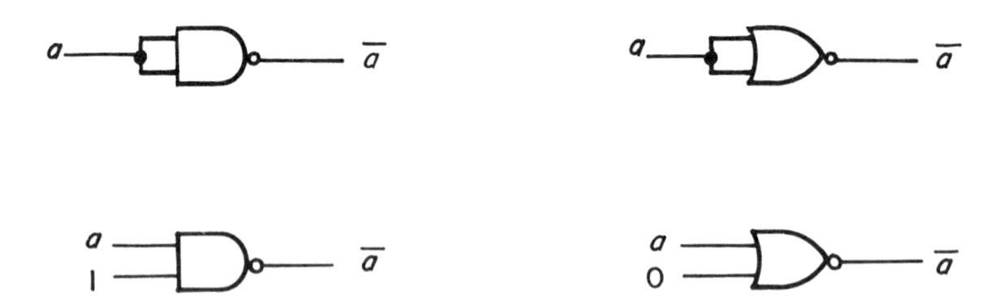

Figure 5.3 Use of gates to perform inversion operation.

NAND gate and its logical equivalent using two-input gates, and a six-input NOR gate and its logical equivalent using two-input gates.

The logical equations of all the devices of Fig. 5.2, in the order in which they appear, are

(a) $c = a \wedge b$

(b) $c = a \vee b$

(c) $c = a \overline{\wedge} b$

(d) $c = a \overline{\vee} b$

(e) $c = \bar{a} \wedge \bar{b}$

(f) $c = \bar{a} \vee \bar{b}$

(g) $c = \bar{a} \overline{\wedge} \bar{b} = a \vee b$

(h) $c = \bar{a} \overline{\vee} \bar{b} = a \wedge b$

(i) $b = \bar{a}$

(j) $c = a \oplus b$

(k) $e = (\overline{a \wedge b \wedge c \wedge d})$ (four-input NAND)

(l) $e = a \wedge b \wedge c \overline{\wedge} d = (\overline{a \wedge b \wedge c \wedge d})$ [same as (k)]

(m) $g = (\overline{a \vee b \vee c \vee d \vee e \vee f})$ (six-input NOR)

(n) $g = a \vee b \vee c \vee d \vee e \overline{\vee} f = (\overline{a \vee b \vee c \vee d \vee e \vee f})$ [same as (m)]

The multiple-input gates and their equivalents indicate the ease with which multiple-input functions can be realized. A number of multiple-input gates are available as ICs, and should be used whenever possible to reduce number of ICs and number of interconnections between ICs.

Gates can be used as inverters, as shown in Fig. 5.3. The only reason for using gates as inverters is that in some designs the designer may be pressed for space and may have an IC with a NAND or NOR gate free (there are four two-input gates per 14-pin IC). The principle is simple: use NAND or NOR in one of the ways which always results in inversion of the input: $a \overline{\wedge} a = \bar{a}$, $a \overline{\wedge} 1 = \bar{a}$, $a \overline{\vee} a = \bar{a}$, or $a \overline{\vee} 0 = \bar{a}$.

The use of gates to enable or disable transmission of a signal, inverted or noninverted, depending on the value of a control or *gating* signal, is illustrated in Fig. 5.4. This control of signal flow is referred to as *gating* the signal. The choice of

Figure 5.4 Use of gates to gate signals.

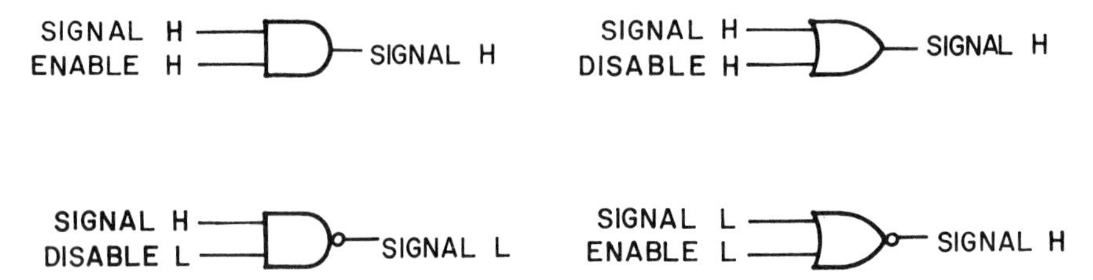

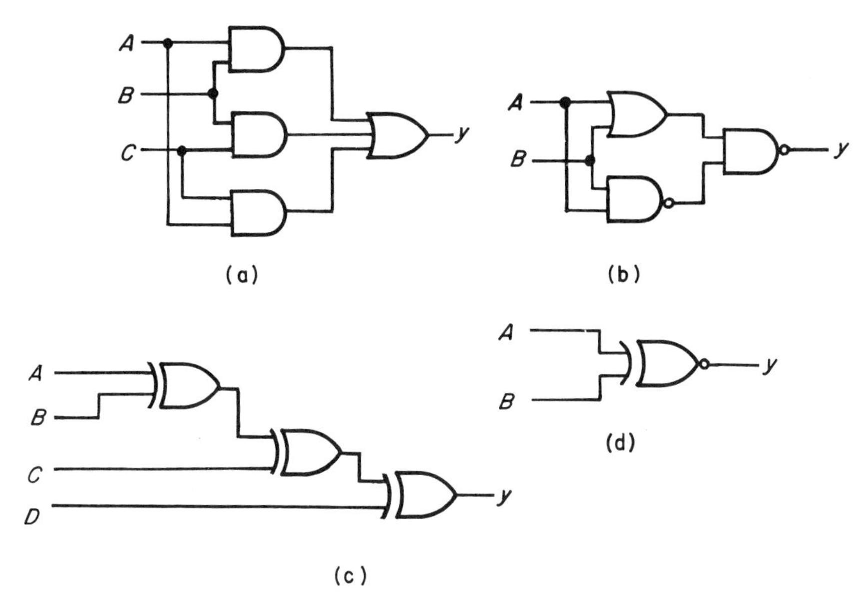

Figure 5.5 Gate implementation of **(a)** $y=(A\wedge B)\vee(A\wedge C)\vee(B\wedge C)$. **(b)** $y=A=B$. **(c)** $y=A\oplus B\oplus C\oplus D$. **(d)** $y=A=B$.

signal names is worth explaining: in general, signal names are chosen which best define the functions of the signals. Thus, the word SIGNAL refers to any signal, and the words ENABLE or DISABLE refer to the functions of enabling the gate to pass the signal or disabling it from passing the signal. The H and L are appended to indicate when the function occurs, on a 1 or a 0, respectively. Thus, ENABLE H enables the gate when it is a logical 1 or disables it when it is a logical 0: it could just as well be called DISABLE L. When a signal is inverted it changes from SIGNAL H to SIGNAL L or SIGNAL L to SIGNAL H.

Consider first the use of the AND gate. Since SIGNAL H$\wedge$ENABLE H$=$ SIGNAL H if ENABLE H$=1$, or SIGNAL H$\wedge$ENABLE H$=0$ if ENABLE H$=0$, the gate passes SIGNAL H without inversion if ENABLE H$=1$, and is held at 0 regardless of the value of SIGNAL H if ENABLE H$=0$. Similarly, an OR gate will pass a signal only if its enable input is 0. If its enable input is high, the output is forced to 1 regardless of the value of the signal input. The use of NAND and NOR gates is exactly the same as for AND and OR gates, except they also perform inversion of the signal.

Complex functions are realized by simply connecting gates together. Thus, the circuits of Figs. 5.5a–5.5d correspond to the following equations:

$$Y=(A\wedge B)\vee(A\wedge C)\vee(B\wedge C) \tag{a}$$

$$Y=A=B=(A\vee B)\wedge(A\,\overline{\wedge}\,B) \tag{b}$$

$$Y=A\oplus B\oplus C\oplus D \tag{c}$$

$$Y=A=B \tag{d}$$

The examples are (a) practical implementations for detecting the condition where at least two of three inputs are on; (b) implementation of $A=B$ from OR, AND, and

NAND; (c) logic for controlling a stairwell light from four landings (see Example 5.3); (d) logic for detecting the event $A = B$.

5.5.2 Gate-Level Design

The important rule in gate-level design is to state the described logical function as a logical equation. This logical equation usually can be written directly because the designer knows the function he wishes to realize. However, the function is not always obvious, as was the case with multiswitch control of a stairwell light discussed in earlier examples. Then it must be deduced from a truth table describing the desired output as a function of the inputs. There are a number of types of graphical methods used to derive a logical equation from a truth table. The derivation of *minimal* logic required to implement a function, that is, the systematic reduction of a logical equation to a minimum of terms, has been developed into a well-defined discipline. In general, the designer does not really need to minimize his logic beyond what he can accomplish by inspection. As we shall describe later in this chapter, beyond a certain degree of complexity it is simpler to use medium-scale and large-scale integrated circuits to synthesize arbitrary input–output logical functions.

Consider the truth table of Table 5.24. We first describe the logic function adequate for each line. Starting at the top, we can write 10 logical equations corresponding to the 10 lines in which $Y = 1$, each equation being a string of noninverted or inverted terms connected by ANDs. Thus, line by line, the equations are

$$Y = \bar{A} \wedge \bar{B} \wedge C \wedge \bar{D}$$

$$Y = \bar{A} \wedge \bar{B} \wedge C \wedge D$$

$$Y = \bar{A} \wedge B \wedge C \wedge \bar{D}$$

$$Y = \bar{A} \wedge B \wedge C \wedge D$$

$$Y = A \wedge \bar{B} \wedge C \wedge \bar{D}$$

$$Y = A \wedge \bar{B} \wedge C \wedge D$$

$$Y = A \wedge B \wedge \bar{C} \wedge \bar{D}$$

$$Y = A \wedge B \wedge \bar{C} \wedge D$$

$$Y = A \wedge B \wedge C \wedge \bar{D}$$

$$Y = A \wedge B \wedge C \wedge D$$

The entire truth table is obtained by ORing the right-hand terms for all ten lines:

$$\begin{aligned} Y = &(\bar{A} \wedge \bar{B} \wedge C \wedge \bar{D}) \vee (\bar{A} \wedge \bar{B} \wedge C \wedge D) \vee (\bar{A} \wedge B \wedge C \wedge \bar{D}) \\ &\vee (\bar{A} \wedge B \wedge C \wedge D) \vee (A \wedge \bar{B} \wedge C \wedge \bar{D}) \vee (A \wedge \bar{B} \wedge C \wedge D) \\ &\vee (A \wedge B \wedge \bar{C} \wedge \bar{D}) \vee (A \wedge B \wedge \bar{C} \wedge D) \vee (A \wedge B \wedge C \wedge \bar{D}) \\ &\vee (A \wedge B \wedge C \wedge D) \end{aligned}$$

This solution is referred as being in *canonical minterm form*. That is, each term in parentheses describes one line of the truth table, wherein each variable appears in

Table 5.24 A Truth Table of a Function to Be Realized in Hardware

Inputs				Output
A	B	C	D	Y
0	0	0	0	0
0	0	0	1	0
0	0	1	0	1
0	0	1	1	1
0	1	0	0	0
0	1	0	1	0
0	1	1	0	1
0	1	1	1	1
1	0	0	0	0
1	0	0	1	0
1	0	1	0	1
1	0	1	1	1
1	1	0	0	1
1	1	0	1	1
1	1	1	0	1
1	1	1	1	1

inverted or noninverted form, and all variables are ANDed; the terms for all lines are ORed. This is a very unwieldy expression: upon reduction it almost inevitably must become a simpler expression. Just as in algebra, logical equations can be simplified. Although this can be done at the equation level, as in algebra, there are simpler diagrammatic techniques.

5.5.3 The Veitch Diagram

A variety of diagrammatic techniques can be used to realize complicated logical operations as minimal equivalent operations—this results in a smaller number of circuit components and faster operation. Perhaps the most popular is the Karnaugh map; we will describe the Veitch diagram instead, because it is easier to use. The Veitch diagram is a form of truth table which is so arranged that minterms (the expressions for individual lines in conventional truth tables) which can be combined (in a way which we shall explain) are adjacent. The formats for Veitch diagrams of 2-, 3-, 4-, 5-, and 6-variable functions are shown in Fig. 5.6. The numbers in the cells correspond to line numbers in corresponding truth tables, where the entries in all columns on the top line are zero, the entries in the rightmost column alternate once per line, the entries in the second column from the right alternate every other line, the entries in the third column from the right alternate every fourth line, etc. This is equivalent to listing all 2^n numerical values for n-bit binary numbers in ascending numerical order, where n is the number of input variables. Table 5.24 is in this format, where A alternates every 8 lines, B every 4, C every 2, and D every line.

In each of the Veitch diagrams, the brackets indicate the areas where the indicated variables are true. Thus, in Fig. 5.6a, A is true in the left column and B is true in the top row. In Fig. 5.6c, A ("8" bit) is true in the two leftmost columns, B ("4" bit) is true in the two top rows, C ("2" bit) is true in the middle two columns, and D ("1" bit) is true in the middle two rows.

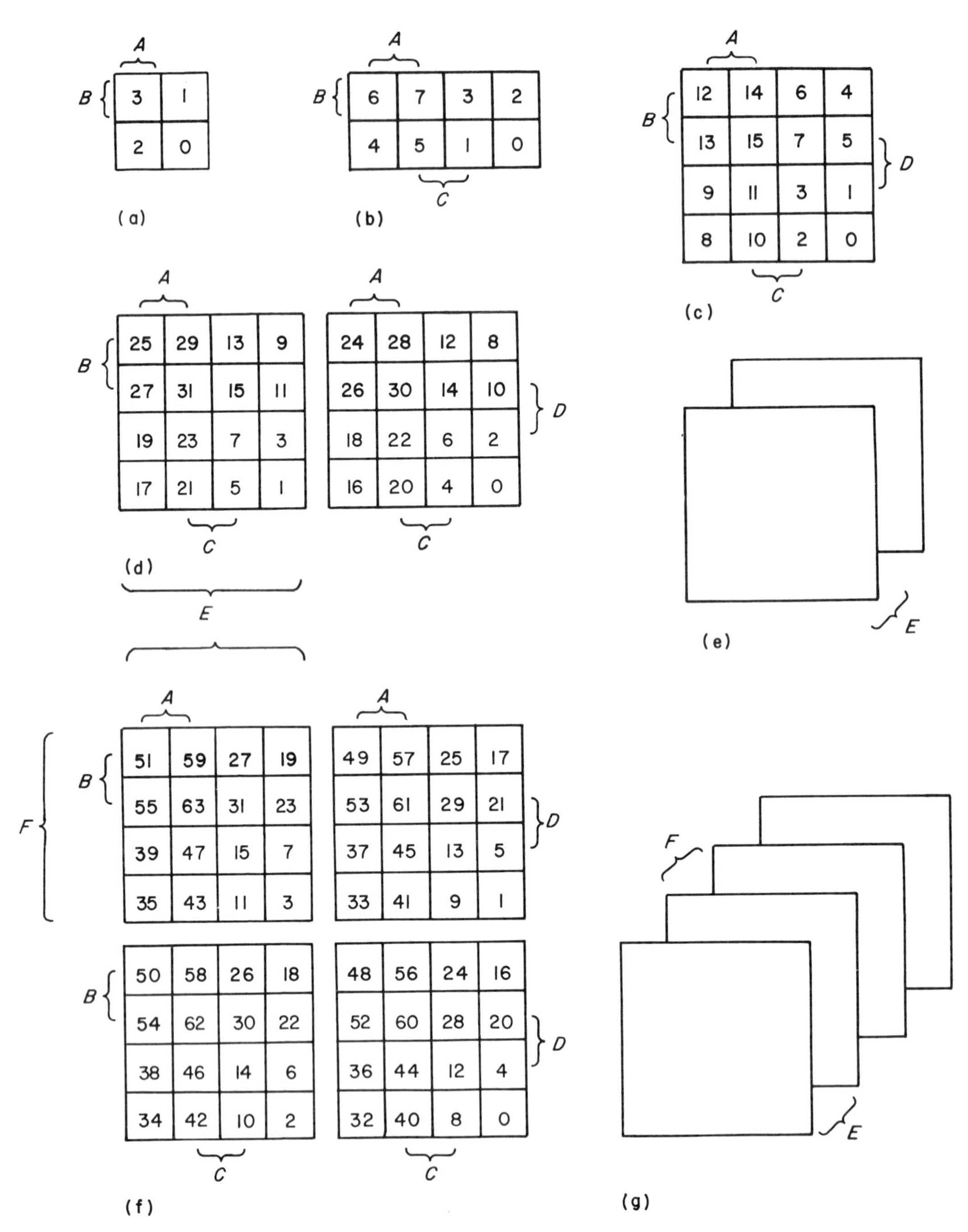

Figure 5.6 Veitch diagrams for two through six variables (with midterm numbers): **(a)** Two variables. **(b)** Three variables. **(c)** Four variables. **(d)** Five variables. **(e)** Visualization of **(d)**. **(f)** Six variables. **(g)** Visualization of **(f)**. (After Blakeslee, *Digital design with standard MSI and LSI*.)

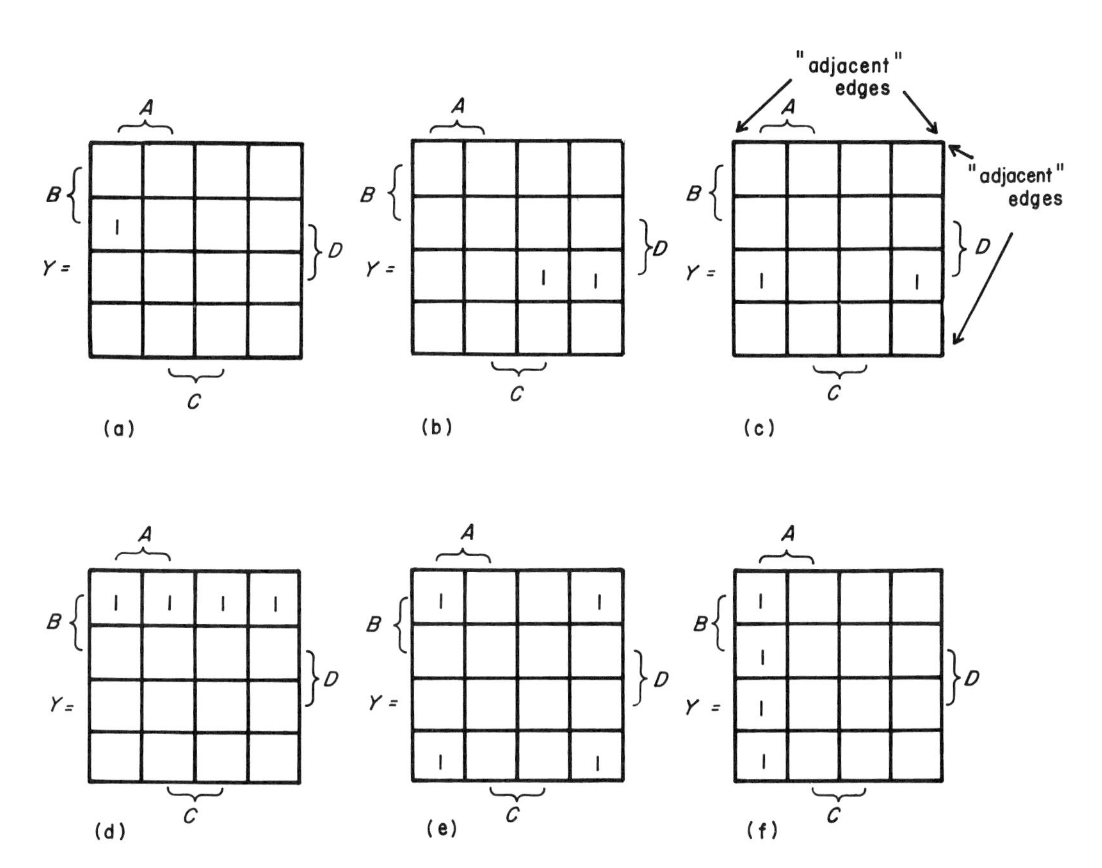

Figure 5.7 **(a)** Function $y=A\wedge B\wedge\bar{C}\wedge D$. **(b)** $y=(\bar{A}\wedge\bar{B}\wedge C\wedge D)\vee(\bar{A}\wedge\bar{B}\wedge\bar{C}\wedge D)=\bar{A}\wedge\bar{B}\wedge D$. **(c)** $y=(A\wedge\bar{B}\wedge\bar{C}\wedge D)\vee(\bar{A}\wedge\bar{B}\wedge\bar{C}\wedge D)=\bar{B}\wedge\bar{C}\wedge D$. **(d)** $y=B\wedge\bar{D}$. **(e)** $y=\bar{C}\wedge\bar{D}$. **(f)** $y=A\wedge\bar{C}$.

To map any logic function, a map corresponding to the appropriate number of input variables is used. For any row of the truth table where the output is 1, the corresponding cell in the Veitch diagram is filled with a 1; all other cells are 0. Thus, in the example of Fig. 5.7a, only row 13 of the truth table is true: hence the variables A, B, C, and D have values 1, 1, 0, 1 (the binary value for 13) for a minterm $A\wedge B\wedge\bar{C}\wedge D$, and that is the only minterm in the logical function. Hence, $Y=A\wedge B\wedge\bar{C}\wedge D$. In the function of Fig. 5.7b, there are two adjacent 1s, corresponding to two minterms:

$$Y=(\bar{A}\wedge\bar{B}\wedge C\wedge D)\vee(\bar{A}\wedge\bar{B}\wedge\bar{C}\wedge D)=\bar{A}\wedge\bar{B}\wedge D$$

That is, C may have either value for Y to be true: it does not contribute to the value of Y, and Y is thus a function only of the three variables A, B, and D. In fact, Y is a function of only three variables if there is only a pair of adjacent terms in the Veitch diagram. Figure 5.7c illustrates that two cells on opposite edges (left and right, or top and bottom) are also "adjacent": the function is

$$Y=(A\wedge\bar{B}\wedge\bar{C}\wedge D)\vee(\bar{A}\wedge\bar{B}\wedge\bar{C}\wedge D)=\bar{B}\wedge\bar{C}\wedge D.$$

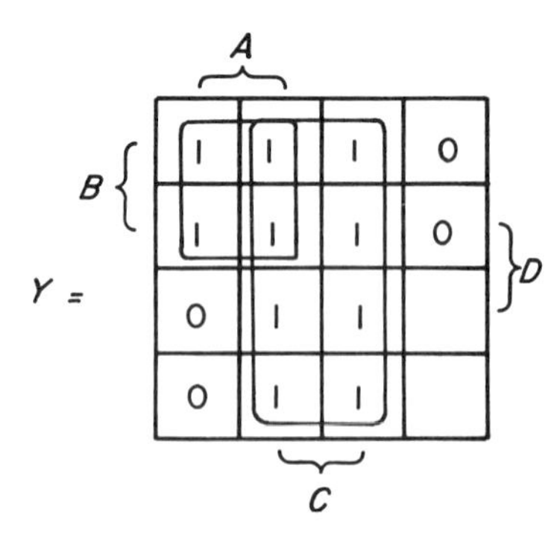

Figure 5.8 Veitch diagram of Table 5.22.

Products of only two variables give four adjacent minterms, as illustrated in Figs. 5.7d–5.7f. In Fig. 5.7d,

$$Y=(A\wedge B\wedge\bar{C}\wedge\bar{D})\vee(A\wedge B\wedge C\wedge\bar{D})\vee(\bar{A}\wedge B\wedge C\wedge\bar{D})$$
$$\vee(\bar{A}\wedge B\wedge\bar{C}\wedge\bar{D})=B\wedge\bar{D}$$

in 5.7e,

$$Y=(A\wedge B\wedge\bar{C}\wedge\bar{D})\vee(\bar{A}\wedge B\wedge\bar{C}\wedge\bar{D})\vee(A\wedge\bar{B}\wedge\bar{C}\wedge D)$$
$$\vee(\bar{A}\wedge\bar{B}\wedge\bar{C}\wedge\bar{D})=\bar{C}\wedge\bar{D}$$

in 5.7f,

$$Y=(A\wedge B\wedge\bar{C}\wedge\bar{D})$$
$$\vee(A\wedge B\wedge\bar{C}\wedge D)\vee(A\wedge\bar{B}\wedge\bar{C}\wedge D)\vee(A\wedge\bar{B}\wedge\bar{C}\wedge\bar{D})=A\wedge\bar{C}.$$

Products of one variable would have ones in all the cells for that variable, and in no others, or in all cells where that variable is zero and in no others.

Consider the example of Table 5.24: it can be mapped as a Veitch diagram, yielding the map of Fig. 5.8. The simplification procedure consists of reading the map off in groups of 1s as large as possible. Thus, in Fig. 5.8 the largest group of 1s corresponds to the eight 1s in the two middle columns corresponding to C. The remaining two 1s are part of a larger group of four corresponding to $A\wedge B$. Therefore $Y=(A\wedge B)\vee C$. These groups are circled in the diagram. The examples of Fig. 5.9 are treated similarly. The larger a group enclosed by a rectangle (each rectangle encloses a space 1, 2, or 4 elements long by 1, 2, or 4 elements wide), the simpler the term it represents. Rectangles may overlap, to result in simpler terms. Thus, in Fig. 5.9a the rectangles join "adjacent" top and bottom edges, and overlap: $Y=(C\wedge\bar{D})\vee(\bar{A}\wedge\bar{D})=(\bar{A}\vee C)\wedge\bar{D}$. In Fig. 5.9b, no elements are adjacent:

$$Y=(A\wedge\bar{B}\wedge C\wedge D)\vee(A\wedge\bar{B}\wedge C\wedge D)\vee(\bar{A}\wedge B\wedge C\wedge D)\vee(\bar{A}\wedge B\wedge\bar{C}\wedge\bar{D})$$

However, the expression can be simplified because the elements in the upper right and lower left corners share the trait $\bar{C}\wedge\bar{D}$, the middle two elements share the trait $C\wedge D$, the leftmost two share the trait $A\wedge\bar{B}$, and the rightmost two share the trait $\bar{A}\wedge B$. Therefore $Y=[(A\wedge\bar{B})\vee(\bar{A}\wedge B)]\wedge[(C\wedge D)\vee(\bar{C}\wedge\bar{D})]$. Note the consider-

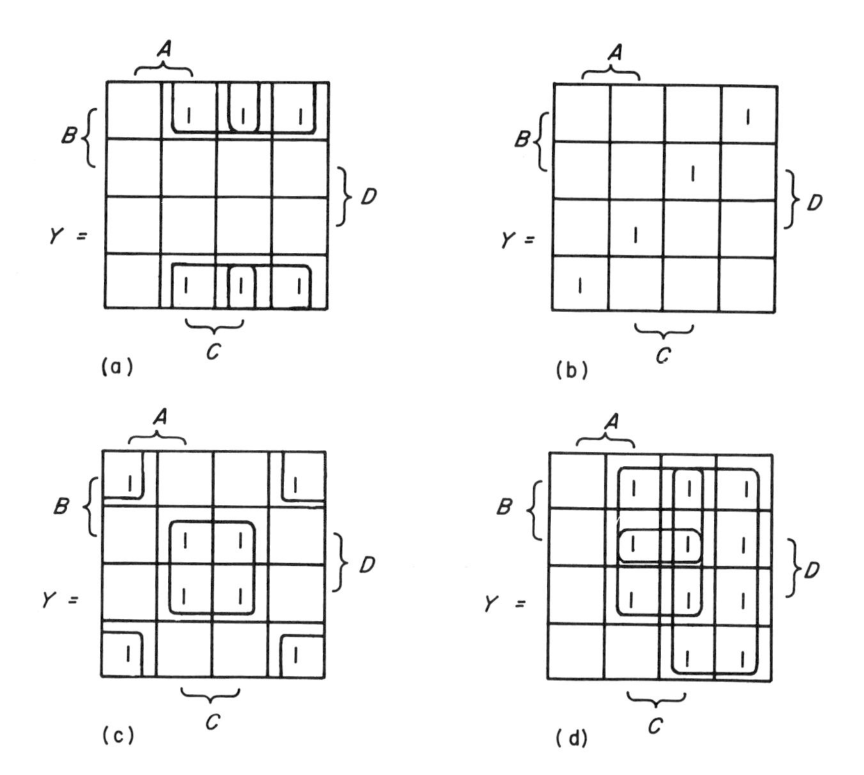

Figure 5.9 Mapping examples.

able simplification possible even after grouping from the Veitch diagram. This is not always the case, but the designer should be alert for such possibilities. Generally further economies beyond the elimination of minterms from the Veitch diagram result from factoring the resulting logical equation.

In Fig. 5.9c, there are two major groupings, yielding $Y=(\bar{C}\wedge\bar{D})\vee(C\wedge D)=\overline{C\oplus D}$. In Fig. 5.9d, there are three groups, yielding $Y=\bar{A}\vee(B\wedge C)\vee(C\wedge D)=\bar{A}\vee(C\wedge(B\vee D)$. The logic implementations of the functions for Figs. 5.9a–5.9d are given in Fig. 5.10.

5.5.4 Simplification of Logical Equations

Another application of the Veitch diagram is the simplification of a known logical equation. For example, consider the design of a circuit to determine whether a four-digit binary number ABCD is prime. The primes are 0001, 0010, 0011, 0101, 0111, 1011, 1101. These can be entered directly into the Veitch diagram by locating corresponding cells. For example, 0001 is in the only cell in which only D is true: the second from the bottom in the last column; 0010 goes in the cell where only C is true: in the third cell of the fourth row. The diagram of Fig. 5.11 was generated this way. The canonical minterm equation is

$$Y=(\bar{A}\wedge\bar{B}\wedge\bar{C}\wedge D)\vee(\bar{A}\wedge\bar{B}\wedge C\wedge\bar{D})\vee(\bar{A}\wedge\bar{B}\wedge C\wedge D)\vee(\bar{A}\wedge B\wedge\bar{C}\wedge D)$$
$$\vee(\bar{A}\wedge B\wedge C\wedge D)\vee(A\wedge\bar{B}\wedge C\wedge D)\vee(A\wedge B\wedge\bar{C}\wedge D)$$

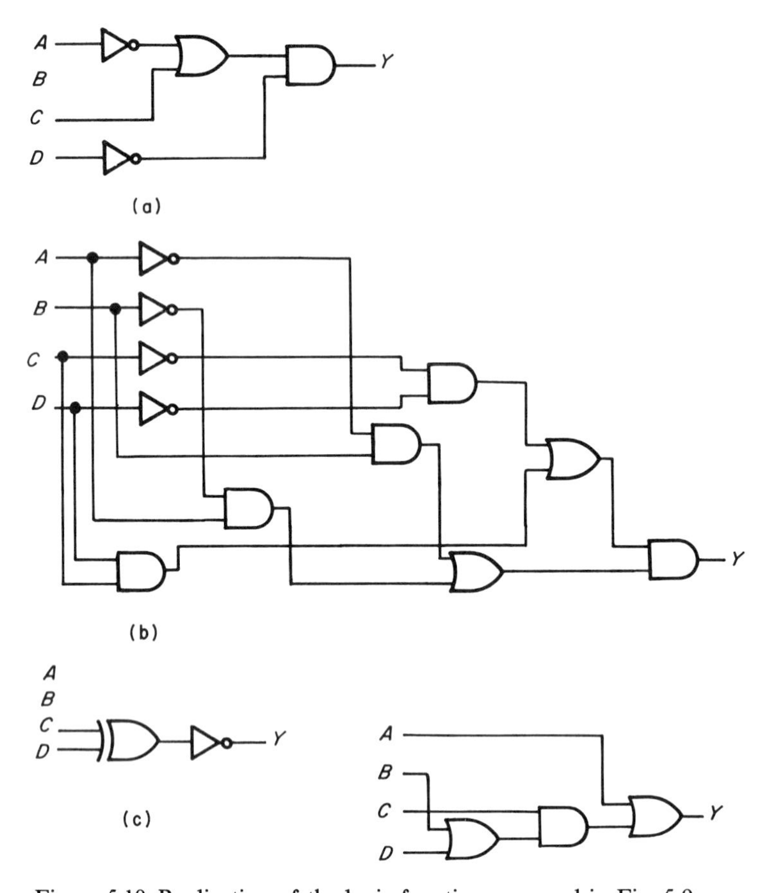

Figure 5.10 Realization of the logic functions mapped in Fig. 5.9.

From the Veitch diagram we get

$$Y=(\bar{A}\wedge D)\vee(\bar{A}\wedge\bar{B}\wedge C)\vee(\bar{B}\wedge C\wedge D)\vee(B\wedge\bar{C}\wedge D)$$

which is much more quickly obtained than it could be by derivation from the canonical minterm equation. It can be further simplified by factoring:

$$Y=\big((\bar{B}\wedge C)\wedge(\bar{A}\vee D)\big)\vee\big(D\wedge\big(\bar{A}\vee(B\wedge\bar{C})\big)\big)$$

5.5.5 *Treatment of "Don't Care" Situations*

The Veitch diagram also allows the designer to make use of "don't care" conditions: that is, *if certain bit combinations at the input are known never to occur, the device can be designed using either a* 0 *or a* 1, *depending on which gives a simpler implementation.*

For example, consider the detection of prime numbers in a BCD digit. The Veitch diagram of Fig. 5.11 would work in this situation as well as for the range 0–15, but we know that the numbers 10–15 never occur. Therefore we can place Xs in their squares, to get the Veitch diagram of Fig. 5.12.

By circling the largest possible groups, we get

$$Y=(\bar{B}\wedge C)\vee D$$

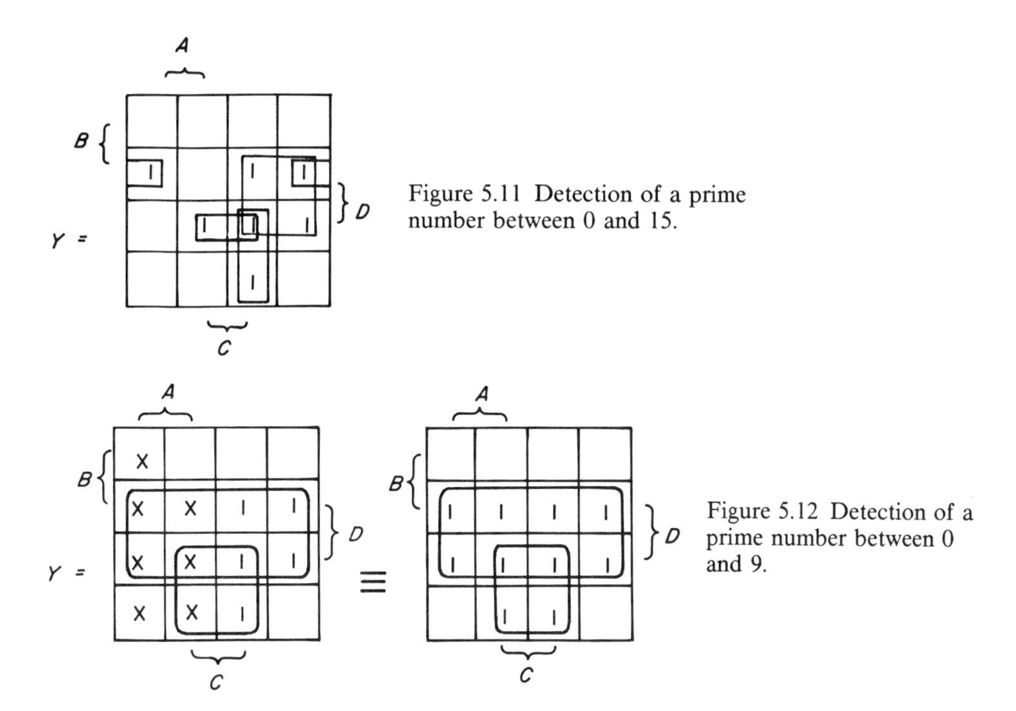

Figure 5.11 Detection of a prime number between 0 and 15.

Figure 5.12 Detection of a prime number between 0 and 9.

Here all the Xs except the lower left corner were counted as 1s, and the X in the lower left corner was counted as a zero. This is indicated in the righthand Veitch diagram. Note that this circuit will not only detect all primes in the range 0–15, but it will also detect the numbers 10, 11, 12. Since these are out of the range of legal inputs, they presumably could never lead to false detection of a prime.

5.6 MSI Combinatorial Logic Devices

The gates and inverters we have presented so far are examples of small-scale integration (SSI): that is, they are built from a relatively small number of transistors. Larger-scale devices, using larger numbers of transistors, are referred to as medium-scale integration (MSI) and large-scale integration (LSI). We shall introduce a few MSI circuits which produce outputs that are invariant functions of their inputs in this section; these are called *combinatorial logic* devices, as are gates. In the next section we shall introduce MSI devices which generate outputs that are functions of their inputs and of *time*.

A *multiplexer* selects data from one of a number of inputs and gates it onto the output. The desired input is selected by presenting the multiplexer with a unique *address*, in binary code, corresponding to the data input desired. The general form of a multiplexer is indicated in Fig. 5.13. For n inputs, m address bits will be required, where m is the smallest integer such that

$$n \leqslant 2^m$$

Multiplexers can be cascaded, as in Fig. 5.14, to address any number of data inputs. The example shows the use of four multiplexers, with five address lines, for a 24:1 multiplexing: Note that, on the last multiplexer, there is an unused input, corresponding to addresses 24–31. If this input is grounded, the output will go to zero for any of these addresses.

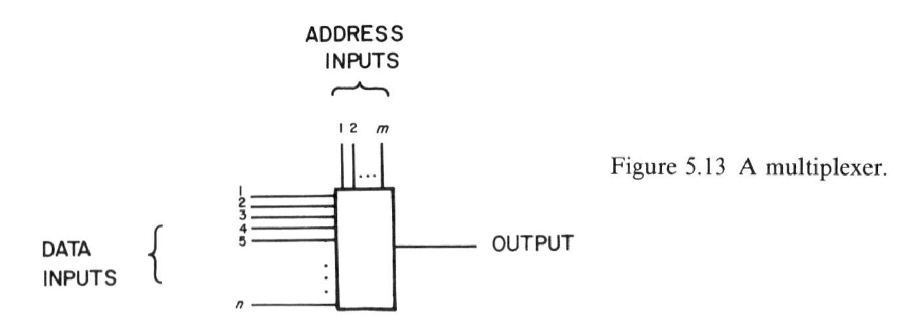

Figure 5.13 A multiplexer.

The obvious application for multiplexers, and the one for which they were originally designed, is to enable a digital device to operate on one of many inputs under some kind of digital control. Thus, a *controller* could cycle through all of the input signals one at a time, to permit a *processor* to process the signals on all the inputs, one at a time. This is practical where the processor is fast enough to follow all the incoming data on all the input data lines without missing something on one line while processing another. This application is block diagrammed in Fig. 5.15. The processor might be sufficiently expensive so that a substantial savings results from having all the inputs *share* the processor, rather than having one processor per input. Certainly, there is a saving of space taken up by the electronics, and probably in

Figure 5.14 Use of three 8-input multiplexers and one 4-input multiplexer to multiplex 24 input channels.

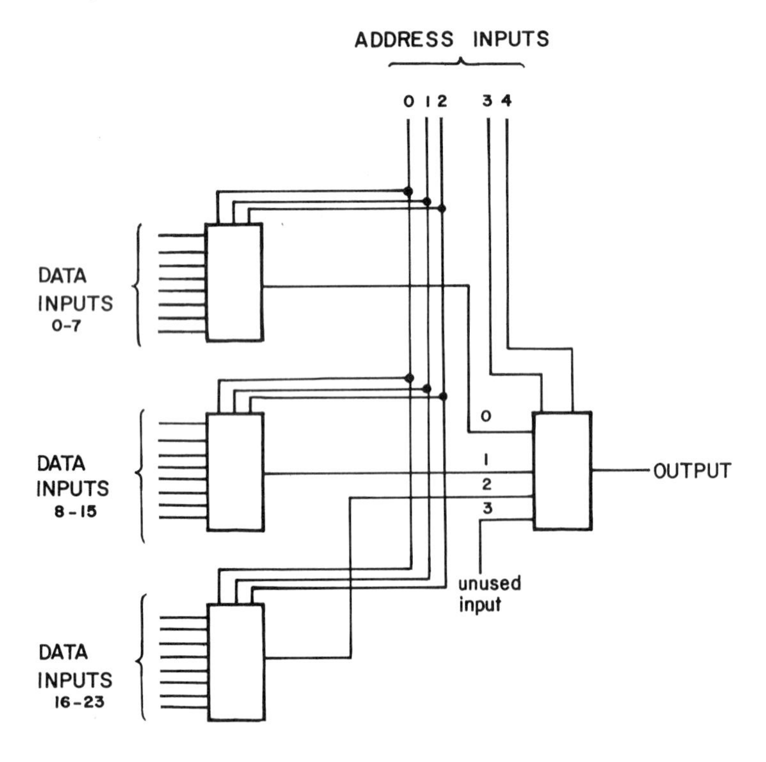

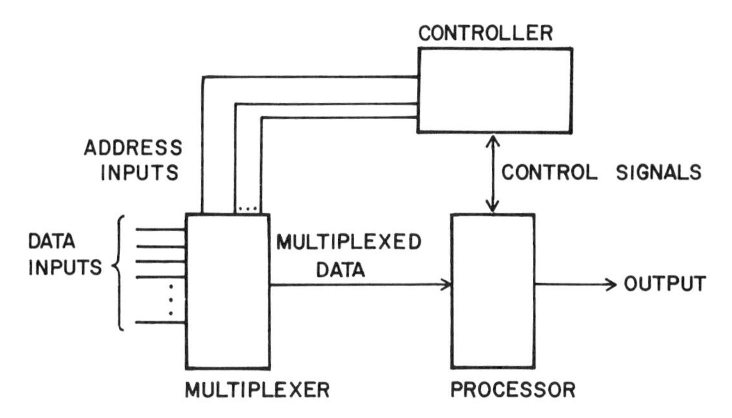

Figure 5.15 Use of a controller and multiplexer to allow a single processor to handle multiple data inputs.

power consumption. The multiplexer permits *interation in time*, rather than *interation in space*. This is illustrated in Fig. 5.16. In Fig. 5.16a, each signal is processed in exactly the same fashion by an identical processor. This is iteration in space, also called *parallel processing*: the operation is repeated over a number of inputs by dedicating one processor to each input. Alternatively, the multiplexer and controller can be used to convert simultaneous signals on different lines to sequential signals on one data line, as in Fig. 5.13 and 5.14, and the processor can then produce a series of outputs on its one output line. This is called *serial processing*, or iteration in time: the operation is repeated over a number of inputs by having them all share a processor, which processes each input identically, at a different time.

Figure 5.16 Iteration **(a)** in space, **(b)** in time.

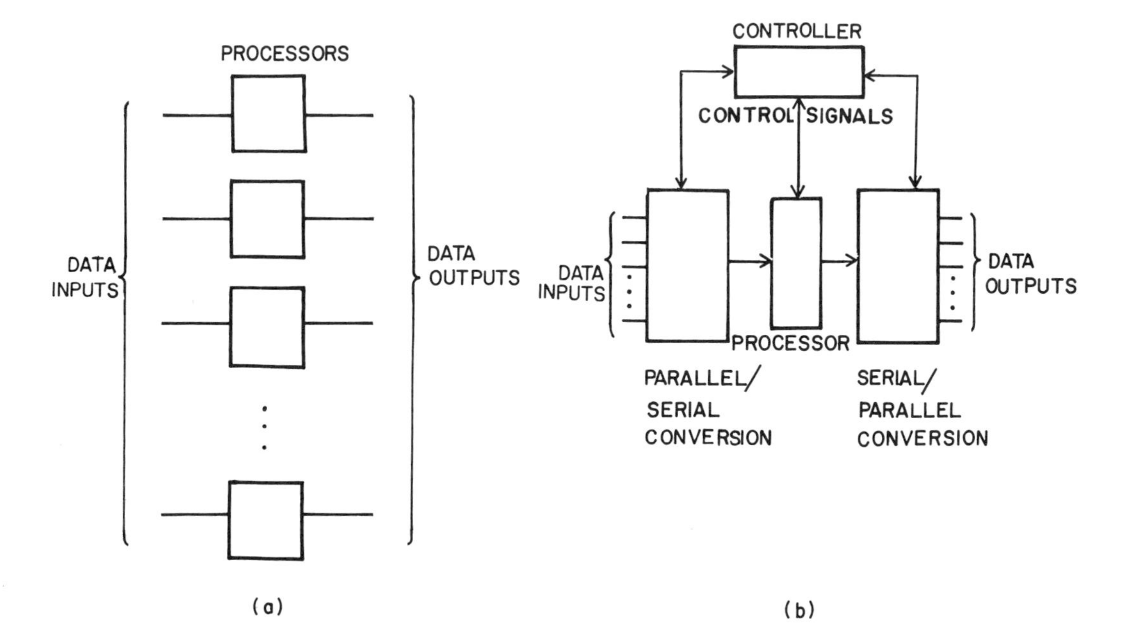

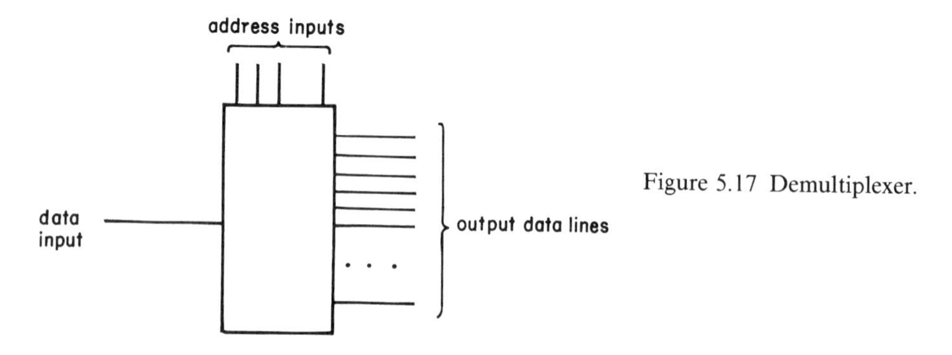

Figure 5.17 Demultiplexer.

If a number of output lines are desired, we can *demultiplex* the processor output. As the name implies, demultiplexing is the reverse of multiplexing: a signal on a single input line is placed onto one of a number of output lines. The desired output line is selected by an address, as illustrated in Fig. 5.17. Adding this to the iteration-in-time diagram of Fig. 5.16b, we can provide multiple input data lines, serial processing of the data, and multiple data output lines, as illustrated in Fig. 5.18. Of course, as the circuit stands, the processor looks at each input only a fraction of the time, and each output line has data on it only a fraction of the time. We will present methods for holding data on output lines using digital storage devices called *registers*, later in this chapter.

The internal design of multiplexers and demultiplexers is obvious from the equations describing their operations. For a 4-input multiplexer,

$$Y = (\bar{A} \wedge \bar{B} \wedge \text{input } 0) \vee (\bar{A} \wedge B \wedge \text{input } 1)$$
$$\vee (A \wedge \bar{B} \wedge \text{input } 2) \vee (A \wedge B \wedge \text{input } 3)$$

That is, the output is described by the sum of minterms, each ANDed with the corresponding input.

Figure 5.18 Use of multiplexer and demultiplexer to process multiple input data lines, using a single processor and placing results on corresponding output data lines.

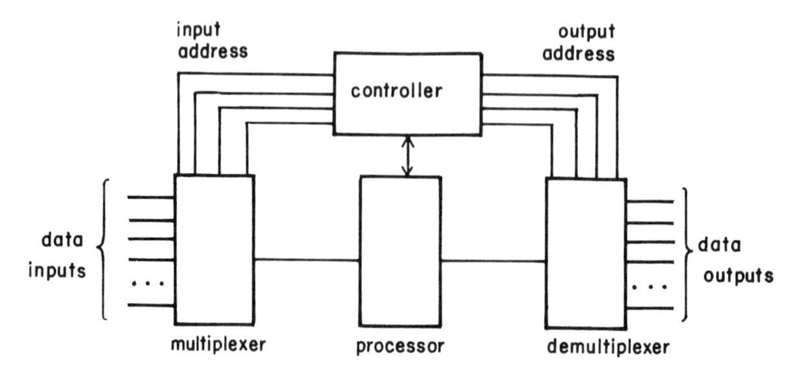

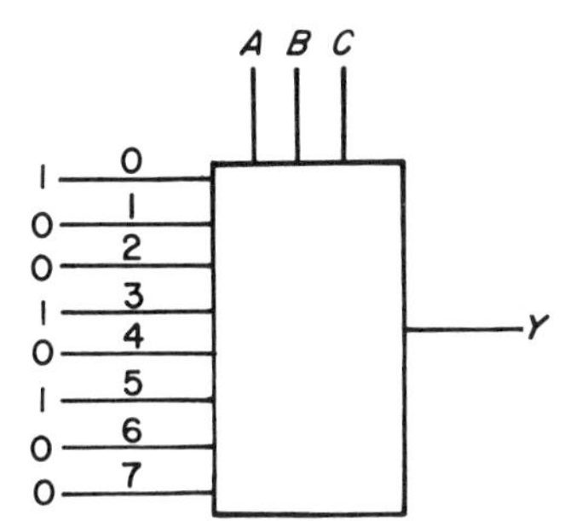

Figure 5.19 Use of a multiplexer to realize a logical function.

Each output of a 4-output demultiplexer is

$Y_0 = \bar{A} \wedge \bar{B} \wedge \text{input}$

$Y_1 = \bar{A} \wedge B \wedge \text{input}$

$Y_2 = A \wedge \bar{B} \wedge \text{input}$

$Y_3 = A \wedge B \wedge \text{input}$

These functions, or functions for larger multiplexers and demultiplexers, can be implemented with gates and inverters. The designer should never build a multiplexer from discrete gates, because they are available as ICs.

Multiplexers and demultiplexers can also be used to realize any logical function. Specifically, a multiplexer or demultiplexer is sometimes the preferred device to produce a single output from a number of input variables. The general method for using a multiplexer for this purpose is illustrated in Fig. 5.19. The input variables are connected to the address inputs, and other logical variables or constants are connected to the data inputs. Depending on the address (logic variables) input, the multiplexer produces a 0 or 1 at its output. For example, if the inputs are $A=1$, $B=0$, $C=0$, the address is 4; data line 4 has a 0 on it, so $Y=0$ if $A, B, C, = 1, 0, 0$. For different address inputs, Y has values indicated in the truth table of Table 5.25. Note that *each input corresponds to a line of the truth table, the input address corresponding to the line in the truth table*. Thus, given a truth table, *any* logical function can be realized using a multiplexer. There are more gates inside the multiplexer than required for minimal design, and propagation time is usually

Table 5.25 The Truth Table for the Design of Figs. 5.19 and 5.20

	A	B	C	Y
0	0	0	0	1
1	0	0	1	0
2	0	1	0	0
3	0	1	1	1
4	1	0	0	0
5	1	0	1	1
6	1	1	0	0
7	1	1	1	0

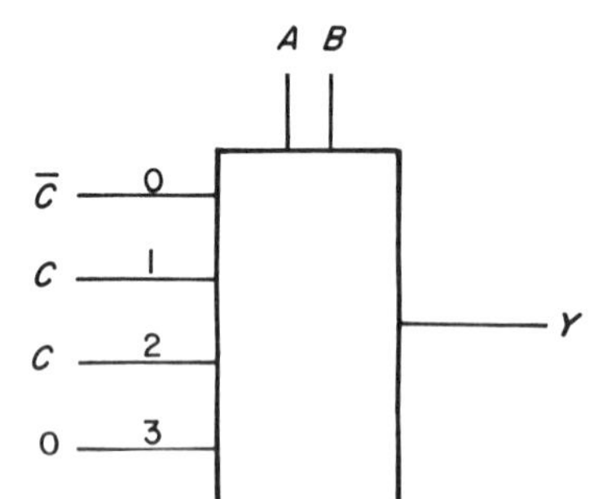

Figure 5.20 Elimination of one address input in the design of Fig. 5.19.

slightly slower than for minimal design, but for all but the most trivial logic functions the multiplexer design is the simplest and requires the fewest ICs.

Note that, for the multiplexer of Fig. 5.19, the output Y can be stated as a function of its address inputs and its data inputs:

$$Y=(\overline{A}\wedge\overline{B}\wedge\overline{C}\wedge\text{input }0)\vee(\overline{A}\wedge\overline{B}\wedge C\wedge\text{input }1)$$
$$\vee(\overline{A}\wedge B\wedge\overline{C}\wedge\text{input }2)\vee(\overline{A}\wedge B\wedge C\wedge\text{input }3)\vee(A\wedge\overline{B}\wedge\overline{C}\wedge\text{input }4)$$
$$\vee(A\wedge\overline{B}\wedge C\wedge\text{input }5)\vee(A\wedge B\wedge\overline{C}\wedge\text{input }6)\vee(A\wedge B\wedge C\wedge\text{input }7)$$

This is identical to a canonical minterm expression, where each input defines the output for a minterm, with one input per minterm.

The design of Fig. 5.20 is equivalent to that of Fig. 5.19, but uses a smaller multiplexer. The variable C has been eliminated as an address input by evaluating the value of Y for each (A, B) combination. Thus, for the first two lines of the truth table $(A, B)=(0,0)$, for address 0. For an input combination of $A=B=0$, $Y=\overline{C}$. Therefore $\overline{C}$ is attached to the address 0 input. For the next two lines of the truth table, $(A, B)=(0,1)$, corresponding to address 1, $Y=C$. For address $(A, B)=(1,0)=2$, $Y=C$. For the last two lines, $(A, B)=(1,1)$, $Y=0$.

To generalize, for any logical function of n variables, a multiplexer with $n-1$ address lines and 2^{n-1} data inputs is adequate. This is possible because, instead of having one input per minterm for all three input variables, each input corresponds to a minterm for the variables A and B, and each input is allowed to be 0, 1, C, or $\overline{C}$.

To give some notion of the savings in ICs, two 4-input multiplexers with common address lines come in a single 14-pin IC, permitting the realization of any two functions of the same three variables (or of the same two variables and any two other variables) with one integrated circuit. Of course, it may be necessary to use one inverter for the third variable to provide inverted inputs, e.g., the $\overline{C}$ input at address 0 in Fig. 5.20. For a function of three logical variables such as that of Table 5.25, there are three realizations of the type in Fig. 5.20. The other two for this function are given in Fig. 5.21. The designer can use any of the possible configurations equally easily: one factor which might determine choice of configuration is whether the inverse of one of the variables is already available elsewhere. This variable could then be used on the data lines, saving an inverter.

Since 2-, 4-, 8-, and 16-input multiplexers are available, single ICs can be used to generate functions of up to five variables. By using multiplexer trees like that of Fig.

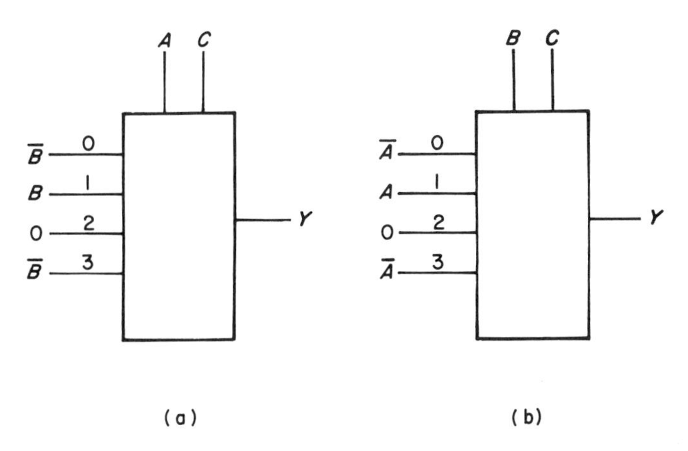

Figure 5.21 Two other implementations of the logic function of Fig. 5.19 and 5.20.

5.14, arbitrarily large multiplexers can be synthesized to handle functions of arbitrarily large numbers of variables.

Multiplexers can be used to handle a part of a logical function; e.g., a multiplexer could be used to handle three out of four variables of a function, with a gate or two to handle the last variable. This can also be done to simplify an otherwise unwieldy Veitch diagram.

Demultiplexers can also be used to implement logic functions. They are useful in situations where a large number of output functions are to be generated from a small number of input variables, and where each output function is true (or false) for only one or a small number of input combinations. For example, if the designer wishes to illuminate one of ten numbered lights to indicate which of ten digits is represented by a BCD code, he could use the circuit of Fig. 5.22, implementing the truth table of Table 5.26.

Very complex functions of large numbers of variables are difficult to implement even with multiplexers and demultiplexers. Such implementations are better accom-

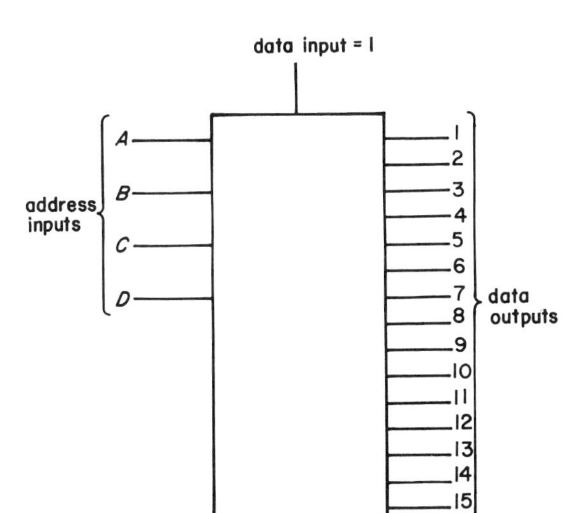

Figure 5.22 Use of a demultiplexer to generate a decimal code from a binary code. Outputs 10–15 are never selected because there are no corresponding BCD codes.

Table 5.26 Generation of a Decimal Code from a BCD Code

BCD		Input		Decimal Output					
A	B	C	D	0	1	2	34	56	789
0	0	0	0	1	0	0	00	00	000
0	0	0	1	0	1	0	00	00	000
0	0	1	0	0	0	1	00	00	000
0	0	1	1	0	0	0	10	00	000
0	1	0	0	0	0	0	01	00	000
0	1	0	1	0	0	0	00	10	000
0	1	1	0	0	0	0	00	01	000
0	1	1	1	0	0	0	00	00	100
1	0	0	0	0	0	0	00	00	010
1	0	0	1	0	0	0	00	00	001

plished with digital memories, which are LSI (large-scale integration) devices. We shall discuss logic function implementation with digital memories later in this chapter.

5.7 Busses

When assembling a large digital machine, it is the custom to interconnect parts of the machine so they can communicate via digital signals. When many subunits are connected together, each with a large number of signals, and each "talking" to a large number of other subunits, the interconnection requirements become burdensome. In many cases, only one device really needs to talk at one time, so it is possible to connect all their outputs to a single set of conductors in such a way that only one device can transmit at a time. Such a communication system is called a *bus*.

The general scheme for a bus is diagrammed in Fig. 5.23. Each device has a set of five inputs and/or five outputs attached to the five wires in the bus. That is, each device sends or receives signals through the bus, or does both. Of course, devices that send must somehow identify themselves, and must also identify the device to which they are sending. Also, each device which transmits must be able to tell whether the

Figure 5.23 A bus.

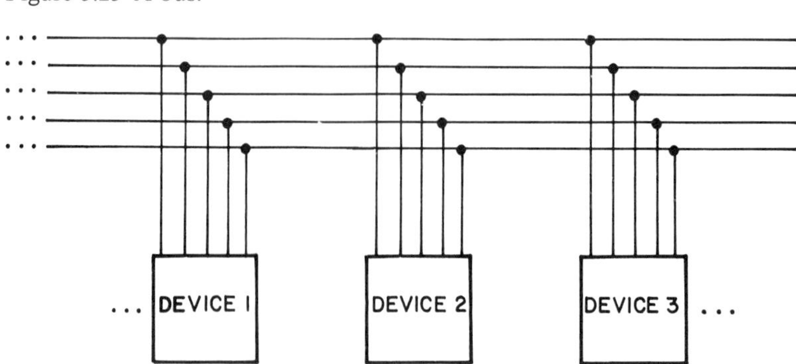

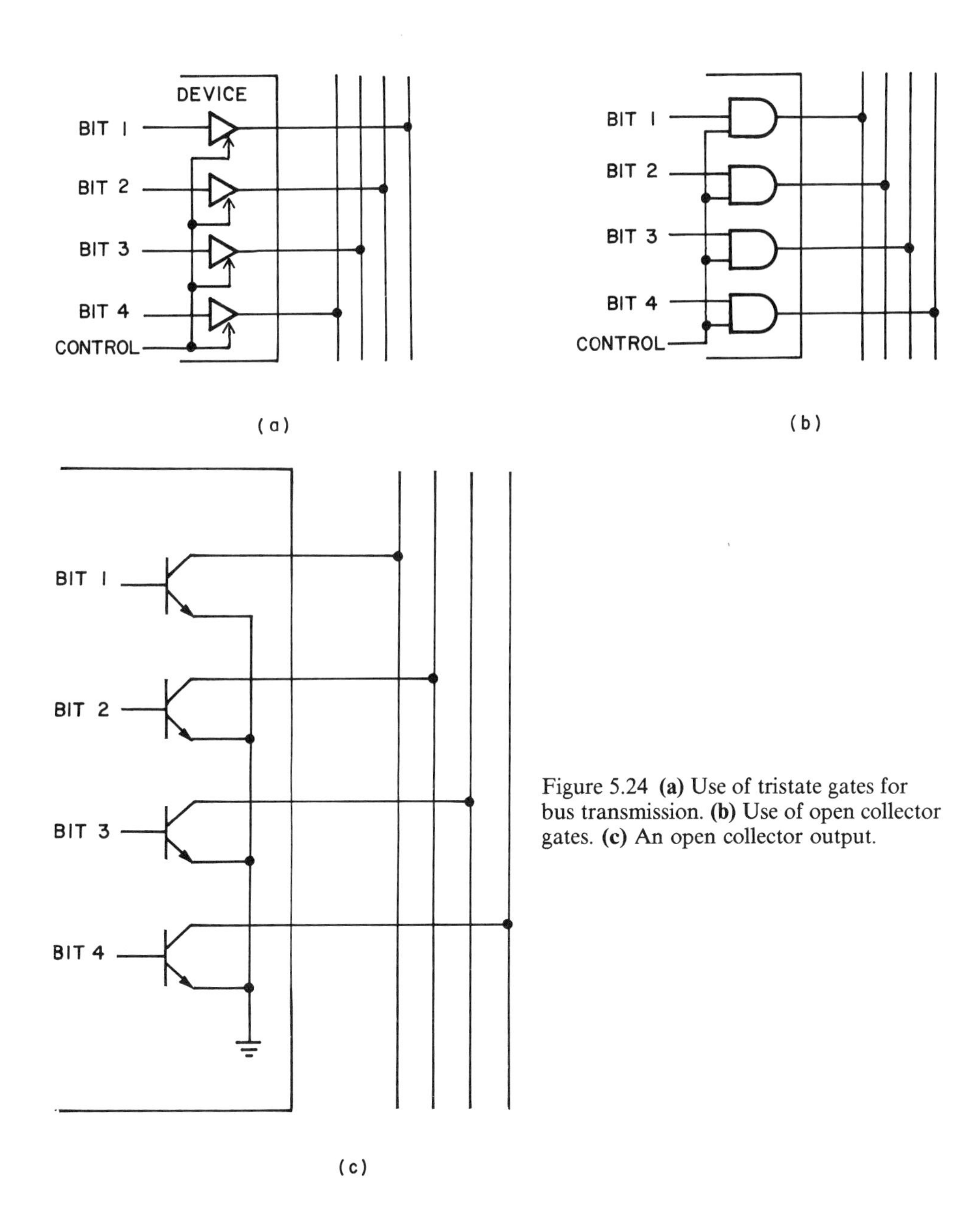

Figure 5.24 **(a)** Use of tristate gates for bus transmission. **(b)** Use of open collector gates. **(c)** An open collector output.

bus is "busy," that is, whether another device is transmitting; and if the bus is busy, it must hold off transmission until the bus is free.

To hold off transmission, the simplest method would be to use *tristate* output gates such as those in Fig. 5.24a. These have three output states: 1, 0, and OFF. The input passes to the output if CONTROL is on, bringing the bus line to a 0 or 1 level, with a low output impedance. If CONTROL is OFF, however, the output presents a high impedance to the bus, permitting some other tristate device to drive the bus. Of course, the bus will work properly only if no more than one device has a low output impedance.

The other method for driving data busses is to use open collector output devices, typically gates. This method is illustrated in Fig. 5.24b. Here, the output is from a transistor collector (Fig. 5.24c). If the output is logical 1, the transistor is not conducting, and presents a high impedance to ground. If the output is logical 0, the transistor is conducting, presenting a low impedance to ground. If all open collector devices tied to a bus are logical 1, there is a high impedance to ground; if one or more is logical 0, there is a low impedance to ground; typically, the bus is tied to +5 V through a resistor to pull it up to logical 1 in the absence of a conducting gate; if a gate conducts, the bus is pulled down to logical 0.

The open collector gate is a means of wired AND/OR operation. For low-level logic (low level = logical 1), also called negative logic, the logic level on the bus is the ORed value of all the gate outputs; for high-level logic (high level = logical 1), also called positive logic, it is the ANDed value of all the gate outputs.

5.8 Flipflops and Registers

The combinatorial logic devices we have been discussing so far permit the generation of any logical function, but they cannot be used for the generation of *sequences* of *logical states*; that is, they passively follow the inputs and do not produce outputs which are time varying (except insofar as the inputs are time varying). Also, combinatorial circuits cannot "remember" previous states of their inputs, and cannot change their properties as a function of time. All of these additional properties are important for the manufacture of devices which perform sequences of operations on their inputs; such devices are themselves essential for most digital applications.

In this section we shall develop the one-bit memory called the flipflop, and show how the flipflop can be applied to applications requiring memory and automatic sequencing.

5.8.1 Flipflops

Figure 5.25 illustrates a single flipflop. If we assume both inputs are initially 0, the outputs Q and $\overline{Q}$ (here, $\overline{Q}$ is not necessarily the inverse of Q, a problem we cannot avoid because this is standard notation) are

$$Q = Reset\ \overline{\vee}\ \overline{Q} = 0\ \overline{\vee}\ \overline{Q} = -\overline{Q}$$

$$\overline{Q} = Set\ \overline{\vee}\ Q = 0\ \overline{\vee}\ Q = -Q$$

Note that if $Q = 1$, $\overline{Q} = 0$, and if $Q = 0$, $\overline{Q} = 1$. The equations do not specify which of the two combinations exists, and either is possible.

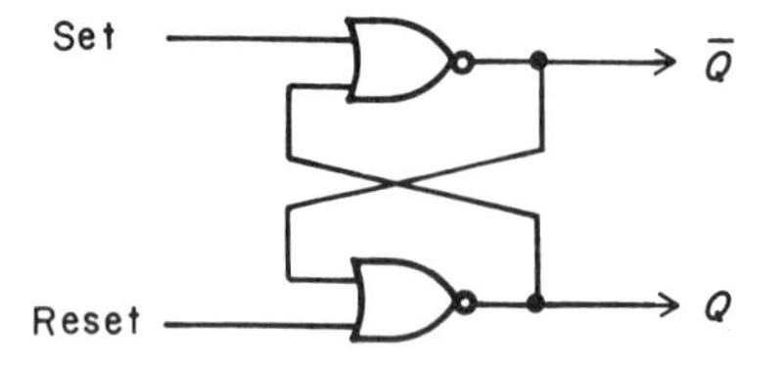

Figure 5.25 A set/reset (S/R) flipflop.

If one input goes high, the corresponding output goes low:

$$Q = Reset \,\overline{\vee}\, \overline{Q} = 1 \,\overline{\vee}\, \overline{Q} = 0$$

or

$$\overline{Q} = Set \,\overline{\vee}\, Q = 1 \,\overline{\vee}\, Q = 0$$

and the other output, with zero input, goes high:

$$Q = Reset \,\overline{\vee}\, \overline{Q} = 0 \,\overline{\vee}\, 0 = 1$$

or

$$\overline{Q} = Set \,\overline{\vee}\, Q = 0 \,\overline{\vee}\, 0 = 1$$

Thus, $Set = 1$, $Reset = 0$, produces $Q = 1$, $\overline{Q} = 0$; and $Reset = 1$, $Set = 0$ produces $Q = 0$, $\overline{Q} = 1$. If, after Q or $\overline{Q}$ goes to 1, both inputs return to zero, then Q and $\overline{Q}$ remain unchanged. Thus, either

$$Q = Reset \,\overline{\vee}\, \overline{Q} = 0 \,\overline{\vee}\, 1 = 0$$

and

$$\overline{Q} = Set \,\overline{\vee}\, Q = 0 \,\overline{\vee}\, 0 = 1$$

or

$$Q = Reset \,\overline{\vee}\, \overline{Q} = 0 \,\overline{\vee}\, 0 = 1$$

and

$$\overline{Q} = Set \,\overline{\vee}\, Q = 0 \,\overline{\vee}\, 1 = 0$$

In addition, if $Set = Reset = 1$,

$$Q = Reset \,\overline{\vee}\, \overline{Q} = 1 \,\overline{\vee}\, \overline{Q} = 0$$

$$\overline{Q} = Set \,\overline{\vee}\, Q = 1 \,\overline{\vee}\, Q = 0$$

But with $Set = Reset = 1$, and then *Set* or *Reset* goes to 0, then

$$Q = Reset \,\overline{\vee}\, \overline{Q} = 0 \,\overline{\vee}\, 0 = 1$$

and

$$\overline{Q} = Set \,\overline{\vee}\, Q = 1 \,\overline{\vee}\, 0 = 0$$

or

$$Q = Reset \,\overline{\vee}\, \overline{Q} = 1 \,\overline{\vee}\, 0 = 0$$

and

$$\overline{Q} = Set \,\overline{\vee}\, Q = 0 \,\overline{\vee}\, 0 = 1$$

That is, if *Set* goes to zero first, $\overline{Q}$ goes high and Q stays zero; and if *Reset* goes to zero first, Q goes high and $\overline{Q}$ stays zero.

To summarize: If both inputs are zero, Q and $\overline{Q}$ will be opposite polarities; they will reflect the most recent entry of data via the *Set* and *Reset* inputs. If *Set* most recently went high with *Reset* low, Q is high and $\overline{Q}$ is low *regardless of subsequent changes* at the *Set* input. If *Reset* most recently went high with *Set* low, Q will be low and $\overline{Q}$ high, *regardless of subsequent changes* at the *Reset* input. If both *Set* and *Reset*

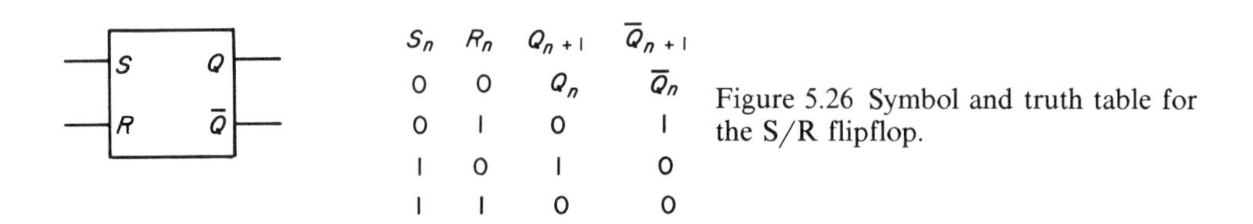

S_n	R_n	Q_{n+1}	$\overline{Q}_{n+1}$
0	0	Q_n	$\overline{Q}_n$
0	1	0	1
1	0	1	0
1	1	0	0

Figure 5.26 Symbol and truth table for the S/R flipflop.

go high, Q and $\overline{Q}$ go low until one of the inputs go low. Q will go high if *Reset* goes low first, and $\overline{Q}$ will go high if *Set* goes low first.

Thus, the flipflop, called a *Set/Reset* or S/R flipflop, can be set (Q goes high) or reset ($\overline{Q}$ goes high) by positive transitions (change from 0 to 1) at the *Set* or *Reset* input, respectively, and as long as both inputs are not simultaneously high, the flipflop will hold this state until reset or set, respectively. It is thus a form of *digital memory*.

The symbol for the S/R flipflop and its truth table are given in Fig. 5.26. The states Q_{n+1} and $\overline{Q}_{n+1}$ represent the states subsequent to inputs S_n, R_n, and previous states Q_n and $\overline{Q}_n$.

Symbols and truth tables for D flipflops and J–K flipflops are given in Fig. 5.27. The D flipflop transfers the D (data) input to the output on a positive transition at its C (clock) input; changes at D are ignored between positive transitions at C. Such a flipflop can be used to "freeze" a logical signal at a certain time, a feature which is useful in circuits which need to be able to hold a digital logic value until the processor can handle it.

The J–K flipflop is a *master–slave* flipflop: Inputs at J and K are sampled on positive transitions at the T (toggle) input, and the sampled J and K values are used

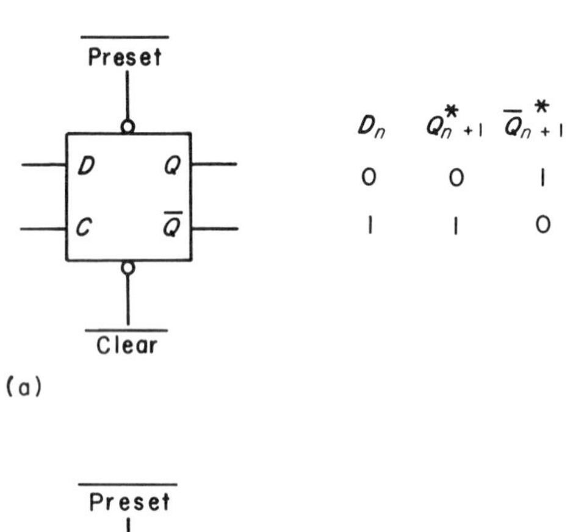

D_n	Q^*_{n+1}	$\overline{Q}^*_{n+1}$
0	0	1
1	1	0

(a)

$\overline{\text{Preset}}$

J T K Q $\overline{Q}$

$\overline{\text{Clear}}$

J_n	K_n	Q^*_{n+1}	$\overline{Q}^*_{n+1}$
0	0	Q_n	Q_n
0	1	0	1
1	0	1	0
1	1	$\overline{Q}_n$	Q_n

(b)

Figure 5.27 **(a)** D flipflop and **(b)** J–K flipflop. *$\overline{\text{Preset}}$ = 0 forces Q to 1, $\overline{Q}$ to 0; $\overline{\text{Clear}}$ = 0 forces Q to 0, $\overline{Q}$ to 1; $\overline{\text{Clear}}$ = $\overline{\text{Preset}}$ = 0 produces undefined output. $\overline{\text{Clear}}$ = $\overline{\text{Preset}}$ = 1: truth tables apply.

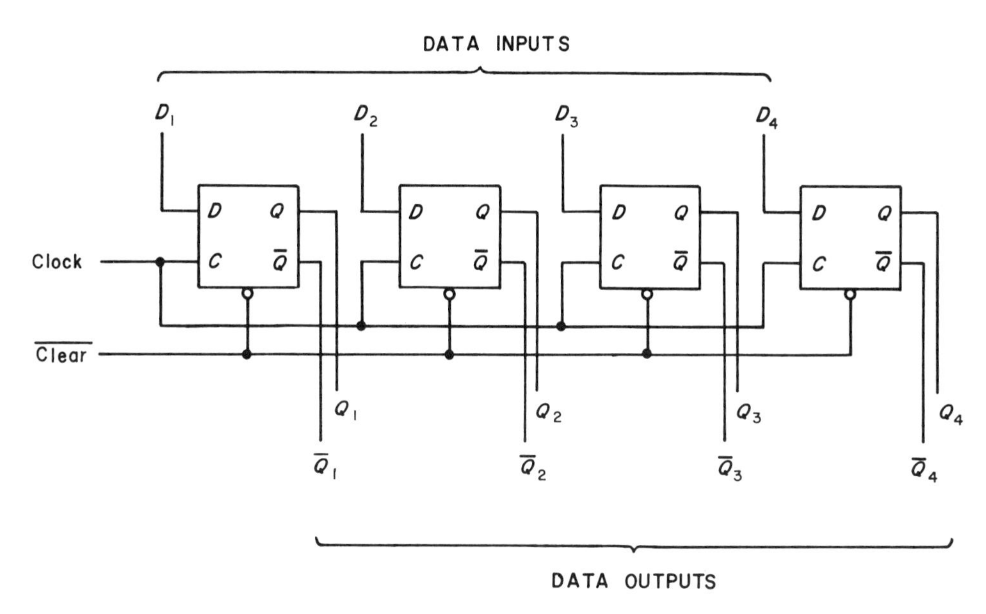

Figure 5.28 A D register.

to determine the Q output on the negative T transition. The device is called a master–slave device because the master flipflop, whose outputs are not available to the user, holds the sampled data (at a positive T transition) and the contents of the flipflop are *jammed* into the slave flipflop (at a negative T transition), whose outputs are Q and $\overline{Q}$. If J is high and K is low, Q goes high; if J is low and K is high, $\overline{Q}$ goes low; if J and K are both low, Q and remain unchanged and if J and K are both high, Q and reverse state (go from 0 to 1 or from 1 to 0)—this last type of behavior is called *toggling*. Toggling is essential to the sequencing of many automatic devices.

In both the D and $J-K$ flipflops, Q and $\overline{Q}$ are of opposite polarities. The $\overline{Set}$ (also called $\overline{Preset}$) and $\overline{Reset}$ (also called $\overline{Clear}$), inputs, when low, force $Q=1$ or $Q=0$, respectively, regardless of the inputs.

5.8.2 Registers

A register is a collection of flipflops, all of which are synchronously driven by the same clock. Figures 5.28–5.32 illustrate some common registers using flipflops. These devices, or devices like them, are all available as MSI integrated circuits. In Fig. 5.28 a simple D register is illustrated. Each flipflop Q output takes on the value at its D input whenever the clock goes positive; this value is held until the next positive clock transition, regardless of changes at the D input between positive clock transitions. Such a device is used to hold a multidigit binary *word*. For example, it could be used to hold data output values on each of a set of demultiplexed data lines, refreshing each time the shared processor recomputes its value. D registers can be the basis for digital computer semiconductor memories, where each register represents one word in memory, and each word is accessed by a specific *memory address*, using address selection similar to that used in multiplexers. We shall discuss such memories in the section on the use of LSI circuits for digital design.

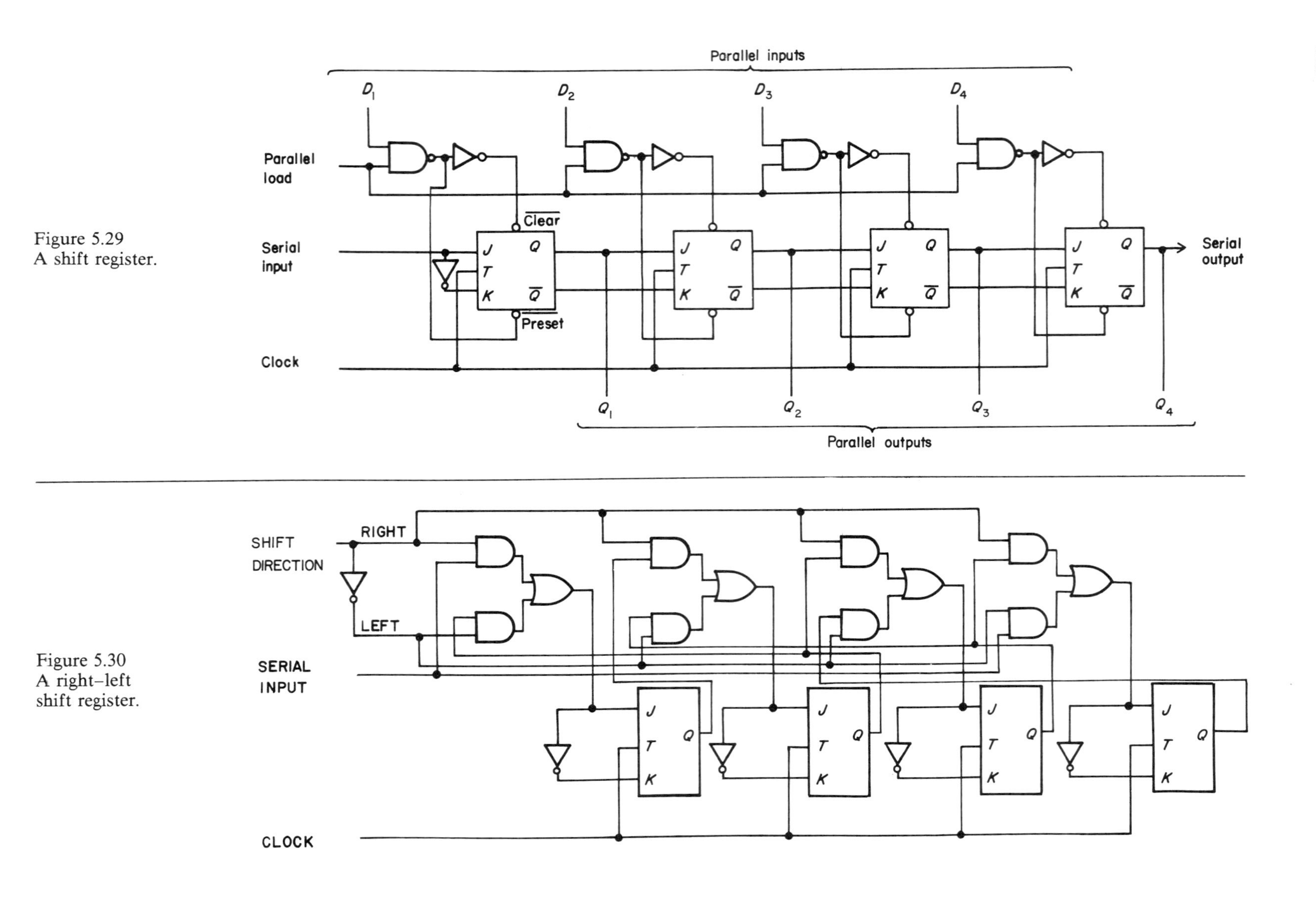

Figure 5.29
A shift register.

Figure 5.30
A right–left
shift register.

The $J-K$ register of Fig. 5.29 is a *shift register*. Data are loaded either serially (one bit per clock cycle) via the serial input, or in parallel (all at once) using the gates to the *set* (*preset*) and *clear* inputs. Using the serial input, data are loaded into the first flipflop on each cycle (sampled on positive transition, transferred to output of first flipflop on negative transition). Using the parallel load, the gate outputs are held at 1 while the parallel load signal is zero, and the data inputs are passed to the *preset* and *clear* inputs of the $J-K$ flipflops whenever *parallel load* is high: thus, a 0 parallel data input clears the corresponding flipflop and a 1 input sets it during *parallel load* = 1.

Regardless of how the device is loaded, the contents of each flipflop are changed on each clock cycle, in synchrony with the other flipflops. On a given clock cycle, the following events occur: (a) on a positive transition, the serial input is sampled by the leftmost, and each of the other flipflops samples the flipflop to its left; (b) on a negative clock transition, the sampled values appear on the outputs. Thus, on each clock cycle, the serial input goes to the output of flipflop 0, the output of flipflop 0 goes to the output of flipflop 1, and so forth: that is, *all of the flipflop contents shift right one flipflop*. For this reason, the device is called a *shift register*. The serial input is loaded into the first flipflop and the content of the last flipflop is lost on each shift.

Figure 5.30 illustrates a left–right shift register. For simplicity, only a serial input is indicated. The shift direction is controlled by a SHIFT DIRECTION signal, such that the register's contents shift to the right if SHIFT DIRECTION = 1, and to the left if SHIFT DIRECTION = 0. This is accomplished using gates to determine whether input to a flipflop is from the flipflop to the right or from the flipflop to the left.

Note that the use of $J-K$ flipflops eliminates a factor which could be a serious problem if D flipflops were used. Each flipflop changes state only after all flipflops have sampled and internally stored in their master flipflops the data coming into them. This is important because if the change of state and sampling of input data both occurred at the same time, as would be the case with D flipflops, those flipflops sampling the outputs of other flipflops would be changing state at the same time as the flipflops whose outputs they were sampling. Since true simultaneity cannot be guaranteed because of inherent differences in response times of the flipflops, a flipflop might sample the preceding one's output after it had changed rather than before. Thus, the contents of a four-bit right–shift flipflop might be 0110, which should be followed by X011 after a shift, with X = the serial input. But due to the synchronization problem, it could be XX11, if the first flipflop changed state before it was sampled by the second. Such synchronization problems, called *race conditions* (the devices can be viewed as *racing* with each other) are avoided by separating input sampling and changing of output state in each flipflop into two phases, which can be done with master–slave devices like $J-K$ flipflops. There are other methods for eliminating race conditions, some of which will be discussed in the next few chapters.

Shift registers are available commercially in a number of different configurations: with different lengths, with and without parallel inputs and outputs, as uni- or bidirectional devices, even with more than one shift register per IC. They have a number of interesting applications in data communication, arithmetic computation, and digital signal generation, processing, and acquisition. We shall discuss each of these types of applications in later chapters.

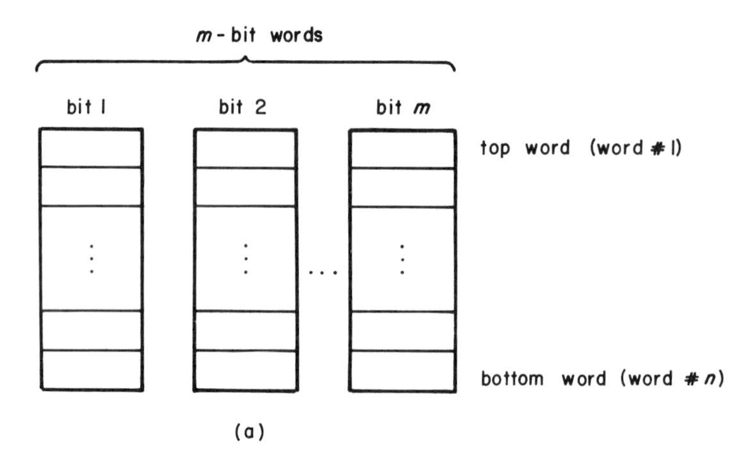

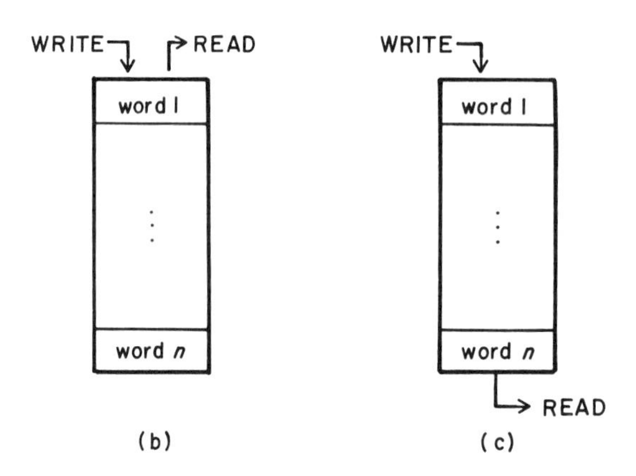

Figure 5.31 Stacks. **(a)** An m-bit by n-word stack. **(b)** Last-in-first-out (LIFO) stack. **(c)** First-in-first-out (FIFO) stack.

Some specialized types of shift registers are useful for short-term storage of a number of digital words. Two of these are so-called *stacks*: words can be placed on the top of a stack of storage word locations, and depending on the type, they can be read off the top or the bottom of the stack. They are usually arranged as specialized shift registers, as illustrated in Fig. 5.31a, where each shift register represents one bit of the m-bit words to be stored, and hence there are m shift-registers in all. These operate strictly in parallel with each other, so all the bits in a word are in the same location in all the corresponding shift registers. To store n words, the shift registers are n bits long.

The last-in-first-out (LIFO) shift register allows words to be written onto the top word of the shift register, with all previous words shifted one place to make room for the new word. Whenever a word is read, it is read from the top, and all words are shifted back up one place. Thus the most recently entered word which has not yet been read is available at the top of the stack for the next read operation. As long as, from the time the stack is empty, the number of writes minus the number of reads is less than or equal to the number of words the stack can hold, the stack will not overflow.

The LIFO operates much like a stack of cafeteria trays: When a tray is placed on top, it weights the pile down the depth of one tray, and when a tray is lifted off the top, the pile pops up one tray height. For this reason, the LIFO is often called a *push-down–pop-up stack*.

As an exercise in the use of flipflops and shift registers, a possible design for a LIFO is given in Fig. 5.32. (The designer should always use LIFOs as MSI integrated circuits rather than construct his own, however.) The register consists of one right–left shift register per bit, plus an extra right–left shift register used to determine which bits contain information and which are empty. The extra register is required because it is not possible to determine whether a flipflop bit contains information or not and also have it code a binary bit, because this would require the flipflop to indicate one of three possible conditions: (1) no information contained in the bit; (2) information contained $=1$; (3) information contained $=0$. Since a binary flipflop can have one of only two possible states, an extra shift register is needed to code which bits (levels in the stack) contain data words. Thus, for an m-bit word width, $m+1$ shift registers are needed. Our example illustrates the design for a 1-bit word with a stack depth of three words.

$\overline{\text{CLEAR}}$ signal clears all flipflops, a useful feature for initialization. When $\overline{\text{CLEAR}}=1$, the device may be used for data storage. Data is presented as a binary 0 or 1 at INPUT, and the top bit in the stack is available at OUTPUT. As can be seen from the figure, data paths go into and out of the upper right–left shift register, so the lower right–left shift register is used to keep track of which bits are occupied: a 1 in a bit of the lower register indicates the corresponding bit of the upper register has data in it; a 0 indicates that stack level is empty. DATA PRESENT gives state of output.

A WRITE/READ $=1$ signal enables one set of AND gates, so that INPUT goes to the J input of the upper left flipflop, and a 1 goes to the lower left flipflop. Also, the output of each flipflop (except the rightmost one) goes to the J input of the next flipflop to the right. Thus, WRITE/READ $=1$ enables both shift registers to be used as right shift registers. WRITE/READ $=0$ enables the other set of AND gates, so that each flipflop (except the two rightmost) receives its J input from the next flipflop to the right. Thus, WRITE/READ $=0$ enables both shift registers to be used as left shift registers.

In right shift (WRITE) mode, INPUT goes to the top (left) flipflop in the data shift register, and a 1 goes to the corresponding flipflop in the status shift register. A clock pulse (transition from 0 to 1 and from 1 back to 0) loads these two flipflops, transfers their previous contents to the next (middle) flipflops, and transfers the middle flipflops' contents to the last (right) flipflops. The contents of the last flipflops are lost.

In the left shift (READ) mode, the rightmost two flipflops are loaded on a clock pulse with zeros, clearing the last word and its status bit; the middle flipflops are loaded from the rightmost ones; and the left flipflops are loaded from the middle ones.

Thus, for each clock pulse, a new data word is loaded if WRITE/READ $=1$, and the previously loaded data words are shifted right. If WRITE/READ $=0$, all the data words are shifted left. Thus, WRITE/READ $=1$ is the same as a "push," and WRITE/READ $=0$ is the same as a "pop." Clearly, a specific sequence of opera-

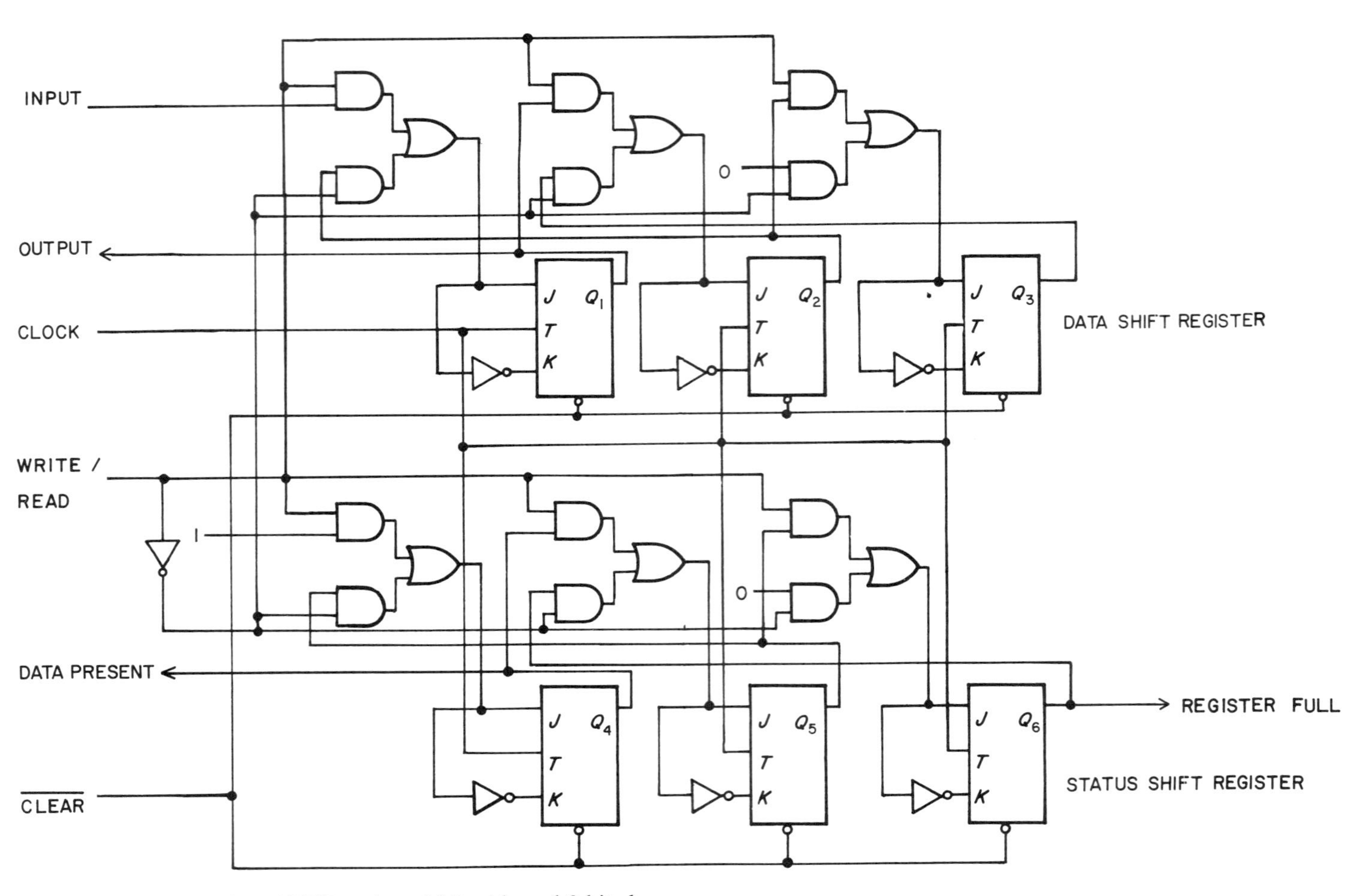

Figure 5.32 A design for a LIFO register 1 bit wide and 3 bits long.

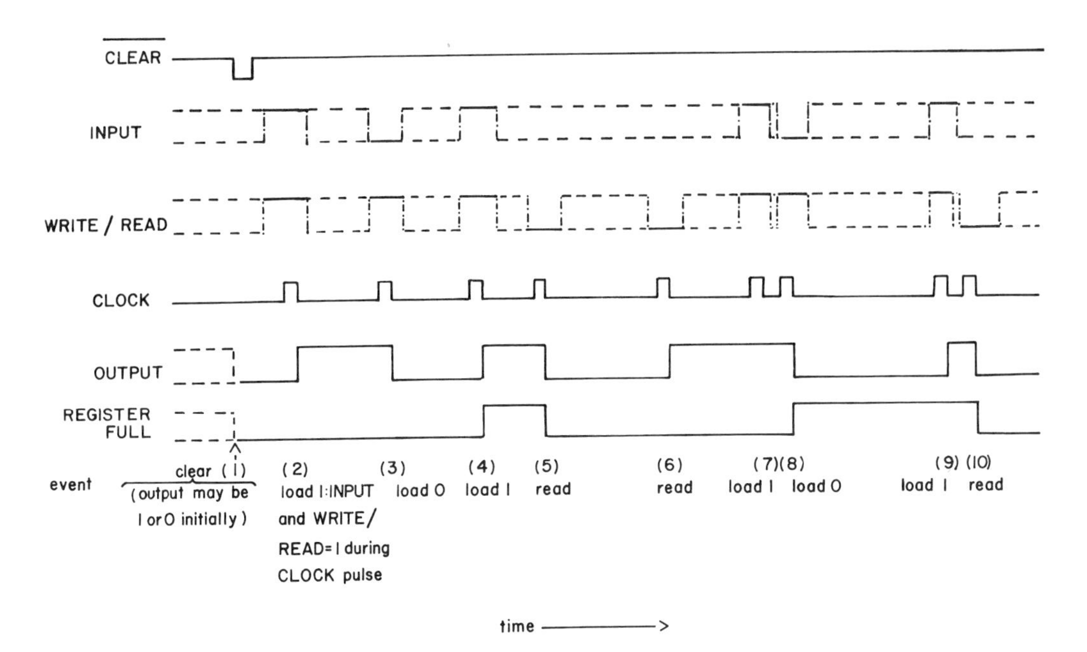

Figure 5.33 Timing diagram for the LIFO.

tions at the inputs is necessary: (a) first, the INPUT is set up to the proper (0 or 1) value, or the OUTPUT is read or used to control some process; (b) the WRITE/READ bit must also be set up for proper shift direction; (c) when (a) and (b) are done (in either order), a CLOCK pulse causes the shift operation. For detection of a full stack, the Q output of the last status bit is available; it can only go on if all levels of the stack are occupied.

To diagram such a complicated sequence of events, it is handy to represent all the relevant signals in time. Such a *timing diagram* is presented in Fig. 5.33, which illustrates the following sequence of operations: (1) clear the LIFO; (2) load a 1; (3) load a 0; (4) load a 1; (5) read the 1; (6) read the previous 0; (7) load a 1; (8) load a 0; (9) load a 1; (10) read the 1. The numbered steps are indicated in the figure. Note that INPUT and WRITE/READ need only be specified during the CLOCK pulse. (Actually, they should be set up early enough to avoid races, and retain the desired values long enough to avoid races with the CLOCK pulse.)

The contents of the LIFO at steps (1)–(10) can also be represented with a truth table, as in Table 5.27. Here, the logical variables INPUT, WRITE/READ, OUTPUT, and REGISTER FULL are listed, as well as the states of Q_1, Q_2, Q_3, Q_4, Q_5, and Q_6. Of course, $Q_1 = \text{OUTPUT}$, and $Q_6 = \text{REGISTER FULL}$. This method is very handy, and is in fact a useful intermediate step in the design of a logical device. Each of the ten "states" is defined as the time at the end of a $\overline{\text{CLEAR}}$ pulse or CLOCK pulse, neglecting transition delays. Later in this chapter we shall discuss the use of such a *state logic design* method for constructing sequential logic machines. The LIFO is, of course, a type of sequential logic device in that the states of the logical variables are functions not only of the logic levels at the inputs and within the machine but also of the previous states of the input and internal logical variables.

Table 5.27 States of Input Variables and Internal Variables of the LIFO of Figs. 5.32 and 5.33 at State Times 1–10[a]

State time		1	2	3	4	5	6	7	8	9	10
Clear		0	1	1	1	1	1	1	1	1	1
Input[b]		x	1	0	1	x	x	1	0	1	x
WRITE/READ[b]		x	1	1	1	0	0	1	1	1	0
OUTPUT ($=Q_1$)		0	1	0	1	0	1	1	0	1	0
REGISTER FULL ($=Q_6$)		0	0	0	1	0	0	0	1	1	0
Q_1	data bits	0	1	0	1	0	1	1	0	1	0
Q_2	data bits	0	0	1	0	1	0	1	1	0	1
Q_3	data bits	0	0	0	1	0	0	0	1	1	0
Q_4	status bits	0	1	1	1	1	1	1	1	1	1
Q_5	status bits	0	0	1	1	1	0	1	1	1	1
Q_6	status bits	0	0	0	1	0	0	0	1	1	0

[a]Arrows indicate (arrows crossing boundary from one state to another) flow of information from one flipflop or another or (arrows confined to one state time) loss or capture of information due to shift out of or shift into shift register).

[b]x = value may be 0 or 1, without affecting values of other variables.

This is true of shift registers, LIFOs, FIFOs, counters, pocket calculators, computers, and a wide range of other digital devices.

The arrows in Table 5.27 allow the designer to follow data up and down in the LIFO stack. For each clock with WRITE/READ$=1$, a datum enters the stack (INPUT) and $Q1 \leftarrow$ INPUT, where $\leftarrow$ is read "assumes the value of." At the same time, $Q_2 \leftarrow Q_1$, $Q_3 \leftarrow Q_2$. The status bits on a WRITE clock (WRITE/READ$=1$) become $Q_4 \leftarrow 1$, $Q_5 \leftarrow Q_4$, $Q_6 \leftarrow Q_5$. Thus, the data and status shift registers shift right one bit: the contents of Q_3 and Q_6 before the shift are lost.

With a READ clock (WRITE/READ$=0$), $Q_1 \leftarrow Q_2$, $Q_2 \leftarrow Q_3$, $Q_3 \leftarrow 0$, $Q_4 \leftarrow Q_5$, $Q_5 \leftarrow Q_6$, and $Q_6 \leftarrow 0$. Thus, everything shifts left and the contents of Q_1 and Q_4 are lost. In the table, entry from the left (WRITE) into a shift register (either the data or status shift register) is indicated by an arrow in the upper left of a table entry which does not originate in the lower right of the preceding row and column. Data shifted in a register during a WRITE are represented by an arrow originating in one cell and crossing diagonally into the cell one row down and one column to the right. Data lost on a WRITE are indicated by an arrow in the lower right of a cell which does not enter another cell. During READs, arrows point up and to the right, rather than down and to the right.

For example, consider the data bit written into the data shift register during state 2. It appears as a 1 in Q_1 in state 2 (WRITE), and can be followed into Q_2 and Q_3 as it is pushed down during the next two WRITEs during states 3 and 4. It then pops up through Q_2 into Q_1 during the next two READs, in states 5 and 6, but is not popped out, because in states 7 and 8 it is pushed down through Q_2 and into Q_3 by two more WRITEs. In state 8, a WRITE forces the bit out the bottom of the stack and it is lost.

The reader should familiarize himself with this notation: it is a convenient way to

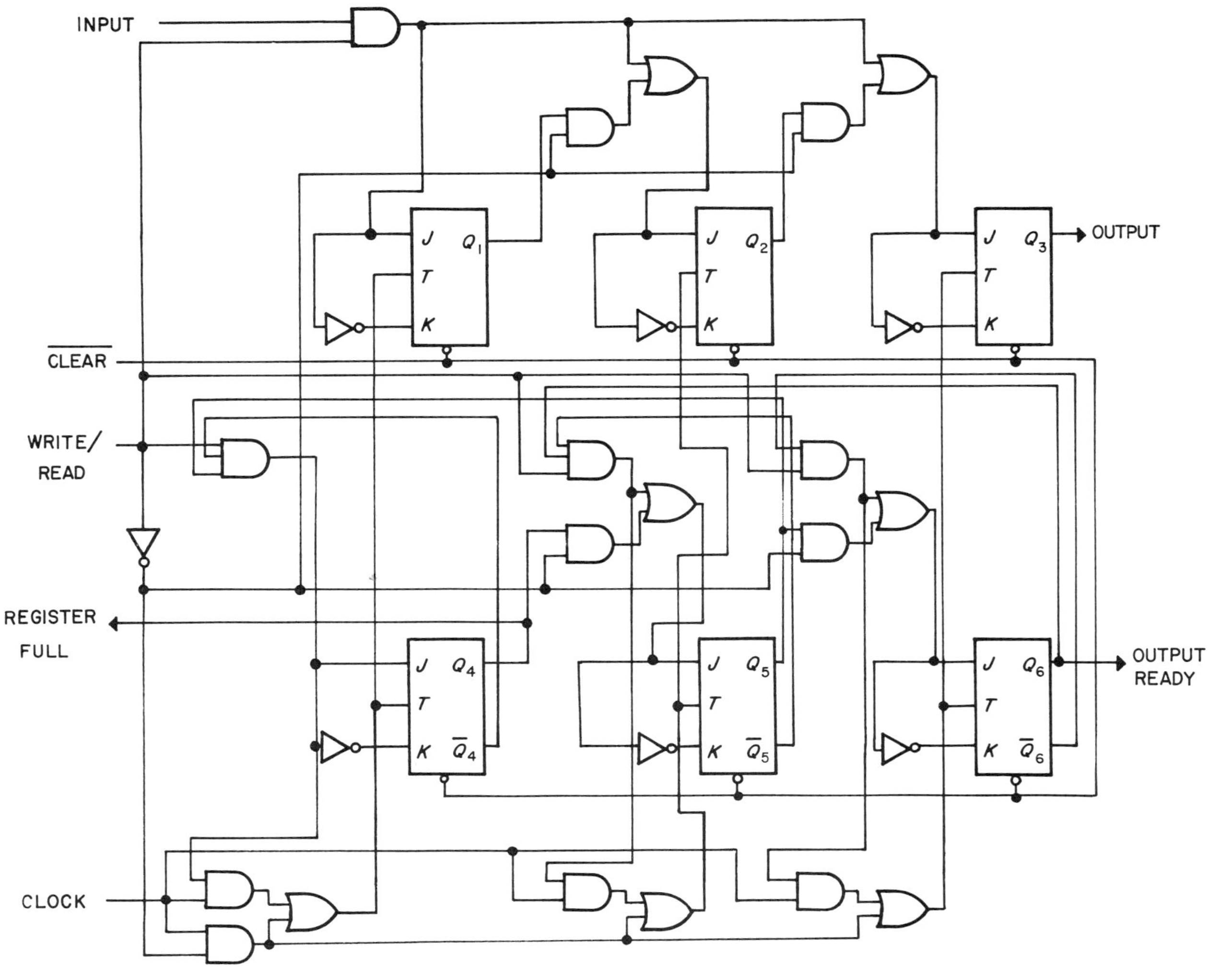

Figure 5.34 A first-in-first-out (FIFO) register.

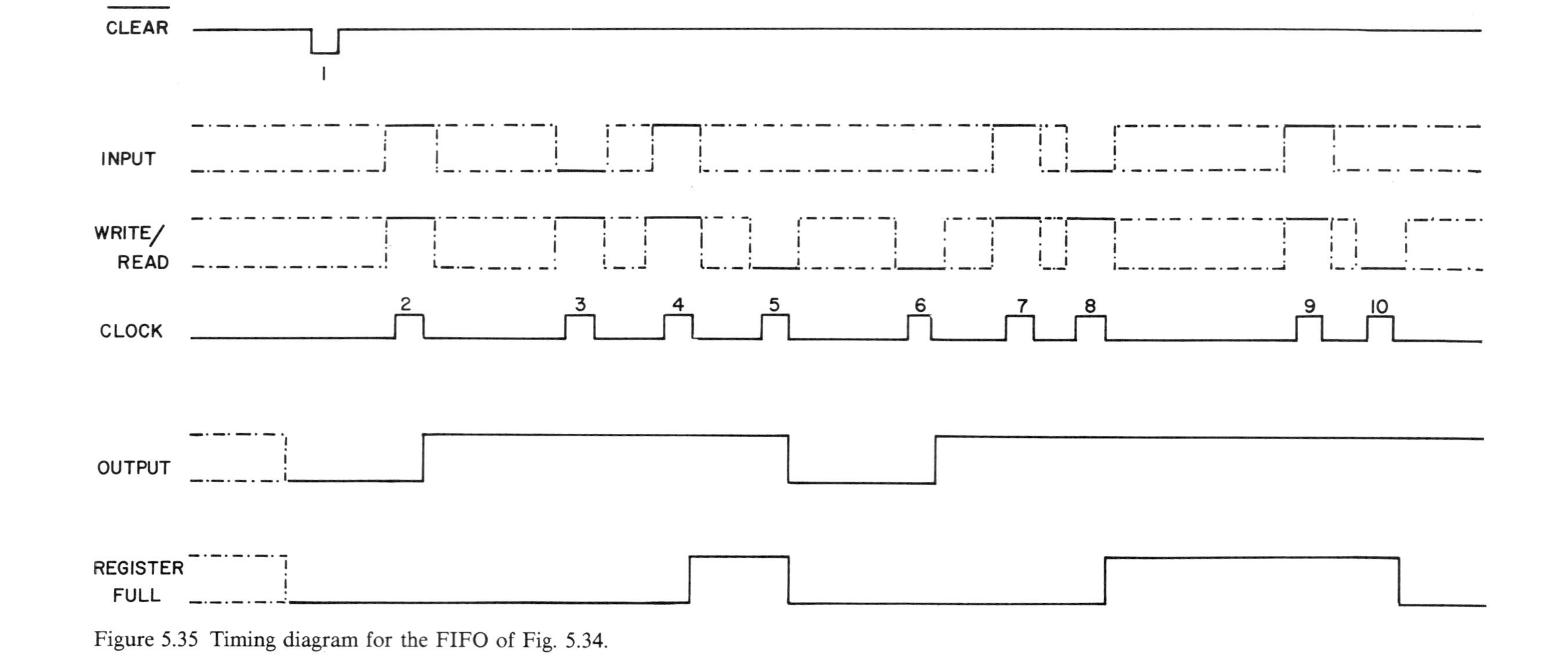

Figure 5.35 Timing diagram for the FIFO of Fig. 5.34.

block diagram the sequential states of sequential logic machines. Note that in each cell, state n (1 or 0) of an internal bit (OUTPUT through Q_6) is the state resulting after the nth clock pulse, whereas the input bits ($\overline{\text{CLEAR}}$ through WRITE/READ) are those present during the clock pulse.

A design for a FIFO is given in Fig. 5.34. Here, on a CLOCK pulse with WRITE/READ$=1$, input data is directly loaded into the rightmost unoccupied flipflop. The criterion for choice of flipflop is as follows: The rightmost flipflop is loaded if it is empty; otherwise that flipflop which is empty and for which the flipflop to the right is filled, is loaded. This is accomplished with gates; WRITE/READ$=1$ enables the gate for the first level (Q_1 and Q_4), and the upper AND gates for the second and third levels (Q_2, Q_3, Q_5, and Q_6). For these gates to produce a 1, they must also receive 1s from the $\bar{Q}$ of the flipflops they are loading and, in the case of Q_1–Q_4 and Q_2–Q_5 pairs, a 1 from the Q of the next flipflop to the right. Thus, for example, the AND gate used to load the J–K inputs of flipflop 4 will generate a 1 (to indicate the first level is occupied) if READ/WRITE$=1$, AND $\bar{Q}_4=1$, AND $Q_5=1$; the top AND gate feeding flipflop 5 will generate a 1 if READ/WRITE$=1$, AND $Q_5=1$, AND $Q_6=1$, the top AND gate feeding flipflop 6 will generate a 1 if READ/WRITE$=1$ AND $\bar{Q}_6=1$.

These AND gates not only provide 1's through the OR gates to the J–K inputs of the lower flipflops; they also (1) pass INPUT to the corresponding data flipflop; and (2) pass the CLOCK to the T inputs of the data and status flipflops. Thus, (1) INPUT goes to the input of the rightmost unoccupied data flipflop; (2) a 1 goes to the corresponding status flipflop; and (3) the CLOCK pulse goes to the toggle inputs of both flipflops. Other flipflops are unaffected.

The reader can quickly verify that when READ/WRITE$=0$, a right shift is executed, with 0 loaded into flipflops 1 and 4. REGISTER FULL$=Q_4$. A timing diagram is given in Fig. 5.35, and a table of the states corresponding to the clock cycles is given in Table 5.28. Note that the FIFO differs from the LIFO in that (a) data enter the "deepest" unoccupied position directly; and (b) data are read from the "bottom" instead of the "top," with shifts on READ/WRITE$=0$ downward rather than upward.

Table 5.28 States of Input Variables and Internal Variables of the FIFO of Figs. 5.34 and 5.35, at State Times 1–10

State time		1	2	3	4	5	6	7	8	9	10
$\overline{\text{CLEAR}}$		0	1	1	1	1	1	1	1	1	1
INPUT		x	1	0	1	x	x	1	0	1	x
WRITE/READ		0	1	1	1	0	0	1	1	1	0
OUTPUT (Q_3)		0	1	1	1	0	1	1	1	1	1
REGISTER FULL (Q_4)		0	0	0	1	0	0	0	1	1	0
Q_1	data bits	0	0	0	1	0	0	0	0	0	0
Q_2		0	0	0	0	1	0	1	1	1	0
Q_3		0	1	1	1	0	1	1	1	1	1
Q_4	status bits	0	0	0	1	0	0	0	1	1	0
Q_5		0	0	1	1	1	0	1	1	1	1
Q_6		0	1	1	1	1	1	1	1	1	1

LIFOs and FIFOs should never be assembled by the designer. Such devices are available as integrated circuits, and are quite inexpensive. They are more sophisticated than the ones presented here, providing such features as an OUTPUT READY signal, indicating presence of data in the register, and separate IN/OUT pulse lines for asynchronous loading and emptying of a FIFO. We shall illustrate the use of such devices in later sections.

5.9 Counters

One specialized type of register is the counter. In such a device, the outputs of some or all of the flipflops are used to represent binary bits in some form of numeric representation. These bits could be the bits of an unsigned positive binary number, the binary-coded decimal digits of a decimal number, or any of a number of other binary representations of numbers.

A binary counter is illustrated in Fig. 5.36. The first flipflop toggles once for each cycle of the input. The second flipflop toggles once per cycle of the Q output of the first flipflop; the third toggles once per cycle of Q_2; and the fourth toggles once per cycle of Q_3. A timing diagram illustrates the relations among input and Q_1–Q_4, in Fig. 5.37. Note that Q_1 cycles half as often as INPUT; Q_2 cycles half as often as Q_1; . . . , etc. That is, each flipflop acts as a frequency divider, with output frequency equal to half the input frequency. This is a property which is very useful in waveform generation, as will be seen in later chapters.

Table 5.29 summarizes the values of Q_1–Q_4 during states 1–18. Note that the outputs Q_4, Q_3, Q_2, Q_1 code number of input cycles, in binary, modulo 16. In order to start the count at zero, the $\overline{\text{CLEAR}}$ inputs of the J–K flipflops are used. By use of appropriate interconnections among flipflops and by using gates to determine when the J, K, and T inputs are 0 or 1, it is possible to design a counter which produces any sequence of bits desired. The sequence of Table 5.30 determines the output of a *BCD* counter: for example, the count goes from 9 to 0 and starts counting up to 9 again. Since this defines the difference between a binary counter and a BCD counter, we can use this rule to modify a binary counter to make a BCD counter. This is done in Fig. 5.38.

Figure 5.36 A binary counter.

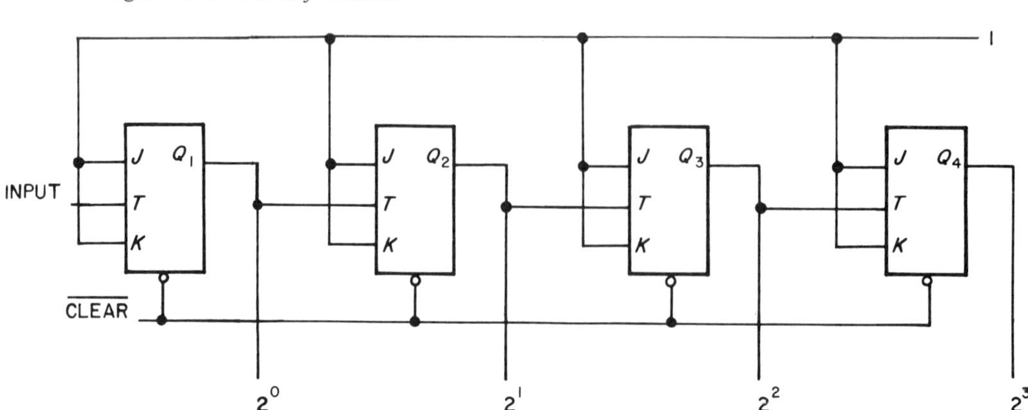

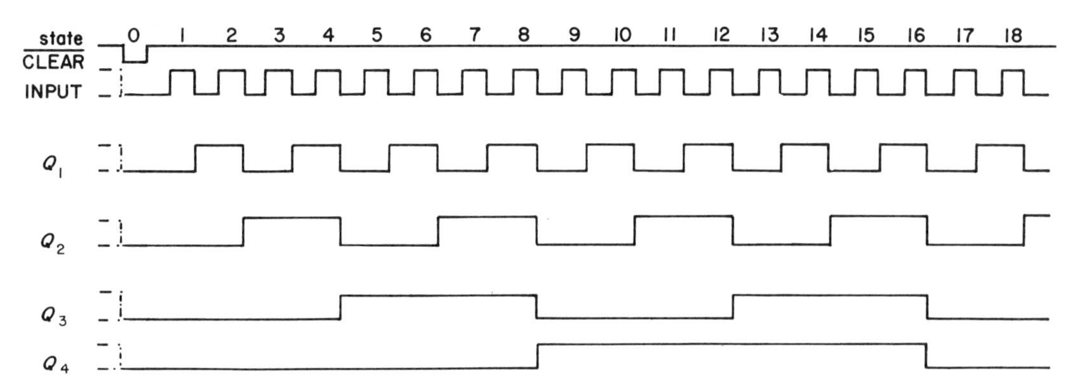

Figure 5.37 Timing diagram for the counter in Fig. 5.36.

If the D output of a BCD counter is fed to the A input of a second BCD counter, the second counter will increment each time the first goes to zero: thus, a series of BCD counters can be used to count "in decimal," where each BCD counter represents a decimal digit in a multidigit count. This is illustrated in Fig. 5.39.

Binary and BCD counters, as well as modulo 12 counters (useful in clock circuits) and other types of counters, are available as ICs, and the average designer need never synthesize his own. In some advanced applications, other count sequences must be synthesized, and straightforward procedures have been developed to make their design a painless operation (see, for example, Blakeslee, 1975).

Table 5.29 States of Q_1–Q_4 at States 0–18, for Counter of Figs. 5.36 and 5.37

State	Q_4	Q_3	Q_2	Q_1
0	0	0	0	0
1	0	0	0	1
2	0	0	1	0
3	0	0	1	1
4	0	1	0	0
5	0	1	0	1
6	0	1	1	0
7	0	1	1	1
8	1	0	0	0
9	1	0	0	1
10	1	0	1	0
11	1	0	1	1
12	1	1	0	0
13	1	1	0	1
14	1	1	1	0
15	1	1	1	1
16	0	0	0	0
17	0	0	0	1
18	0	0	1	0

Table 5.30 Outputs of BCD Counter

State	$Q_D(8)$	$Q_C(4)$	$Q_B(2)$	$Q_A(1)$
0	0	0	0	0
1	0	0	0	1
2	0	0	1	0
3	0	0	1	1
4	0	1	0	0
5	0	1	0	1
6	0	1	1	0
7	0	1	1	1
8	1	0	0	0
9	1	0	0	1
0	0	0	0	0

The binary and BCD counters we have presented belong to a class called "ripple counters": at any change of state caused by an input toggle, the output of flipflop 1 changes first followed by a change of flipflop 2, and so forth, because flipflop 1 is the input to flipflop 2, flipflop 2 is the input to flipflop 3, etc. Therefore, the outputs cannot change simultaneously. In some application, however, the outputs must change simultaneously because even a momentarily incorrect count would result in improper function. For example, if the count output produces a proportional voltage, as in some waveform generators, as the count goes from 7 to 8, the output would go from 0111 to 0110 (since Q_1 changes first), to 0100 (Q_2 changes second), to 0000 (Q_3 changes third), to 1000. Admittedly, this all occurs in a fraction of a microsecond, but during that time a transient "glitch" would appear on the voltage waveform, as illustrated in Fig. 5.40. The dotted line shows the desired waveform.

Figure 5.38 A BCD counter.

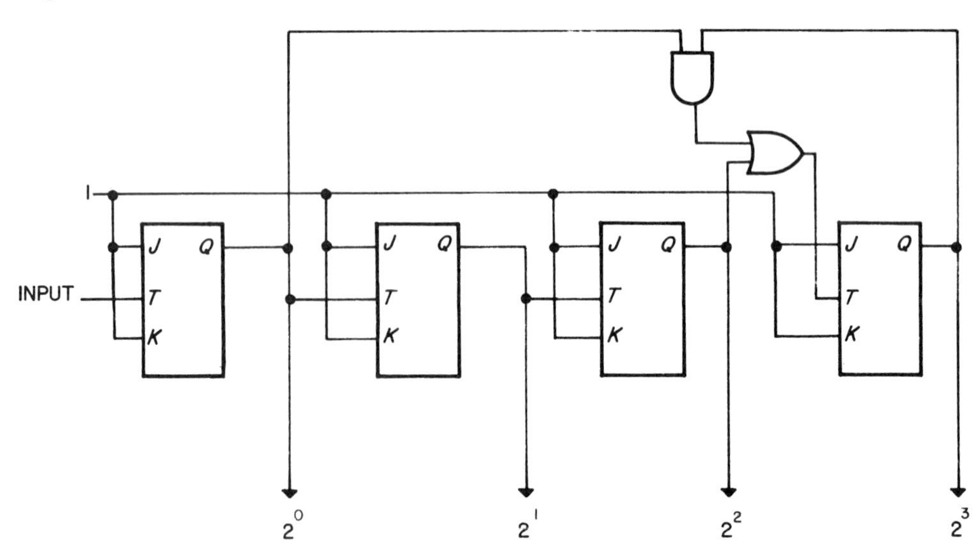

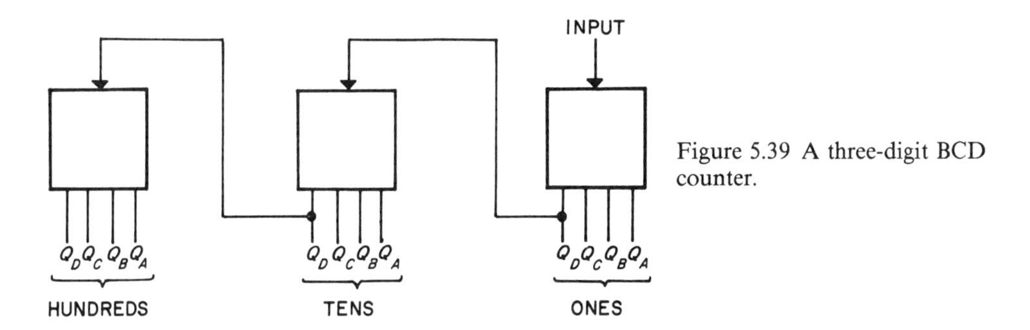

Figure 5.39 A three-digit BCD counter.

To circumvent this problem, a class of counters has been devised, called *synchronous* counters. They are designed to use one phase of the input signal to determine the next output state by loading the next output state into the master flipflops of master–slave flipflops (just as $J-K$ flipflops sample their inputs to determine the next output state on the rising edge of the T input); then the alternate phase is used to force all the outputs to their new states simultaneously (just as the falling edge of the T input causes the $J-K$ flipflop output to go to its new, predetermined state).

One approach to synthesizing a synchronous binary counter is illustrated in Fig. 5.41. The rule for the design is: if all lower-order bits are 1, a flipflop should toggle on the next state.

Thus, the "1" flipflop toggles on every cycle, so that $J_1 = K_1 = 1$. The "2" flipflop toggles if $Q_1 = 1$, so that $J_2 = K_2 = Q_1$. When $Q_1 = 1$, Q_2 toggles; when $Q_1 = 0$, Q_2 remains unchanged. The "4" flipflop (Q_3) toggles if Q_1 AND $Q_2 = 1$, so that $J_3 = K_3 = Q_1 \wedge Q_2$. When $Q_1 \wedge Q_2 = 1$, Q_3 toggles; otherwise Q_3 remains unchanged. Similarly, $J_4 = K_4 = Q_1 \wedge Q_2 \wedge Q_3$. Since all Qs assume their new states when the input goes from 1 to 0, they all change synchronously. The reader can verify from Table 5.30 that this rule, or *algorithm*, will generate a binary count sequence. Other

Figure 5.40 A waveform "glitch" caused by a ripple counter output being used to generate a proportional voltage output.

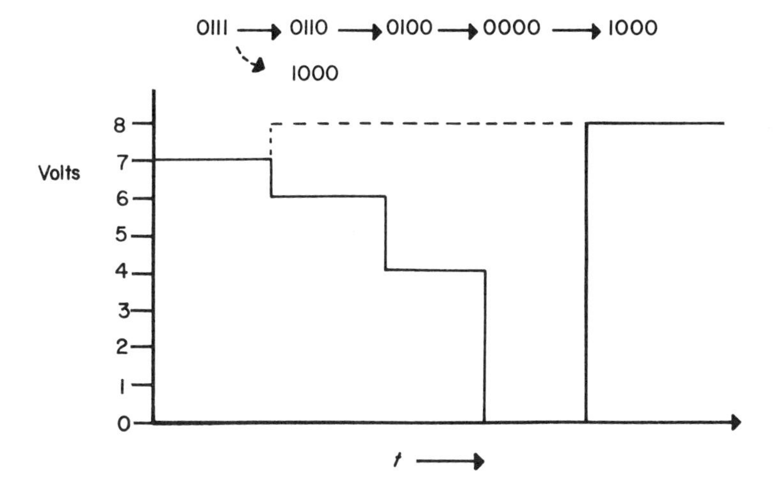

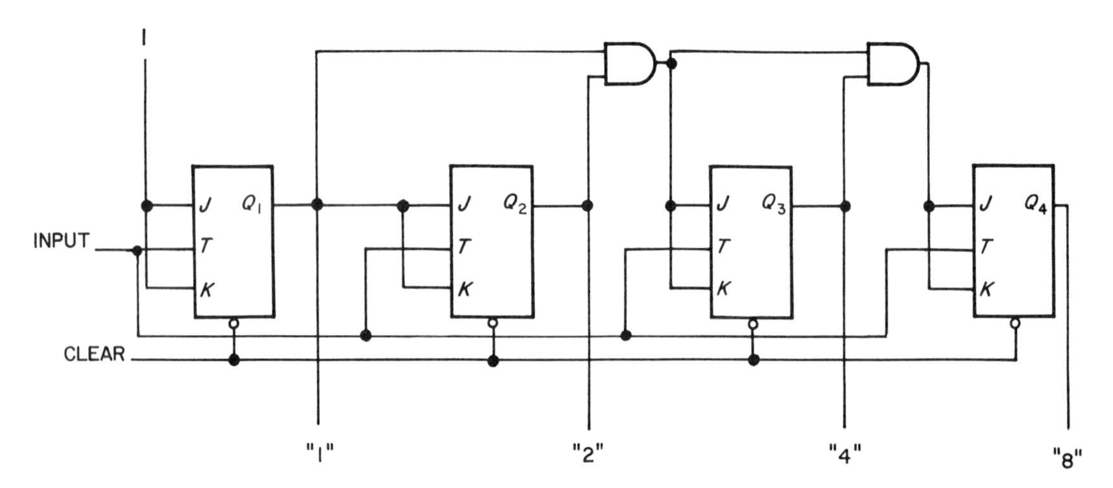

Figure 5.41 A synchronous binary counter.

Figure 5.42 Cascading asynchronous binary counters: **(a)** Modification to permit cascading. **(b)** Two cascaded 4-bit counters.

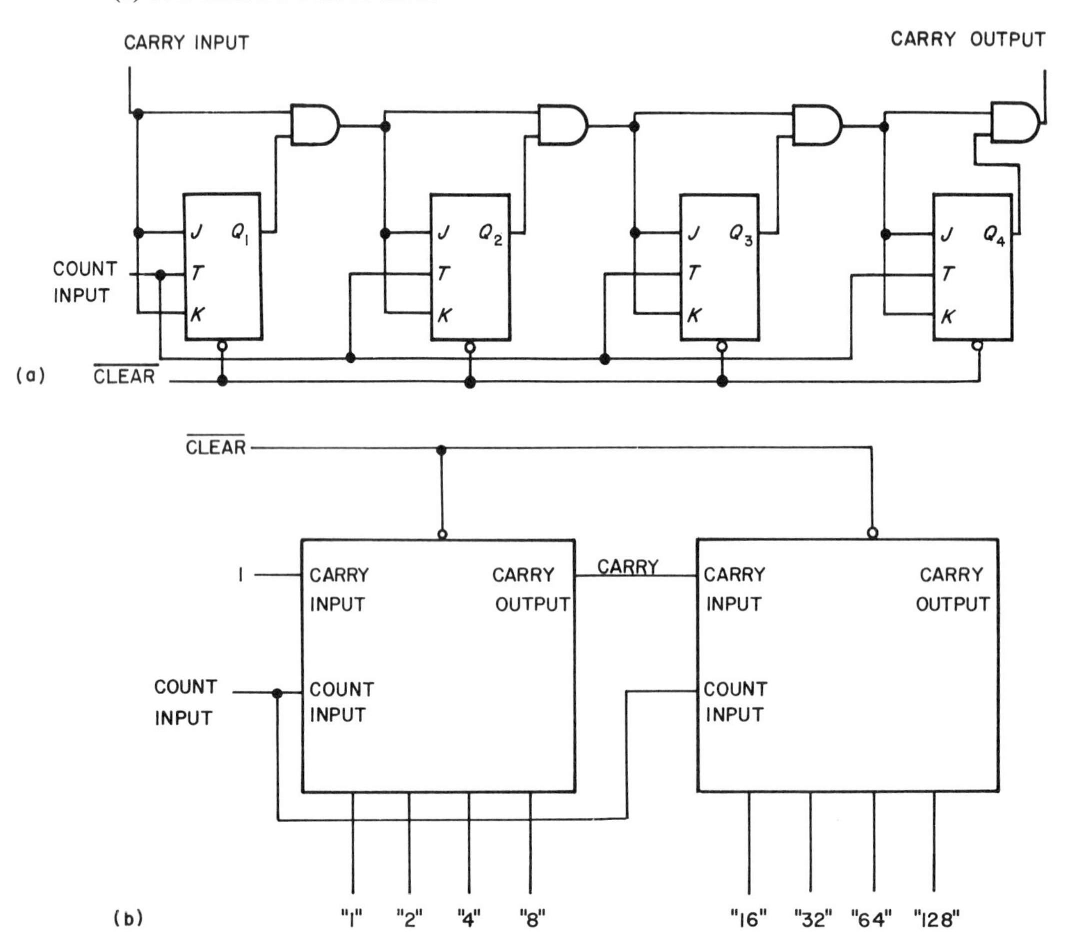

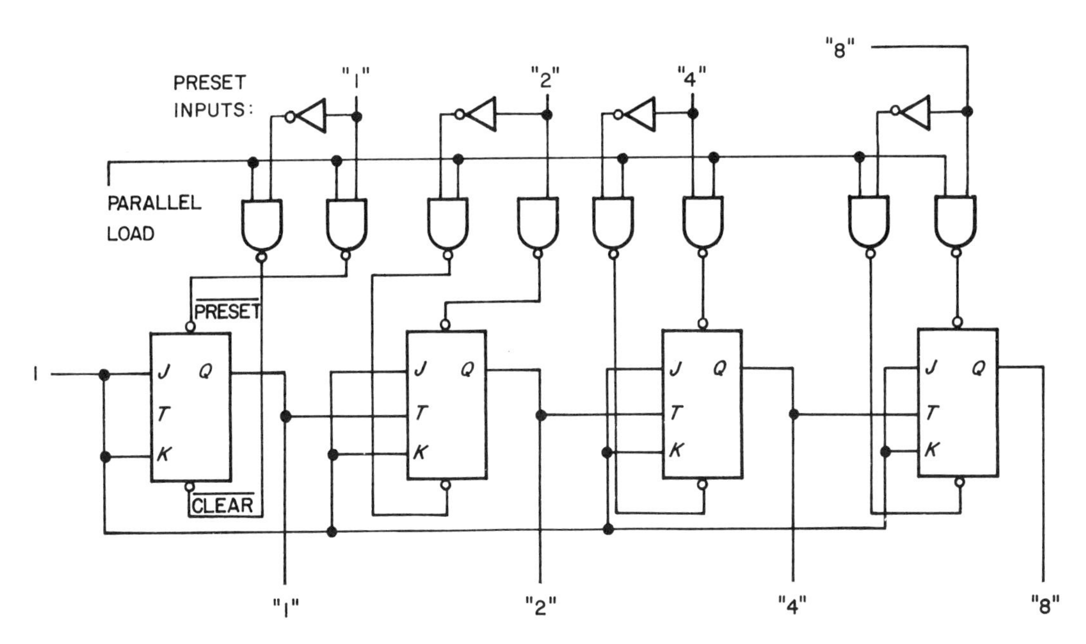

Figure 5.43 Incorporation of presettability into a binary counter design.

algorithms are used to generate other count sequences, with the INPUT signal clocking the synchronous transition of all the bits.

Clearly some form of "carry" information must be available at the output of a synchronous counter IC, and synchronous counter ICs must have carry inputs, in order to cascade ICs to make larger counters. Figure 5.42a shows how a 4-bit synchronous binary counter design can include a carry input and output, and Fig. 5.42b indicates the cascading of two such 4-bit counters. Additional AND gates have been added so cascading ICs is equivalent to lengthening the sequence of gates and flipflops. The carry inputs and outputs and additional gates permit extension of the algorithm to any number of bits. Note that the carry input of the first 4-bit counter must be connected to logical 1 to enable the first flipflop to toggle.

There are two other modifications of the basic counter design which are of some importance: (a) the ability to preset the counter to some initial count value (other than zero, which $\overline{\text{CLEAR}}$ provides); and (b) the ability for the counter to count up (increment) or down (decrement). Figure 5.43 illustrates one way to include presettability in a counter design: it is similar to the parallel load feature of a shift register, consisting of gates which pass the preset inputs into the J–K flipflops via the asynchronous $\overline{\text{CLEAR}}$ and $\overline{\text{PRESET}}$ inputs when enabled by PARALLEL LOAD = 1. When PARALLEL LOAD = 0, the $\overline{\text{PRESET}}$ and $\overline{\text{CLEAR}}$ inputs are disabled.

A binary counter which counts down is illustrated in Fig. 5.44. It differs from an "up" counter in that the $\overline{Q}$ output of each flipflop toggles the next flipflop, rather than the Q output. As a consequence, when a flipflop Q-output goes to 1, the next flipflop toggles. Recall that toggling occurs when Q outputs go to 0 in the up counter. Thus, a carry to the next, more significant, bit occurs in the up counter when a bit goes from 1 to 0; a borrow occurs in the down counter when a bit goes from 0 to 1. The output of the down counter is tabulated in Table 5.31. Note that

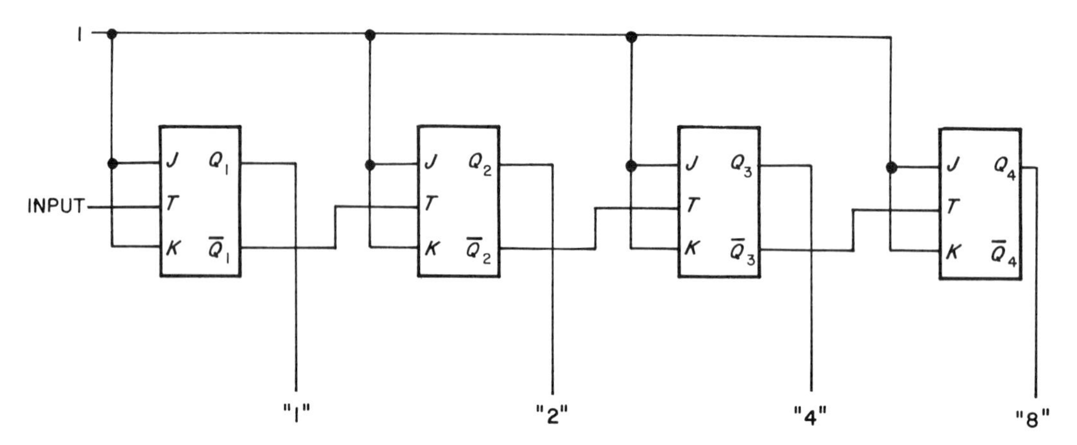

Figure 5.44 A down counter.

the outputs of the flipflops constitute a backward count from zero through 15 to 1, 0, 15, etc. That is, the counter counts down, modulo 16. The outputs are also, therefore, the twos complement of the state number: thus, for example, state number 7 is 1001, or 9, or the 4-bit twos complement of 7.

The reader by this time will be anticipating the implementation of an up–down counter, illustrated in Fig. 5.45. Gates are used to pass the Q output of a flipflop to the T input of the next flipflop when UP/DOWN = 1, and to pass the $\overline{Q}$ output to the next T input when UP/DOWN = 0. Synchronous UP/DOWN counters are also available: the design of one is left to the reader as an exercise in the problems at the end of this chapter.

Possible implementations of a variety of registers have been illustrated here only

Table 5.31 The Outputs of the Down Counter of Fig. 5.44

State	Q_4	Q_3	Q_2	Q_1
0	0	0	0	0
1	1	1	1	1
2	1	1	1	0
3	1	1	0	1
4	1	1	0	0
5	1	0	1	1
6	1	0	1	0
7	1	0	0	1
8	1	0	0	0
9	0	1	1	1
10	0	1	11	0
11	0	1	0	1
12	0	1	0	0
13	0	0	1	1
14	0	0	1	0
15	0	0	0	1
16	0	0	0	0
17	1	1	1	1
18	1	1	1	0

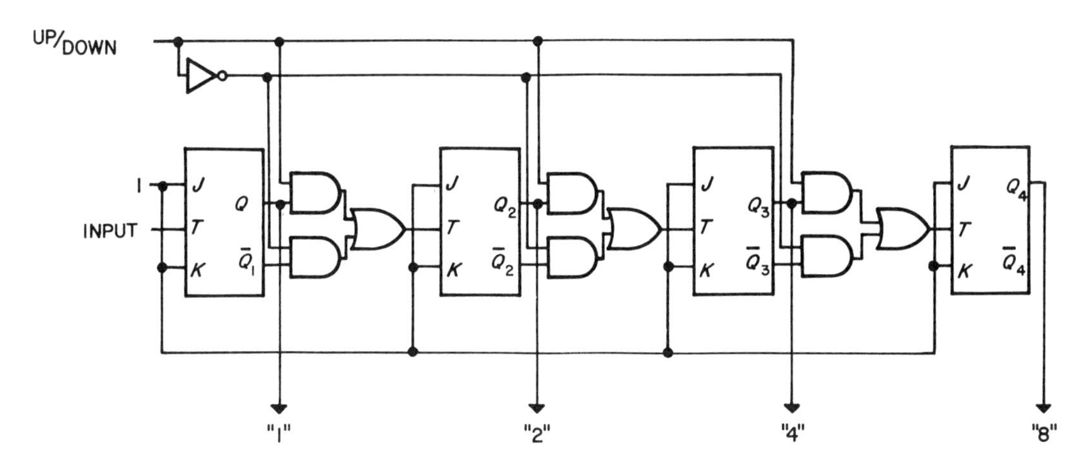

Figure 5.45 An up/down counter.

because the designer occasionally needs to make his own special-purpose registers, and to provide the reader with some familiarity with the use of flipflops and gates in digital design. However, the designer should always begin his design process by block diagramming his system into major components such as registers: then he can usually find components which closely match his requirements, such as counters, shift registers, D registers, LIFOs, and FIFOs. It is almost never necessary to actually build a register, although it is not uncommon to have to add gating functions to register inputs.

Thus, the designer should generally use gates only for enabling or disabling signals, to perform combinatorial logic functions, to modify larger devices, and to interface signals to busses and cables. Similarly, flipflops should be used as 1-bit status indicators, or for other purposes requiring storage of single bits. For more complex functions, MSI circuits should be used. This approach requires that the designer be familiar with *all* of the commercially available ICs, rather than rely on just a few types.

Section 5.10 is for advanced students.

5.10 Some Other MSI Devices

Medium-scale integrated circuits include such devices as registers and counters, multiplexers, and demultiplexers. They also include a number of arithmetic elements, which we will describe in the next section of this chapter. There are a number of other MSI devices which should be briefly mentioned at these points, as useful building blocks of which the reader should be aware. These include (a) parity generators, (b) comparators, (c) display decoders, and (d) arithmetic functions. We shall discuss the first three in this section; Section 5.12 reviews arithmetic functions.

5.10.1 Parity Generators

When transmitting binary data under circumstances in which they may be degraded by interference, it is customary to transmit extra bits which serve as a check on the data. If data are transmitted as a series of m words, each n bits long, an extra bit can be transmitted with each to indicate whether an odd or even number of 1's appears in the word, as the $(n+1)$th

Table 5.32 Parity Generation

Bit	1	2	3	4	Parity
	0	0	0	0	0
	0	0	0	1	1
	0	0	1	0	1
	0	0	1	1	0
	0	1	0	0	1
	0	1	0	1	0
	0	1	1	0	0
	0	1	1	1	1
	1	0	0	0	1
	1	0	0	1	0
	1	0	1	0	0
	1	0	1	1	1
	1	1	0	0	0
	1	1	0	1	1
	1	1	1	0	1
	1	1	1	1	0

bit of the word. If desired, a final $(m+1)$th word can be transmitted, each bit of which indicates whether an odd or even number of ones was transmitted in that bit location of all the data words: e.g., the ith bit of the last word indicates whether the total number of 1's in the ith bits of all the data words was odd or even. Such an error-checking code is called a *parity* code, and is the simplest and therefore most economical error code.

With a low and random error rate, only occasional words will have an inverted bit. The probability of two inversions in the same word, which would cause a correct parity check (an even number of 1's would still be even after two bit inversions, for example), is very low. Once an error is detected, an individual word or the whole block of words can be retransmitted, whichever is easier. If the error rate gets high enough to imply occasional double inversions, the system should be repaired or redesigned.

Parity generation is illustrated in Table 5.32, for a 4-bit word. For an odd number of 1's, the parity bit is 1; for an even number, it is 0. This is known as even parity. The inverse rule (even = 1, odd = 0) is called odd parity. The formal definitions are as follows: An *odd* parity bit is adjusted so the total number of 1 bits (data plus parity bits) is odd. For *even* parity, the parity bit is adjusted so the total number of 1 bits is even. Either will work, obviously.

To obtain a parity bit, we can add all the bits of a word together, modulo 2, at the transmitter. The resulting sum is an even parity bit which can be transmitted at the end of the word. The receiver can add all the word bits plus the parity bit, and if transmission is correct (or if an even number of bit inversions has occurred), the sum should be 0; if incorrect (one or an odd number bit inversions) it will be 1.

Example 5.5. Transmission of a 4-bit code

Consider transmission of the data word 0101. It has an even number of ones, which when added modulo 2 is $0+1+0+1 \pmod 2 = 0$, so the parity bit is 0, (using even parity). At the receiver, adding the four data bits plus the parity bit gives 0: $0+1+0+1+0=0$. If in transmission, one bit is inverted, we could have 1101, 0001, 0111, or 0100 with the parity bit = 0, or the parity bit itself could be the inverted bit, with the data transmitted correctly as 0101. In all of these cases, the sum including the parity bit will be odd, signaling an error. If two bits are inverted, the sum will be even and no error will be detected.

More complex error codes, requiring more than one error bit, can be used to determine which bit actually is in error, obviating the need for retransmission but requiring transmission of a larger number of bits in the first place

---

5.10.2 Comparators

A comparator has as input two binary numbers A and B, and it generates three output bits, labeled $A < B$, $A = B$, and $A > B$. The output whose condition is true equals 1, and the other two, which are false, equal 0. Commercially available comparators typically accept two 4-bit binary numbers and $A < B$, $A = B$, and $A > B$ input from a comparator of the next less significant 4 bits, and generate $A > B$, $A = B$, and $A < B$ outputs. They can thus be cascaded together to compare numbers with any number of bits. The comparator handling the least significant 4 bits has a 1 applied to its $A = B$ input and a 0 to its $A < B$ and $A > B$ inputs. Design of such a comparator is given as an exercise in the problems at the end of the chapter.

5.10.3 Decoder/Drivers

To display the contents of a BCD, octal, or hexadecimal number, seven-segment displays are commonly used, as illustrated in Fig. 5.46. These are either light-emitting diodes (LEDs) arranged so that a row of diodes forms each segment on the display, each row being turned on and off by appropriate digital logic, or liquid crystal displays (LCDs) are used, with similar driving logic. Liquid crystal segments become opaque or transparent when a polarizing voltage is applied, for use with transmitted or reflected light.

Other types of numeric displays which predate the LED and LCD are still in use, but have no advantage over the newer forms. We shall concentrate on LED displays here because a wider selection is presently available.

The process of generating signals to drive the numeric segments is diagrammed in Fig. 5.47. A 4-bit BCD or a 3-bit octal count from a TTL counter is fed to the inputs of a decoder/driver. The decoder/driver uses gates to generate the appropriate bar pattern for the LED display. These seven different signals are appropriately amplified and led to the LED display.

A variety of devices are available. Some systems accommodate hexadecimal numbers. Some have internal *latches* (storage logic) which can hold and display a number while the

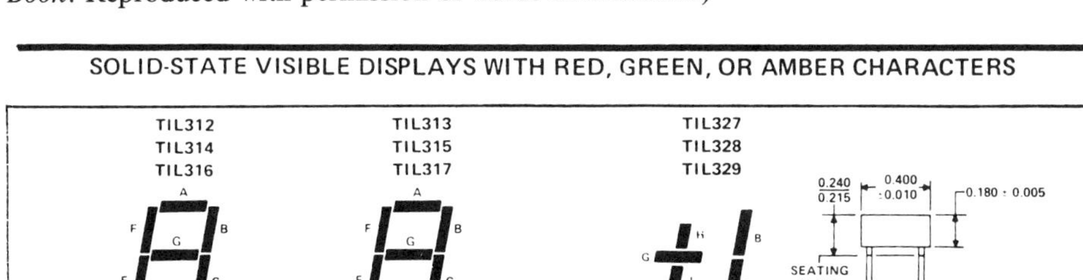

Figure 5.46 Seven-segment displays. (From Texas Instruments, *The Optoelectronics Data Book*. Reproduced with permission of Texas Instruments.)

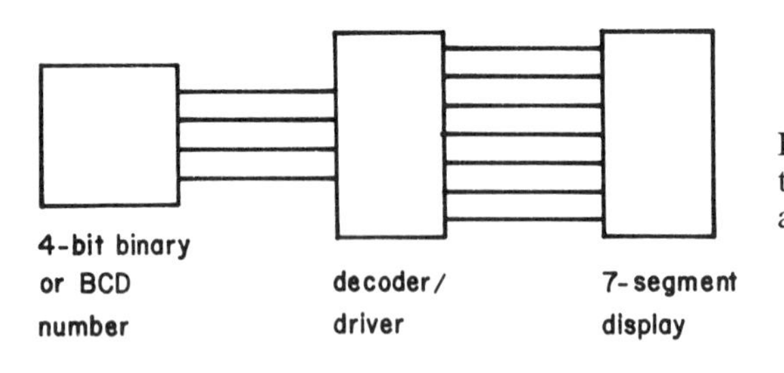

Figure 5.47 Use of a decoder/driver to generate a 7-segment display of a 4-bit binary or BCD number.

counter starts its next count. Multiplexing can be used, so that a single driver/decoder can illuminate a number of LED displays at once: each is lit for a brief time, then another, then the next, so that the displays *time share* the decoder/driver.

5.11 Preset Counters

There is a large number of applications for which it is necessary to count the number of TTL events on a signal line, and to signal the occurrence of n events, where n is a predetermined count value: such a counter is a *preset counter*; its output consists of a TTL level change or pulse which occurs n events after initiation of the count.

One approach to a preset counter implementation is that of Fig. 5.48. A regular binary counter is used, with external $\overline{\text{CLEAR}}$ and gated input. Once GATE goes on, the input is counted. (Assume $\overline{\text{CLEAR}}=0$ sometime before GATE $\leftarrow 1$, and that $\overline{\text{CLEAR}}=1$ during counting.) When the count of n is reached, a positive TTL level appears on the output. The count is determined by the input bits to the comparator: when $A = B$, TTL OUT goes high.

The preset count value is the output either of some digital logic device or of the poles of a switch with 0's and 1's at their inputs. There exist a variety of 10-position four-pole switches with specially designed layout, which connect the wiper to the appropriate poles so the poles corresponding to the BCD count are shorted to the wiper for each of the 10 positions: thus, if the wiper is tied to logical 1 and the four outputs are tied through appropriate resistors to logical zeros, the outputs will have the appropriate patterns of 1's and 0's to encode in binary the 10 switch positions corresponding to 0–9 (Fig. 5.49). Of course, BCD switches are generally used for manual programming of counters, although switches could be used for octal programming of binary counters in some applications.

Alternatively, the parallel load inputs of a down counter can be loaded with the preset

Figure 5.48 A preset counter.

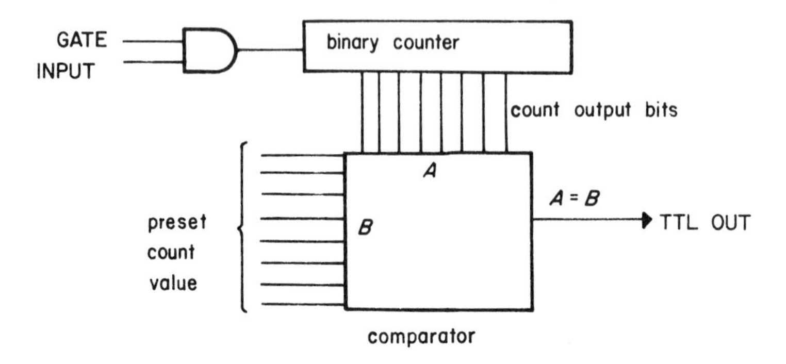

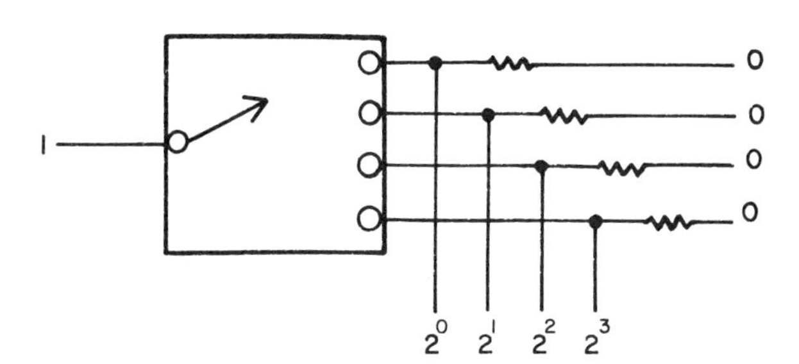

Figure 5.49 A 10-position BCD switch.

count before the beginning of the count, and the counter can be decremented on each TTL input event. When the count is equal to zero, the input will have toggled *n* times. Alternatively, if an extra TTL pulse is applied to the input just before counting, *n* counts will bring the output through zero to minus one, with a borrow generated at the borrow output. Either approach will work.

A preset counter is said to *free run* if it reinitializes and starts a new count on each completion of a count. The configuration for a free-running preset counter is given in Chapter 6, in the discussion of frequency division.

5.12 Implementation of Binary Arithmetic

5.12.1 Computation Methods Using Counters

A number of methods exist for implementing arithmetic functions. For example, counters can be used for addition, subtraction, multiplication and division. Figure 5.50 illustrates the use of counters for addition and subtraction.

Four-bit counters can be cascaded for any number of bits. Arbitrarily large words are symbolized by the striped arrows. The addend or subtrahend is loaded into the up/down counter by the controller upon receipt of a signal to begin addition or subtraction (LOAD). At the same time the augend minus one or minuend minus one is loaded into the down

Figure 5.50 An adder/subtractor using counters.

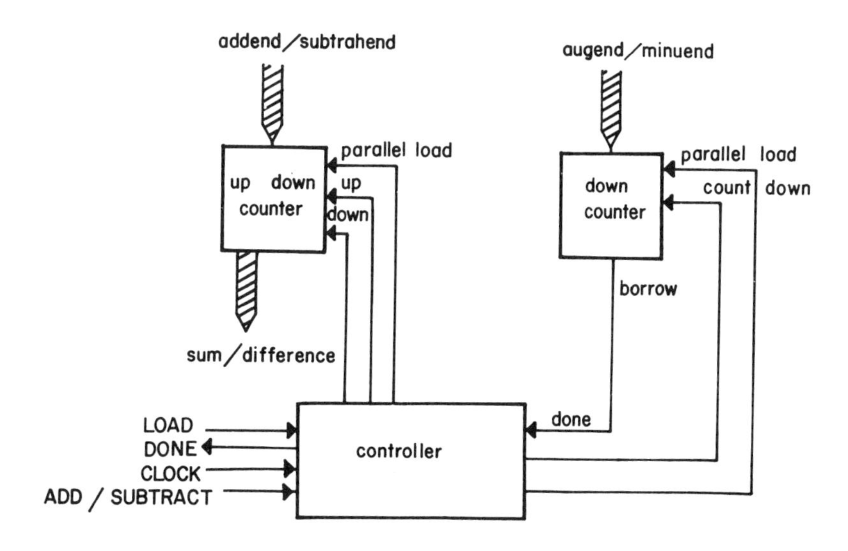

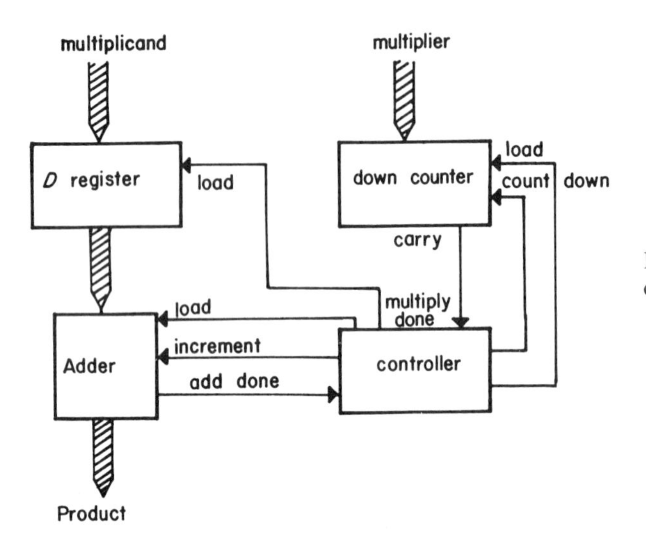

Figure 5.51 A multiplier using counters.

counter. Then the controller sends pulses to the up/down counter to increment (if adding) or decrement (if subtracting) the counter. With each increment or decrement of the up/down counter, the down counter is decremented once. Therefore, if the augend/minuend is n, there will be n increments or decrements of the up/down counter before a borrow signal is generated at the down counter: this is the desired number of increments or decrements, so the borrow output from the down counter is used to signal completion (done). At the end of this sequence, the resultant sum or difference is in the up/down counter.

Figures 5.51 and 5.52 show how counters can be used for multiplication and division. This highly schematized circuit functions by adding the multiplicand to zero n times, where $n =$ the multiplier. The adder is the same as the one in Fig. 5.50: the initial addend is zero, and the sum from each addition is the addend for the next, and the product after the last addition. This number is in the up/down counter of the adder (since only addition is done, it can be just an up counter). The down counter of the adder is refreshed each time with the multiplicand, whose value is stored in the D register. The *add done* signal notifies the

Figure 5.52 A divider using counters.

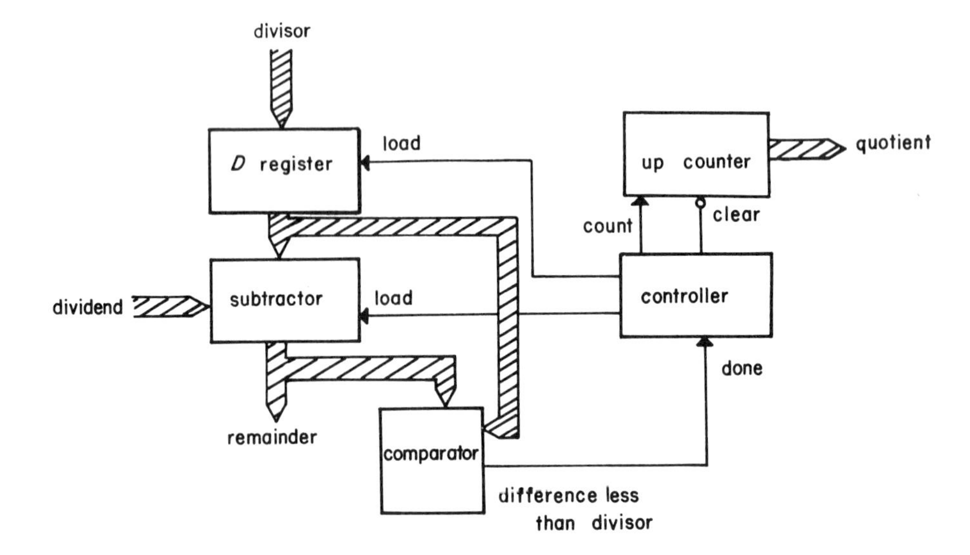

controller that an addition is done, at which time, if *multiply done* is zero, another addition is begun. After the nth addition, *multiply done* goes to 1, signaling the last addition. When both *add done* and *multiply done* equal 1, the multiplication is finished.

Division is accomplished by successive subtractions, as indicated in Fig. 5.52. Here, the divisor is stored in a D register to serve repeatedly as a minuend in iterative (repeated) subtraction. The dividend is loaded into the up/down counter of an adder–subtractor like that of Fig. 5.50. The contents of the D register are repeatedly subtracted from the up/down counter of the subtractor. After each subtraction, the up counter is incremented, and if the comparator finds the contents of the subtractor counter are still larger than the divisor, another subtraction is initiated. Once the subtractor counter contents are less than the divisor, subtraction ceases. Then the remainder is in the subtractor counter and the quotient, which is the count of the number of subtraction, is in the up counter.

With one exception, counters are rarely used in arithmetic computation. Cheaper, faster arithmetic elements are available which use combinatorial logic, as will be described below. The exception to this rule lies in the fact that BCD arithmetic is not easily implemented with combinatorial logic, but by using BCD counters in the devices of Fig. 5.50–5.52, BCD computation is easily accomplished.

5.12.2 Computation Methods Using Combinatorial Logic

The binary arithmetic tables are truth tables, so it should be possible to use Veitch diagrams to design adders, subtractors, multipliers and dividers. This is illustrated in Fig. 5.53 for addition, 5.54 for subtraction, and 5.55 for multiplication. Note that the adder and subtractor

Figure 5.53 Addition using gates.

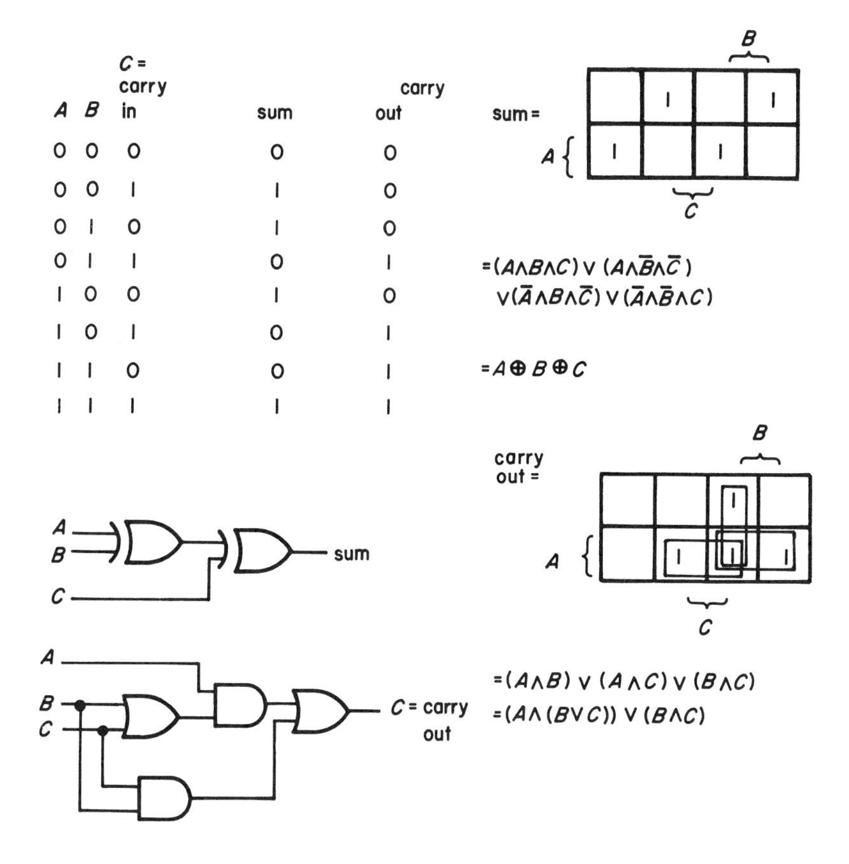

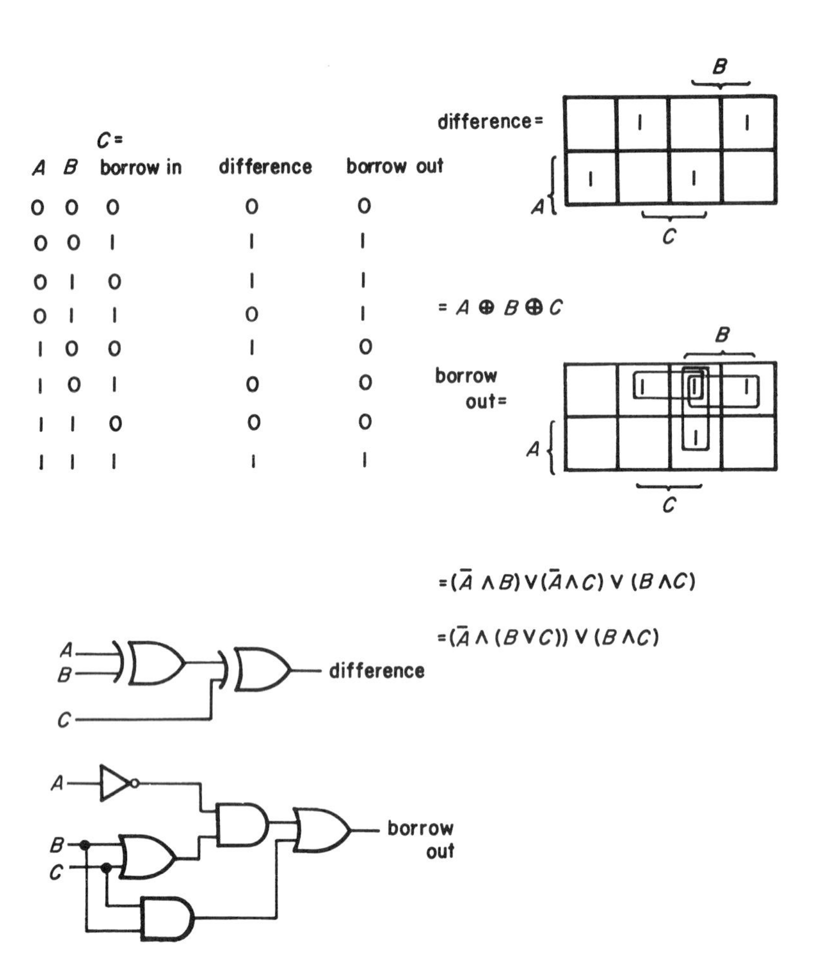

Figure 5.54 Subtraction using gates.

are very similar: *sum* and *difference* are identical and *borrow* out differs from *carry out* only by an inversion of A. Clearly the devices can be cascaded for any number of bits. For particularly high-speed applications, special carry look-ahead circuitry is available, so the carry need not "ripple" through the adder. These are not needed for most custom applications, but they are moderately easy to add if required.

Generally, subtraction is implemented by generating the twos complement of the minuend and adding to the subtrahend. This is accomplished by inverting the minuend, adding it to the subtrahend, and adding one by placing a 1 on the carry input of the least significant bit's carry input.

Multiplication can be accomplished by any of a number of methods using combinatorial logic. For example, gates can be put together to realize the logical function underlying each output bit. This is conceptually the simplest, and probably operates the fastest. However, to multiply two 12-bit binary numbers and generate a 24-bit binary output, 24 output functions of 24 inputs are needed, with $2^{24} \simeq 1.6 \times 10^7$ output states. With discrete gates, this is too cumbersome a task; even with multiplexers and demultiplexers, a large number of ICs would be required. If a read-only memory (ROM) were used, where the 24 inputs represent addresses, and there are 24 outputs, the memory would be organized as 2^{24} 24-bit words, a very unwieldy device indeed, even with the 1 bit$\times 2^{14}$ word memory ICs currently available. Instead, combinatorial logic devices exist which provide high-speed multiplication at reasonable cost, by performing 2 bit×2 bit or 4 bit×4 bit multiplications; for larger input words, these are combined to provide a full set of partial products, which are then summed with special devices to provide the final products.

This method is not illustrated here because the implementation is too complex to explain in the available space; however, clear instructions are available from manufacturers for the implementation of multipliers for any word lengths. This type of logic can also be used for division, by successively approximating the quotient bit by bit and determining whether the quotient–divisor product is larger than the dividend.

5.13 State Logic Design

We have presented methods for using combinatorial logic to generate logical functions, and a number of sequential devices such as shift registers and counters have been examined. In Sections 5.11 and 5.12 we briefly explained the use of counters and combinatorial logic for implementation of binary arithmetic functions.

Since arithmetic functions are available as MSI devices, the question arises, What do we need sequential devices (devices which change state at clock transitions and remain unchanged between transitions) for? Assuming they have some utility beyond the obvious purposes of shifting, counting, and stacking bits, another question arises: How does one design with sequential logic?

Of course there are uses for sequential logic: it is incorporated into signal generators, used in digital signal analysis, and built into computers, calculators, and a variety of process control devices. (The latter do what they sound like: they control processes, ranging from simple sequencing logic for washing machines to complex robots working in factories.) There exists a variety of techniques which can be used to design sequential machines, all of which can be loosely subsumed under the phrase *state logic design*.

All sequential devices which are deterministic (not ruled by chance) can be said to use an *algorithm*, which is a set of rules governing their operation. An algorithm can also be defined in terms of the function of a sequential device: An algorithm is a procedure which, when followed, gives the desired result.

The design process for a machine which operates according to an algorithm can be subdivided into three stages: (a) statement of the algorithm; (b) flow diagram; and (c) black box design. Of course, all three phases are restricted to available operations, and *all three are really restatements, in different terms, of the machine's design*.

To define the stages in more detail, we shall take the arithmetic process of multiplication through the three design stages.

5.13.1 Multiplication Using Iterative Bit-by-Bit Multiplication and Addition of Partial Products

We know that the truth table for a 1×1 bit multiplication is the same as the truth table for the AND operation. Therefore if we multiply each bit of the multiplier by the entire multiplicand using AND gates and add the partial products as we do in long multiplication, we can do unsigned binary multiplication by an iterative method using relatively simple logic components.

The algorithm:

1. Load 12-bit multiplier and multiplicand. Clear 24-bit product and set the bit counter to a value of 12.
2. Multiply multiplicand by rightmost bit of multiplier.
3. Add result to bits 1–12 (leftmost bit to 13th from left) of product.
4. Decrement the bit counter.

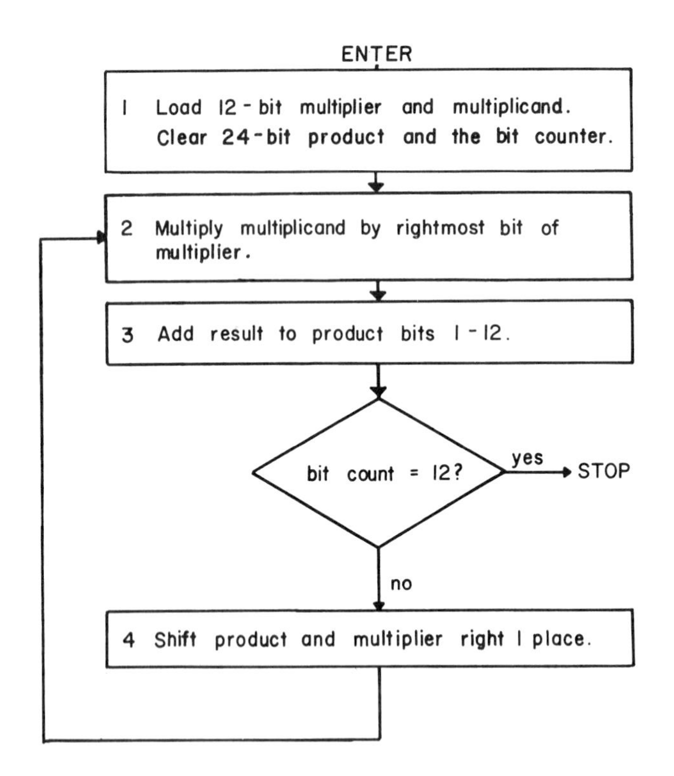

Figure 5.55 Multiplication algorithm using addition, counting, and shifting.

5. If bit counter=0, go to step 7. If bit counter>0, go to step 6.
6. Shift product and multiplier right 1 bit, go to step 2.
7. Stop. Product is complete.

The flow diagram is given in Fig. 5.55. Although it is comparatively trivial in this case, flow diagrams enhance the clarity of algorithms by clearly indicating "program" flow, especially for complex algorithms.

The block diagram is given in Fig. 5.56. Note that one shift register is used as product and multiplier; as multiplier bits are shifted right and lost, they make room for product bits.

There are a number of ways a controller could be designed. Consider a state logic approach. Here, for every operation by the controller (clear, shift, load, etc.), we define a numbered state. The CLOCK can be used to drive a counter whose output is the state number. START can be used to force the counter to an initial state to start the state sequence, and the sequence can stop in a final state when multiplication is done by gating CLOCK with a signal which is one except in the done state. The following states are used:

0. State entered by START pulse, which resets counter. Subsequent states are entered on negative CLOCK transitions. DONE, which is on during state 26 (final state), goes off, enabling CLOCK. Clear shift register (CLEAR SR = 1).

1. Load multiplier into rightmost 12 bits of shift register; load multiplicand into D register (LOAD = 1).

2 to 25. On even numbered states, load sum into 13 leftmost bits of shift register. (LOAD SUM = 1). On odd numbered states, shift right (SHIFT = 1).

26. DONE = 1. Inhibit clock.

A controller consisting of a counter, a demultiplexer, and some gates, is illustrated in Fig. 5.57. Positive logic is assumed for all inputs and outputs, although many inputs and outputs are actually negative logic in commercial ICs. Although the OR gates at the outputs of the

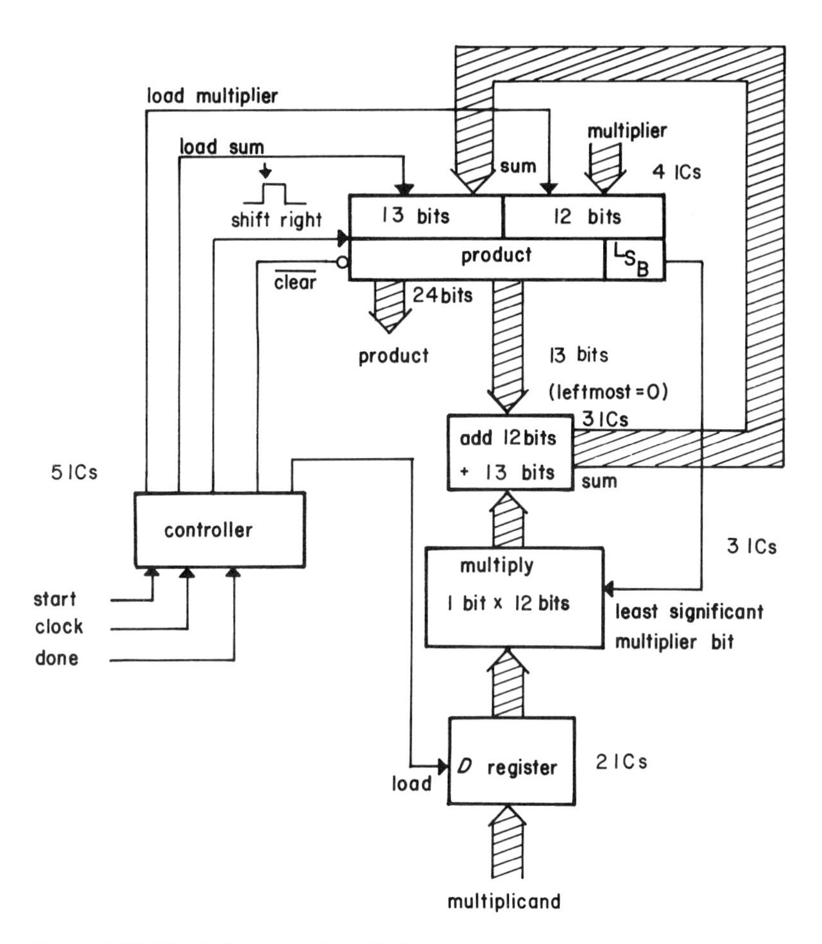

Figure 5.56 Block diagram of multiplier.

demultiplexers look formidable, use of open collector demultiplexers with negative logic permits use of wired OR methodology.

The controller illustrated in Fig. 5.57 requires approximately 5 ICs; the rest of the multiplier (Fig. 5.56) requires approximately 12 additional ICs if MSI adders and shift registers are used. By contrast, the next most efficient multiplier in terms of number of ICs is the type using MSI multipliers to generate partial products (9 ICs), and adders (17 ICs) to sum the partial products and produce the output product. The partial products method, requiring approximately 26 ICs, generates the output in about 100 nsec. The iterative shift-multiply-add scheme, requiring approximately 17 ICs, requires 27 clock cycles, or perhaps 2.7 μsec if a 10-MHz clock is used. The iterative method is therefore preferable in this example, as long as a 2.7-μsec multiply time is acceptable.

In general, iteration in time saves components by sacrificing time; iteration in space saves time. The trade-off does *not* apply for small enough numbers of iterations (e.g., a 4 bit$\times$4 bit multiply, which would require one MSI multiplier vs. a four-step iteration in time which would require about six ICs).

Most important, state logic design is an easy design methodology for implementing algorithms in general. Since all sequential machines can be described by algorithms, state logic design methods can be used for all sequential machines. Before going on to a description of LSI devices, let us consider one more example of the use of state logic design.

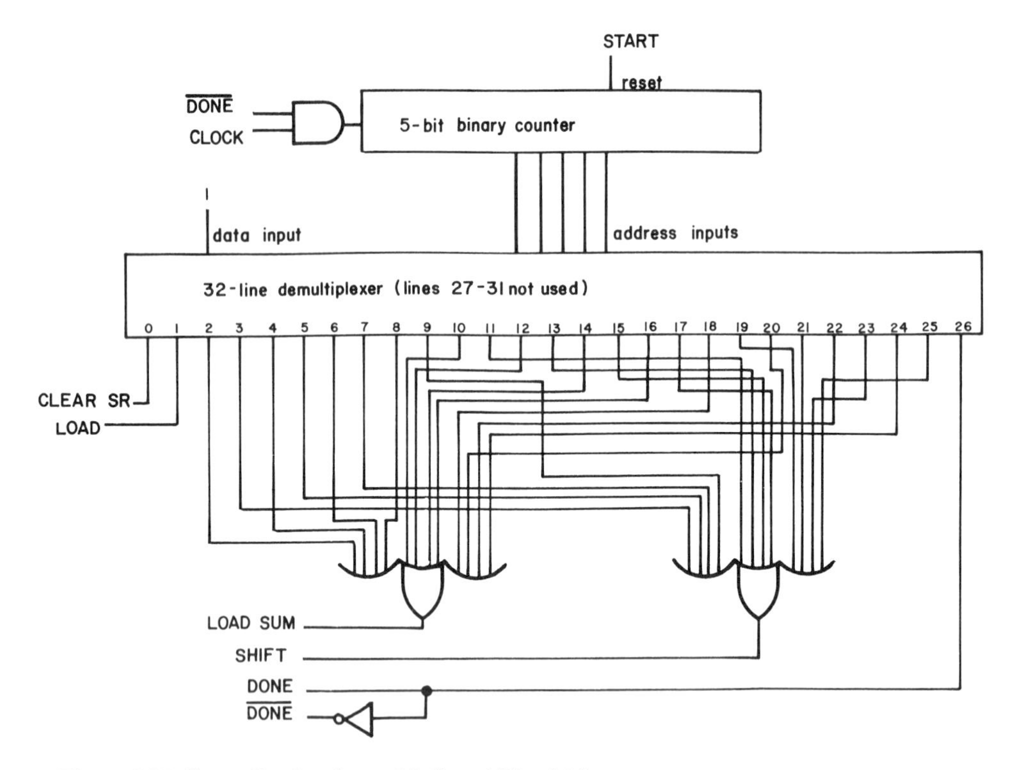

Figure 5.57 Controller for the multiplier of Fig. 5.54.

5.13.2 *Finding the Maximum Number in a Set of 16 12-Bit Numbers*

We can easily find the maximum of two numbers, using arithmetic comparators. By repeatedly comparing the maximum of the previous comparison with a new number, we can find the maximum number of a set of numbers. We shall design a machine which will accept from 1 to 16 numbers, store them, find the maximum and present the maximum at its output when done.

The algorithm:

1. Wait for signal to read a number (READ = 1) from the input.
2. When READ = 1, Set DONE = 0, and write the input into an input FIFO.
3. If LAST = 1, go to step 4. If LAST = 0, go to step 1.
4. After reading all input numbers, shift FIFO into D register.
5. If FIFO is empty, go to step 8.
   If FIFO contains data, go to step 6.
6. Insert FIFO output into D register if FIFO number greater than D register number.
7. Shift FIFO, go to step 5.
8. DONE $\leftarrow$ 1, go to step 1.

The flow diagram is given in Fig. 5.58. There are a number of loops: the first waits for READ = 1, when the "program" is entered. After each input word except the last, the "program" returns to this waiting loop. The program also returns to this loop after finding the maximum. Once the last datum is entered, the program enters another loop, processing the numbers one by one until the last one is processed.

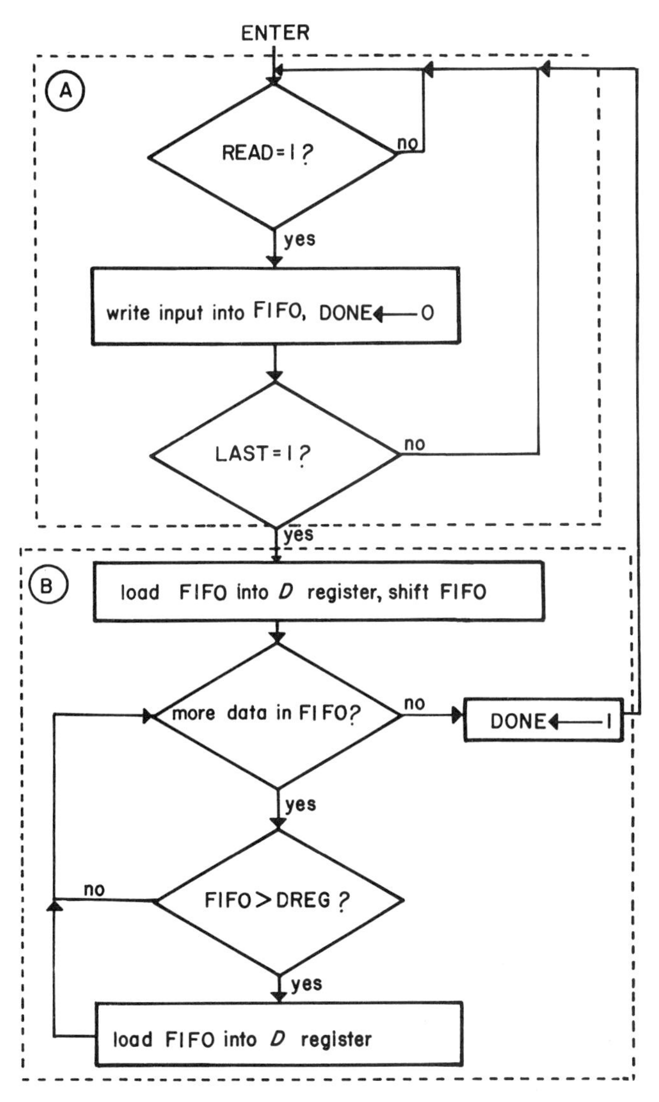

Figure 5.58 Flow diagram for finding maximum of a set of numbers.

The machine is diagrammed in Fig. 5.59. The comparator generates a single logical variable called FIFO > DREG, used to determine whether the FIFO contents will be LOADed into the D register.

A controller for the device is diagrammed in Fig. 5.60. In terms of state numbers, the algorithm is written

0. Entered from state 3 (DONE = 1), when READ = 1. For each READ, issue a WRITE to FIFO while in this state.
1. Entered from state 0 when READ$\wedge$LAST = 1. While in this state, use CLOCK to SHIFT (on negative clock transition) and LOAD (on positive clock transition) if FIFO > DREG = 1. First operation may be either SHIFT or LOAD, so it is not necessary to ensure that positive or negative transition occurs first.
2. Entered from state 1 when SHIFT$\wedge$FIFO OCCUPIED = 1. DONE = (STATE 2 = 1).

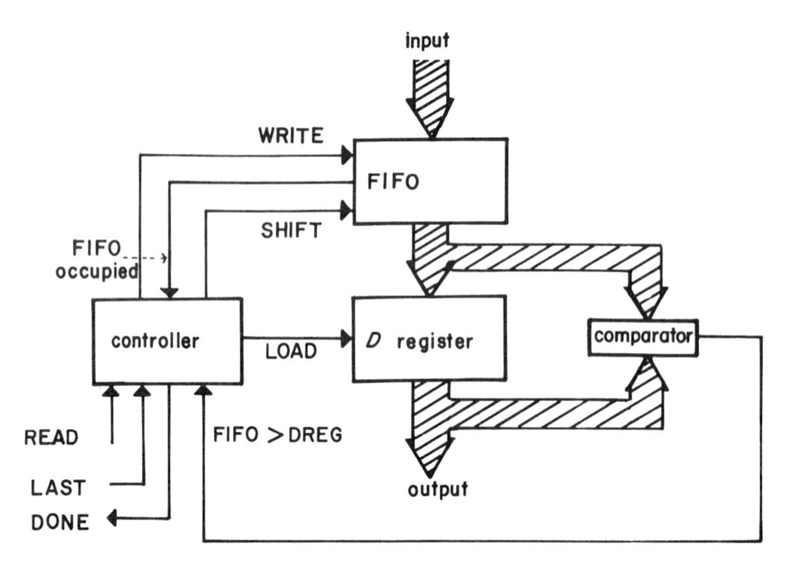

Figure 5.59 Machine to find the maximum of a set of numbers. All numbers are 12-bit unsigned numbers; the FIFO is 16 words deep.

Here the smaller loops were executed without change of state, e.g., repeated reads in state 0, repeated shift/loads in state 1. We were able to shift and load using positive and negative transitions of the same clock pulses; if this had not been possible, two states would have been required, or state 1 could have enabled a substate counter, which would cycle through any desired number of substates on clock transitions, using each substate to accomplish a different task.

Note that, since there are only three states, the demultiplexer has been eliminated in order to simplify design. State 0 is defined as $2^0 \overline{\vee} 2^1$.

Figure 5.60 Controller for maximum finder.

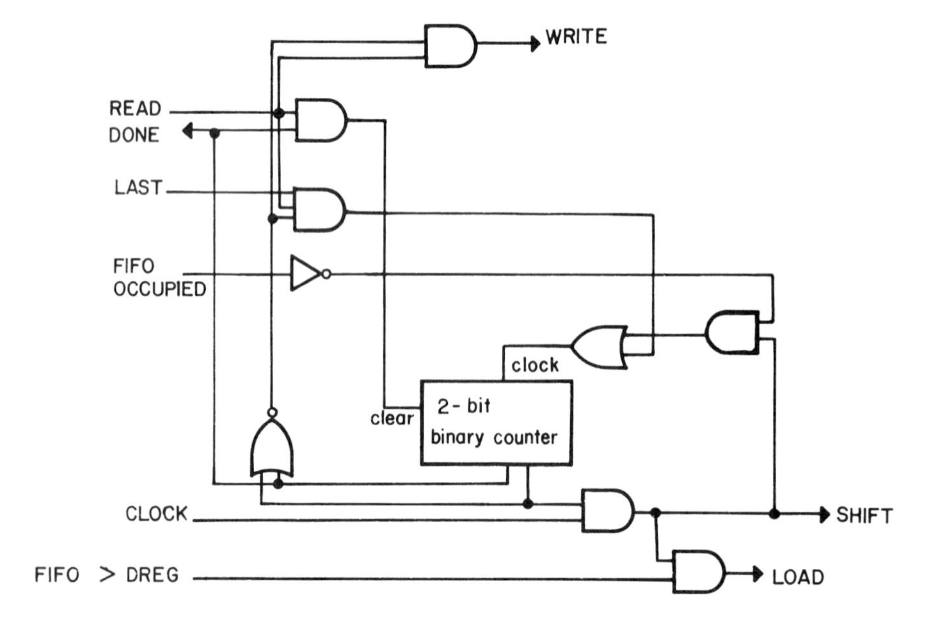

Clearly if an algorithm has much complex branching in the flow diagram which cannot be handled easily by substate counters, it will be necessary to go to noncontiguous states, no matter how cleverly state number assignments are made. The use of up/down counters, interpolated dummy states, and even preset counters (noncontiguous states can be jammed into the counter with the parallel load feature) sometimes solve these problems. However, the design eventually becomes too cumbersome: at this point, the designer should consider using a microcomputer.

5.14 Large-Scale Integration: Memories

LSI includes some of the larger arithmetic elements such as multipliers, accumulators (special registers used in computers), and other components for computers. The latter include complete microcomputers on single chips, semiconductor memories, and input/output handling chips for such machines.

Other LSI devices include programmable logic arrays (a form of custom-designed special-purpose memory) and calculator chips.

In this section we will briefly describe semiconductor memories and some of their applications outside computers.

Random access memories, or RAMs, are semiconductor or other devices which can store digital data under external control, with a specific location corresponding to each memory address. Data can be written into or read out of RAMs. Generally, the number of addresses in a RAM is a power of two, and sufficient address lines are provided so each location is addressable and each address corresponds to a memory location. RAMs may be used in groups: for example, a set of 12 RAMs with their address lines in parallel and separate data lines, each holding $2^{10} \times 1$ bits, could be used to configure a memory that would hold 2^{10} 12-bit words. Similarly, by using an 8-line demultiplexer to select RAMs (most RAMs have a select input which must be low to enable their address lines—such an input is called a *chip select*), the upper 3 bits of a 13-bit address can be used to select RAMs and the lower 10 bits to select addresses within RAMs giving a total of $8 \times 10 = 2^{13}$ addresses. In this configuration the data lines are connected in parallel for all 8 RAMs, giving a memory organized as 2^{13} 1-bit words.

Figure 5.61 illustrates the configuration for a $2^{13} \times 8$ bit memory: it has 2^{13} addresses corresponding to 2^{13} 8-bit words. It is configured from $2^{10} \times 1$ bit elements. Since $2^{10} = 1024 \cong 10^3 = 1\text{k}$, 1024 is referred to commonly as 1k. Thus, this is an $8\text{k} \times 8$ bit memory.

There are a number of types of memory which can be used as ICs in digital electronics. These include

a. *Random Access Memory* (RAM), wherein each memory word is accessed for reading or writing via a specific address, access time being approximately equal for any combination of successive locations.
b. *Serial Access Memory* (SAM), wherein the memory consists of a circular shift register (serial output connected to serial input). A counter keeps track of the "address" of the bit available for reading and writing (the serial output and input bits). To read or write a given address, the register is shifted until the counter matches the desired address: clearly a large change of address takes longer than a short one. With the advent of cheap RAMs this type of memory is falling into disuse.
c. *Read-Only Memory* (ROM), in which the binary contents are wired in at the factory as a step in the IC manufacturing procedure. These act like RAMs, except it is not possible to change the contents. Although it is possible to obtain custom-designed ROMs, they are too expensive for production in small quantities.
d. *Programmable ROMS* (PROMs), which can be written using special equipment. These hold their contents until erased with high-intensity ultraviolet light and reprogrammed.

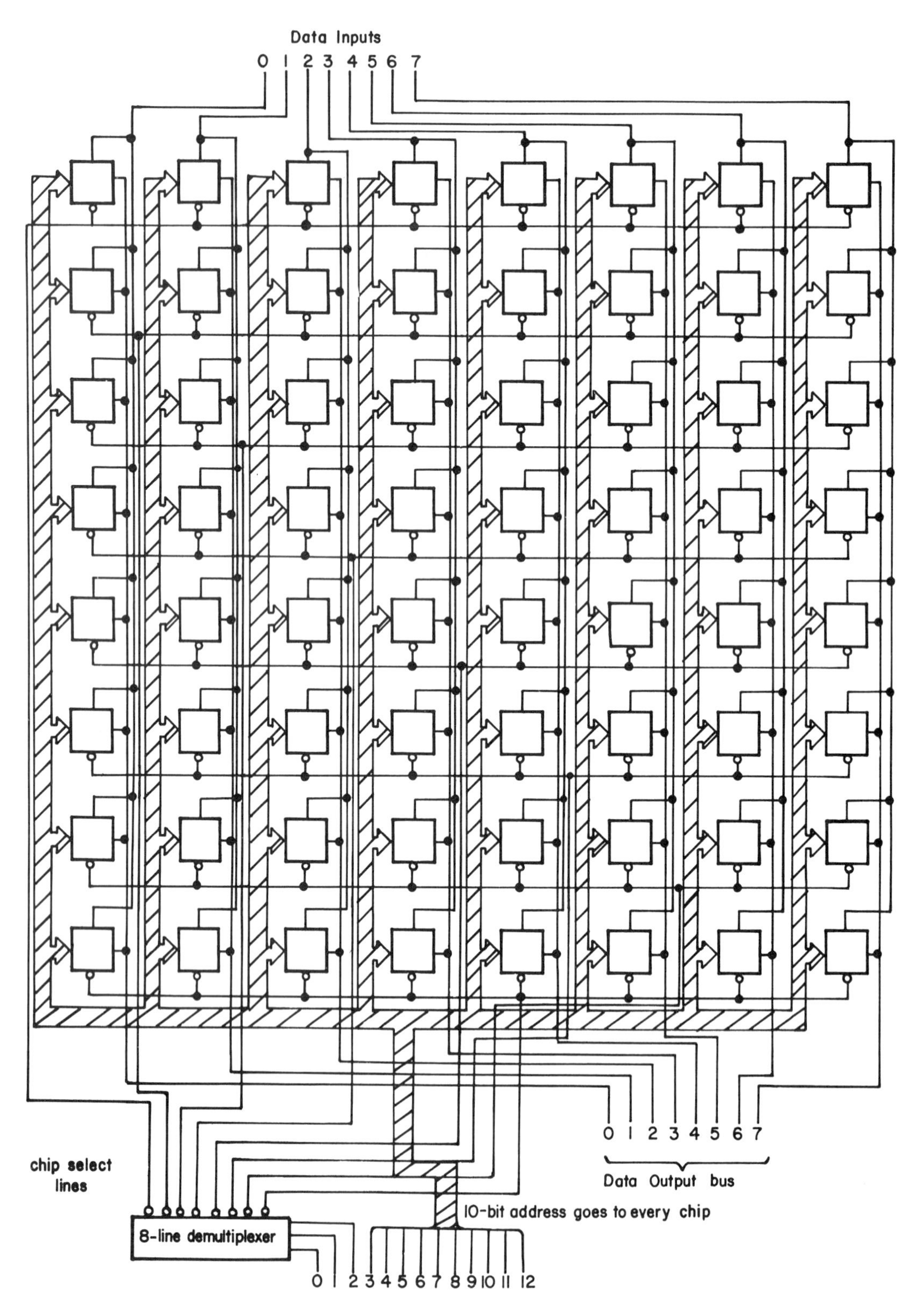

Figure 5.61 Configuration of a $2^{13} \times 8$-bit memory, using $2^{10} \times 1$-bit elements.

e. *Programmed Logic Arrays* (PLAs), some of which can also be programmed, do not have a full-scale memory complement, but are an inexpensive way of making a ROM-like device, in which not all inputs codes correspond to defined outputs, and a given output can be specified by more than one input code.

RAMs are used for temporary data storage because they are volatile: that is, their contents are lost if power is removed. ROMs, PROMs, and PLAs are nonvolatile, but cannot be written on during normal operation. The old core memories have the advantage of allowing RAM-like operation, without the inherent volatility of RAMs but they are inconvenient for the applications described in this book.

A number of new memory types have recently appeared. We can expect corelike RAMs to become available in the near future.

RAMs, being used for temporary data storage, are good "scratch pads" for digital devices; they are used as a computer memories for the full range of computer sizes, often in a mixture of ROM, RAM, and core memory.

ROMs and PROMs are used for permanent storage, such as the programs in microcomputers, and start-up programs in larger machines. They are also used to sequence sequential machines from one state to the next, and they are very useful for data conversion (e.g., for conversion from the matrix code corresponding to the location of a keyboard key to USASCII, the digital standard code for alphanumeric characters), table lookup (trigonometric tables, for example), and generation of complex logical functions. A PROM is used to test a new memory content: if it is correct, a ROM is manufactured with the same content if the number of devices or speed requirements (ROMs are faster) justify the expense; PROMs are used for slower devices produced in smaller quantities.

5.14.1 Code Conversion and Table Lookup

We shall consider two examples here: the generation of USASCII code for a keyboard character, and the design of a high-speed multiplier.

Figure 5.62 illustrates the code conversion process: The device is a 90-key keyboard encoder. It accepts the key coordinate of a 10 column $\times$ 9 row keyboard (pressing a key sets a column bus and a row bus to 1: each key has a unique row–column location) and generates a 10-bit ASCII output.

First, in Fig. 5.62a, the 9 row inputs and 10 column inputs are fed as 19 inputs to an "address generator." The 19 bits must be converted to a more economical form of representation, such as a 7-bit binary or 8-bit BCD number. We do not want to use a ROM directly because it is uneconomical: most memory addresses would never be used and a large ROM (10 bits by 2^{19} words) would be needed. Similarly, a ROM for address generation would be uneconomical: 7 bits by 2^{19} words. After address generation the ROM is manageable size: 10 bits by 2^7–2^8 words.

The coordinates are most easily converted to a 2-digit BCD number, since the coordinates themselves can be expressed as a two-digit decimal number directly. We will have the column digit represent units, and the row digit will be decades, since this is more efficient: resultant numbers are 00–89, rather than (if reversed) 00–98, with numbers ending in 9 missing. By converting rows and columns into separate BCD digits, rather than using all 19 bits to generate a single 7- or 8-bit (binary or BCD) number, the number of degrees of freedom is reduced, and design is greatly simplified.

Conversion with multiplexers would require four 8-input multiplexers and four 9-input multiplexers; this is clearly impractical. Demultiplexer design would require a similarly large number. Using gates, however, results in an economical design, illustrated in Fig. 5.62b. Each bit of the address is generated by ORing all the numbered lines whose line numbers would have 1 in that bit. Thus, all odd address lines are ORed for the 1-bit. Numbers 2, 3, 6, and 7 are ORed of the 2-bit. Numbers 4, 5, 6, and 7 are ORed for the 4-bit. The 8-bit is just the 8-line for rows, since there is no 9-line; for columns, it is the 8-line ORed with the 9-line.

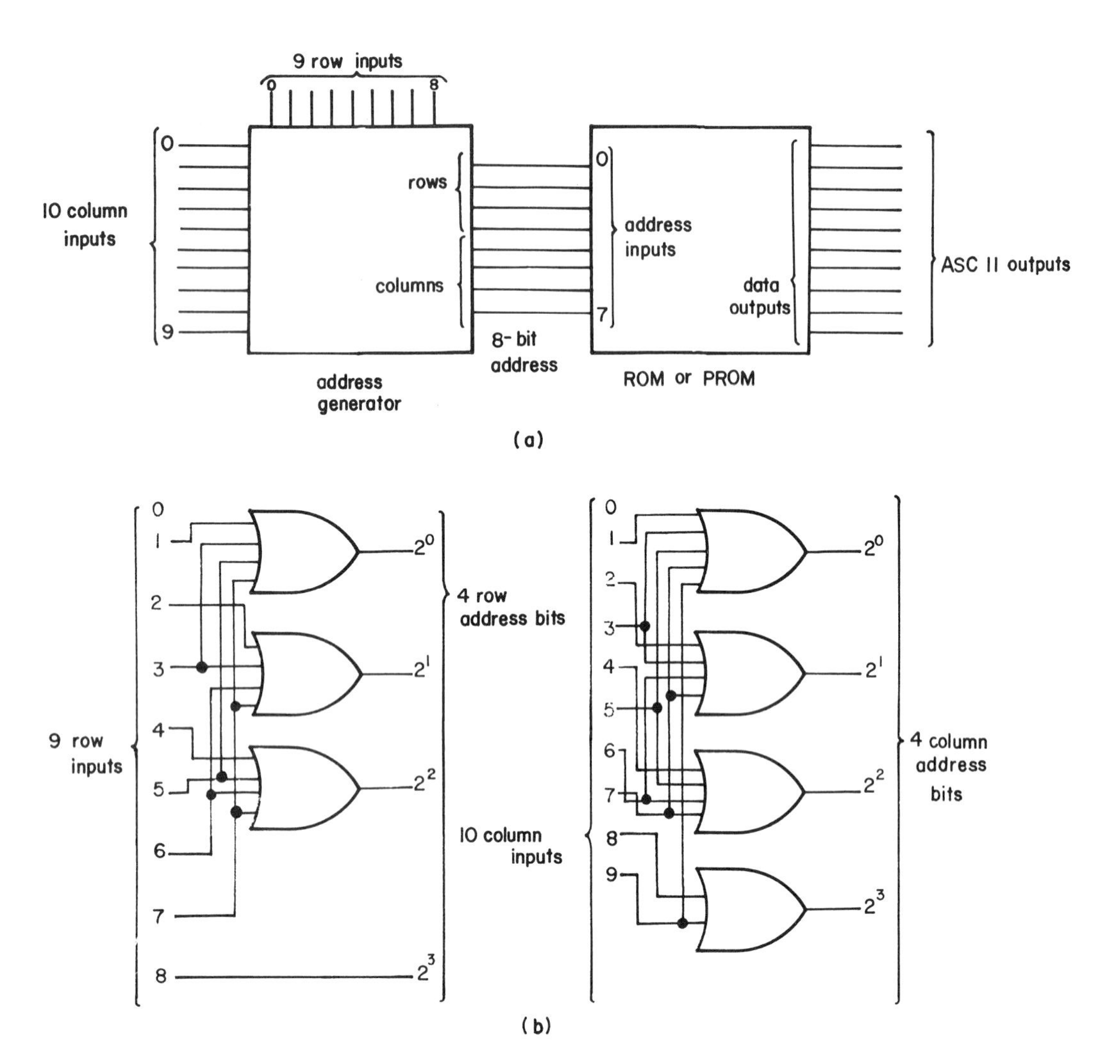

Figure 5.62 Generation of ASCII code from row×column keyboard input.

The 8-bit address goes to the address inputs of the ROM or PROM: in each address, from location 0 to 89, is stored the ASCII code of the corresponding keyboard character. Therefore, the correct ASCII code appears at the output for any keyboard character. (Note: this scheme is not really practical because it cannot correctly code simultaneously depressed pairs of keys.)

5.14.2 Generation of Logical Functions

We have described the use of MSI devices such as multiplexers and demultiplexers to generate logical functions. ROMs and PROMs can be used for this purpose as well. Consider the device of Fig. 5.63. The output functions are 7 logical signals used to drive a 7-segment numeric display. Seven output bits are required, as functions of 4 BCD input bits. Seven multiplexers would be required, one for each bar on the 7-segment display. Alternatively, one 8×32 bit ROM or PROM would do the job, as illustated in Fig. 5.63. Four of the 5 input address bits, 16 of the 32 addresses, and 7 of the 8 output bits are used. The contents of all 32

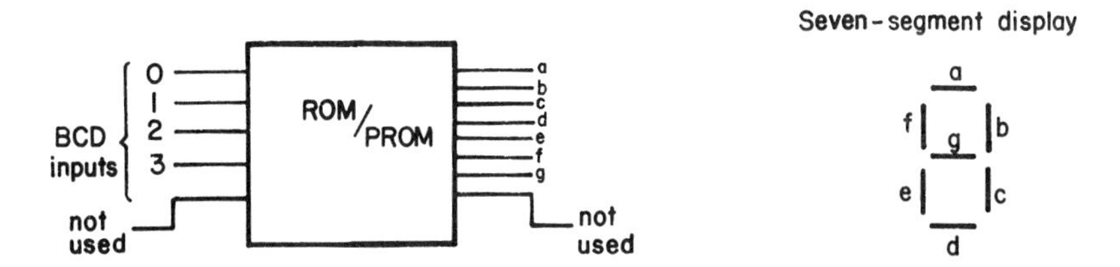

Figure 5.63 Use of a ROM or PROM to generate the driver signals for a 7-segment BCD numeric display.

addresses are given in Table 5.33. Of course, in this particular case, the ROM–PROM approach is impractical, because decoder/drivers are available as MSI ICs, and some displays even have their own decoder/drivers built in. However, the example illustrates the use of ROMs for generation of logical functions.

5.14.3 Control of Sequential Machines

A ROM can be used to control the sequence of operations of a sequential logic machine. Selected input and output variables and the current address being accessed in the ROM are used to determine the next address. In a properly designed device, each current state unambiguously defines a next state: thus the device is a state logic device.

A ROM-controlled state logic device is illustrated in Fig. 5.64. The memory has 8 input bits ($2^8=256$ addresses), and a 9-bit output. The output consists of a 4-bit *link*, used to partially determine the next address, and 5 bits of output. On each clock cycle the link is loaded into the *D* register, and forms part of the address input to the ROM (either by some combinatorial algorithm, or directly, as 4 of the 8 address bits). The newly selected address corresponds to a new set of output bits, and a new link, which in conjunction with the inputs determines the address to accessed on the next clock cycle.

A concrete example of a ROM-controlled device is illustrated in Figs. 5.65 and 5.66, and in Table 5.34. Figure 5.65 is a flow diagram of the sequence of steps in a device which accumulates the total length of a line, as computed from serially input (x, y) coordinates along the line. The algorithm is started by RSTRT=1; this condition forces the algorithm to

Table 5.33 Contents of a ROM or PROM Used to Decode a Seven-Segment Display of a BCD Digit

Address	Segments							
	a	b	c	d	e	f	g	h (not used)
00000	1	1	1	1	1	1	0	X
00001	0	1	1	0	0	0	0	X
00010	1	1	0	1	1	0	1	X
00011	1	1	1	1	0	0	1	X
00100	0	1	1	0	0	1	1	X
00101	1	0	1	1	0	1	1	X
00110	0	0	1	1	1	1	1	X
00111	1	1	1	0	0	0	0	X
01000	1	1	1	1	1	1	1	X
01001	1	1	1	0	0	1	1	X
01010–11111:	0	0	0	0	0	0	0	

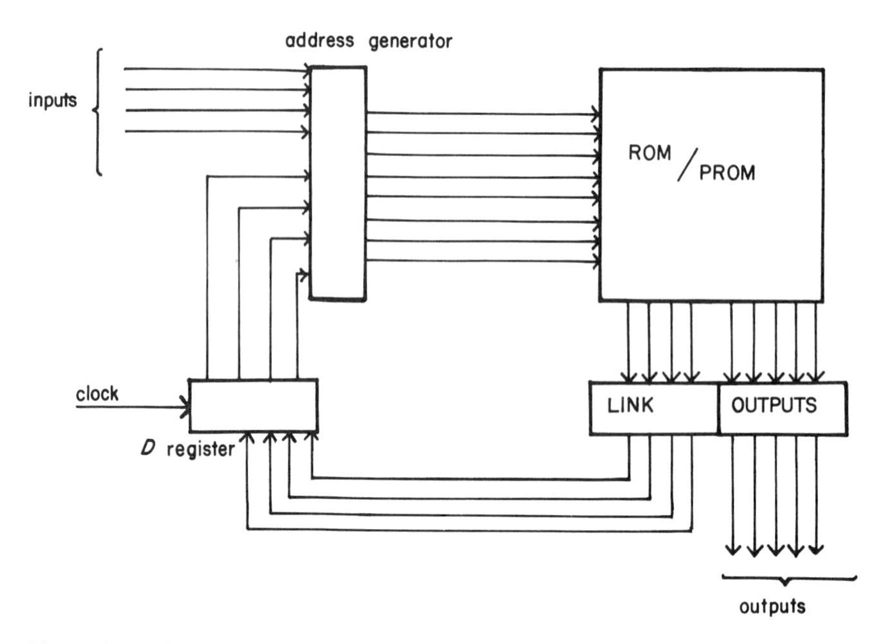

Figure 5.64 Use of a ROM or PROM to sequence a state logic machine.

step 00, during which CLSUM = 1, and LOADXY = 1, to clear the sum register and load the first x and y coordinates (register X is loaded with XIN; register Y is loaded with YIN). The algorithm goes to step 01 and checks if RSTRT = 1. If so, it goes back to state 00; otherwise, it goes on to step 10, and checks whether XYCOM = 1. If XYCOM = 1, either the x or the y input (XIN or YIN) has changed, i.e., is different from X or Y, respectively. If so, the program goes on to state 11; if not, it returns to state 01. Whenever state 11 is entered, the

Table 5.34 ROM Contents for Line-Measuring Machine

ROM address inputs			ROM data outputs			
Link 12	XYCOM 3	RSTRT 4	Link 12	CLSUM 3	LOADXY 4	ADSUM 5
00	0	0	01	1	1	0
00	0	1	01	1	1	0
00	1	0	01	1	1	0
00	1	1	01	1	1	0
01	0	0	10	0	0	0
01	0	1	00	0	0	0
01	1	0	10	0	0	0
01	1	1	00	0	0	0
10	0	0	01	0	0	0
10	0	1	00	0	0	0
10	1	0	11	0	0	0
10	1	1	00	0	0	0
11	0	0	01	0	1	1
11	0	1	00	0	1	1
11	1	0	01	0	1	1
11	1	1	00	0	1	1

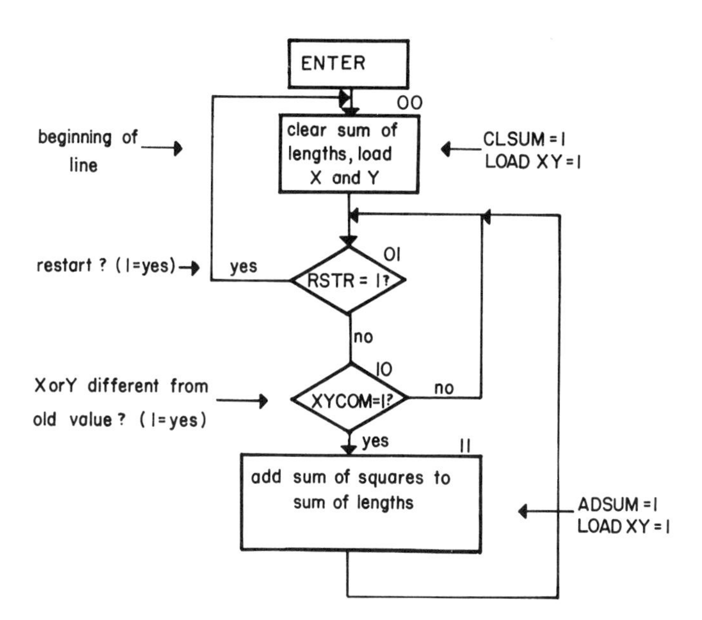

Figure 5.65 Flow diagram of a state logic machine used to determine the length of a line.

Figure 5.66 Block diagram of the machine of Fig. 5.65 and Table 5.34.

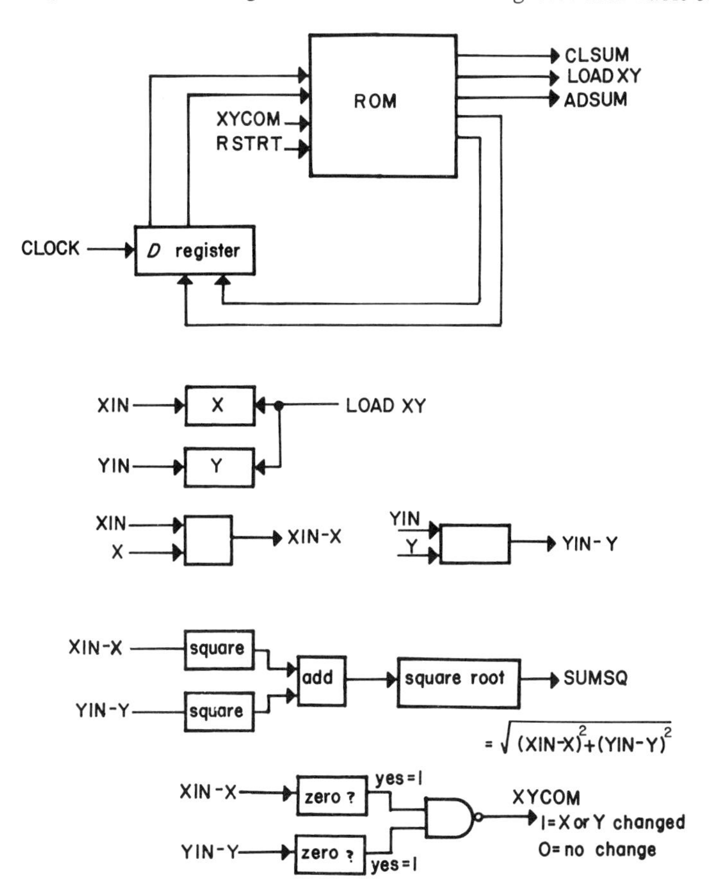

sum of squares is computed (details not shown here) to determine the length of the new line segment, and added to the sum register: control then passes to state 01.

Since there are only four states, 00–11, the link part of the ROM address input and data output need only be 2 bits. The rest of the ROM address input is formed by XYCOM and RSTRT (the two input control variables) and the rest of the output bits are CLSUM (to clear the sum), LOADXY (to load new x and y values), and ADSUM (to add a new line segment length to the sum).

Table 5.34 illustrates the contents of the ROM, which is organized as 16 words$\times$5 bits. State 00 generates a link of 01, regardless of inputs, as well as LOADXY = CLSUM = 1. State 01 generates a link of 10 if RSTRT = 0, or a link of 00 if RSTRT = 1. State 10 generates a link of 11 if XYCOM = 1, or 01 if XYCOM = 0 if RSTRT = 0. State 11 turns bits 4 and 5 (LOADXY and ADSUM) on, and generates a link of 01 if RSTRT = 0.

For each state, there are four possible ROM addresses, depending on the conditions of XYCOM and RSTRT. Thus, there are four substates for each state. A variety of techniques exists for optimization of ROM use by more sophisticated address generation algorithms, but we cannot go into these here.

The ROM-controlled logic machine is an elegant solution to execution of algorithms. It also is, by some people's definitions, a rudimentary computer.

Problems

1. Using gates and inverters, design minimum logic (schematics with as few logic devices as possible) which produces a 1 at its output:

a. if any 8 of its 10 inputs are on, but not if any other number of inputs is on;
b. if any 2 nonadjacent inputs of 10 inputs are on: assign the 10 inputs positions in a row;
c. if its output is Y:
(1) $Y=(A\wedge B)\vee((C\wedge B)\vee(A\wedge C))$
(2) $Y=(A\wedge B)\wedge((C\wedge B)\vee(A\wedge C))$
(3) $Y=(A\wedge B\wedge C)\vee(A\wedge B\wedge D)\vee(B\wedge C\wedge D)\vee(A\wedge B)$
(4) $Y=(A\wedge B\wedge C)\vee(A\wedge B\wedge D)\vee(B\wedge C\wedge D)\vee(A\wedge B)$

2. Implement designs of Problem 1 using multiplexers or demultiplexers.

3. Implement the designs of Problem 1 using ROMs: tabulate the ROM contents for each.

4. **a.** Convert the following binary numbers to octal, decimal, and hexadecimal: 00100110, 100111101, 0110110, 1111, 111011110111000, 01.100011, 1011101/1101110.
b. Convert the following octal numbers to binary, decimal, and hexadecimal: 0134, 2375, 1476, 4327, 3246.725, 247/375.
c. Convert the following decimal numbers to binary, octal, and hexadecimal: 1247, 328, 65, 4629, 921.384, 467/921.
d. Convert the following hexadecimal numbers to binary, octal, and decimal: AEF4, 341, 829E, C16F, 824A1, 321.86, F124/82.

5. Perform the following computations in long-hand, using the number system in which the numbers are represented. Check your answers by converting to decimal, computing, and converting back to the original number system.

128_{16} $+321_{16}$ $+643_{16}$ $+82F_{16}$	$428F_{16}$ -821_{16}	321.4_{16} $\times 81.E_{16}$	$112F_{16}\overline{)821.000_{16}}$
1121_8 432_8 746_8 517_8	724.32 -25.62_8	1146_8 $\times 121.6_8$	$731_{82}\overline{)567.4_8}$
11001101_2 1101011_2 101110_2 10111001_2 1010111_2	11011001_2 -11011.01_2	110011100.1_2 $\times 001.001_2$	$110011.0011_2\overline{)1101101.00000_2}$

6. Write the simplest logical equations from the following Veitch diagrams. Further simplify by factoring, where possible. $\times$s are "don't care" conditions.

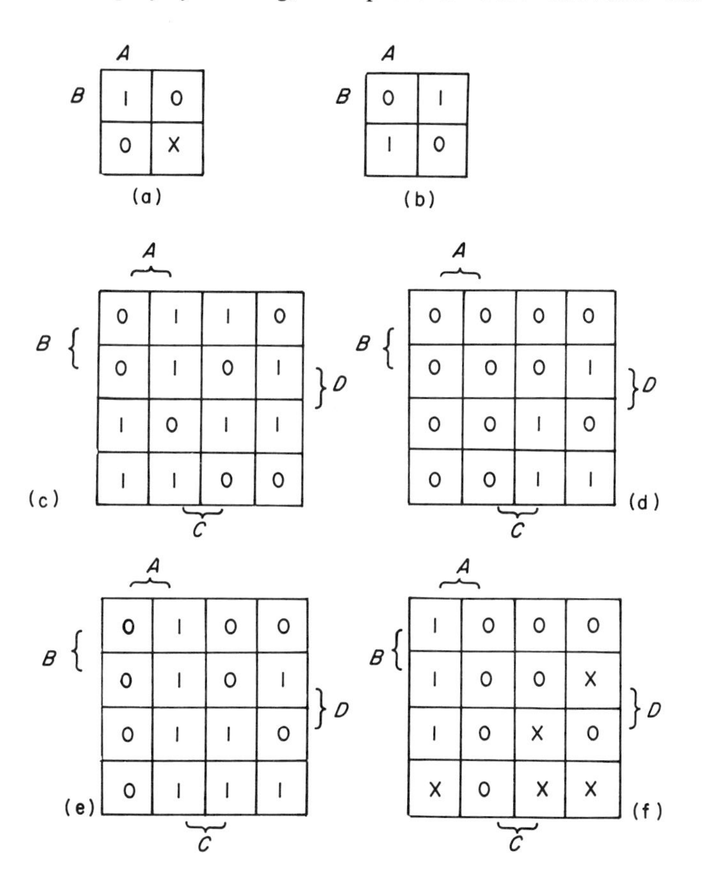

7. Using gates, design
   a. a $J-K$ flipflop;
   b. a D flipflop.

8. Using flipflops and gates, design
   a. a right–left parallel-in, serial-in/parallel-out, serial-out shift register;
   b. a combined LIFO/FIFO, which operates as a LIFO when a signal LIFO/FIFO=1, as a FIFO when LIFO/FIFO=0.

9. Characterize the registers in Problem 8 with (a) timing diagrams, (b) a truth table designating successive states.

10. a. Design a synchronous up/down binary counter.
    b. Design a synchronous up/down BCD counter.

11. Design a 4-bit comparator with two 4-bit inputs A and B, and outputs $A<B$, $A=B$, and $A>B$.

12. Modify the comparator design of Problem 11 for cascading 4-bit comparators: this means adding $A<B$, $A=B$, and $A>B$ inputs for propagation from less significant bits.

13. Design a state logic machine to accept up to 16 7-bit numbers, storing them in a FIFO, and then go through the numbers selecting the number with the largest number of 1's (e.g., 1101110 has more 1's than 1101100). Block diagram the device, and present two alternative designs for the controller: a counter-based design and a ROM-based one.

Selected References

Barna, Arpad, and Dan I. Porat. *Integrated Circuits in Digital Electronics*. John Wiley, New York, 1973.

Blakeslee, Thomas R. *Digital Design with Standard MSI and LSI*. John Wiley, New York, 1975.

Clare, Christopher R. *Designing Logic Systems Using State Machines*. McGraw-Hill, New York, 1973.

Kintner, Paul M. *Electronic Digital Techniques*. McGraw-Hill, New York, 1968.

Lenk, John D. *Handbook of Logic Circuits*. Reston Publ. Co., Reston, VA, 1972.

Peatman, John B. *The Design of Digital Systems*. McGraw-Hill, New York, 1972.

6

Waveform Generation and Signal Conditioning

Paul B. Brown

In this chapter we shall consider the major application of analog circuitry: the generation of waveforms and the modification of waveforms, usually referred to as signal conditioning or signal processing. We shall also extend the concept of processing analog signals by analog computation methods and hybrid computation methods, to show how simple analog and digital building blocks can be used to measure, detect, and record *specific features* of signals, such as their peak sizes, average voltage level and power, and other, more sophisticated features, such as the occurrence of specific waveforms.

In the next chapter we shall continue this subject, using digital and hybrid (combined analog/digital) methods. There, we shall emphasize the advantages of *discrete time*, *continuous voltage*, and *discrete time*, *discrete voltage* representation of signals. In this chapter, we are dealing with *continuous time*, *continuous voltage* (i.e., analog) signals, such as the outputs of real-world devices, which may change at any time (hence continuous time), which produce all voltage levels within a finite operating range (hence continuous voltage); and *continuous time*, *discrete voltage* signals, such as the binary outputs of level detectors, which can vary at any time (hence continuous time), but occupy only a finite number of states (on and off, 0 and 1: hence discrete voltage).

A number of analog waveform-generating and waveform-processing devices rely on level detection (the detection of the crossing of a voltage *threshold* by an analog signal) and switching (the nearly instantaneous transition of an impedance from a high to a low level or *vice versa*) for their proper functions. We shall therefore first develop these concepts.

6.1 Level Detection

In Chapter 4, we introduced the concept of a device which produces a binary output which is equal to 1 when the analog input voltage exceeds a comparison voltage and equal to 0 when the input voltage is less than the comparison voltage. When they are exactly equal, the output is defined as equal to the value it held when the two voltages were most recently unequal. This device is referred to as a *voltage comparator*, a *threshold detector*, a *level* or *voltage discriminator*, or a *trigger circuit*. These terms are usually used in different contexts: A *voltage comparator* compares two voltages A and B, and the output indicates $A > B$ by logical 1 and $A < B$ by logical 0. A *threshold detector*, *level detector*, or *voltage discriminator* is a voltage comparator which compares a signal A with a fixed threshold level B, again indicating $A > B$ and $A < B$ by logical 1 and 0, respectively. A *trigger circuit* is a threshold detector whose output is used specifically to initiate (*trigger*) some event, such as an oscilloscope sweep, or the turning on of a building's heating furnace or air conditioner.

Such a device would be an ideal op amp in open loop configuration with input voltage attached to positive input A and comparison voltage attached to negative input B. The output will be $\pm V_{max}$, for voltages above or below the comparison voltage, respectively. This circuit is illustrated in Fig. 6.1a. In such a circuit, $-V_{max}$ is logical zero and $+V_{max}$ is logical 1. An approximation to the ideal device is obtained using a nonideal op amp, with its frequency limitations (it cannot change state instantaneously or follow infinitely fast input signals) and gain limitations (output will be equal to m times $(V^+ - V^-)$ for inputs within the range $-V_{max}/m$ to $+V_{max}/m$).

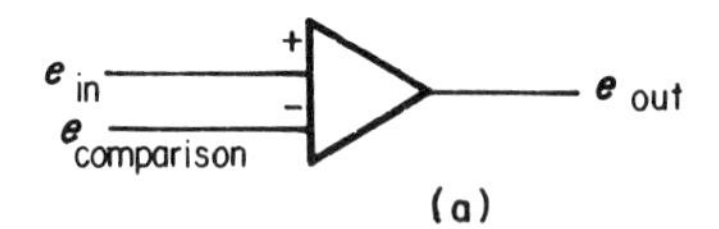

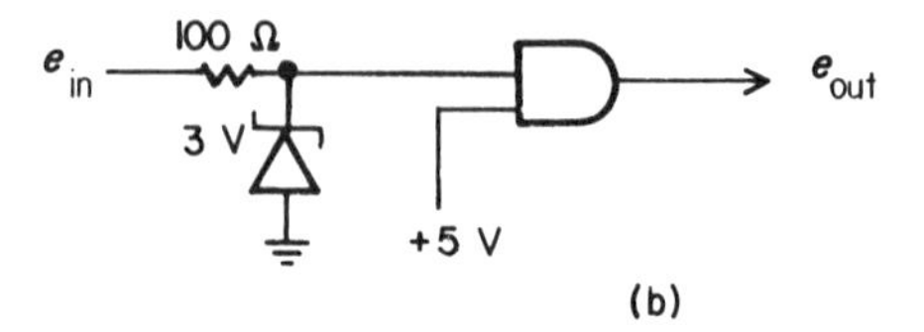

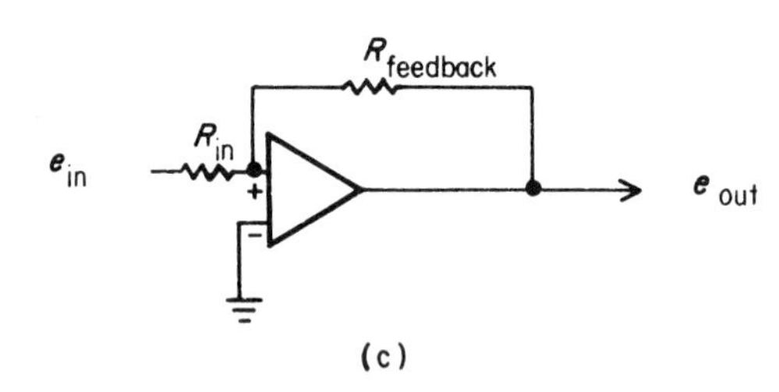

Figure 6.1 Three methods of voltage comparison.

A second alternative is to use a single-input TTL device (such as an inverter), with input voltage limiting to protect the input (Fig. 6.1b). The device will have an output of logical 0 if the input voltage is below the threshold for a 1, or a 1 if the input is above the threshold for a 0.

Since the two thresholds are generally different in real-world devices for noise protection purposes, the device will change its output from 0 to 1 at a higher input voltage than is required to change the output from 1 to 0. This property is referred to as *hysteresis*, a property we shall discuss in more detail with regard to the third way of making a discriminator: an op amp in *positive feedback* configuration. In a single-input TTL device the thresholds are fixed, so it would be necessary to vary the offset of the input signal in order to vary the effective threshold.

All of these devices have the property of comparing a signal A with a threshold B, to produce a logical 1 if $A > B$ and a logical 0 if $B > A$, neglecting hysteresis.

The third of the three alternatives, the one most commonly used, is to utilize the positive feedback configuration introduced in Chapter 4 (Fig. 6.1c). For such a device, often called a *Schmitt trigger*, it will be recalled [Eqs. (4.48), (4.49)] that

$$V^{+}_{\text{threshold}} = -V^{-}_{\text{max}} \frac{R_{\text{in}}}{R_{\text{feedback}}} \tag{6.1}$$

$$V^{-}_{\text{threshold}} = -V^{+}_{\text{max}} \frac{R_{\text{in}}}{R_{\text{feedback}}} \tag{6.2}$$

The difference in negative and positive threshold is referred to as hysteresis: this can be a useful feature in level detection, as will be explained shortly.

To vary the threshold, it is necessary only that the desired comparison voltage be applied to the negative input, as in Fig. 6.2. Here, for e^{+} to be less than the

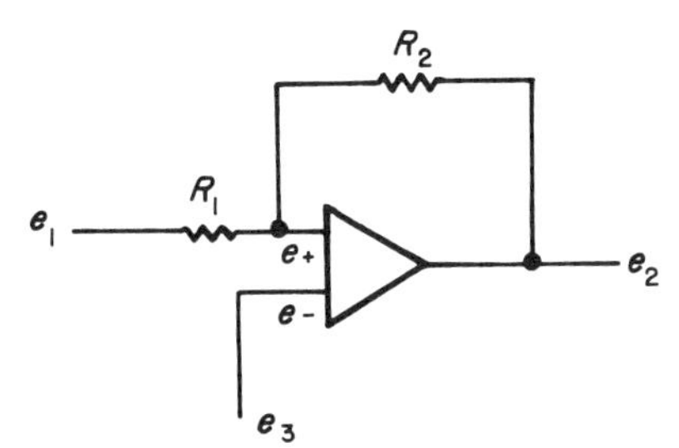

Figure 6.2 Positive feedback level detector with variable threshold.

comparison voltage e_3, with the output equal to $V^+_{\max}$,

$$e^+ = e_1 - \frac{e_1 - e_2}{R_1 + R_2} R_1 < e_3$$

$$= \frac{e_1 R_2 + V^+_{\max} R_1}{R_1 + R_2} < e_3$$

Solving for e_1, when $e^+ = e^- = e_3$, at the negative threshold $V^-_{\text{threshold}}$, gives

$$e_1 = \frac{e_3(R_1 + R_2) - V^+_{\max} R_1}{R_2}$$

Similarly, the positive threshold is

$$V^+_{\text{threshold}} = \frac{e_3(R_1 + R_2) - V^-_{\max} R_1}{R_2}$$

Generalizing,

$$V^{\mp}_{\text{threshold}} = \frac{e_3(R_1 + R_2) - V^{\pm}_{\max} R_1}{R_2} \tag{6.3}$$

where $V^+_{\max}$ is used for negative threshold and $V^-_{\max}$ is used for positive threshold.

What is the advantage of hysteresis in a threshold detector? Figure 6.3 graphically illustrates the advantage. When a signal is contaminated by noise (Fig. 6.3a) multiple triggering can occur without hysteresis (Fig. 6.3b); with hysteresis (Fig. 6.3c), the multiple triggering is suppressed, although the actual times at which the output rises and falls are changed somewhat. Even with filtering, this kind of contamination with noise is not uncommon, so that hysteresis is very useful in producing single triggered pulses from each analog excursion which crosses threshold.

The circuit of Fig. 6.4 illustrates a simple and effective method of modifying the output of an op amp-based discriminator for TTL compatibility and variable polarity. The output of the op amp is clamped to a range of 0–3.3 V, which is TTL compatible. Since the rise and fall time may be slow enough for multiple triggering of conventional TTL, a Schmitt trigger input inverter is used: note the miniature hysteresis loop drawn inside the symbol of the first inverter, to signify Schmitt trigger input. A single inversion produces logical 1 whenever the signal is *below* (more negative than) threshold (and hence goes on during a negative slope at the input); a double inversion produces an output which goes on when the signal goes *above* (more positive than) the threshold (and hence goes on during a positive slope at the input). Therefore the selector switch which selects single or double inversion is called a *slope selector*, and is similar to the slope selector on an oscilloscope trigger

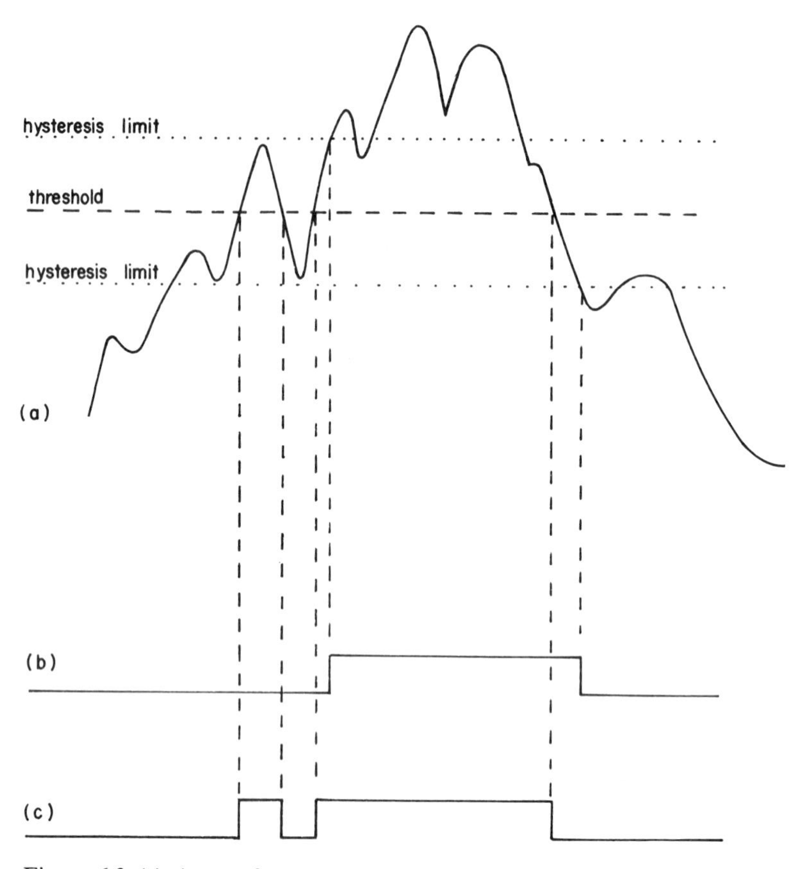

Figure 6.3 **(a)** A waveform contaminated with high frequency noise. **(b)** Output of a discriminator with hysteresis. **(c)** Output of a discriminator without hysteresis.

circuit. Note that the *slope* polarity is independent of the *voltage* polarity of the threshold: either a negative or positive threshold voltage may be associated with either a negative or positive slope polarity to provide all four possible combinations.

The most obvious application of the Schmitt trigger is to produce a pulse each time an analog signal exceeds a threshold. Such circuits are used as inputs to TTL circuits for analysis of timing of analog pulses, to count analog pulses, or to initiate

Figure 6.4 A complete discriminator with TTL output.

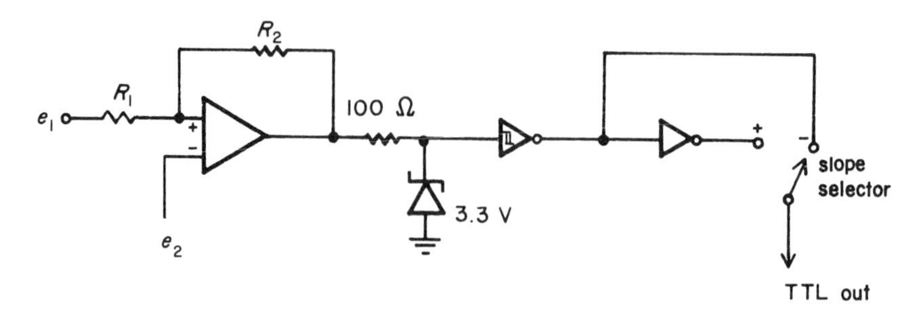

action of digital machines on the occurrence of analog pulses; or they may be used to trigger oscilloscope sweeps or to alter the behavior of analog circuits, as we shall illustrate in later sections of this chapter.

6.2 Switching Techniques

Another important analog-processing tool, often used in conjunction with voltage discrimination, is the *analog switch*, a device which can act much like a mechanical switch by very rapidly increasing or decreasing the conductance of a path by several orders of magnitude. Such switches are controlled by electric control signals, much like electromechanical relays. Since the semiconductor switches are very fast, have longer lifetimes than electromechanical relays, and are not subject to contact bounce (relay contacts can bounce, producing multiple make–break switch actions when they open or close), they have largely superseded relays, except in some high-power or high-voltage applications.

The symbol that we shall use for an analog switch is given in Fig. 6.5a. This particular switch is a single-pole, single-throw (SPST) switch, normally open (N.O.): that is, when the CONTROL signal is off (logical 0) there is a very high ohmic resistance (typically $>10^{10}\ \Omega$ between the "switch contacts"), and when CONTROL is on (logical 1) the resistance is low (typically $<10^{2}\ \Omega$). CONTROL is, for most semiconductor switches, a TTL signal.

Since these are semiconductor devices, they typically exhibit such nonidealities as noise, leakage current, and dc offset, but these are generally of no great importance.

Analog switches are obviously used to vary impedances: therefore they can be used to select among a number of analog signals, a process called *analog multiplexing*; or they can be used to select digitally component values in a circuit, e.g., to change the R or C of an RC network using digital control. Such uses are illustrated in Figs. 6.5b and 6.5c, respectively.

6.3 Pulse Generators

In a number of applications, it is useful to be able to cause a pulse of fixed amplitude and duration to occur upon arrival of some external input: typically a pulse is initiated when the input goes from logical 1 to logical 0, or vice versa. We say the device is *triggered* on a *negative* or *trailing edge* if a $1 \rightarrow 0$ transition causes a pulse, or it is triggered on a *positive* or *leading edge* if a $0 \rightarrow 1$ transition causes a pulse.

Pulse generators are one type of multivibrator (MVB). The flipflop, which has two stable states which can be held indefinitely, is known as a *bistable* MVB. The pulse generator, which has only one stable state which can be maintained indefinitely (the off state), is a *monostable* multivibrator. A digital oscillator, which cannot maintain either state forever but flips back and forth between them, is an *astable* MVB. The monostable MVB, or monostable for short, is also called a *one-shot*.

One way to generate a pulse is to ac couple the input to a gate or inverter, as in Fig. 6.6. This is adequate if the desired output pulse is brief and precise duration is not needed. An input pulse of sufficient duration will cause the capacitor to charge, generating a transient at the inverter input. For as long as the transient is above the logical 1 threshold, the inverter output will be off. Note that for input pulses longer than this time, determined by the RC time constant, the output pulse will always be

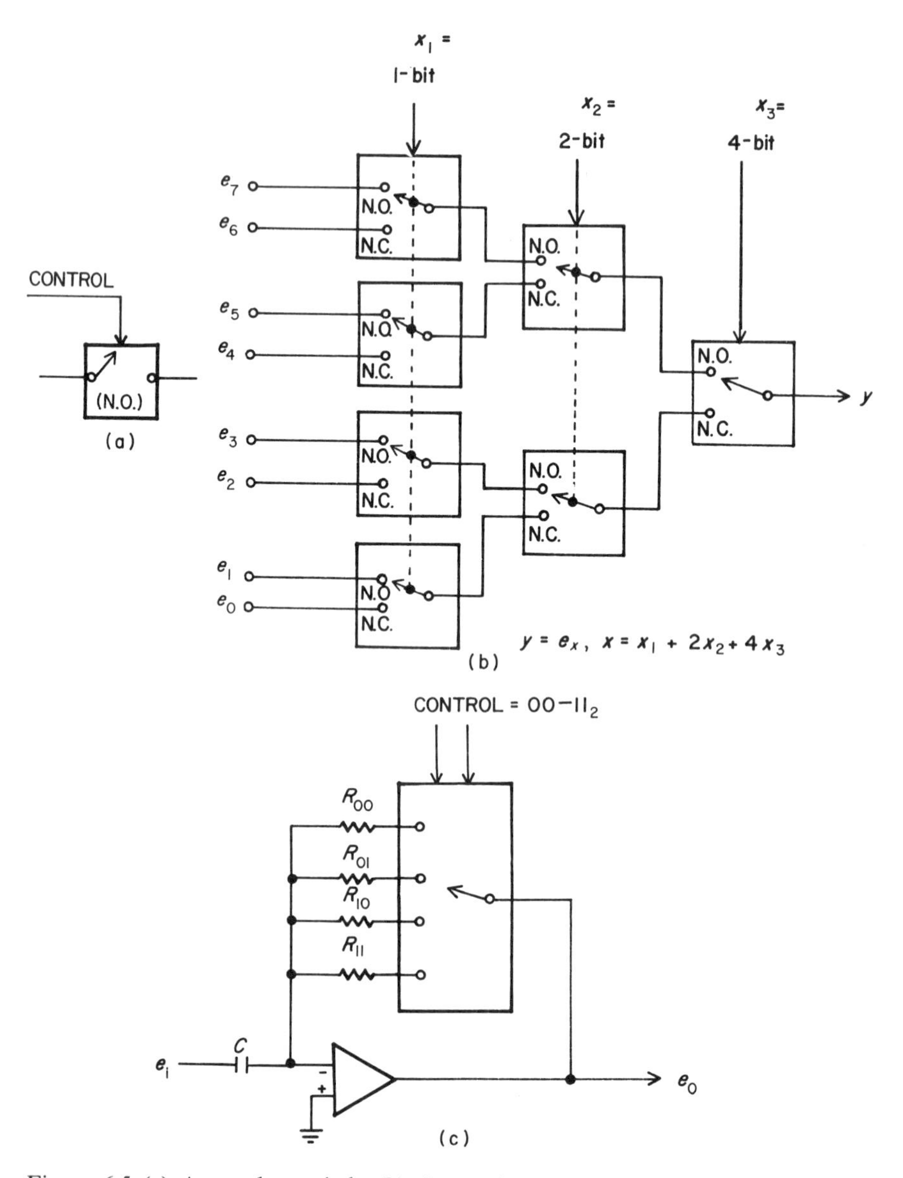

Figure 6.5 **(a)** An analog switch. **(b)** An analog multiplexer. **(c)** Selection of analog component values: control of *RC* time constant of an integrator.

the same length; for shorter input pulses, the positive transient will be cut short, reducing the output pulse duration.

The pulse generator of Fig. 6.7 is a true monostable multivibrator. Regardless of the length of input pulse, the output pulse is of fixed duration, determined only by *RC*, because as soon as the upper gate goes off, the lower gate goes on, holding the upper gate off regardless of any events at the input. The lower gate can go back off again only when the *RC* transient goes back above its threshold. In fact, whereas the one-gate device of Fig. 6.6 would produce two brief pulses in a time shorter than its normal output pulse if the input pulses were sufficiently short and close together, the

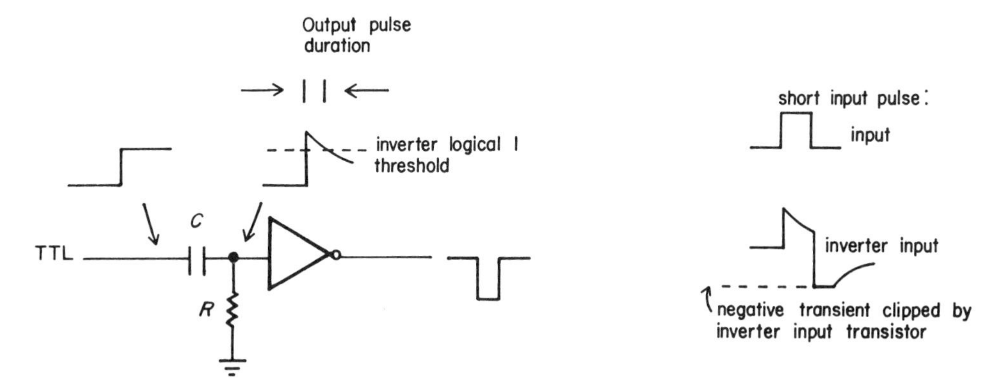

Figure 6.6 A primitive pulse generator.

monostable multivibrator of Fig. 6.7 would ignore the second input pulse if it occurred during the output pulse.

The monostable of Fig. 6.7 has one disadvantage: if the capacitor is not given adequate time to discharge before a second input pulse, it may not generate an adequate transient to trigger a second output pulse, or if it does, the transient may be decreased in size, resulting in a briefer output pulse. This means that only if the input pulses are sufficiently far apart in time will the output reliably produce pulses of a fixed duration. The time required can be expressed as the ratio of pulse duration to pulse interval which, if not exceeded, guarantees that the output pulse will be within a specified percentage of the duration. This ratio of pulse duration to pulse interval is referred to as the device's *duty cycle*. Modern IC pulse generators commonly have duty cycles in excess of 90%.

Triggering on the trailing edge of an input pulse can be accomplished by using NOR gates instead of NAND gates, or by placing an inverter in front of the input of the NAND gate MVB, as illustrated in Fig. 6.8.

The designer may wish to assemble a monostable if no IC versions are available and one is needed in a hurry, but normally it is better to buy one of the many commercial IC versions, which have sophisticated designs permitting them to be retriggered at any interval, and even allowing pulses to be prolonged by retriggering

Figure 6.7 A monostable multivibrator.

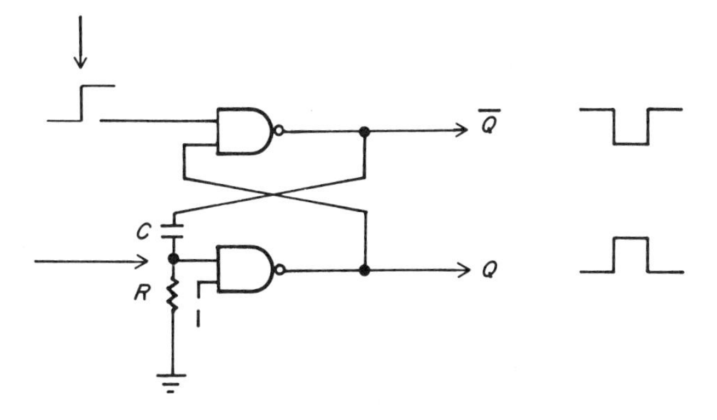

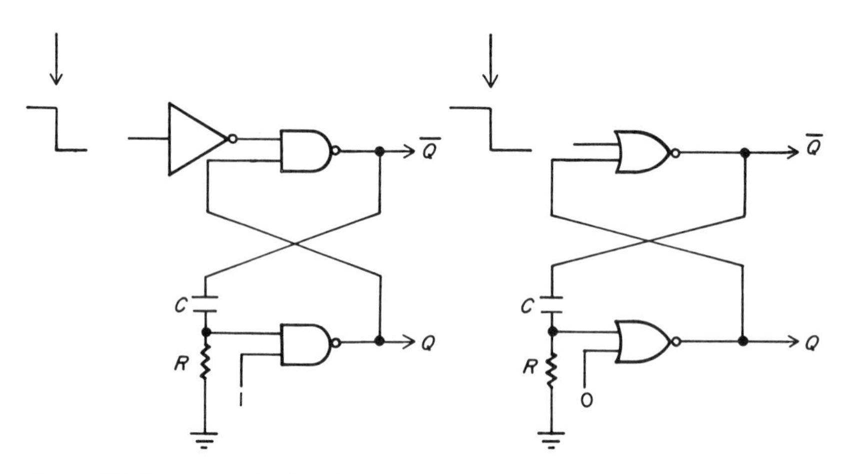

Figure 6.8 Two ways of triggering on a trailing edge.

during an output pulse or allowing them to be cut short by clearing them during the output pulse. These devices often have multiple inputs for triggering from a number of different sources. Their pulse durations are controlled by use of external R and C.

Pulse generators may be cascaded to produce pulses in any temporal pattern. For example, a pair of multivibrators, each triggered on the trailing edge of the other, can be used to make a square wave oscillator. Trains of pulses (a train of pulses is a set of two or more pulses occurring on a single signal line) can be generated by enabling such a square wave for a specified time. The following example incorporates these principles.

Example 6.1. Production of a pulse train with complex timing properties

Figure 6.9 illustrates the generation of two bursts of pulses with controllable pulse frequency within each burst, burst durations, and pulse width. The output sequence is triggered by the leading edge of START (Fig. 6.9a). After a delay t_1, a sequence of pulses begins at the output. The interval between pulses in each burst is equal to t_2, the pulse width is equal to t_3, the two bursts have duration t_4, and there is a delay between bursts equal to t_5.

Delay t_1 is accomplished with the t_1 MVB in Fig. 6.9b: the one-shot is symbolized as a box with the trigger polarity, duration, and output polarity (+output$=Q$, −output$=\overline{Q}$). The t_1 MVB fires on the leading edge of START and generates a pulse of duration t_1. The delay is accomplished by using the trailing edge of t_1 to trigger the t_6 MVB, which in turn is used to gate the entire output-generating process. At the same time, interval t_4 is initiated by triggering the t_4 MVB via a gate which is enabled by the t_5 MVB negative output, which is on at the beginning of the sequence. The turning on of t_6 and t_4 and the already high output of the t_3 MVB cause the input to the t_2 MVB to go high, triggering it and starting the first output pulse. At the end of each t_2 pulse, t_3 goes on, which turns off the input to the t_2 MVB; when the t_3 pulse goes off, the t_2 pulse is retriggered. The t_2–t_3 pair thus constitutes a pair of reciprocating one-shots, each triggering the other: This continues until t_4 goes off, forcing a halt to t_2 pulse generation. When t_4 goes off, t_5 is triggered. T_4 and t_5 are a reciprocating pair which alternate until t_6 goes off. Thus, when t_4 goes on a second time a second pulse train at t_2 is generated. When t_4 goes off a second time, t_5 triggers again; however, t_5 cannot retrigger t_4 a second time because by the second time t_5 goes off, t_6 is off, preventing the input to t_4 from going low because it is already low.

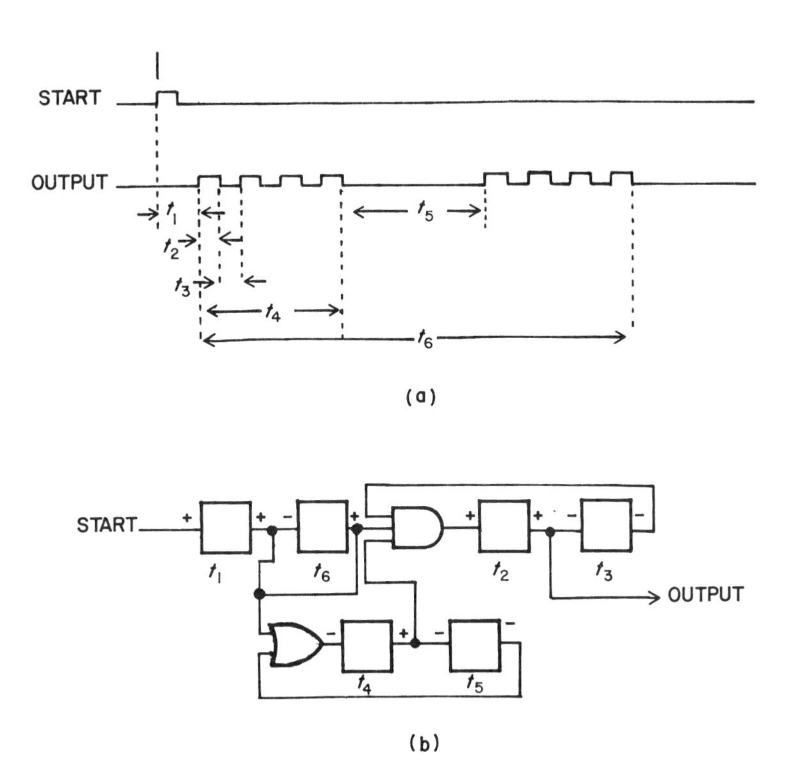

Figure 6.9 A complex sequence of pulses generated with a combination of pulse generators.

The use of pulse generators in combination is one of two ways to produce complex sequences of pulses. The other way is to use flipflops, or devices based on flipflops such as counters and shift registers, to derive such pulses from a clock. The method using pulse generators is thought of as an analog control method because the pulse durations of the one-shots may be continuously varied by varying the R of the RC pulse duration control circuit. The clock-derived pulses are digitally controlled because their durations are multiples of the clock period and cannot take on intermediate values between these multiples.

6.4 Oscillators

The designer of signal-generating circuits is generally faced with one of three types of design problems, the first two of which are (a) design of a circuit which generates a periodic waveform or (b) design of a circuit which, upon occurrence of a specific input waveform, produces an output waveform of predetermined shape. The former devices are, broadly speaking, called *oscillators*: the latter are *triggerable* waveform generators. Often one type will include the other as a component. A third type of waveform generator produces a continuously varying output which is not periodic, but random. This is called a *noise generator*. We shall discuss all three categories of waveform generator in this chapter.

The most common periodic wave forms generated by oscillators are illustrated in Fig. 6.10. The sine wave is commonly used for frequency response testing of analog circuits; as a waveform to drive transducers, to produce pure sound tones, mechanical vibrations, etc; and as a carrier for amplitude or frequency modulation. Of

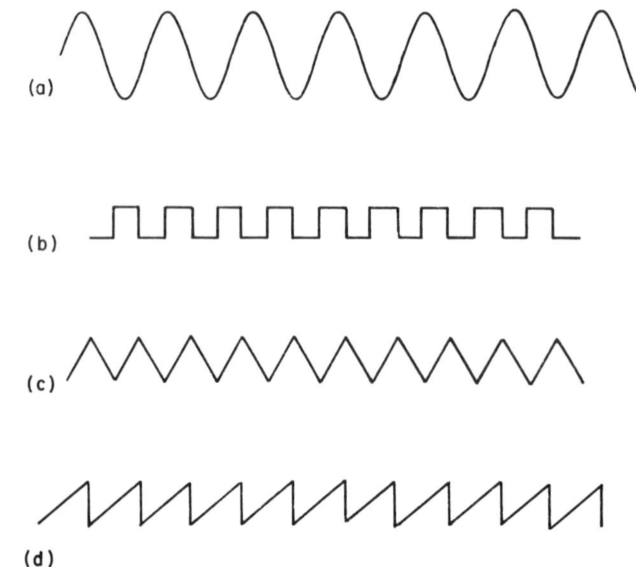

Figure 6.10 The most commonly used periodic waveforms: **(a)** sine wave; **(b)** square wave; **(c)** triangle wave; **(d)** sawtooth.

course, sine waves have the advantages in all these applications of possessing only one Fourier component and possessing all orders of derivatives. The square wave is convenient for calibrating gains of circuits, for testing rise times, and to provide the basic clock frequency for sequential digital machines. The triangle wave is used not only as a test stimulus but also as an intermediate waveform in some types of waveform generators. Note that the time derivative of a triangle wave is a square wave and that the time integral of a square wave is a triangle wave. The sawtooth is most commonly used as the horizontal deflection voltage on oscilloscopes: the slow ramp deflects from left to right during display of the trace, and the fast ramp returns the trace to the left side of the screen.

The design and analysis of oscillators is a science in itself. We are fortunate in having available a number of ICs and modules which make fabrication of oscillators simple for all but the most exacting purposes, e.g., sine wave generators of great spectral purity (very little harmonic distortion). If very pure sine waves are desired (say, less than 0.1% harmonic content) it is best for the amateur designer to spend a sizable sum on a precision sine wave oscillator with regulated frequency and amplitude and minimal distortion. Since the construction of a precision sine wave oscillator is so difficult, and since the construction of less precise ones is so easy, we shall describe only the circuits which are easy to build. Square wave, triangle wave, and sawtooth oscillators are very easy to build, so we shall present their designs here as well.

6.4.1 Square Wave Oscillators

A simple square wave oscillator requiring only one op amp is illustrated in Fig. 6.11. The positive feedback circuit ensures that the output is either V_{max}^{+} or V_{max}^{-}, and only occupies intermediate states during transitions from one extreme to the other. The negative feedback charges the capacitor C through resistor R_2, until $e^{-} = e^{+} = V_{max}^{\pm}/2$, at which time the output changes state. Then the capacitor charges in the

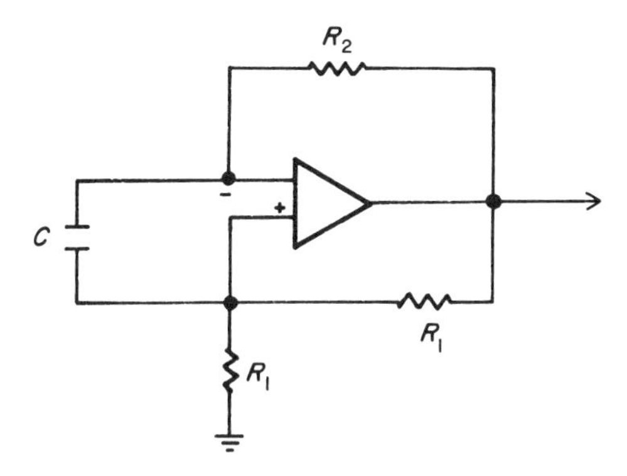

Figure 6.11 A square wave oscillator.

other direction until once again $e^- = e^+ = V_{max}^{\pm}/2$. By varying R_2 or C, the charging rate, and therefore the frequency of oscillation, can be varied. A continuously variable R_2 and decade variable C permit easy construction of an oscillator which can cover a very wide frequency range.

Quartz crystals are used as capacitive elements in precision oscillators because when cut in the proper plane, the quartz is dimensionally very stable over a wide temperature range: even greater stability is achieved by putting a crystal oscillator in a temperature-controlled oven. Quartz oscillators can be built from discrete components, as in Fig. 6.12, for about \$20. For about twice that cost, they can be purchased as DIP ICs, with any specified oscillation frequency.

If less precision is acceptable, a TTL oscillator can be built using TTL ICs such as the 555 multivibrator (Fig. 6.13), which will oscillate at a frequency determined by an external resistance and capacitance. The designer simply chooses R and C according to a graph of R and C versus desired frequency, and the device can be assembled and running in minutes, for a total cost of \$1–2.

The 555 (and similar devices) is inexpensive, easy to use, and can be wired as either a monostable or astable multivibrator. It has a range of supply voltage which permits use as an analog or TTL device. A number of other IC multivibrators are available, with such features as pulse reset (aborts a pulse), retriggerability (pulse can be extended by triggering again before it is terminated), and even voltage-controlled oscillation rate.

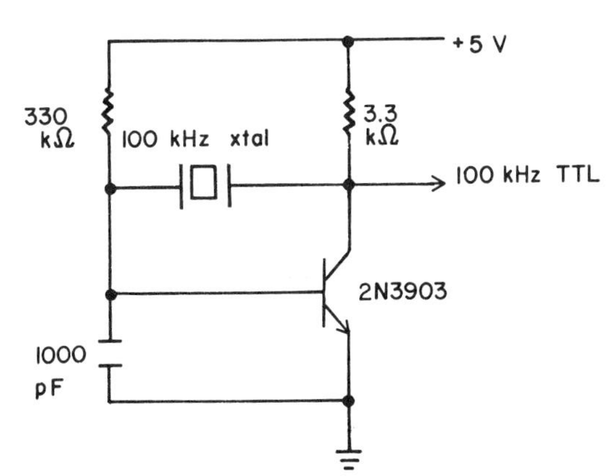

Figure 6.12 A crystal-controlled TTL square wave oscillator.

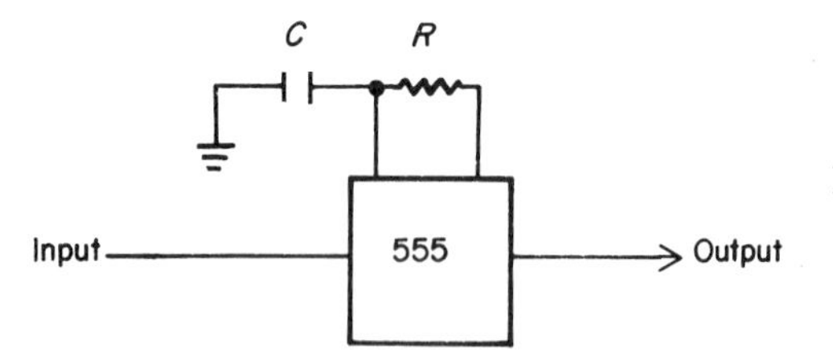

Figure 6.13 The 555 timer.

6.4.2 Triangle Wave Oscillators

The circuit of Fig. 6.14 has a triangle wave and a square wave output. The first op amp is a positive feedback threshold detector with threshold $= V_{\max}^{\pm} \times (1\ \mathrm{k}\Omega/10\ \mathrm{k}\Omega) = V_{\max}^{\pm}/10$. Its output $V_{\max}^{\pm}$ is fed to an inverting integrator with time constant RC, whose output is fed back to the threshold detector.

When the square wave output is at $V_{\max}^{+}$, the integrator, regardless of its initial value, begins integrating, to produce a negative-going ramp of constant slope:

$$e_{\text{triangle}} = -\frac{1}{RC}\int V_{\max}^{+}\, dt = -k^{-}t \tag{6.4a}$$

When the threshold for the Schmitt trigger is exceeded (e.g., $e_{\text{triangle}} = V_{\max}^{+}/10$), the Schmitt trigger output flips to $V_{\max}^{-}$, and the triangle wave immediately reverses to produce a ramp of constant slope:

$$e_{\text{triangle}} = \frac{-1}{RC}\int V_{\max}^{-}\, dt = -k^{+}t \tag{6.4b}$$

If

$$V_{\max}^{+} = -V_{\max}^{-},\ k^{+} = -k^{-}$$

For the entire triangle wave, $e_{\text{peak-to-peak}}$ in either direction will take the form of a ramp whose duration is half the period of the oscillation $T/2$:

$$e_{\text{peak-to-peak}} = -\frac{V_{\max}^{+}}{RC}\int_{0}^{T/2} dt = \frac{-V_{\max}^{\pm}T}{2RC}$$

We know the total ramp traverse to be equal to

$$e_{\text{peak-to-peak}} = V_{\max}^{+}/10 - V_{\max}^{-}/10 = 2V_{\max}^{+}/10$$

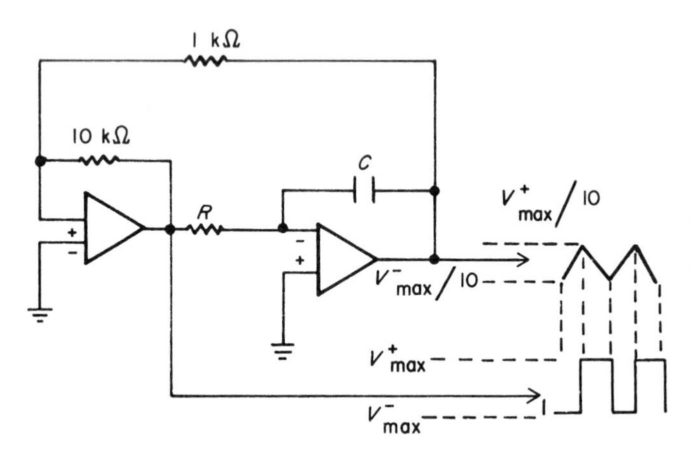

Figure 6.14 A square wave/triangle wave generator.

Equating these two expressions and solving for T gives

$$\frac{2V_{\max}^{+}}{10}=-\frac{V_{\max}^{\pm}T}{2RC}$$

$$T=\frac{4RC}{10}=0.4RC$$

Since $f=1/T$, $f=2.5/RC$.

The frequency can be varied by changing R or C: note that changing $V_{\max}^{\pm}$ will *not* vary the frequency; hence the behavior of the device should be relatively insensitive to changes in supply voltage, which is the primary determinant of $V_{\max}^{\pm}$. However, the peak-to-peak amplitude will vary with $V_{\max}^{\pm}$.

The up and down ramps can be made independent by having different R and C for each ramp: this is accomplished with diodes, as in Fig. 6.15. Whether a single RC or separate RC_{up} and RC_{down} is used, frequency is usually varied continuously by using a potentiometer for R and in decade steps by switching C in steps of ten.

The main use of triangle waves is as an intermediate in the synthesis of sine waves. There are a number of ICs on the market, such as those produced by Exar, which have simultaneous square wave, triangle wave, and sine wave outputs. We shall examine some of these in the section on amplitude and frequency modulation. Triangle waves are also used to sweep some test parameter up and down, for example, to vary frequency of a sine wave to determine a system's frequency response.

6.4.3 *Sine Wave Oscillators*

Sine waves are produced from triangle waves with nonlinear waveform shaping circuits such as the ones described in Chapter 4. Such circuits are not simple to design, but they are available commercially. However, the designer will usually do best to buy an IC function generator which produces sine waves from internally generated triangle waves.

Figure 6.15 Modification of triangle wave generator for asymmetric slopes.

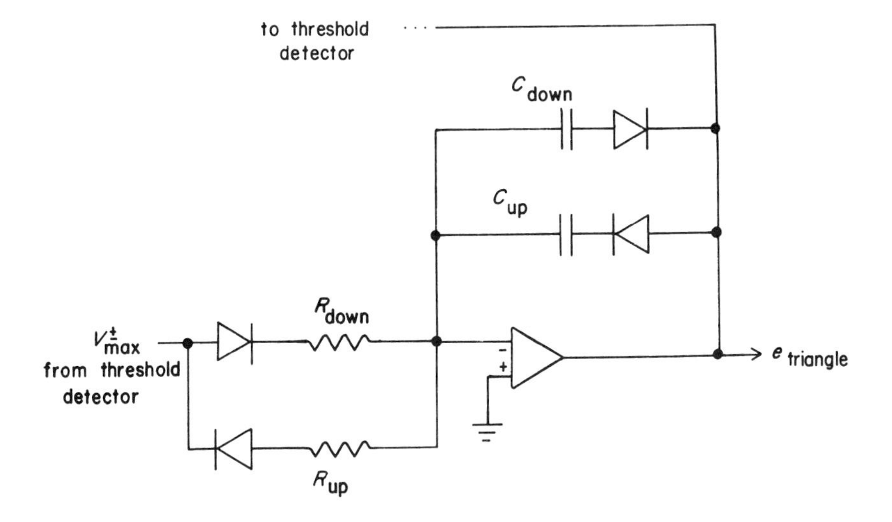

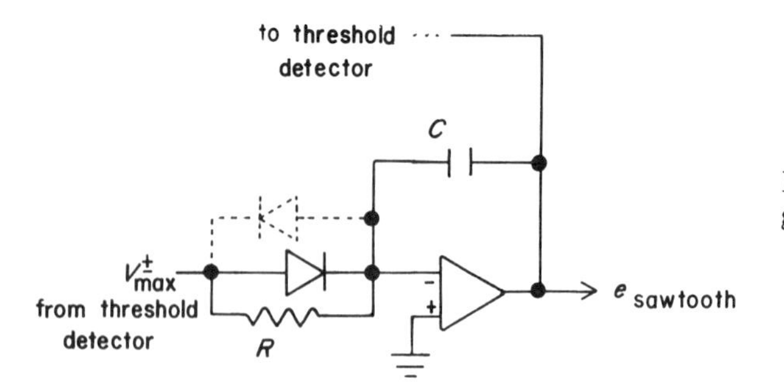

Figure 6.16 A sawtooth generator.

6.4.4 Sawtooth Oscillators

A sawtooth is a triangle wave with one ramp (up or down) several orders of magnitude faster than the other. One method for sawtooth generation is to modify the triangle wave generator so that R is nearly zero for one polarity of $V^{\pm}_{\max}$, as in Fig. 6.16. Here a diode, in one direction or the other, is placed in parallel with R. The ramp in one direction will have a slope determined by RC, and in the other direction the diode will act as a very low R, for very high slope.

The sawtooth is used for a linear sweep in one direction and fast return, for example on oscilloscope traces, raster scans on television sets and some types of computer graphics devices, and to linearly vary some test variable when testing a system's response to that variable.

6.5 Amplitude and Frequency Modulation

Although the most obvious use of amplitude modulation (AM) and frequency modulation (FM) is in radio frequency applications, a number of scientific applications benefit from the use of AM and FM. In general terms, it is useful to be able to modulate the amplitude or frequency of an oscillator. This is because it is often useful to be able to test a system with a sinusoid or other periodic signal which is systematically varied in amplitude or frequency over a whole range of values.

6.5.1 Amplitude Modulation and Demodulation

An amplitude modulator is illustrated in Fig. 6.17. The carrier signal, which is of fixed frequency and amplitude, is multiplied by the modulator signal, which should have no frequency components greater than one-half the carrier frequency, but which otherwise has any desired frequency distribution.

If we multiply one sinusoid by another, the output frequency composition is, from trigonometry

$$(A\sin\omega_A t)\cdot(B\sin\omega_B t)=\frac{AB}{2}\left[\cos(\omega_A-\omega_B)t-\cos(\omega_A+\omega_B)t\right] \tag{6.5}$$

That is, half the output power is contained in the frequency equal to the sum of the input frequencies, and half is contained in that frequency equal to the difference in the input frequencies. If the modulator signal has a richer frequency composition, say $g(f)$, then the output frequency composition is $\frac{1}{2}g(f+c)+\frac{1}{2}g(f-c)$, where c is

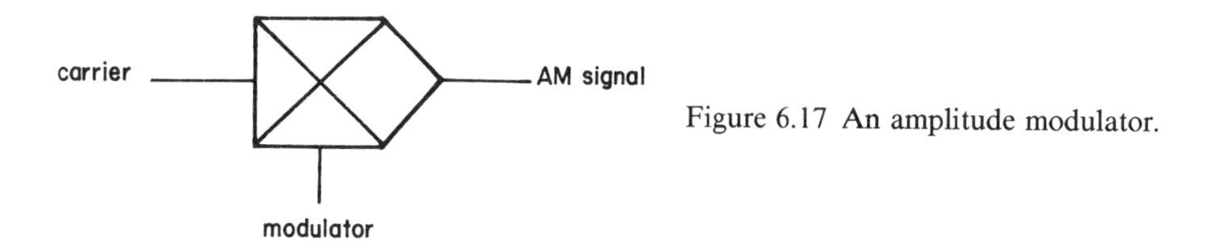

Figure 6.17 An amplitude modulator.

the carrier frequency. Note that there is no trace of the input frequencies $g(f)$ or c in the output spectrum $g(f+c)$, $g(f-c)$.

If dc components (A and B below) are present in the input signals, then the output is

$$[A+g(f)]\cdot[B+c]=AB+Bg(f)+Ac+\tfrac{1}{2}g(f+c)+\tfrac{1}{2}g(f-c) \qquad (6.6)$$

Note that, if there is a dc component in the modulator, the carrier frequency c appears at the output: if there is a dc component in the carrier, the modulator spectrum $g(f)$ appears at the output. If there is a dc component in both inputs, there is a dc component at the output as well as the two input spectra. A great deal of effort has gone into the development of designs which *supress* the input frequencies, which are not desired in most applications: the simplest is to ac couple these inputs to the modulator and to balance internally all offsets in the modulator. Alternatively, or in addition, notch and bandpass filters can be used to remove the unwanted frequencies $g(f)$ and c from the output.

As will be described in the chapter on radio frequency circuits, sometimes one band of frequencies, $g(f+c)$ or $g(f-c)$ is also removed: the two bands are called *sidebands*, and use of only one is called *single-sideband suppression*. All of the required information in the input signal is retrievable if the original carrier frequency is known; hence all the transmission power can be concentrated in one of the sidebands, resulting in greater transmission range and a more economical use of the radio frequencies available.

The designer can use a multiplier for his AM applications, or for some applications he may wish to use one of a variety of balanced modulators available as modules or ICs on the market.

Amplitude demodulation, or recovery of the modulator signal from the AM signal, can be accomplished by dividing the AM signal by the carrier, but this is rarely done because the carrier is not usually available at the receiver. Instead, demodulation is usually accomplished by rectifying the signal and low-pass filtering the output, sometimes using *translation filter* techniques (*heterodyning*). The details of amplitude demodulation and frequency demodulation will be developed in Chapter 8, where radio frequency techniques are described.

6.5.2 Frequency Modulation

Frequency modulation is less easily understood on an intuitive basis than AM. We can easily visualize an AM waveform, where the frequency of oscillation appears to be fixed but the *envelope* is a replica of the modulator. (The envelope is the outline formed by connecting the positive peaks with one smooth curve and the negative peaks with another.) In FM, the amplitude of the signal carries no information. Instead, the frequency varies with time as a function of the amplitude of the

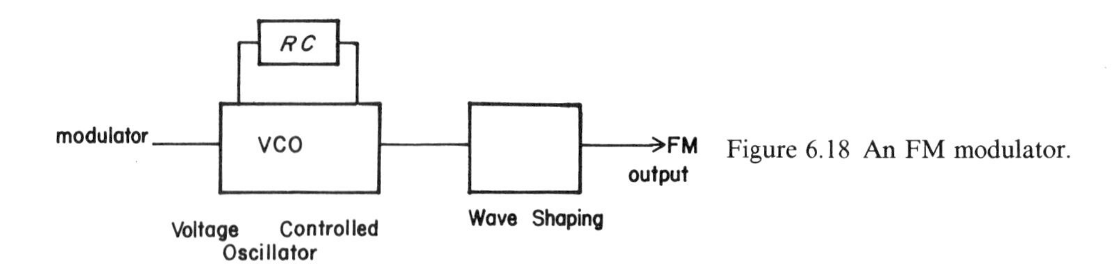

Figure 6.18 An FM modulator.

modulator. Thus, for a carrier frequency c and a modulator $m(t)$, the output frequency $g(f)$ is

$$g(f) = g(c + km(t)) \tag{6.7}$$

That is, the frequency of the output is equal to the carrier c plus $m(t)$ multiplied by a conversion constant k. If $m(t)$ is zero at a particular time, the output frequency is c at that time. If $m(t)$ is negative, the output frequency is less than c; if $m(t)$ is positive, the output frequency is greater than c. Thus, all the input information which is present as an amplitude variation in time $m(t)$ becomes encoded in the output as a frequency variation in time f, and the output amplitude does not carry any information.

However, FM required, until recently, greater circuit complexity in terms of numbers of circuit components, and it still requires a greater bandwidth than AM for effective use. With modern electronics components the complexity problem has been beaten.

Figure 6.18 illustrates an FM modulator of the type used in modern IC function generators. The modulator is used as a control signal to vary the frequency of an oscillator, called a *voltage controlled oscillator* (VCO). The center frequency (the frequency at which the VCO oscillates if the modulator input is zero) is adjustable with external R and C.

The VCO itself is usually a triangle wave oscillator with a means of varying its frequency under control of an external voltage. In principle, any resonant circuit could be used, including a sine wave oscillator, with some means of varying the effective R or C of the oscillator under the control of an external signal. This kind of direct production of a spectrally high quality sine wave with direct control of its frequency is not yet practical, although highly desirable. Instead, a VCO similar to that of Fig. 6.19 is generally used. It is a triangle wave generator similar to that of Fig. 6.14, with a multiplier used to amplitude modulate the square wave. Since the ramp slope depends on the voltage into the integrator,

$$e_{\text{triangle}} = -\frac{1}{RC}\int V_{\text{int}}\,dt = -\frac{V_{\text{int}}}{RC}\int dt$$

The period T is solved for as before:

$$e_{\text{peak-to-peak}} = -\frac{V_{\text{int}}}{RC}\int_0^{T/2} dt = -\frac{V_{\text{int}}T}{2RC} = \frac{2V_{\max}^{\pm}}{10}$$

Solving for T gives

$$T = \frac{4RC}{10}\cdot\frac{V_{\max}^{\pm}}{V_{\text{int}}}$$

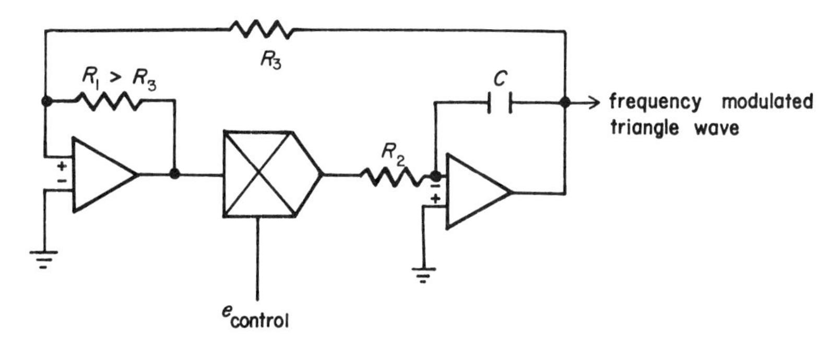

Figure 6.19 A voltage-controlled oscillator (VCO).

Since $V_{max}^{\pm}/V_{int} = 1/e_{control}$,

$$T = \frac{0.4RC}{e_{control}}$$

and

$$f = \frac{1}{T} = \frac{e_{control}}{0.4RC} \tag{6.8}$$

Thus, frequency is proportional to $e_{control}$. R and C are adjusted for the desired center frequency f.

In an FM modulator of the type described here, the triangle wave goes through a shaping circuit, usually a multiple breakpoint transconductance amplifier of the sort described in Chapter 4, to convert the triangle wave to a sine wave. The errors in this conversion are responsible for the relatively poor spectral purity of such devices, although more expensive modules have 0.1% distortion (−60 dB total harmonic content).

ICs and modules, available from a number of manufacturers, contain all the circuitry necessary for both AM and FM operation, so the designer need not make his own. With one or two such devices, costing only a few dollars each, all the commonly used waveforms can be generated, with a nonlinearity of about 1% for triangles and ramps and about 2.5% for sine waves. The designer can thus devote his energies to the more productive black box design of analog systems, and less to the relatively unproductive design of the black boxes themselves.

6.6 Automatic Gain and Frequency Control

Given the possibility of controlling the frequency or amplitude of a signal with a control voltage, it should be possible to sense frequency or amplitude of a signal, generating a dc level which is proportional to one or the other, and feed this voltage back as control voltage to a frequency or amplitude modulator. It would then be possible to automatically maintain a steady frequency or amplitude of a signal under conditions of variable load, temperature or supply voltage, which would cause them to vary. Such control is referred to as *automatic gain control* (AGC) or *automatic frequency control* (AFC), and can be used to maintain a steady signal in the presence of influences which tend to cause it to vary.

6.6.1 Automatic Gain Control

The gain (or, strictly speaking, the amplitude) of a signal is most prone to variation if it is a recorded or transduced signal from an uncontrolled source, such as the reception of a radio transmission or the recording of a person's voice as the person moves around a room with a stationary microphone. Also, some types of signal generators are prone to amplitude fluctuation as the frequency of oscillation is changed.

It is also often helpful to have a circuit which maintains constant mean signal power when the bandpass of filters applied to that signal is changed: for example, if the bandwidth is reduced, and significant portions of the signal spectrum are filtered out.

An automatic gain control (AGC) is block diagrammed in Fig. 6.20. The control signal for a fixed amplitude would be a dc level. From this is subtracted a dc level corresponding to the rms voltage of the output: the difference is the error, which is multiplied by the input signal. If the output signal becomes too small, the error becomes more positive, increasing the gain of the multiplier; if the output signal becomes too large, the error signal becomes less positive, decreasing the gain of the multiplier.

The difference circuit is diagrammed in Fig. 6.21. In order to control independently the set point and feedback sensitivity, separate amplifiers control the gains of both signals.

Modules with rms can be bought already assembled, or the designer can full-wave rectify and low-pass filter the output signal. The purpose of the rms circuit is to transduce the signal power into a proportional voltage and use that voltage as a monitor of signal power. Alternatively, it is possible to control the gain by some other signal parameter, such as heights of peaks, or power of a particular frequency component. The latter would be accomplished by interposing a bandpass filter between the output and the rms circuit. The peak-regulating mode would be implemented by interposing a peak detector, which samples successive peaks and smooths them, between the output and the rms circuit.

There are a number of AGC ICs available: the designer should check their prices and specifications before building his own.

Figure 6.20 An automatic gain control (AGC) circuit.

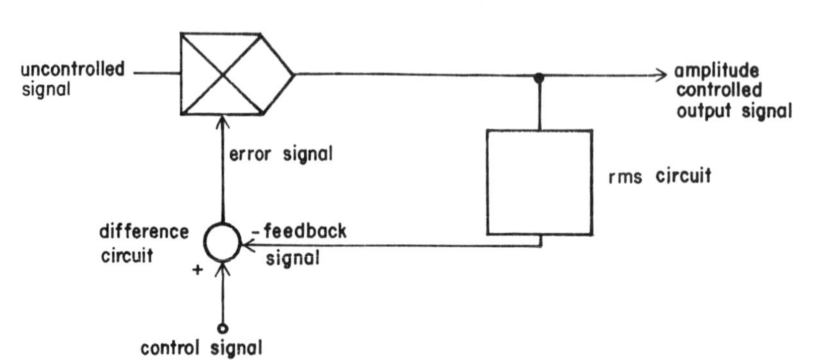

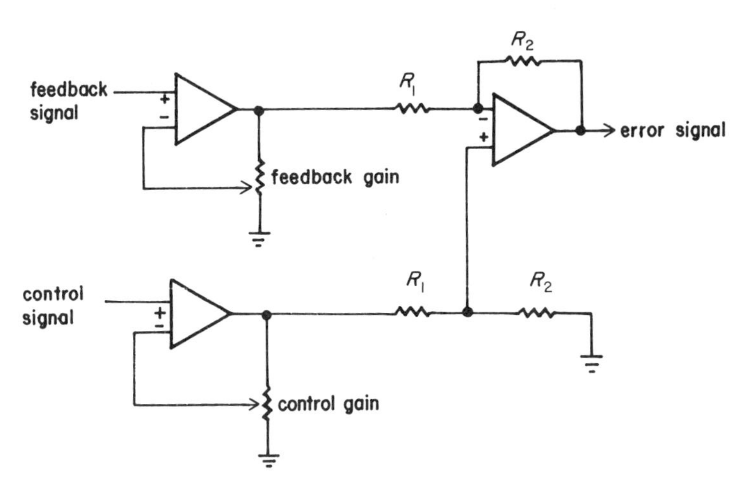

Figure 6.21 The difference circuit for the AGC of Fig. 6.20.

6.6.2 Automatic Frequency Control

There are a number of ways to control frequency: probably the best is with phase-locked loops, which we shall describe in Section 6.7. All must involve the use of a variable frequency oscillator whose oscillation frequency can be controlled by a control signal. Figure 6.22 illustrates a simple automatic frequency control (AFC) circuit. A voltage-controlled oscillator is driven by a signal which is derived from an input control signal and a feedback signal whose voltage is proportional to frequency. The circuit can be operated without the feedback, but then the VCO must be very stable in its frequency output.

Note the similarity of this circuit to an op amp in the negative feedback configuration, with the interpolation of a voltage-to-frequency and a frequency-to-voltage converter. This basic configuration can be used to control essentially any aspect of a signal (e.g., its frequency, amplitude, power, or other more subtle characteristics) or of any physical variable (position of a mechanical activator, intensity of a light, etc.). This general process is often referred to as *servo control*, and is a subject of widespread interest among engineers and scientists because of its enormous utility.

Figure 6.22 An automatic frequency control (AFC).

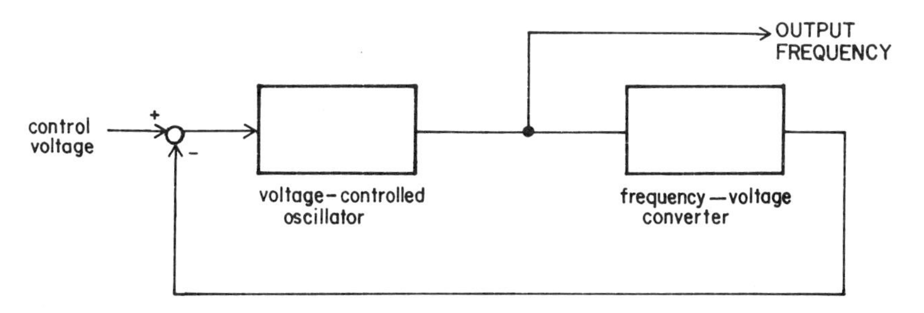

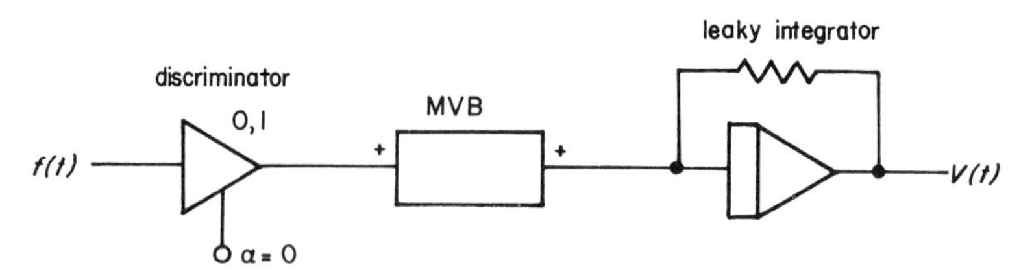

Figure 6.23 A frequency-to-voltage converter.

A frequency-to-voltage converter which could be used in Fig. 6.22 is diagrammed in Fig. 6.23. A discriminator produces uniform positive pulses of uniform duration. These are low pass filtered by a leaky integrator: the integration provides an output whose amplitude is proportional to the number of pulses per unit time, which is equal to the input frequency. The integrator is a leaky one to prevent infinite gain for dc, that is, to permit the output to fluctuate with frequency rather than cumulate total number of input oscillations.

6.7 Frequency Division and Multiplication

We have already described a method for frequency addition and subtraction, namely, amplitude modulation. Also, we have discussed means for the generation of an output frequency proportional to an input voltage, in the section on frequency modulation. There also exist techniques for the generation of a frequency which is a predetermined integral fraction or multiple of an input frequency, and hence for production of a frequency which is any rational multiple of an input frequency.

6.7.1 Frequency Division: TTL Signals

Figure 6.24 illustrates the process of frequency division. An input TTL signal is fed to a preset counter with count n: the output consists of a pulse every n input cycles. In many applications, however, a symmetric output is required, for which each phase of the output is the same number of cycles in length. To accomplish this, the frequency is first divided by $n/2$ with a preset counter set to a value of $n/2$, as in Fig. 6.25. The output consists of a pulse once each $n/2$ cycles. This output is not symmetric, one phase of the cycle typically lasting one input cycle or less and the other phase lasting $n/2-1$ cycles. Therefore this signal is fed to a divide-by-two counter to produce a symmetric square wave (each phase is the same number of input cycles in length).

But what if the frequency is to be divided by an n that is an odd number? The circuit of Fig. 6.26 permits generation of a symmetric divide-by-n output for both odd and even frequencies: pulses are triggered on both leading and trailing edges of the input, to double the number of cycles per second. Since each MVB fires on the leading or trailing edge, its pulse must be finished before the next trailing or leading edge: generally a minimal pulse width, such as 100 nsec, is used, which is sufficiently

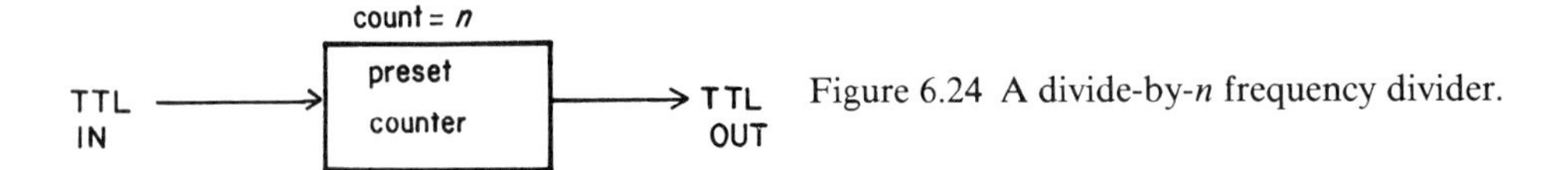

Figure 6.24 A divide-by-n frequency divider.

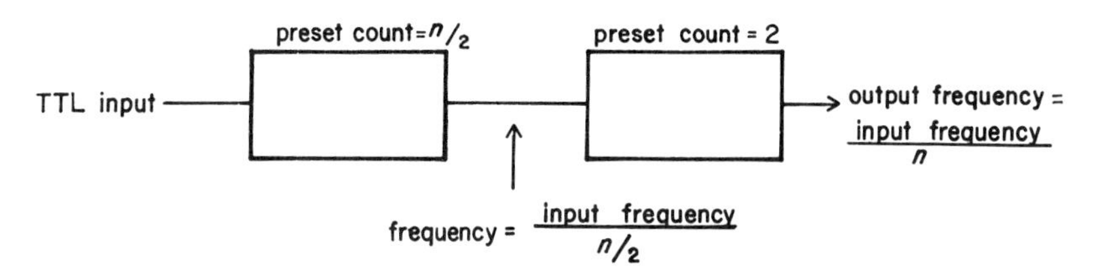

Figure 6.25 A symmetric frequency divider for TTL signals.

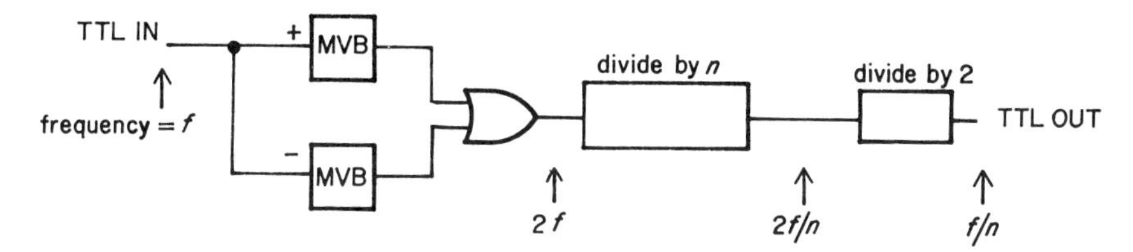

Figure 6.26 A symmetrical divide-by-n circuit for odd or even n.

long to trigger the divide-by-n preset counter and sufficiently brief to serve for very high input frequencies. For example, a 100-nsec pulse would be adequate for a symmetric square wave at the counter input, up to frequencies of $1/(4\cdot 100\text{ nsec})=2.5$ MHz. Then the pulse frequency is first divided by n to provide a divide by $n/2$, and then by 2, to provide an output frequency which is equal to the input frequency times $1/n$.

The preset counters used in these designs are free running: that is, they immediately begin a new count upon completion of a count. Figure 6.27 presents a

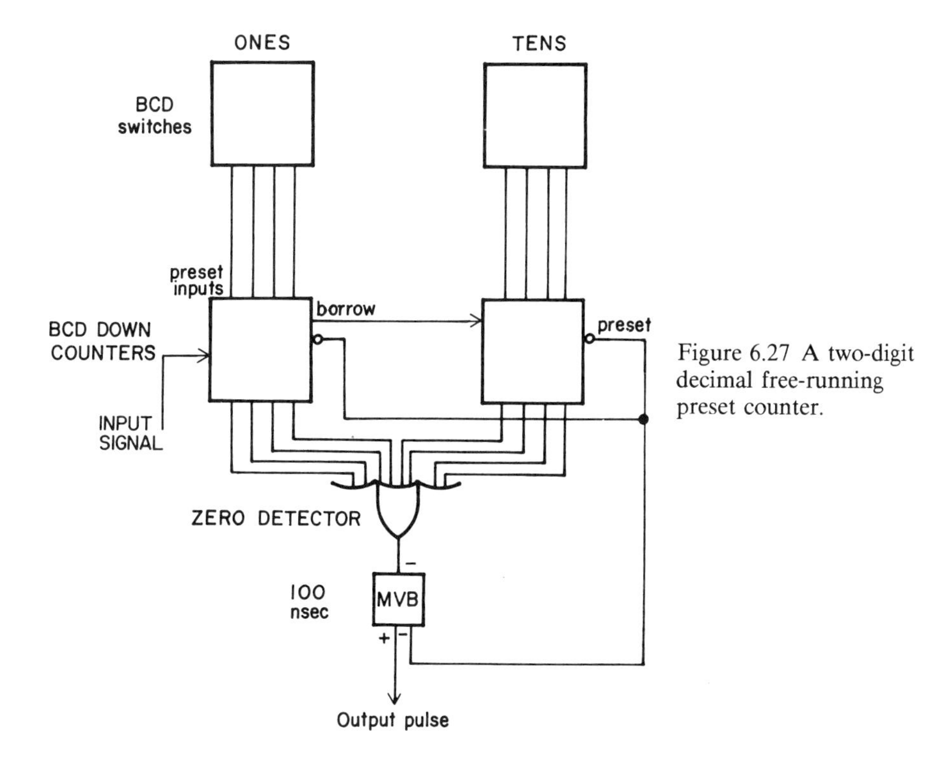

Figure 6.27 A two-digit decimal free-running preset counter.

configuration for a free-running preset counter which can be used for most applications. The specific device shown is a two-digit counter, but the design can be used for any number of digits. When the down counter reaches zero, the MVB loads the counter with the desired preset count n. Subsequent input events decrement the counter n times to zero, at which time the zero detector fires the MVB, starting a new preset count. The MVB pulse is used as the output as well as for resetting the count. The 100-nsec pulse is sufficiently long to reset the counter and sufficiently brief not to interfere with counting, for most applications.

6.7.2 Phase-Locked Loops

Phase-lock techniques have been in use since the 1930s, but only in the last decade or so have they been in common use in inexpensive consumer and laboratory products, and available in a number of varieties as ICs. They are instrumental in a number of processes, including generation of rational multiples of oscillator frequencies, FM demodulation, and extraction of low-level signals from noise.

A basic phase-locked loop is illustrated in Fig. 6.28. The device consists of a VCO, a TTL frequency divider, a phase comparator with inputs from the VCO and an external input, and a low-pass filter which removes ripple from the phase comparator output and feeds the resulting dc level to the control input of the VCO. The VCO has an RC pair used to determine its *neutral frequency*, i.e., that frequency at which it will oscillate if zero volts is applied to its input. It will deviate from this neutral frequency if a nonzero input voltage is applied: frequency deviation is proportional to voltage. Generally additional components are added to control the relation between frequency deviation and voltage.

The VCO typically has sine, triangle, and square wave outputs: the square wave is generally TTL compatible, or can easily be modified to be so.

The frequency divider counts the TTL square wave and produces a square wave whose frequency is $1/n$ times the VCO frequency and symmetric. This is compared with the input frequency by the phase comparator.

A number of designs for phase comparators exist. If both inputs are digital, which is easily arranged, an EOR gate can be used: When the inputs match, there is no output; during mismatch, the output is high. The more mismatch there is (the greater the difference in phase or frequency), the more time the output is high, since the

Figure 6.28 A phase-locked loop.

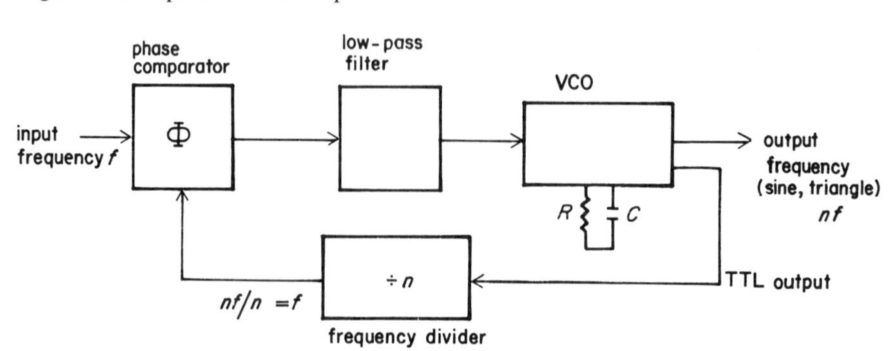

percentage of time the output is high is proportional to the difference in phase or frequency of the two inputs. Therefore, filtering the comparator output should produce a dc voltage whose level is proportional to the error in frequency. This actually is true even if the inputs are not symmetric square waves, but most phase comparators work best if their inputs are at least roughly symmetric. Some comparator designs permit the use of virtually any periodic waveform at the reference input.

The phase comparator is thus a means of measuring the difference between two frequencies. The fixed frequency fed into the phase-locked loop is the *reference* frequency, and the output of the frequency divider can be referred to as the feedback frequency. The dc level out of the low-pass filter is proportional to the difference between these, and it forces the VCO frequency to deviate from neutral toward the desired output frequency, which is the input frequency times n. When the output reaches this frequency, the two frequencies into the phase detector are equal, since the output frequency nf is divided by n by the frequency divider to produce $nf/n = f$, the input frequency. There will be a finite phase mismatch between the two comparator inputs, enough to force the VCO to produce an output frequency nf.

The phase-locked loop, or PLL, is thus a feedback loop which uses negative feedback to control frequency, much like an op amp. The PLL can be used to multiply frequencies and, since it has a VCO, it can produce triangle, sine, or square waves with frequency nf. Integrated circuits containing the VCO, phase comparator, and terminals for a low-pass filter are inexpensive, and only a few external components are needed to set the neutral frequency and frequency range of the VCO and to provide the desired frequency cutoff of the low-pass filter.

6.7.3 Frequency Division with PLLs

If a PLL is connected in the configuration of Fig. 6.29, it becomes a frequency divider. Here the input frequency is divided and the PLL is simply used to produce a phase-locked oscillation. Obviously the symmetric divide-by-n counters described earlier are a simpler method of TTL frequency division, but the PLL is perhaps the best device to use if a triangle or sine wave output is desired.

6.8 Frequency Synthesis

Note the effect of an input frequency divider (Fig. 6.29) and a feedback frequency divider (Fig. 6.28) on the ratio of output to input frequency. If the input divider factor is called n_1 or the feedback divider ratio is called n_2, the output frequency f_o is

Figure 6.29 Use of a phase-locked loop for frequency division.

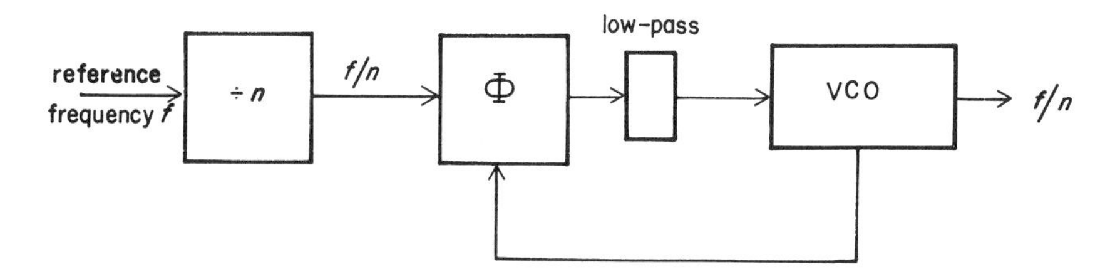

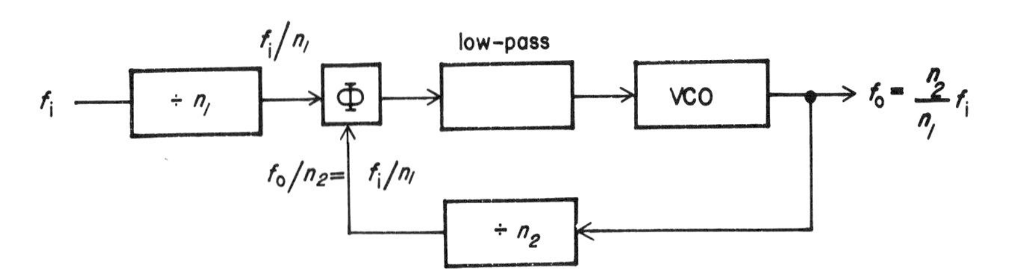

Figure 6.30 A PLL with input and feedback frequency dividers.

(see Fig. 6.29)

$$f_o = f_i/n_1 \tag{6.9}$$

or (see Fig. 6.28)

$$f_o = n_2 f_i \tag{6.10}$$

If both input and feedback frequency dividers are used, as in Fig. 6.30, we can solve for the output frequency in much the same fashion as we solve for output voltage in the negative resistive feedback op amp configuration. We know that, once the PLL is locked, the two input frequencies to the phase comparator are equal. One input is the input frequency f_i divided by the input count n_1 and the other is the output frequency f_o divided by n_2. Equating these gives

$$f_o/n_2 = f_i/n_1 \tag{6.11}$$

Solving for f_o gives

$$f_o = \frac{n_2}{n_1} f_i \tag{6.12}$$

Notice the very close analogy to the expression relating output to input for the negative feedback op amp configuration:

$$e_o = \frac{R_2}{R_1} e_i$$

Thus, we reach the very important generalization:

> The phase-locked loop with input and feedback frequency division can be used to multiply an input frequency by any rational fraction.

Given this rule, in principle we can synthesize any desired frequency using any standard, or reference, frequency, to any desired degree of precision. We shall give an example.

Example 6.1. Generation of frequencies from 1 to 99,999 kHz with 1-Hz increments, using 60-Hz line voltage as reference.

Figure 6.31 presents a circuit which will serve this purpose. The power line voltage is reduced and clipped for TTL compatibility and a divide-by-60 counter is used to produce a 1-Hz

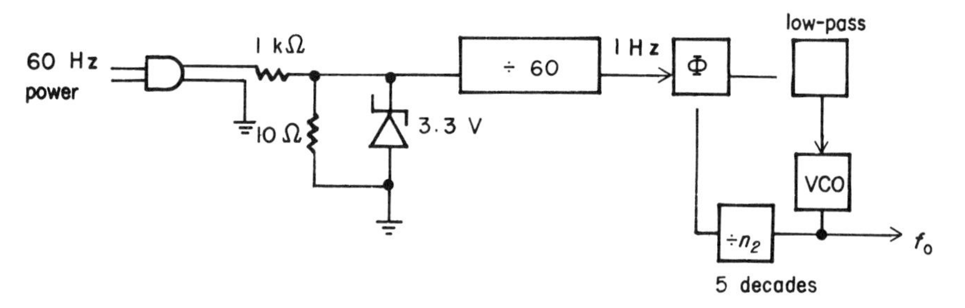

Figure 6.31 A simplified version of a frequency synthesizer for frequencies from 1 to 99,999 Hz with 1-Hz resolution.

reference. The PLL has a frequency divider which can be set to n_2. Any number of decades can be used for n_2: in this example, we want to use five decades, each programmable with a BCD switch for unit increments from 1 to 99,999 Hz.

No attempt is made to diagram the passive components needed for the VCO and low-pass filter. Actually, probably additional switching of these components would be required in order to vary the VCO neutral frequency, since there are probably no VCOs which can be locked over such a wide frequency range without varying the neutral frequency.

6.9 Triggerable Waveforms

In Chapter 5 we introduced the multivibrator, which is triggered on a leading or trailing edge, and which, once triggered, produces a pulse with a duration determined by some R and C. By varying R and C, the duration can be controlled. In an example we showed how one-shots can be interconnected to produce complex sequences of pulses. In this section we shall show how a wide variety of triggerable waveforms such as triangle waves, sawtooth waves, and gated sine waves can be produced easily and efficiently.

6.9.1 A General-Purpose Trigger Circuit

There are a number of modes in which triggerable circuits can be operated. Fig. 6.32 illustrates a simple trigger circuit which operates in six different modes. In the RECURRENT mode, it can be free running, like an oscillator. It can be triggered by a manual pushbutton, either to produce a single output waveform when triggered, or to run recurrently as long as the pushbutton is depressed. We shall refer to these modes as MANUAL SINGLE and MANUAL CONTINUOUS, respectively. If the circuit is to be electrically controlled, e.g., by TTL pulses, it may be triggered on leading or trailing edges (+TRIGGER or −TRIGGER, respectively), or it may be enabled to fire recurrently as long as the TTL input is on: we refer to this as the GATED mode.

The trigger circuit of Fig. 6.32 can be used to control triggerable waveforms, in each of the modes we have defined above. A flipflop is set by the triggering event, producing TTL OUT: this signal is on for the duration of the waveform and enables the generation of one repetition of the waveform. When the waveform is done, a TTL level change signals the end of the cycle, and a 100-nsec MVB is triggered to

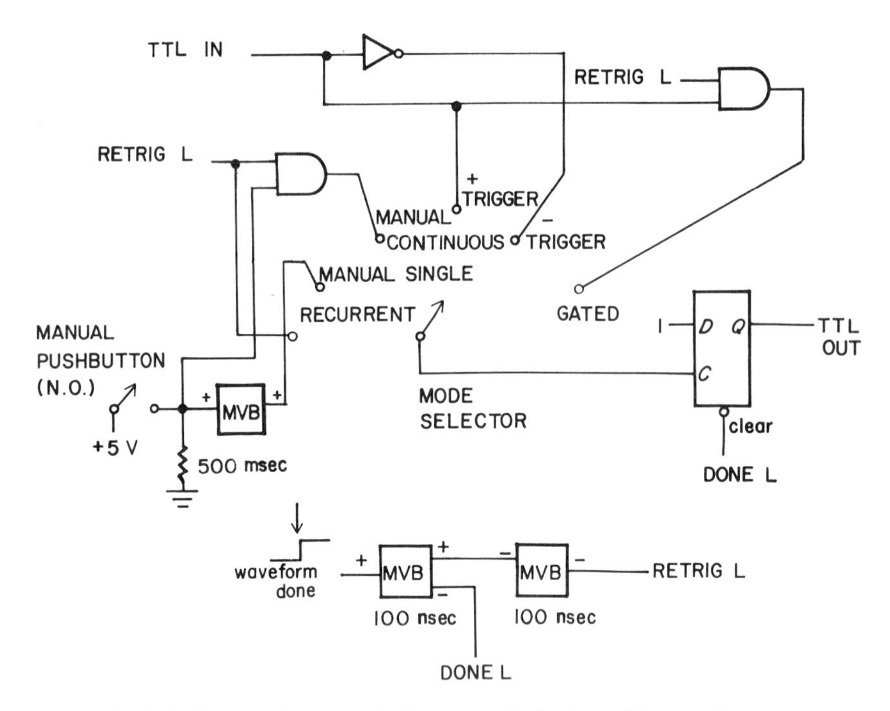

Figure 6.32 A simple trigger circuit for control of triggerable waveforms.

produce DONE L, which clears TTL OUT, and DONE H, which fires a delayed 100-nsec pulse, RETRIG L.

In RECURRENT mode, RETRIG L is connected to the flipflop: When the connection is initially made, the normally high level of RETRIG L triggers the flipflop directly. At the end of each waveform, RETRIG L goes low and then high again, retriggering the flipflop.

In MANUAL SINGLE mode, a normally open pushbutton switch output, held at logical 0 by a resistor to ground, sets the flipflop when the button is pressed: a direct +5-V connection is established, reversing the normally low switch output. Note that a 500-msec MVB is used to prevent retriggering due to switch contact bounce. It is drawn as having a 500-msec pulse duration, but any convenient duration greater than the duration of the buttonpress is adequate.

In MANUAL CONTINUOUS mode, the TTL level out of the pushbutton is ANDed with RETRIG L, which is normally high. After DONE L ends, having cleared the flipflop, RETRIG L is triggered, and passed as a low pulse to the flipflop clock input: the flipflop is triggered on the positive transition at the end of RETRIG L. Thus, the flipflop is triggered by one of two events: initial closure of the pushbutton switch passing a positive transition through the AND gate, or at the end of RETRIG L. Hence the flipflop will be retriggered after each waveform until the pushbutton is released.

In +TRIGGER mode, the flipflop is triggered on each positive transition of TTL IN. In −TRIGGER mode, it is triggered on each negative transition of TTL IN, by interposing an inverter between TTL IN and the flipflop clock.

In GATED mode, the flipflop is triggered by positive transition of TTL IN, and by RETRIG L as long as TTL IN is high.

We have not explained what triggers the DONE MVB. As we describe each type of triggerable waveform, we shall show how the end of each is detected.

6.9.2 Triangle Wave

We shall define a triggerable triangle wave as going from zero initial state to a positive peak, back through zero to a negative peak, and back to zero.

A *serrasoid*, or triangle wave, is generated with the circuit of Fig. 6.33. It is the same as the triangle wave oscillator described earlier except for a few modifications: (1) A switch has been interposed between the discriminator and the integrator, permitting the second op amp to operate as an integrator (N.O. position) or a unity gain follower with zero input (N.C. position). The switch is controlled by the TTL OUT from the trigger circuit of Fig. 6.32. (2) A circuit to detect completion of the triangle wave consists of a discriminator which fires an MVB when the triangle wave crosses zero from positive to negative, setting a flipflop which was initially cleared, which then enables a WAVEFORM DONE level when the triangle wave returns to zero from the negative side. This rather elaborate system ensures that WAVEFORM DONE is not generated at the beginning of the serrasoid, as it first goes positive.

The serrasoid can be put through wave-shaping circuitry to produce a sine wave, if desired.

The variable R is a potentiometer with maximum/minimum ratio of slightly over 10; the variable C is a set of capacitors on a decade switch for tenfold step variations.

The ascending and descending ramps can be independently controlled by using two variable Rs and Cs with steering diodes, as described in the earlier discussion of the serrasoid oscillator.

The triangle wave can be generated singly (MANUAL SINGLE, ±TRIGGER) or recurrently (RECURRENT, MANUAL SINGLE, or GATED).

Figure 6.33 A triggerable serrasoid (triangle wave) generator.

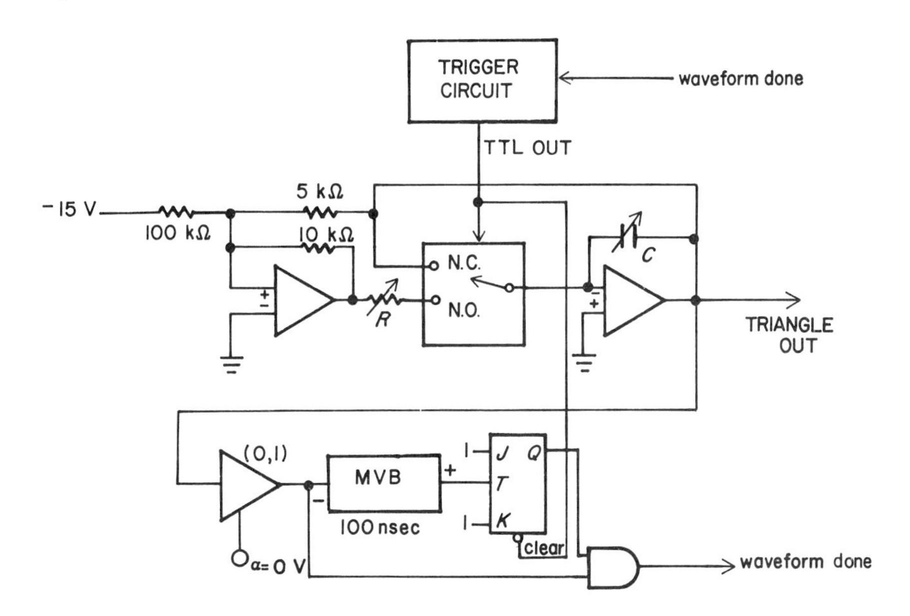

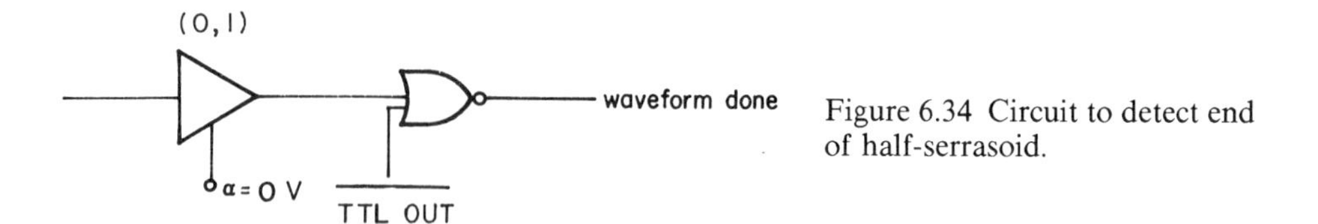

Figure 6.34 Circuit to detect end of half-serrasoid.

To generate only a half-cycle (positive ramp to peak and negative ramp to zero), the WAVEFORM DONE circuit is simplified: the inverse of TTL OUT is NORed with the output of the discriminator. This level is prevented from going high until after the waveform is started, and then goes high as soon as the serrasoid reaches zero from the positive side.

The waveforms at various points in the circuits of Figs. 6.33 and 6.34 should be reconstructed by the student as an exercise.

6.9.3 Sawtooth Wave

The sawtooth wave starts at zero, goes to a maximum, and is reset rapidly to zero. The very simple circuit of Fig. 6.35 accomplishes this. The switch enables the integrator when TTL OUT goes on: the ramp proceeds until a level of +10 V is reached, at which time the WAVEFORM DONE signal is generated by the discriminator. This clears TTL OUT, causing the switch to short the capacitor and force the output rapidly back to zero.

6.9.4 Digital Control of Waveform Timing: Pulses

Digital control of waveforms is becoming more and more popular: with only one timing standard, a clock, any of a wide variety of waveforms can be produced using digital control. Such control methods have the advantage of great stability and precision, ability to interact with other digitally controlled waveforms, and ease of

Figure 6.35 A triggerable sawtooth generator.

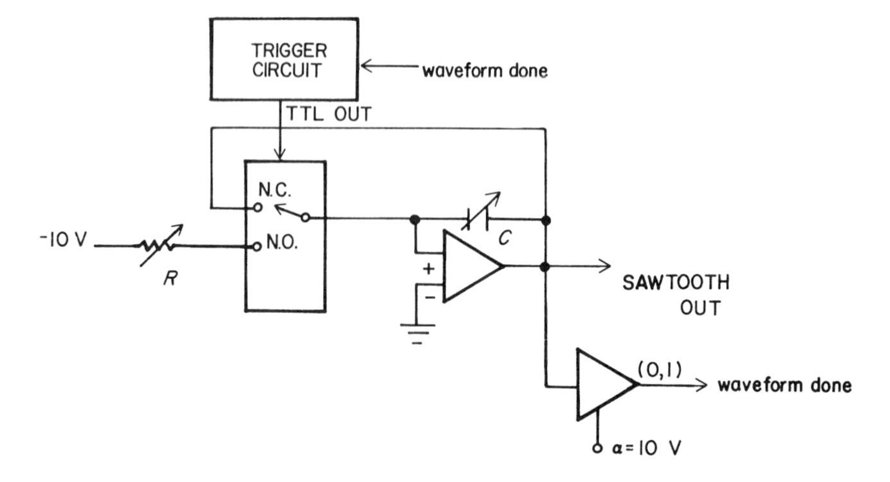

design and maintenance. We shall describe digital control of pulse duration in this section. Digital noise generation will be covered later in this chapter.

A digitally controlled TTL pulse generator is diagrammed in Fig. 6.36. The device consists of the familiar preset counter constructed from programmable down counters and BCD switches. TTL OUT, when on, enables the counters; when off, it presets them to the desired value from the switches. When enabled, the counters count down to zero, at which time *waveform done* goes on, clearing TTL OUT.

The BCD switches control the pulse duration, as a multiple of the clock input to the down counters. The clock input to the counters is itself decade variable, consisting of the input of a decade multiplier switch which selects outputs from any of a string of symmetric divide-by-10 frequency dividers. Thus the pulse duration is any value from 1 to 999, times 1 μsec to 1 sec.

Note that, if an external TTL signal is used as the clock input, durations can be controlled in terms of number of external TTL events. This is very useful in programming the number of events to be produced, for example the number of pulses in a pulse train.

6.9.5 Analog Pulses

The simplest way to generate pulses of controllable duration, amplitude, polarity, and dc offset is to use TTL one-shots to control timing, and use the TTL pulse to generate an analog pulse which can then be conditioned with analog circuitry. The

Figure 6.36 A digitally programmable triggerable pulse generator.

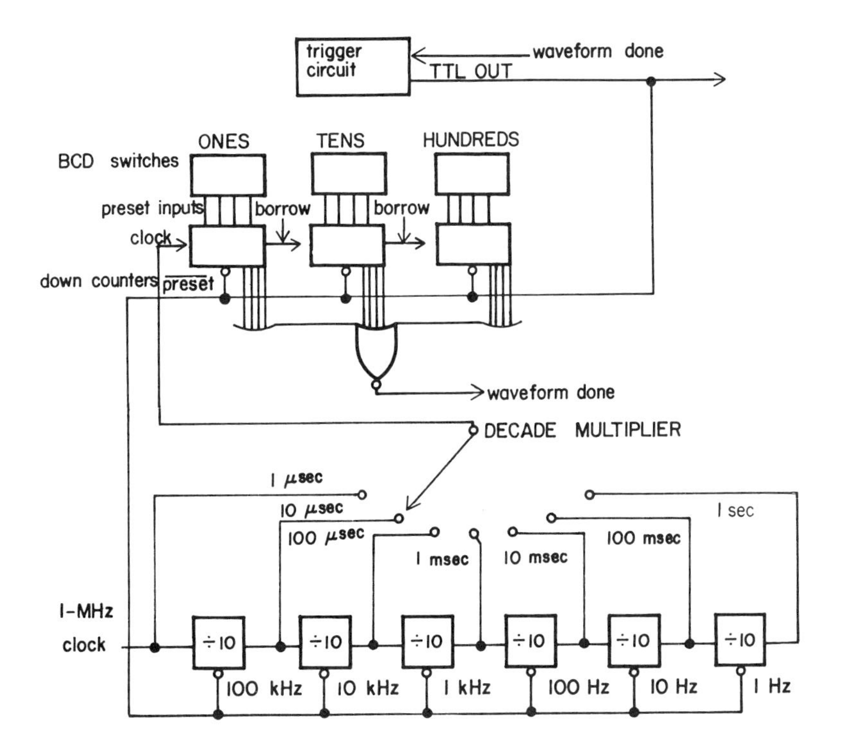

circuit of Fig. 6.37 uses a TTL pulse to generate positive or negative analog pulses, which then can be conditioned for variable amplitude, polarity, and dc offset.

The trigger circuit is simplified over the usual trigger circuit, in that the flipflop is replaced with the MVB with analog control of pulse duration using variable R and C. The TTL OUT generated by this pulse generator triggers RETRIG L on its trailing edge, providing a delayed retrigger pulse for RECURRENT, MANUAL CONTINUOUS, and GATED modes.

The TTL pulse drives a voltage discriminator with $\pm V_{max}$ outputs: Resistances are adjusted for thresholds of about 1.0 V and 2.0 V, compatible with TTL pulses. The $\pm V_{max}$ discriminator output is summed with an offset voltage which is used to bring the quiescent voltage to exactly 0, and a variable negative feedback resistance permits varying pulse amplitude from 0 to $-V_{max}$. This negative pulse is summed with a dc offset voltage for baseline control in a negative feedback op amp configuration, with polarity switching accomplished by bringing the pulse to the negative or positive summing junction of the output op amp.

The zero baseline is adjusted by switching polarity repeatedly while adjusting the zero baseline pot: When polarity flips without influencing the baseline, the zero baseline adjustment is completed. Now amplitude, polarity, and dc offset can be adjusted independently without affecting each other.

This type of noninteracting adjustment of amplitude, polarity, and baseline can be added to the output of any waveform generator. We shall not include it in the discussions of other waveforms, but the reader should keep in mind the possibility of such independent controls, which are often useful.

Figure 6.37 An analog pulse generator with independently controlled duration, amplitude, polarity, and dc offset.

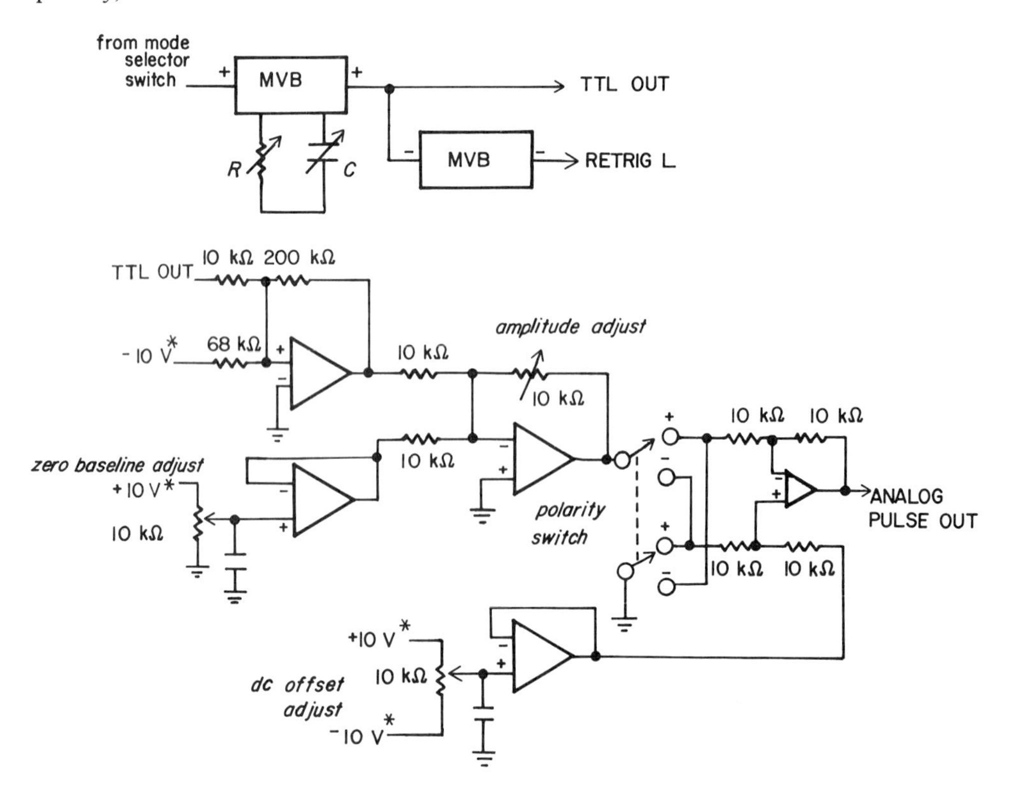

6.10 Complex Waveforms

Waveforms of great complexity can be synthesized by summing simpler waveforms, each triggered at the appropriate time. For example, a trapezoidal waveform can be synthesized using two sawtooth generators and two pulse generators, as shown in Fig. 6.38. The up ramp is a sawtooth; the TTL OUT pulse from the sawtooth generator triggers an analog pulse on its trailing edge, to produce the plateau portion, and the TTL OUT from the pulse generator produces a pulse and sawtooth for the down ramp. Since sawtooth 2 and analog pulse 2 must be the same length, analog pulse 2 should be derived directly from the TTL OUT of the sawtooth, rather than with a separate MVB.

Once amplitudes and offsets of all the trapezoid components are properly adjusted the entire trapezoid can be fed to a waveform-conditioning circuit for independent control of amplitude, polarity, and baseline.

Digital wave forms, e.g., TTL pulse trains, can be synthesized readily using the trigger circuit with its different modes. Figure 6.39 illustrates the use of triggerable preset counters for generation of two signals, each of which consists of a pulse train

Figure 6.38 Synthesis of a trapezoid from sawtooth waves and pulses.

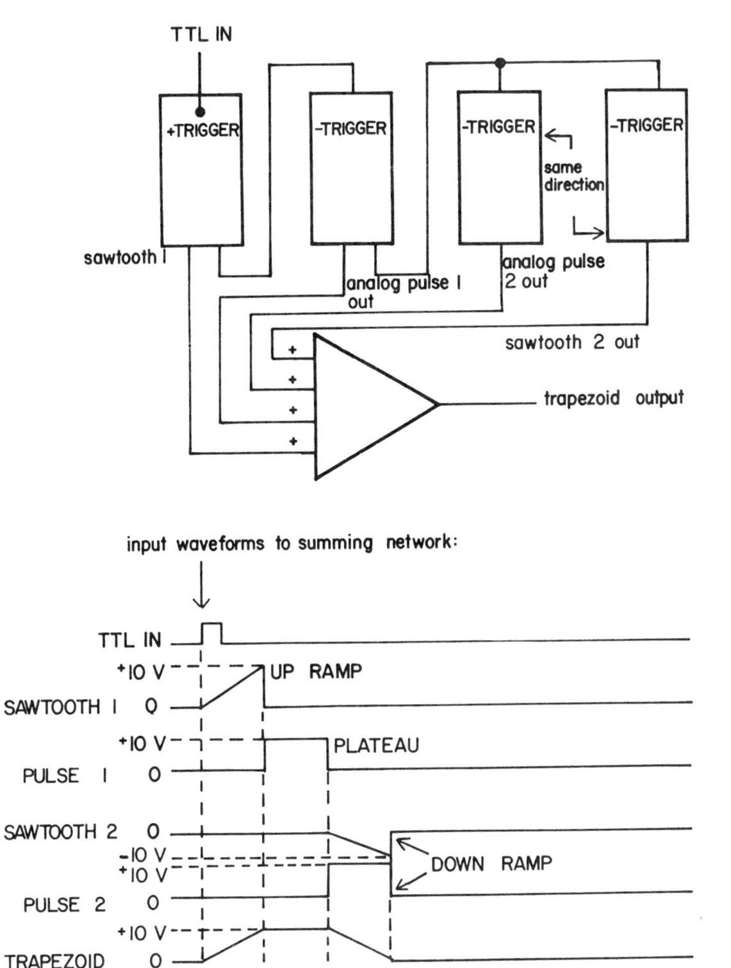

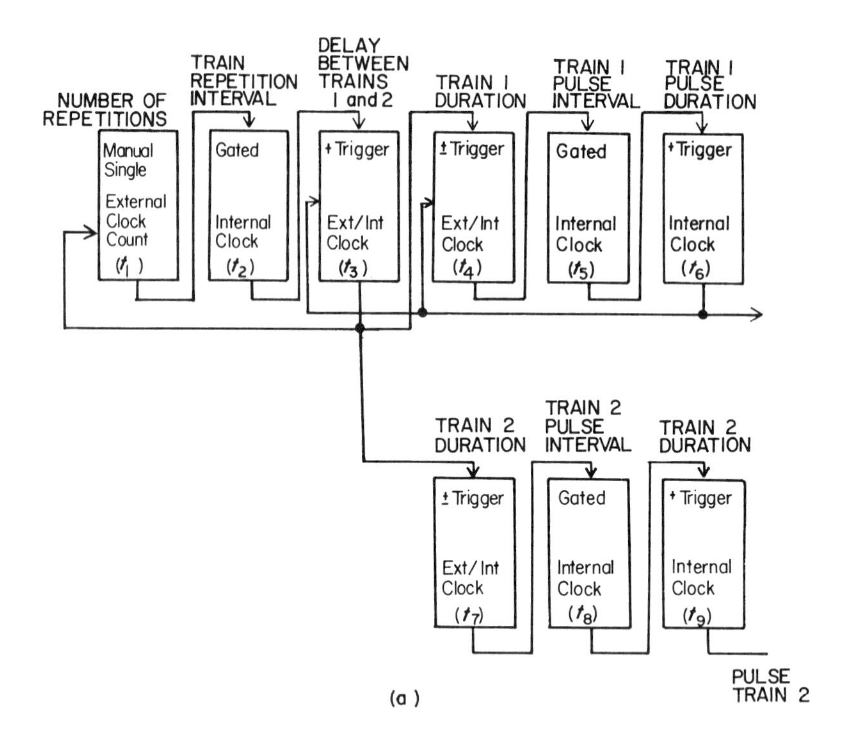

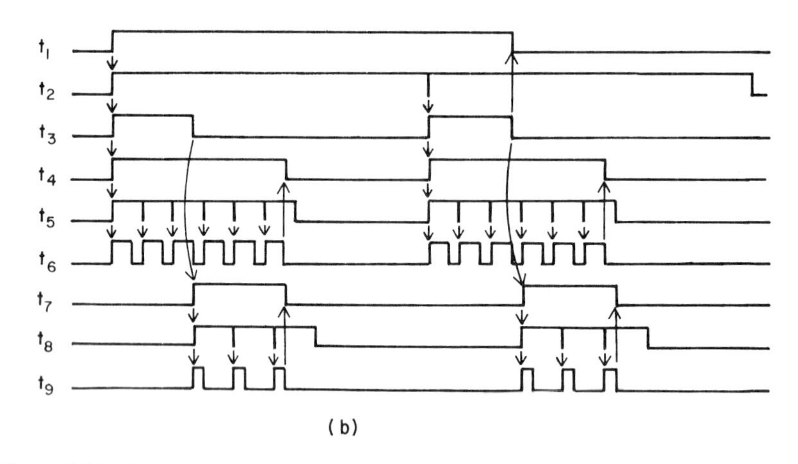

Figure 6.39 Generation of pulse trains with digital counters. Counter configuration of **(a)** produces timing sequences diagrammed in **(b)**. For example of **(b)**, counters t_1, t_4, and t_7 all use external count inputs with preset counts of 2, 5, and 3, respectively. Arrows in timing diagram indicate which events cause triggering and completion of counts.

with independent control of repetition rate, pulse width, and train duration. The two trains are controlled via a series of nine pulse generators, to control number of trains, repetition rate, and delay between trains. The circuit of Fig. 6.39a produces waveforms of the type illustrated in Fig. 6.39b. Note that some pulse generators have EXT CLK or EXT/INT CLK. These are preset counters, similar to those of Section 6.7.4. By using them to count the number of pulses produced by the generators they gate, they can be used to control the number of pulses gated. Thus, the NUMBER OF REPETITIONS counter counts the number of DELAY pulses; after the last desired DELAY pulse, it goes off, stopping generation of further DELAY pulses, and hence further outputs for TRAIN 1 and TRAIN 2. Similarly, DELAY can be timed either as length of time, with INT CLK, or as number of pulses in TRAIN 1. TRAIN 1 DURATION and TRAIN 2 DURATION can both be controlled in the same way, either counting the internal clock or counting number of pulses in TRAIN 1 or TRAIN 2, respectively.

By using ramps or triangle waves to frequency or amplitude modulate sine wave generators, it is possible to generate "swept" amplitudes or frequencies of sinusoidal oscillations, varying their amplitudes or frequencies through a time-varying series of different values. This is particularly valuable in the testing of equipment: If the modulating sawtooth or triangle wave is used to sweep an oscilloscope beam (x axis), the response of a system to the modulated sine wave can be displayed as the vertical deflection (y axis). This allows the display of the system's frequency or amplitude response as the envelope (the waveform consisting of a line connecting

Figure 6.40 A "tone pip" generator, used to trapezoidally modulate a sine wave. Trapezoid ramps and plateau are specified in terms of number of thousandths of a cycle; waveform always starts on a positive zero crossing.

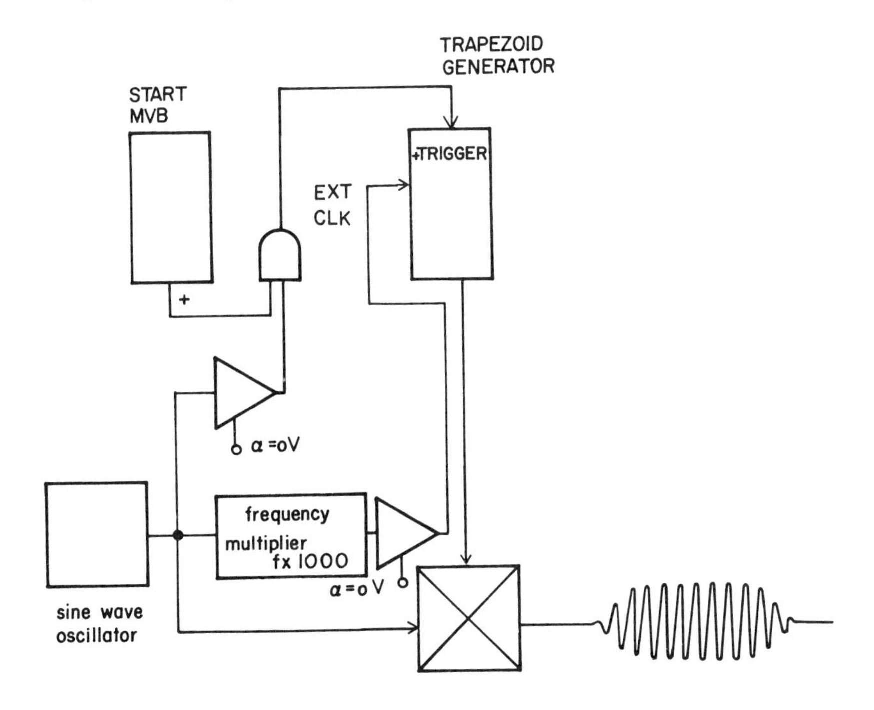

successive positive peaks and another connecting successive negative peaks) of the displayed waveform. This is particularly effective using a storage oscilloscope.

Figure 6.40 illustrates the generation of a "tone pip," a trapezoidally modulated sine wave which always starts on a positive-going zero crossing and ends on a negative-going zero crossing. The number of cycles in the tone pip is digitally controlled. Before implementating such a device in this form, the designer should consider using some of the monolithic IC function generators which are available commercially, since they are simpler. With these waveforms, as well as random noise, the designer has at his or her disposal the great majority of signal waveforms, or the components for their synthesis, used in scientific research.

6.11 Noise

Noise has been discussed in Chapters 2 and 3, where its statistical properties, spectral composition, naturally occurring forms, and use in characterizing systems were discussed. We shall present here methods for generation of gaussian bandpass noise and for generation of Poisson point processes. We shall also describe methods for generating very good approximations of these two types of signal using a totally deterministic device, a seemingly impossible technique which in fact produces a signal which is often more useful for testing systems than true or "natural" random noise.

6.11.1 Analog Noise Generation

6.11.1.1 Gaussian Bandpass Noise

The most common method of generating gaussian bandpass noise is to transduce a naturally occurring random process into a corresponding electric signal, or to use naturally occurring thermal noise. The bandpass characteristics of the resulting signal are thus largely determined by the electronics used for transduction and amplification, since a properly selected natural process will possess a bandpass which is much wider than is needed for most scientific applications.

Examples of the first type include the use of "shot effect" noise, produced by bandpass filtering the output of a Geiger counter or the signal on the plate of a vacuum tube with constant grid voltage. The stochastic point processes, in this case the random decay of radioactive nuclei and the random arrival of electrons at the plate of the vacuum tube, produce a signal which is truly random, has a guassian voltage distribution, is stationary, and has a wide spectrum. With careful amplification and filtering, the spectrum can be shaped to any desired pattern.

Most commonly, thermal noise or the noise across a diode junction is used. Inexpensive modules are now available which provide stable sources of noise with all the properties mentioned above. The designer should therefore simply select such modules according to desired signal power and bandwidth, rather than construct his own.

6.11.1.2 Poisson Pulse Trains

Shot noise, or *shot effect* noise, is the signal (current) produced at the plate of a vacuum tube as it is randomly bombarded by electrons. Shot noise is a *Poisson process*, that is, with a Poisson-shaped interevent interval distribution. Therefore it is only necessary to produce a uniform pulse on each occurrence of a shot effect event. Thus, for example, radioactive decays, detected with a Geiger–Muller tube, can be used to trigger a pulse generator to produce a Poisson pulse train. Such a pulse train has a Poisson-distributed interval distribution, and, as would be expected from the fact that gaussian noise can be generated from the same process, it has a flat spectrum up to a corner frequency which is determined by the width of the uniform pulses. The mean rate of the Poisson train can be varied by varying the rate of events detected by the Geiger counter, i.e., by varying the distance of the radioactive source from the detector or the size of aperture in a shield placed between the source and the detector. *Frequency division of the detector output will not yield a Poisson process*, but rather, the greater the divisor, the closer the process will approach a *normal* process, with a normal distribution of interpulse intervals..

Discrimination of white noise, diagrammed in Fig. 6.41, can be used to generate a good approximation of a Poisson process. The noise must have a high bandpass and the output pulses must be brief relative to the shortest desired interval. The mean rate of the Poisson process is varied by low-pass filtering the noise: the lower the bandwidth, the lower the mean rate. The interval statistics of such a pulse train should be examined to verify a good approximation to Poisson statistics.

6.11.2 Digital Generation of Pseudorandom Noise

There are a number of methods currently in use for production of digital signals which are very good approximations of random telegraph signals (that is, signals with binary voltage and random transition times). Some of these require computer-generated sequences of numbers; while these are the easiest to use for computer modeling of random processes, there is a much simpler and less expensive way to generate pseudorandom binary sequences with a minimum of digital ICs. Such sequences are called pseudorandom because (a) they are generated by deterministic devices and are therefore predictable and eventually must repeat their sequence; (b) they nevertheless display many statistical processes of stationary random processes, such as roughly constant mean and variance and zero serial correlation. This method uses shift registers to generate pseudorandom binary sequences, which in turn may be used to produce bandpass gaussian noise or any of a variety of pseudorandom point processes. The general scheme is illustrated in Fig. 6.42.

The n-bit shift register is driven by a clock with frequency $f = 1/\Delta t$, where Δt is the interval between clock pulses. It can be preset to an initial value by a parallel load pulse, which loads the contents of n switches set to 0 or 1. Any initial value, except all bits equal to zero, is permissible. For certain shift register lengths, it can be shown that at least one feedback algorithm can be found such that all possible combinations of n bits (except all zeros) occur in the shift register before any particular value is repeated. The feedback algorithm simply consists of addition of certain parallel output bits modulo 2. The interested reader should consult Golomb (1967) for a proof of this fact. Zero is an illegal number

Figure 6.41 Production of a Poisson pulse train from gaussian noise using a discriminator and multivibrator: the pulse width must be small relative to the period of the highest frequency component in the bandpass noise.

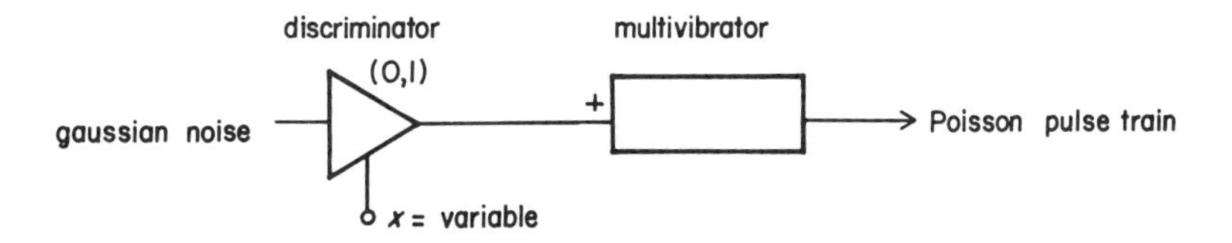

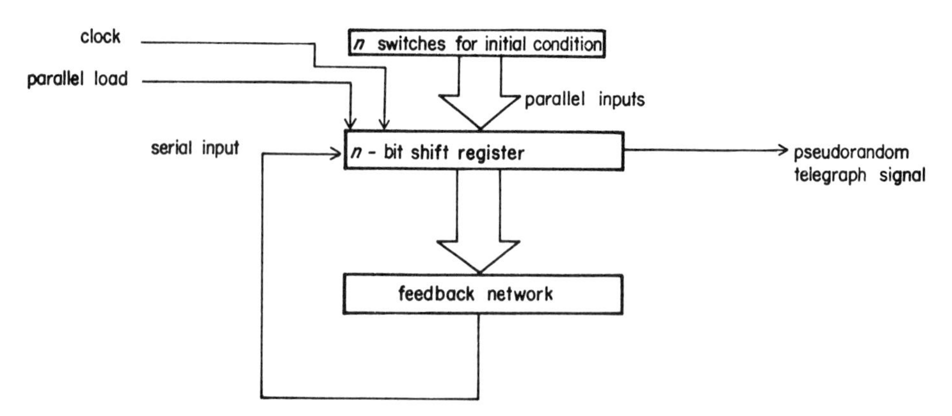

Figure 6.42 A pseudorandom noise generator using a shift register.

because once the shift register contents equal zero the feedback would be zero, and the shift register contents would remain at zero forever.

Since the shift register has n bits, a sequence in which all possible arrangements of 1's and 0's (except all 0's) occur must have a number of steps equal to

$$p = 2^n - 1$$

since there are 2^n possible numbers represented by n bits, and 0 is excluded. Such a sequence is called a maximal length sequence, and has the property that its contents, viewed as an n-bit number, vary randomly from one clock state to the next, with the qualification that once all possible values except zero have occurred, the sequence repeats. This means the signal at the serial output (bit 33) is periodic, with period $(2^n - 1)\Delta t$. The signal is thus referred to as a *pseudorandom binary sequency* or *pseudorandom telegraph signal*. The output bit can be said to simulate a series of *Bernoulli trials*, similar to coin flips. The intervals between transitions approximate the binomial distribution, which for large n approximates a Poisson process. These events can be used to produce pulses, which can be used to approximate a Poisson process. Such a pulse train is called a *pseudorandom process*, specifically a *pseudo-Poisson pulse train*.

Sequence length can be quite large for a relatively short shift register. For example, $p = 8,589,934,591$ for $n = 33$. Thus, even if $\Delta t = 1$ μs, for 1 million shifts per second, the sequence will run for 8,589 sec, or almost $2\frac{1}{2}$ hours, without repeating.

Figure 6.43 presents a design for a 33-bit shift register which will produce a maximum sequence. Feedback is from bit 33 and from bit 13. By gating the clock, the sequence can be started and stopped under external control. Reset to the initial value is controlled either manually or electrically. If desired, the initial value can include a bit which is always equal to 1, to prevent accidental loading of an initial value equal to 0.

The pseudorandom binary sequence produced by the circuit of Fig. 6.43 has the output waveform of Fig. 6.44a and the cross-correlation function of Fig. 6.44b. The corresponding power spectrum is illustrated in Figure 6.44c. The line spectrum has a spacing of $1/(2^n - 1)\Delta t$, or approximately $1/(8 \times 10^9)10^{-6} = 0.00125$ Hz, a spacing so dense that no ordinary passive or active RC filter could separate the individual frequencies, except very near dc. *Therefore it is for all practical purposes a true bandpass white noise, adequate for testing the frequency response of any analog system and many digital signal processors.*

The pseudorandom binary sequence can be bandpass filtered to produce any smaller power spectrum. Typically, the noise is confined to a bandwidth which is greater than the

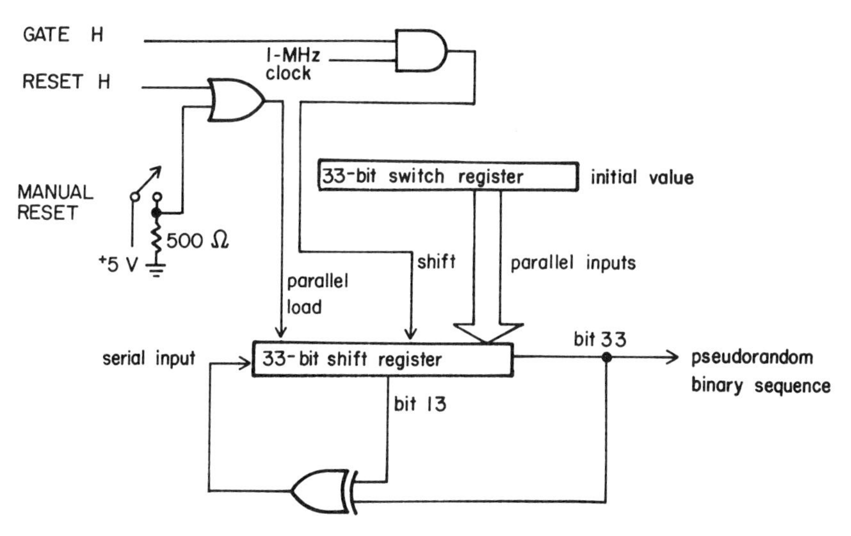

Figure 6.43 A 33-bit maximum sequence shift register noise generator.

Figure 6.44 **(a)** Output of the pseudorandom telegraph signal generator of Fig. 6.42. **(b)** Autocorrelation function of the signal of Fig. 6.43a.

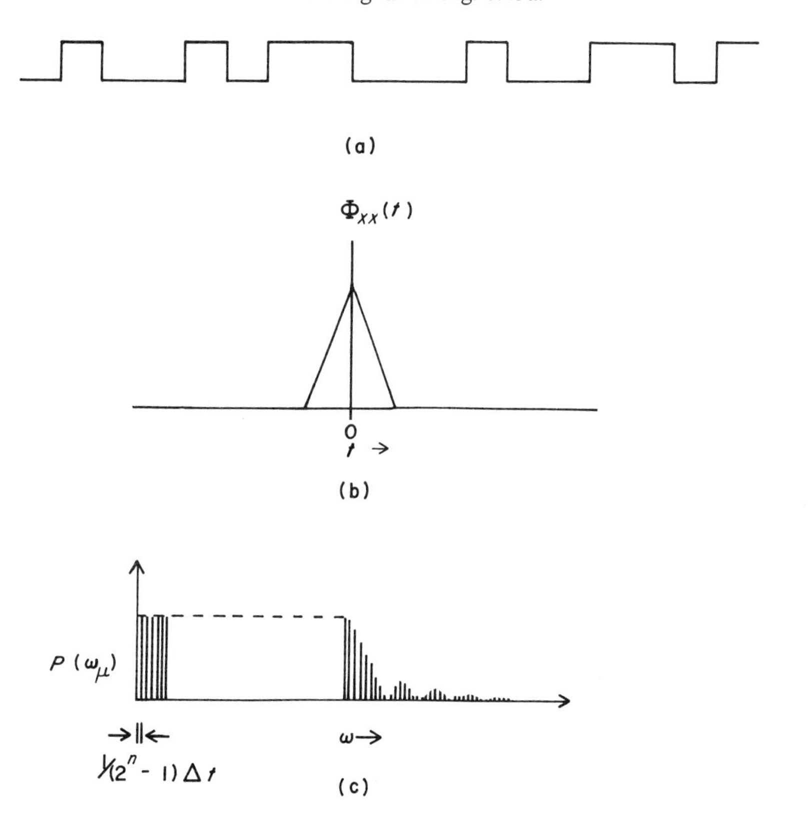

bandwidth of the system being tested, but with most of its power within the system bandwidth. Obviously, as has been emphasized in Chapter 3, the noise test signal has the advantage of testing all frequencies simultaneously. The use of a repeatable noise permits the measurement of an average system response, thus facilitating the determination of the causal and random components of the system response. Noise is also the best input for determining the nonlinear characteristics of a system under test.

6.12 Filters

Generally, filters are thought of as signal-conditioning devices used in measurement and recording apparatus. In fact, signal conditioning is used in both signal generation and signal measurement and recording. Thus, for example, filters are used to control the bandwidth of noise in noise generators, as well as for modification of other waveforms containing more than one frequency component.

All filters can be considered to be means of altering the relative amplitudes and phases of different frequency components of signals. Passive filters have been considered in earlier chapters. In this chapter, we shall limit our discussion to active filters, giving only a few examples of the most commonly used devices. The reader is referred to texts listed at the end of the chapter for more extensive treatments of filter design.

By cascading filters, a wide variety of spectrum-shaping devices are possible. We shall describe low-pass, high-pass, and notch filters here. The reader should use the more advanced references at the end of the chapter if more details are required.

6.12.1 Low-Pass Filters

The idealized low-pass filter for most applications is one which does not modify a signal's frequency components below its cutoff (corner) frequency, and which passes no frequency components above its corner frequency. Such a device has a rectangular frequency response function and is referred to as an ideal low-pass filter.

One approximation to the ideal low-pass filter is the Butterworth filter, illustrated in Fig. 6.45. The filters illustrated are first and second order: for each order n, a 6-dB/octave slope is obtained on the "shift" of the frequency response curve: thus, a fourth-order filter would provide a 24-dB/octave rolloff. Filters can be cascaded (placed in series) to realize any desired order, the resultant being the sum of the orders of the component filters: thus, three identical second-order filters in cascade produce a sixth-order filter.

Note that the first-order filter is just an RC filter with an op amp follower output. R and C are selected according to the desired cutoff frequency ω, where

$$\omega = \frac{1}{\tau} = 2\pi f = \frac{1}{RC} \tag{6.13}$$

The input impedance Z of the circuit is selected as the first step in the design process. Than the values of the components R, R_2, C_1, and C_2, are determined from

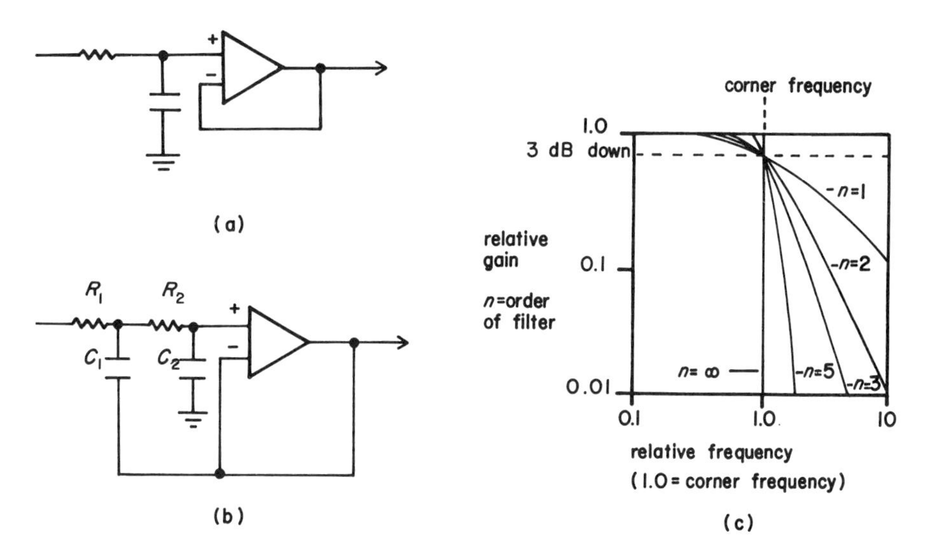

Figure 6.45 The low-pass Butterworth filter: **(a)** first order; **(b)** second order; **(c)** frequency response.

the following formulas:

$$C_1 = \frac{5}{2\pi f_c Z} \quad \text{farads} \tag{6.14}$$

$$C_2 = \frac{1}{2\pi f_c Z} \quad \text{farads} \tag{6.15}$$

$$R_1 = 1.25Z \quad \text{ohms} \tag{6.16}$$

$$R_2 = 0.16Z \quad \text{ohms} \tag{6.17}$$

where f_c is the cutoff frequency.

To vary the cutoff frequency, the capacitors may be switch selectable and/or R_1 and R_2 may be potentiometers, as illustrated in Fig. 6.46. The capacitors are typically selected at decade intervals for decade variable range selection, and the resistors are selected for a 1–11-fold variation by making a series combination of a

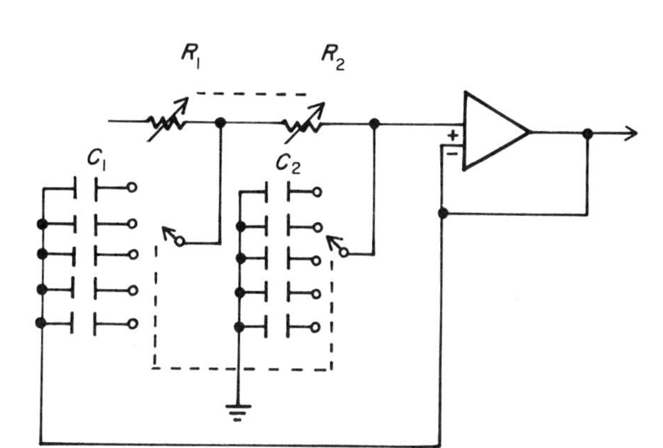

Figure 6.46 Use of ganged potentiometers and decade variable C values to vary f_c, the corner frequency of the Butterworth low-pass filter.

fixed resistor and a potentiometer of ten times as large maximum resistance for both R_1 and R_2. The capacitor switches and the potentiometers are ganged in pairs for covariation of R_1 with R_2 and C_1 with C_2.

If only switches are used to select fixed cutoffs, decade multiples of 1 and 3 or of 1, 2, and 5 are commonly used.

Other approximations to the ideal low-pass filter exist: the references at the end of this chapter go into some detail on the available design.

6.12.2 High-Pass Filters

The high-pass Butterworth filter is illustrated in Fig. 6.47. Once again, the design process begins with selection of appropriate Z. R and C are selected for the one-pole design the same way as for the low-pass filter [Eq. (6.13)]. R_1, R_2, C_1, and C_2 for the 2-pole design selected according to the following formulas:

$$C_1 = C_2 = \frac{1}{2\pi f_c Z} \quad \text{farads} \tag{6.18}$$

$$R_1 = \frac{\sqrt{2}}{2} Z \quad \text{ohms} \tag{6.19}$$

$$R_2 = \sqrt{2}\, Z \quad \text{ohms} \tag{6.20}$$

Variation of f_c can be accomplished, as before, by covarying C_1 and C_2 or R_1 and R_2, as was done for the low-pass filter in Fig. 6.46.

Figure 6.47 High-pass Butterworth filters: **(a)** first order; **(b)** second order; **(c)** frequency response.

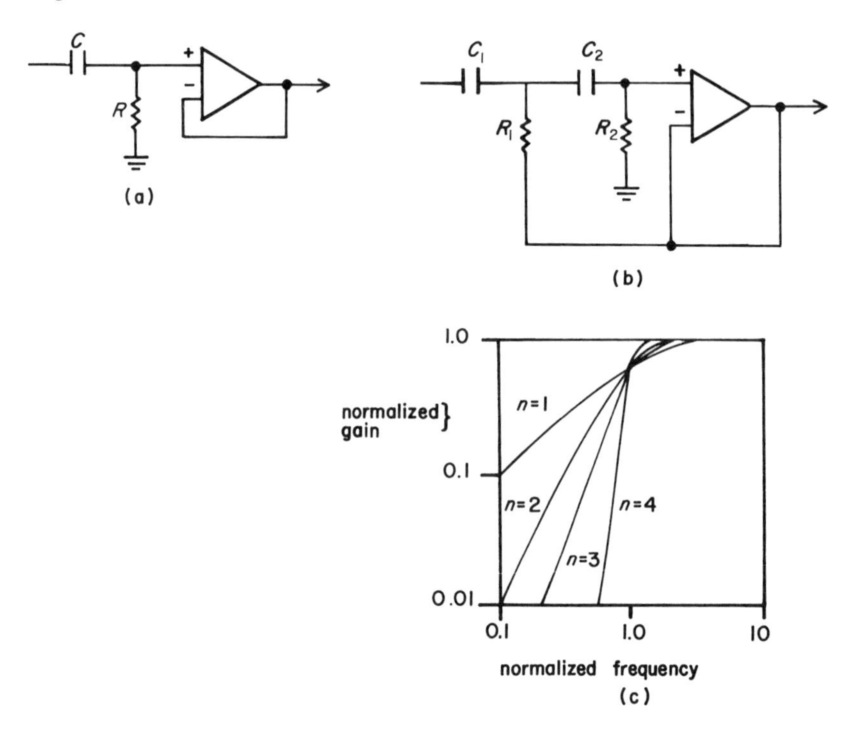

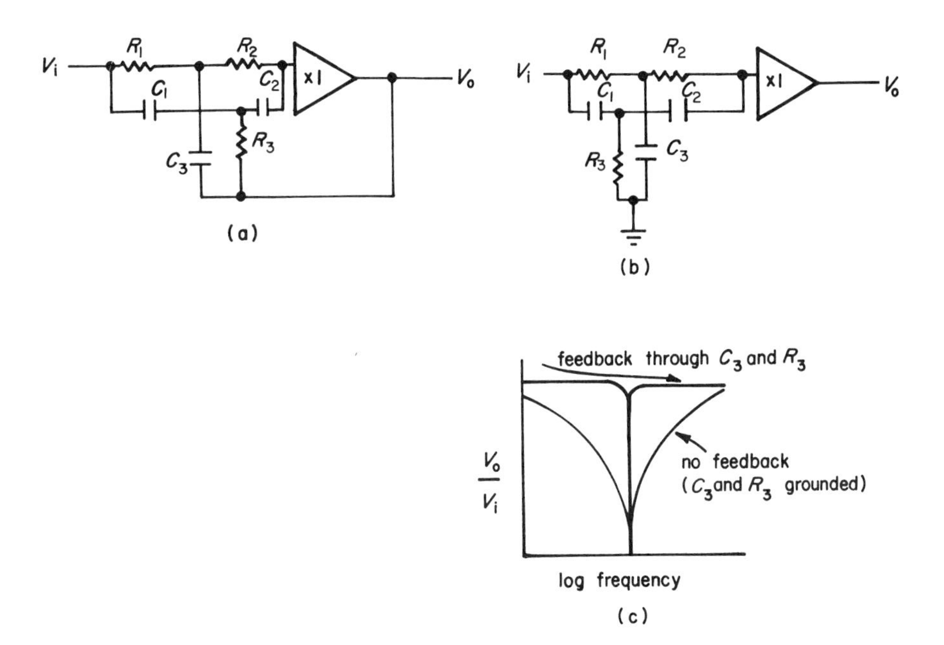

Figure 6.48 The twin-tee notch filter. **(a)** Feedback through R_3, C_3. **(b)** R_3, C_3 grounded. **(c)** Frequency response functions for **(a)** and **(b)**.

6.12.3 Notch Filters

By incorporating a twin-tee network into a filter, a sharply defined notch can be introduced into the frequency response function. The *twin-tee* notch filter is illustrated in Fig. 6.48. Such a sharp notch is commonly used to reject specific frequencies from a signal, usually because they are contamination picked from extraneous sources. Perhaps the most common application is in removing 60-Hz interference from signals. The design equations are

$$\frac{1}{2\pi f_c} = R_1C_1 = R_2C_2 \tag{6.21}$$

$$R_2 = R_1, \qquad C_2 = C_1, \qquad C_3 = 2C_1, \qquad R_3 = R_1/2 \tag{6.22}$$

6.13 Event Detection

Electric waveforms often contain embedded in them events which must be detected automatically in order to analyze them or in order to control some process. Examples of such event detection include level detection (introduced in Section 6.1), peak detection, maximum slope detection, and many more sophisticated paradigms.

6.13.1 Window Discrimination

If a record consists of waveforms of different amplitudes, it is not uncommon to wish to detect all pulses of a specific amplitude range and ignore others. For example, the output of a scintillation counter may consist of a sequence of pulses of different amplitudes, reflecting radioactive decays yielding emissions of different energies. Since each energy level signals a

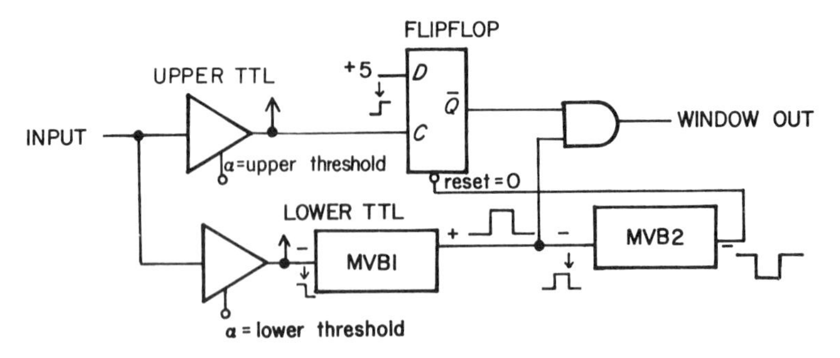

Figure 6.49 A window discriminator.

particular kind of decay, we can count the number of occurrences of decays of a particular energy level and ignore the rest by selectively attending to the pulses of corresponding amplitude and not the rest. Such a technique is referred to as *window discrimination*.

The process is schematized in Fig. 6.49. Whenever an event exceeds the upper threshold (pulse too big), a flipflop is set. When the event amplitude falls below the lower threshold, MVB1 fires a pulse which is passed by the gate if the flipflop is off (e.g., lower threshold was exceeded but upper threshold was not: the event peak fell within the window), and is not passed by the gate if the flipflop was on (e.g., the lower and upper thresholds were both exceeded: the event exceeded the upper window limit).

Sample waveforms are illustrated in Fig. 6.50. The first pulse exceeds lower threshold at A, turning on LOWER TTL; it then exceeds upper threshold at B, turning on UPPER TTL and FLIPFLOP; at C, the upper threshold is recrossed downward, turning off UPPER TTL; at D the lower threshold is recrossed, turning off LOWER TTL and firing MVB1 (pulse width exaggerated for purposes of illustration). MVB1 is not passed by the gate because the

Figure 6.50 Waveforms for the window circuit of Fig. 6.49.

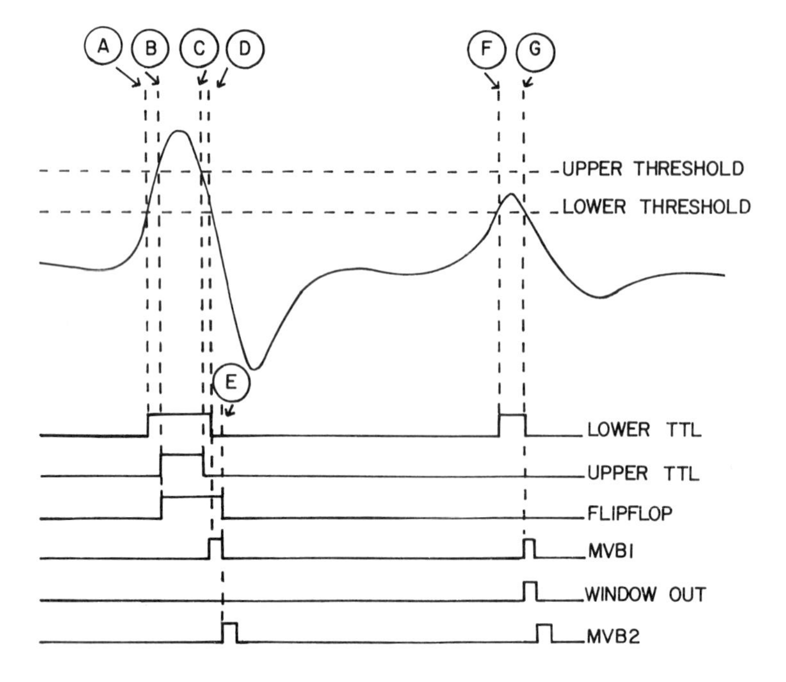

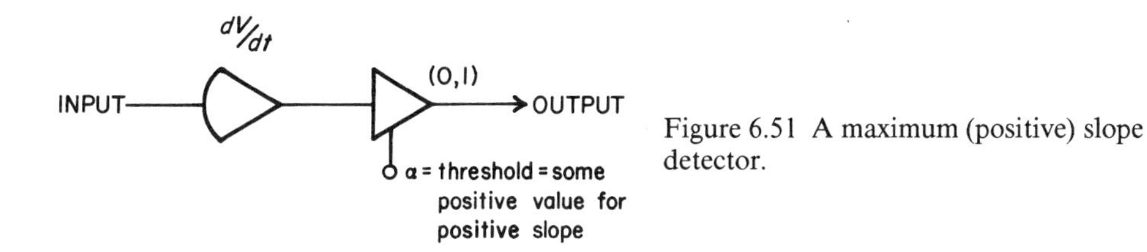

Figure 6.51 A maximum (positive) slope detector.

flipflop is on: hence there is no WINDOW OUT pulse at D, because the upper level was exceeded. At E, MVB1 goes on, and the flipflop is cleared. When the smaller pulse goes on and off at F and G respectively, LOWER TTL goes on and off, triggering the MVB1–MVB2 sequence. Since UPPER TTL does not go on, the gate passes MVB1 as WINDOW OUT.

6.13.2 Detection of Maximum Slope

If precise timing of a variable duration event is required, it is often desirable to detect the event's onset. This might be sufficiently well approximated by detecting its maximum slope, if it has a rapidly rising onset. A simple method of doing this is to differentiate the signal and, if the signal onset is consistently the fastest event in the record, the differentiated signal can be discriminated, as illustrated in Fig. 6.51. If the exact time of maximum slope is desired, then the output of the differentiator can be used as input to a peak detector since a slope has its maximum when the first derivative is at a peak (a maximum).

6.13.3 Peak Detection

The positive peak of a signal can be defined as the point at which the first derivative goes downward through zero. Thus, the simple circuit of Fig. 6.52 can be used to detect positive peaks.

6.13.4 Event Sampling

In many analysis procedures, it is useful to sample a signal at the time that the desired event is detected, and to hold that sample until the next event. The circuit of Fig. 6.53 illustrates

Figure 6.52 A peak detector.

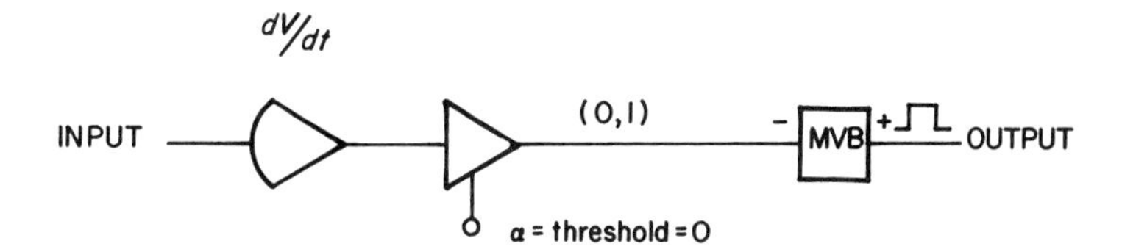

Figure 6.53 A sampling circuit.

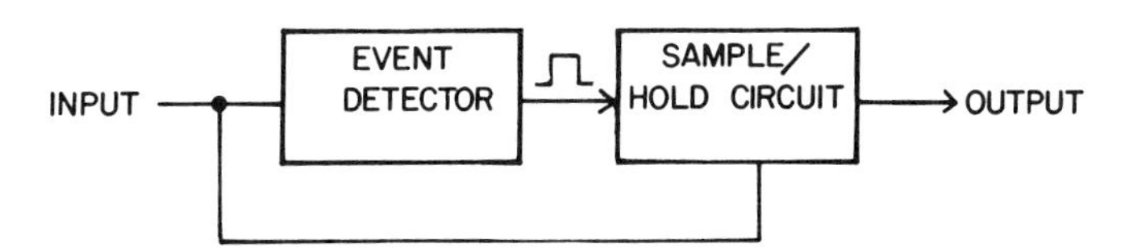

this process. The mysterious SAMPLE/HOLD circuit indicated consists of an analog storage device whose voltage level can quickly be set to the input level, and which will hold this level until it is set to a new input level. We shall discuss the SAMPLE/HOLD circuit in Chapter 7.

6.13.5 Detection of Complex Events

Sequences of events, or complex waveforms, can sometimes be detected by characterizing them as some invariant sequence of the simpler types of events which we have discussed. In general, once one goes beyond relatively simple waveform recognition tasks, it is better to use a computer. In such cases, the analog signal is converted to a series of numbers representing its amplitude as it varies over time. These numbers are then processed by the computer program, seeking patterns to be identified. With the advent of inexpensive microprocessors, this is becoming a relatively inexpensive process.

Problems

1. Design a comparator with $V^{+}_{\text{threshold}} = +1.5$ V, $V^{-}_{\text{threshold}} -2.5$ V, assuming $V^{+}_{\text{max}} = +10$ V and $V^{-}_{\text{max}} = -10$ V.

2. Using MVBs, design a generator to produce a sequence of pulse trains at a single output, with control of (1) interval between trains, (2) train duration, (3) pulse spacing within the train, and (4) pulse widths within the train.

3. Design a square wave oscillator using one op amp to produce a 1-kHz square wave. You will have to derive design equations.

4. Design a 1-kHz triangle wave oscillator.

5. Design a 1-kHz sawtooth oscillator.

6. Design an FM modulator to produce a 1-kHz triangle wave with control input of 0 V, 900 Hz at −1 V, and 100 Hz at +1 V.

7. Diagram a system to multiply an input TTL frequency by 11/7.

8. Using triggerable waveform generators, design a device to produce the waveform of the figure below. Include controls for delay, duration, amplitude, and dc offset.

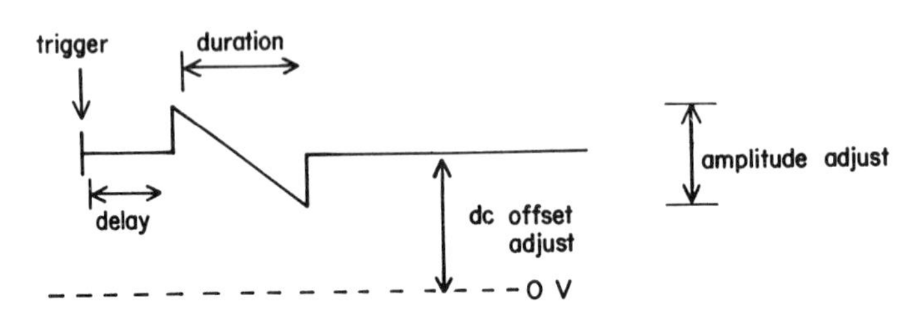

9. Design a bandpass filter with 24-dB/octave rolloff at upper and lower corner frequencies of 100 Hz and 10 kHz. Include a twin-tee notch filter with a notch at 300 Hz.

10. Using standard analog computation elements, build circuits to detect the following events:

a. coincident (simultaneous) occurrence of peaks on each of two separate signals;

b. two successive positive peaks with no intervening crossing of zero volts;

c. signal A greater than signal B (two separate signals).

Selected References

D'Azzo, John J., and Constantine H. Houpis. *Feedback Control System Analysis and Synthesis*. McGraw-Hill, New York, 1966.

DiStefano, Joseph J., III, Allen T. Stubberud, and Ivan J. Williams. *Feedback and Control Systems*. Schaum's Outline Series, New York, 1967.

Eimbinder, J. *Designing with Linear Integrated Circuits*. John Wiley, New York, 1969.

Golomb, S. W. *Shift Register Sequences*. Holden-Day, San Francisco, 1967.

Mitra, S. K. *Analysis and Synthesis of Linear Active Networks*. John Wiley, New York, 1969.

Mitra, S. K. *Active Inductorless Filters*. IEEE Press, New York, 1971.

7

Digital Signal Processing

Paul B. Brown

As we have intimated in earlier chapters, there is a relatively new field, digital signal processing, in which continuous analog signals are converted to a discrete time representation as well as a discrete amplitude representation, and then processed using techniques appropriate to the discrete form, namely, digital techniques. Generally the processed analog signal is "played back" after delays ranging from negligibly short (less than 1 μsec) to arbitrarily long after the original occurrence of the signal. This process is outlined in Fig. 7.1.

This sounds like a complicated way to do simple things, and for processing of single analog signals, this approach is often uneconomical. But digital processing can usually be extended to multiple signals with little extra expense by multiplexing input and output lines and sharing a single processor. Also, the digital representation of an analog signal may be processed in ways which are impossible or impractical using analog techniques, and the digital representation or some extracted feature may be recorded in digital format on an appropriate storage medium such as digital tape or disk. Finally, signals can be generated *de novo* using digital methods. Waveform parameters can be digitally controlled and the most complex of waveforms can be repeated conveniently—even waveforms which closely approximate random noise.

First, we shall introduce the concept of discrete time and discrete amplitude representations of continuous analog signals. We shall next describe digital-to-analog conversion (the conversion of a series of numbers to a corresponding series of voltages in time); analog-to-digital conversion (the generation of a series of numbers to represent a corresponding series of voltages in time); and last of all, commonly used techniques of digital signal processing.

7.1 Discrete Time and Discrete Amplitude Representations of Analog Signals

7.1.1 Discrete Time Signals

In the great majority of cases, the analog signal to be processed is sampled at discrete points in time, separated by equal intervals. The signal represented by such a series of periodic samples is referred to as a *discrete time analog signal*. Note that the amplitude of the signal is still an analog quantity, in that it may occupy any value over a finite operating range.

The sampling of an analog signal is accomplished by either "freezing" it with a *sample/hold*, or *S/H*, circuit each time a periodic sample signal occurs, or by converting it to a digital value so rapidly that it does not have time to change its amplitude appreciably. The latter process is generally considered to be too risky, in that most digitizers are subject to potentially large errors in digitizing signals which are varying during the digitization process. We shall explain this problem in the section on analog-to-digital conversion. We will concentrate here on the use of S/H circuits.

An S/H circuit is illustrated in Fig. 7.2. It consists of a capacitor, an analog switch, and a high-input impedance op amp. Assume for the sake of simplicity that the input impedance of the op amp is so high that it is not an appreciable load on the capacitor/switch circuit. The capacitor C is thus part of an RC low-pass filter applied to the input signal, where R is equal to R_{on} or R_{off}, the switch resistance in the on or off state, respectively. Assume that $R_{\text{on}} \ll R_{\text{off}}$, as is the case with any good switch, and therefore $RC_{\text{on}} \ll RC_{\text{off}}$. Thus, for

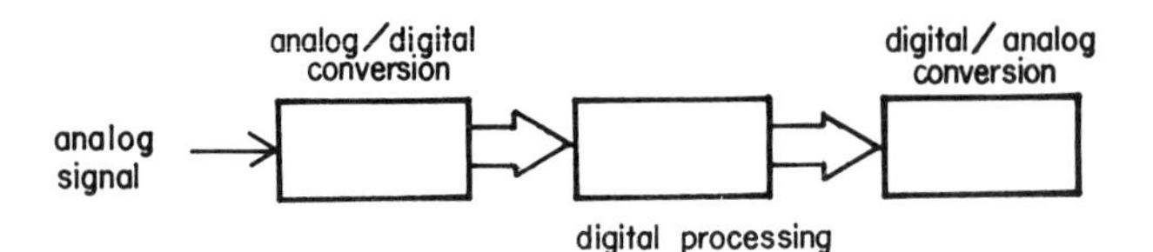

Figure 7.1 The simplest case of digital processing of an analog waveform.

example, if $R_{on} = 100\ \Omega$ and $R_{off} = 10^{10}\ \Omega$, with $C = 100$ pF (10^{-10} F), the S/H circuit has a time constant $\tau_{on} = RC_{on} = 10^{-8}$ sec (10 nsec) when on, and $\tau_{off} = RC_{off} = 10^{0}$ sec (1 sec) when off. If a TTL pulse is applied to the switch which is several on-time-constants long (say, 1 μsec, or $100\tau_{on}$), the capacitor will charge to a value which is essentially equal to the signal value (if the analog signal is not too fast); this signal value will be held on the capacitor essentially unchanged for a fraction of its off-time constant (say, 10 msec, or $10^{-2}\tau_{off}$). Thus, a sample of the signal derived from a 1-μsec sample period can be held for a much longer time, 10 msec in this case. Note that the ratio of τ_{on} to τ_{off} is determined essentially by the ratio of R_{on} to R_{off}, and the actual magnitudes of the time constants can be selected by selecting C. For this reason, all the components except C are incorporated in a single IC; the designer selects C for the desired sample time and hold time, within the constraints of the ratio of R_{on} and R_{off}. Obviously, the ideal S/H has infinitesimal sample time and infinite hold time, whereas in reality the designer must trade off speed of sampling against duration of hold period.

We mentioned that in the sampling mode the capacitor voltage can track the input signal if the input signal is sufficiently slow: this is obvious if the $R_{switch}C$ circuit is appreciated as a low-pass filter. For most purposes it is adequate if the $R_{on}C$ time constant $\tau_{on} \geqslant 100\tau_c$, where $\tau_c = 1/\omega_c = 1/2\pi f_c$, f_c being the corner frequency for the input signal bandpass. In this way, all frequency components of the input signal can be tracked by the S/H. Thus, for a τ_{on} of 10^{-8} sec, a signal with a 6-MHz bandwidth could be adequately tracked.

Given that the S/H is sufficiently fast not to be the frequency limiting factor, the frequency at which the signal is sampled is necessarily the limiting factor determining the upper limit of frequencies which can be represented. This can be understood from the graphs of Fig. 7.3.

The input signal (Fig. 7.3a) is envisioned as being sampled by multiplication with a pulse train of amplitude 1 (during sampling) and 0 (during hold), as depicted in Fig. 7.3b. The resulting waveform, Fig. 7.3c, is similar to the voltage on the output of an ideal switch which is turned on during the pulses of Fig. 7.3b. The frequency spectra of Figs. 7.3a–7.3c are illustrated in Figs. 7.3d–7.3f. The input signal must have a bandwidth (Fig. 7.3d) which is less than the spacing between spectral lines of the pulse train (Fig. 7.3e). Recall that the frequency output of an analog multiplier is the convolution of the two input frequencies: thus, the analog spectrum is repeated, centered about each of the discrete frequency peaks of the periodic pulse spectrum.

The important point of this discussion is that the output spectrum is an unambiguous representation of the input spectrum (and therefore the input waveform) only if the different bandpass regions of the output spectrum do not overlap. If these frequency bands overlap, than the representation is ambiguous.

Figure 7.2 A sample hold (S/H) circuit.

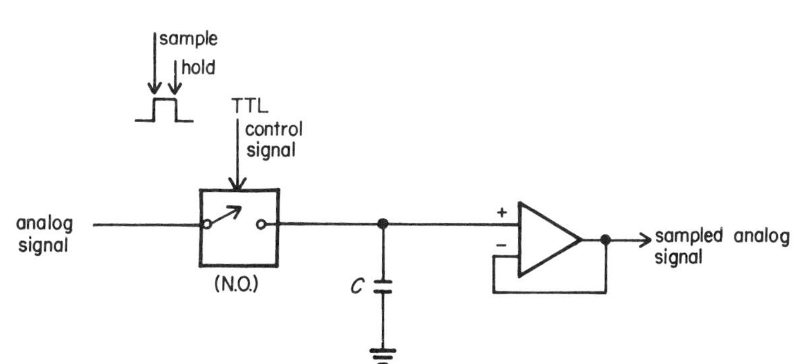

To avoid such ambiguity the input bandpass must be less than one-half the sampling frequency. This is the *sampling theorem*, and the upper frequency so defined is called the *Nyquist frequency*. That is,

$$f_N = \tfrac{1}{2} f_s \tag{7.1}$$

where

f_N = Nyquist frequency

f_s = sampling frequency

This rule is very important. The Nyquist frequency not only defines the highest frequency which is unambiguously represented in a discrete time representation of a continuous time signal. *It also defines the highest frequency which may be present in a continuous time signal without causing ambiguities in its discrete time representation.* Frequencies above the Nyquist frequency are not simply lost; they are represented as lower, *alias*, frequencies. The relation between an input frequency and its alias frequency is depicted in the graph of Fig. 7.4a. Note that the alias frequency is an alternately increasing and decreasing frequency, as a function of the input frequency.

To avoid aliasing, then, it is essential that the input spectrum be low-pass filtered to exclude frequencies at or above the Nyquist frequency. The rolloff frequency and slope of the

Figure 7.3 Spectrum of a signal sampled at discrete times. **(a)** Input signal in the time domain. **(b)** Sampling signal in the time domain. **(c)** The sampled signal in the time domain. **(d)** The input signal in the frequency domain. **(e)** The sampling signal in the frequency domain. **(f)** The sampled signal in the frequency domain. Note that the sampled signal is the product of the input signal and the sampling signal in the time domain, and therefore the convolution of the two signals in the frequency domain. This results in a repetition of the input spectrum at intervals corresponding to the intervals between the impulses in the sampling spectrum in the frequency domain.

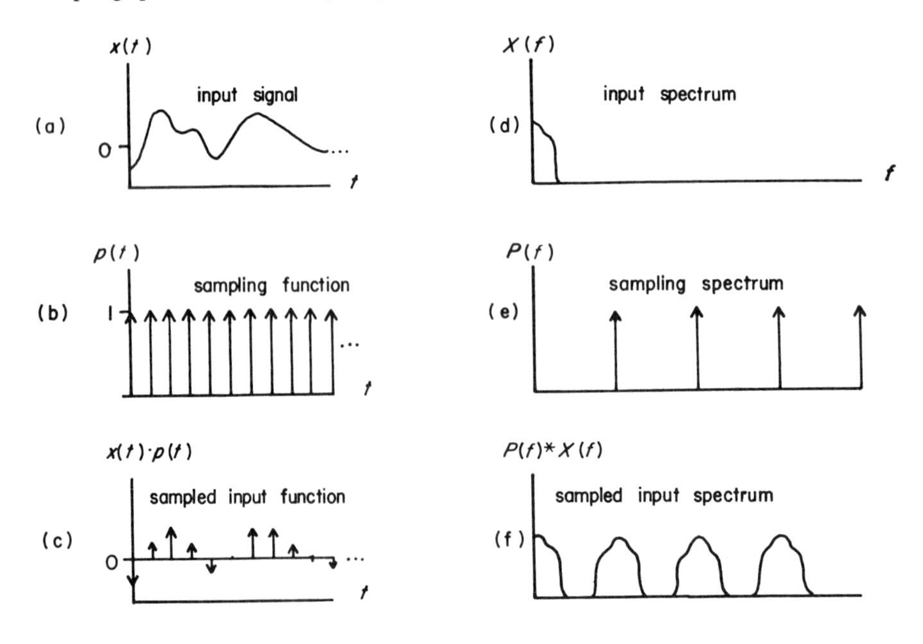

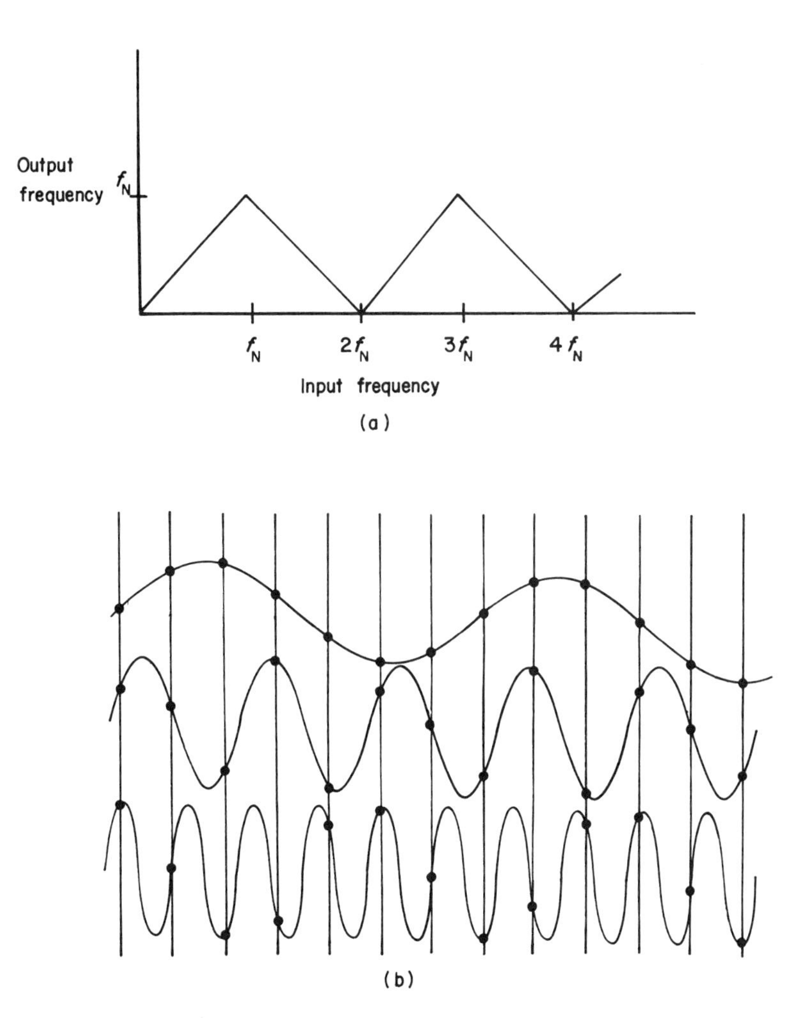

Figure 7.4 Phasing. **(a)** Output frequency as a function of input frequency in conversion of a continuous time, continuous amplitude signal to a discrete time, discrete amplitude representation. **(b)** Example of aliasing. In the top waveform, the sine wave is much lower frequency than the sampling rate, and the sampled waveform (dots) is of the same frequency. The center waveform is higher frequency, but still lower than half the sampling rate. The output frequency (dots) is the same as the input frequency. The lowest waveform has a higher frequency than one half the sampling rate, resulting in a lower-frequency (alias) output (dots).

filter should be selected so the frequency components above the Nyquist frequency are of negligible amplitudes compared to the amplitudes of components below the Nyquist frequency. Examples of aliasing are given in Fig. 7.4b.

7.1.2 Discrete Amplitude Signals

Once a signal has been "frozen" at discrete points in time by an S/H circuit, it can be converted to digital representation, using an *analog-to-digital converter* (A/D converter, ADC). In nearly all applications, such conversions take place each time the signal is frozen by the S/H, at a delay after initiation of the HOLD status of the S/H which is sufficiently long to allow the S/H signal to settle to a steady level. Such a system is diagrammed in Fig. 7.5. Note that the HOLD status must be maintained until conversion is complete.

The A/D converter, after a conversion delay, produces an output binary number, usually in parallel output format (one pin on the A/D converter module for each bit). The magnitude of the binary number codes the magnitude of the input voltage (i.e., the voltage out of the S/H in Fig. 7.5). As long as the A/D can keep up with the S/H, and as long as the range of binary numbers corresponds to the range of input voltages, the A/D will produce a series of binary numbers which accurately reflects (within $\pm\frac{1}{2}$ bit) the input signal.

A reciprocal process exists, called *digital-to-analog* conversion (D/A conversion), using a *digital-to-analog* converter (D/A converter, DAC), which will produce a series of voltages which are proportional to a corresponding series of input numbers. This process is illustrated in Fig. 7.6. Note that the output of the D/A converter only changes at clock intervals, because that is when the digital input to the DAC changes. Note also that the output voltage can only occupy discrete values proportional to the values of the binary numbers used to produce it. These are important points:

a. Once a continuous time signal has been converted to a discrete time signal, it cannot be converted back to the original continuous time signal.
b. Once a continuous amplitude signal has been converted to a discrete amplitude signal, it cannot be converted back to a continuous amplitude signal.

Upon reflection, this makes sense. Obviously a signal cannot be completely represented by sampling it at discrete points, no matter how closely those points are spaced. Of course, if the signal is sufficiently slow or the sampling is fast enough, we can arbitrarily closely approximate it with a discrete time representation: this means that the discrete time representation *converges* to the continuous time representation as the interval between samples becomes

Figure 7.5 S/H, A/D system, with MVBs used to control timing.

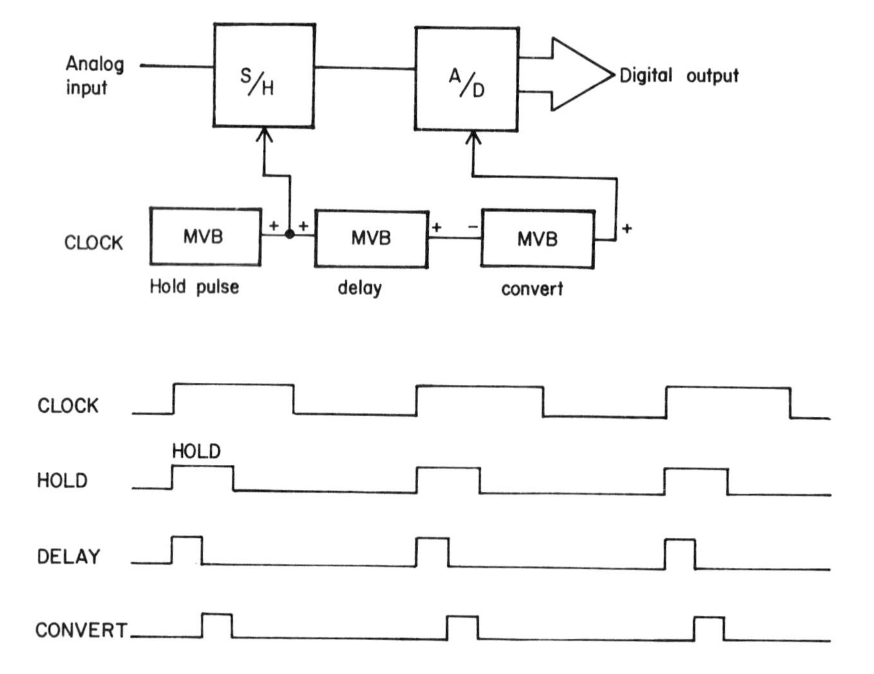

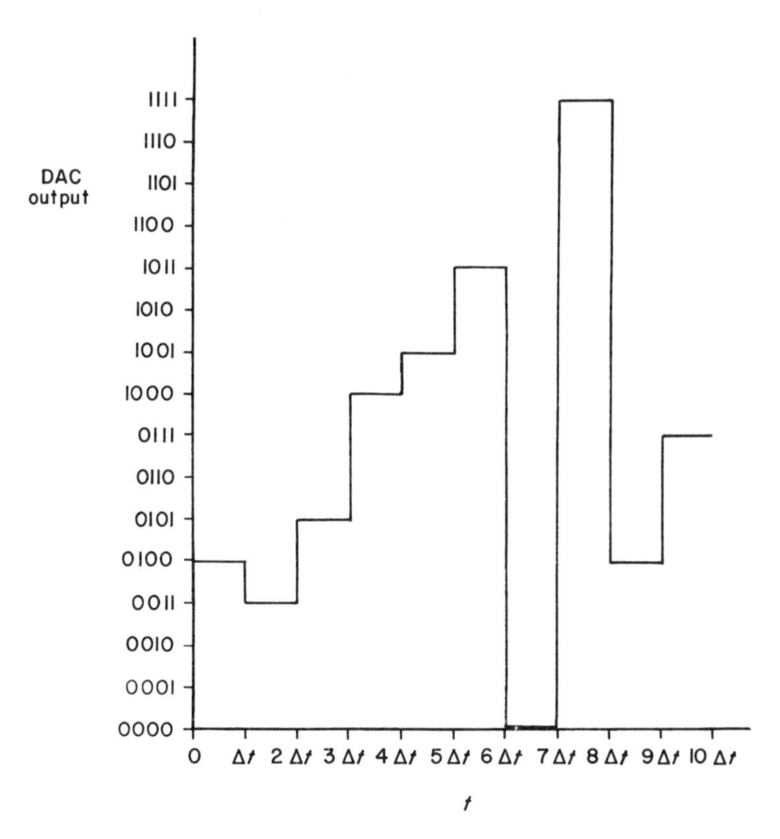

Figure 7.6 D/A conversion. Input sequence: (offset binary) 0100, 0011, 0101, 1000, 1001, 1011, 0000, 1111, 0100, 0111.

vanishingly small. Similarly, a continuous amplitude signal cannot be completely represented by a discrete amplitude number or voltage sequence because there is a finite number of possible values in the discrete form and an infinite number of *actual* values in the continuous form. If we make our numbers more and more precise (extend the number range onto which we map the input voltages), we can arbitrarily closely approximate the continuous amplitude form. This means that the discrete amplitude representation *converges* to the continuous amplitude representation as the voltage difference coded by a given numerical difference (say, one bit) becomes vanishingly small.

In practice, it is generally possible to increase the sampling rate to the point where all desired frequency components are represented. The number of bits used to represent a voltage range determines how closely the discrete amplitude form represents the continuous amplitude input signal of an A/D converter. A/D converters can be obtained with different numbers of bits. Generally, the more bits required, the longer it takes for the A/D converter to finish a conversion, and the more expensive the A/D converter will be. The same trade-offs apply to D/A converters.

7.2 Binary Representation of Discrete Amplitude Signals

These are two principal methods of representing voltages digitally: one is *offset binary*, in which the lowest (most negative or least positive) voltage in the range represented is signalled by a binary 0, and the highest (least negative or most positive) voltage is represented by the

Table 7.1 Comparison of Offset Binary and Signed Bipolar Codes for Different Analog Voltages

Voltage level	Offset binary	Signed bipolar
Most positive	11111111	01111111
Zero	10000000	00000000
Most negative	00000000	10000000

maximum unsigned number, or $2^n - 1$, where n is the number of bits available. Thus, for a voltage range of -5–$+5$ V and for an 8-bit representation, -5 V would be represented as 00000000, 0 V would be 10000000, and $+5$ V would be 11111111. An increment or decrement of 1 would correspond to $(V_{max}^{+} - V_{max}^{-})/2^n$, or 10 V/256=0.0390625 V. Thus, strictly speaking, 00000000 represents any voltage from -5 to -4.9609375 V, 10000000 represents any voltage frnm 0 to $+0.0390625$ V, and 1111111 represents any voltage from $+4.9609375$ to $+5$ V.

Thus, any voltage is represented by a corresponding binary number. Since a *range* of voltages (0.0390625 V wide, in this example) is represented by a given number, the number is ambiguous. For example, 00000111 represents any number in the range -4.765625–-4.7265625 V. This means that our *resolution*, or our ability to distinguish between two

Figure 7.7 The error signal in A/D conversion. **(a)** The input signal. **(b)** The analog representation of the digitized input signal. **(c)** The difference between **(a)** and **(b)**, the error signal.

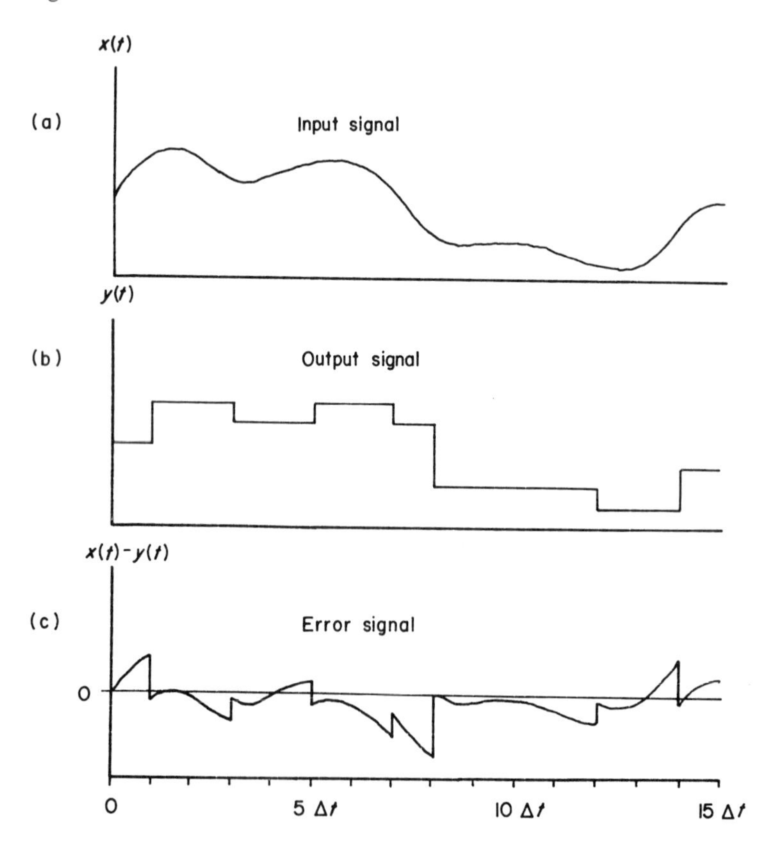

different voltages, is limited on the average to voltages separated by one unit, or 0.0390625 V or more. In general, resolution is specified as

$$V_{res} = \frac{V_{max}^{+} - V_{max}^{-}}{2^n} \tag{7.2}$$

where n is the number of bits available.

Offset binary is the most convenient means of representing voltages in terms of rapid A/D conversion. However, it is usually more convenient to use *signed operation*, where 00000000 represents 0 V, 01111111 is the largest positive number, and 11111111 is the largest negative number. However, signed binary is not quite as convenient in terms of the inner workings of a DAC or ADC. Therefore, most systems use offset binary for actual conversions, and convert to and from signed binary for arithmetic manipulations.

Conversion from one system to the other turns out to be very simple: all that is necessary is to invert the leftmost bit. This is illustrated for a few values in Table 7.1.

Some A/D and D/A converters have built-in inverters so that the most significant bit can be wired according to whether offset binary or signed bipolar operation is desired.

Discrete amplitude systems impose on continuous amplitude signals an *error* which is the difference between the continuous and discrete amplitude forms of the signal. This is graphically illustrated in Fig. 7.7. The continuous amplitude signal is plotted with a corresponding discrete amplitude signal, and the difference between the two is plotted to indicate the error. This error constitutes a distortion of the continuous amplitude signal, and this distortion can be significant in some applications.

7.3 D/A Conversion

We have introduced the concepts of S/H, A/D, and D/A, and the S/H circuit has been described. In this section, we shall describe the methods commonly used to perform A/D and D/A conversions. Even though most scientists never need to know how to build these devices, because they are available as ICs and modules, it is helpful in applying them to know how they work. We shall describe D/A converters first because they are used as integral parts of some types of A/D converters.

A commonly used design for a 4-bit D/A converter is presented in Fig. 7.8. Each bit of the binary input controls a switch. The circuit is arranged in such a way that each switch injects a current into the network when the corresponding bit is on, and serves as a shunt to ground when the bit is off. The amount of current i reaching the op amp negative input is equal to

$$i = V_{max}^{+} \sum_{i=1}^{n} \frac{a_j}{2^j R} \tag{7.3}$$

where a_j is the value of bit j (0 or 1), and there are n bits. This current is equal to the feedback current:

$$i^{-} = i_{feedback} = V_o / R \tag{7.4}$$

Substituting the expression in Eq. (7.3) for i^{-} gives

$$V_{max}^{+} \sum_{j=1}^{n} \frac{a_j}{2^j R} = -V_o / R$$

Rearranging gives

$$V_o = -V_{max}^{+} \sum_{j=1}^{n} \frac{a_j R}{2^j R} \tag{7.5}$$

Thus, the output voltage is proportional to the input digital code.

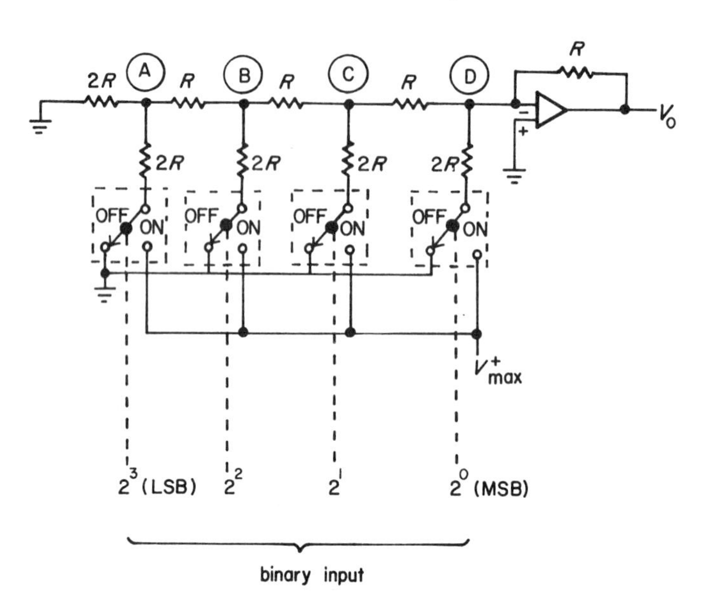

Figure 7.8 A 4-bit D/A converter.

DACs are available with a number of input codes (offset binary, signed binary, BCD, and others), unipolar or bipolar voltage ranges, current or voltage output.

If the input bits to the DAC do not all change simultaneously when the digital value changes from one number to another, there will be a brief interval during which the DAC input is in error: it is not equal to the old or the new number, but to an erroneous value. This was mentioned in Chapter 5, in the discussion of ripple counters and synchronous counters. When such transient erroneous digital inputs occur, they will result in transient erroneous DAC outputs, referred to as "glitches." Obviously, the solution to this problem is to use synchronous inputs, or to "de-glitch" the DAC input with a buffer register (called a "latch"), which is synchronously loaded with new values only after the input value has completely settled.

7.4 A/D Conversion

There are many methods of converting analog voltages into digital numeric representations. In general, the less expensive methods are either slower or less precise than the more expensive ones.

One simple method of A/D conversion is illustrated in Fig. 7.9. The analog input controls the frequency of oscillation of a VCO (voltage-controlled oscillator). Generally, this input should be the output of an S/H. The CONVERT COMMAND initiates a sequence of events, namely: (1) reset the digital counter; (2) enable the counter to count VCO oscillations during a fixed period; (3) latch (store) the counter output in an output register. All of the components are very inexpensive, and with a maximum VCO output frequency of 1.024 MHz, a 10-bit conversion can be completed during a 1-msec count time. The VCO is adjusted so that it oscillates at a minimum frequency for the lower input voltage limit and at its maximum frequency for the higher limit. Since the count of oscillations in a fixed interval is proportional to frequency (frequency=number of oscillations per unit time) and the VCO frequency is proportional to input voltage, the count is proportional to input voltage. There are a few subtleties of design necessary for proper operation, but these need not be discussed here.

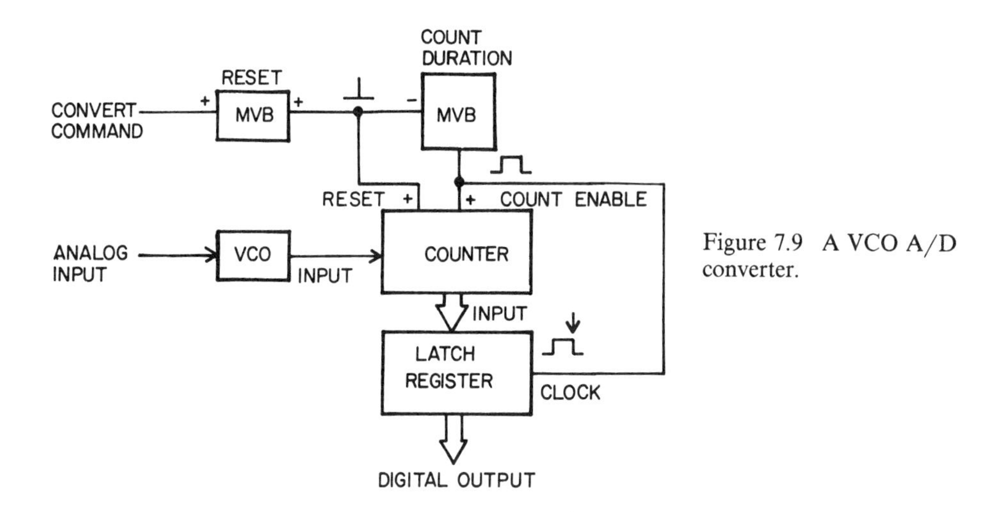

Figure 7.9 A VCO A/D converter.

The VCO method is a medium-speed method, and is limited by the linearity of the VCO. A slower, but highly precise method, appropriate for digital voltmeters, is the dual-slope integration method, illustrated in Fig. 7.10. The input signal is added to a positive reference voltage, V^+_{max}, and inverted, producing a negative signal. This is then integrated over a fixed time period T, charging capacitor C. The capacitor is then discharged by integrating a reference voltage of opposite sign, V^+_{max}, until the integrator output is zero, as detected by the comparator. During this time, a clock is counted to determine how long it takes to discharge the capacitor. The count is proportional to the discharge time, which is proportional to the integrated input voltage. This is in turn proportional to the average value of the input voltage during time T. The method is very accurate, although quite slow. It is inexpensive and reliable, requiring no precision components or timing.

The fastest method commonly used is the successive approximation method of Fig. 7.11. The analog input is compared with the output of a DAC, whose input consists of a D register initially cleared to zero. One bit at a time, trial 1s are loaded into the D register, starting with the MSB. During each trial, the DAC output is compared with the analog input: if the DAC output is greater than the analog input, the digital number in the register is too large, and the

Figure 7.10 The dual-slope integration method of A/D conversion.

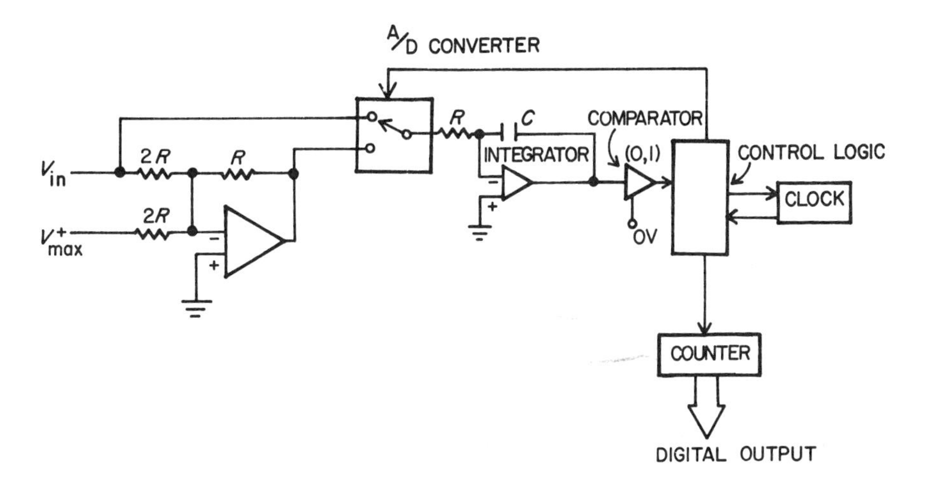

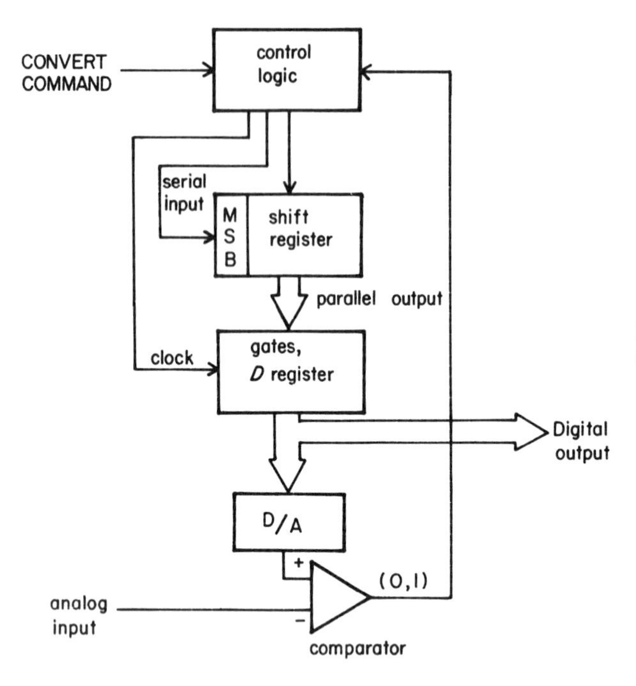

Figure 7.11 A successive approximation A/D converter.

trial bit is reset to zero. If the DAC output is smaller than the analog input, the number is not too large, and the trial bit is kept at a value of 1. This is repeated until all bits of the *D* register have been tried. The final value is the digital code for the analog input.

The successive approximation requires the most circuitry, but it is limited in accuracy only by the accuracy of the D/A converter (18-bit D/As are readily available), and is very fast. Prices for successive approximation A/D converters continue to drop, to the point where they will be used for all but the most cost-limited applications.

Note that, if the signal is changing rapidly, it may cross the boundary between one digital value and another during conversion. If this occurs, large errors in the digital representation can result. For this reason, use of an S/H module is recommended for all applications: S/H modules are so inexpensive compared to ADCs that there is little excuse for omitting them.

7.5 Multiplexing

When multiple analog signals are processed, it is often less expensive to use one digital system to process them all, and to *multiplex* the operation. We have already introduced the concepts of analog and digital multiplexers.

Figure 7.12 illustrates the use of analog multiplexers in a digital signal-processing system. By placing the multiplexer at the earliest point in the system, the eight analog signals share all subsequent electronics, resulting in a considerable economy.

7.6 Digital Signal Processing

Obviously, one does not convert an analog signal into digital form just to convert it back into analog form. Intervening between these two functions is a third one, digital signal processing. In the digital domain, it is possible to treat a series of numbers just as though they constituted a digital representation of the corresponding analog signal. Such processing may consist of straightforward arithmetic processing, such as the addition or subtraction of two signals, or

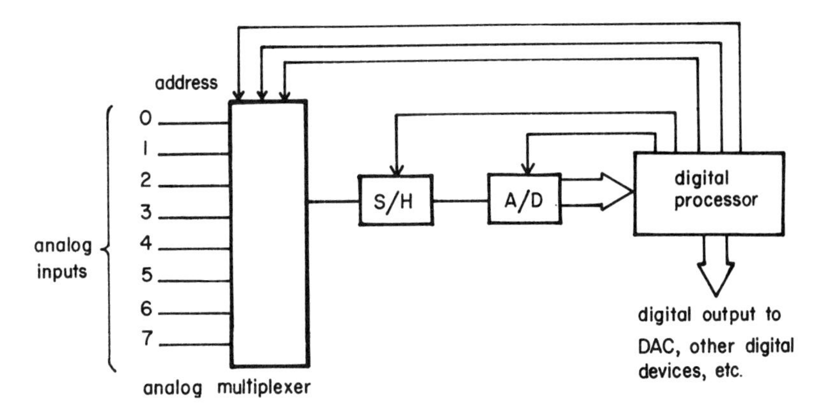

Figure 7.12 Use of an analog multiplexer to process multiple signals with an S/H, ADC, and digital processor.

any of the processes we have introduced in the earlier discussion of analog signal processing: integration, differentiation, multiplication, division, threshold detection, filtering, or the detection of complex events. Signals can be generated *de novo*, in digital signal generators. Or a number of processes can be implemented which are impossible or impractical using analog computation methods, such as filters with unusual frequency response functions, averagers, nonlinear processors, or complex event detectors.

The digital processor itself may be a special-purpose device dedicated to one particular application, or it may be a general-purpose computer which is capable of performing other tasks as well. With the advent of inexpensive microprocessors, it has become more economical to use a microprocessor for many applications in which computers would previously have been much too expensive. In large volumes, a manufacturer can frequently incorporate a microprocessor into a design for $10 or less, replacing SSI, MSI, or even LSI circuits costing many times more.

7.6.1 Signal Averaging

When a signal is repetitive, either because it is intrinsically periodic (e.g., the pattern of gasoline consumption in one cylinder of a car whose engine is running at constant speed) or because it has a causal relation to some other repetitive signal (e.g., the impulse response of a system which is being tested with repeated impulses), the scientist is often more concerned with the waveform of a "typical" response than any individual response. Of course, if the system is entirely deterministic, the two are the same. However, if there are sources of randomness in the response, such as a noise component in one of the inputs to the system, or internally generated noise (e.g., thermal noise), then the two will differ in many systems, such as brain cells responding to small, repeated inputs. The random component of the response may be orders of magnitude greater than the causal component. A technique which is used to emphasize the causal component and de-emphasize the random component is called *signal averaging*, or *averaging*, for short.

The technique relies on the fact that if n samples are summed, random noise will sum its power, and not its voltage, and hence the noise sum will have an amplitude proportional to $\sqrt{n}$. The signal amplitude will be proportional to n, however. Therefore, signal to noise ratio is enhanced by a factor equal to $n/\sqrt{n} = \sqrt{n}$.

Figure 7.13 illustrates the technique. Each waveform is displayed on the same time and amplitude scale. Note that individual waveforms differ substantially from each other (first five responses and last five responses). The averages of the first and last ten responses are

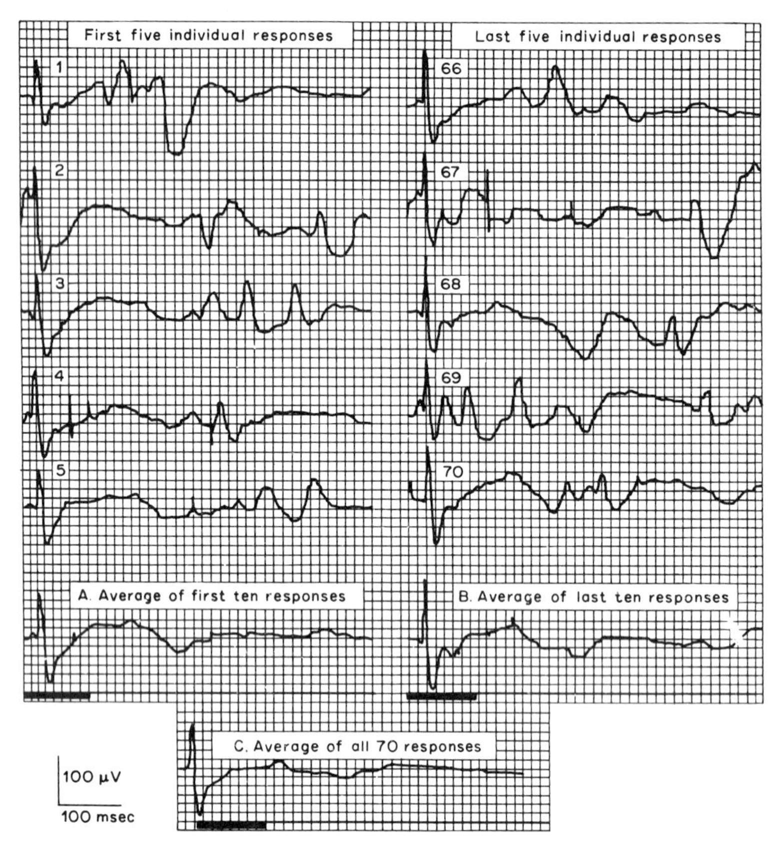

Figure 7.13 Signal averaging. (After Figure 2.3, Rosenblith, 1959. Used with permission of MIT Press.)

much more similar to each other. The average of all 70 responses has eliminated a large fraction of the variable portion of the responses, and what is left consists mostly of the invariant portion of the responses.

The averaging technique consists of summing m signals, all in phase with each other, and dividing by m. Consider the case of Fig. 7.14, where the impulse response of a system is desired, and where the system's response is contaminated by random noise. A digitizing "sweep" consists of 1000 outputs of an A/D converter, clocked at 1000 samples/sec, and each sweep is triggered by the impulse used as input to the system. As each sample is gathered, it is added to the sum of all previous samples corresponding to the same time in the averaging sweeps. Thus, if $a_{m,1}, a_{m,2}, \ldots, a_{m,1000}$ represent the 1st–1000th samples in the mth sweep, and $a_{1,n}, a_{2,n}$, in $a_{m,n}$ represent the nth sample in the 1st–mth sweeps, and if $a_{S,1}, a_{S,2}, \ldots, a_{S,1000}$ represent the 1st–1000th values of the sum across m sweeps, then

$$a_{S,1}, a_{S,2}, \ldots a_{S,1000} = \sum_{i=1}^{m} a_{i,1}, \quad \sum_{i=1}^{m} a_{i,2}, \quad \ldots, \quad \sum_{i=1}^{m} a_{i,1000}$$

Generalizing for n samples per sweep and dividing by m to get the average a_A,

$$a_{A,1}, a_{A,2}, \ldots, a_{A,n} = \frac{1}{m}\sum_{i=1}^{m} a_{i,1}, \quad \frac{1}{m}\sum_{i=1}^{m} a_{i,2}, \quad \ldots, \quad \frac{1}{m}\sum_{i=1}^{m} a_{i,n} \tag{7.6}$$

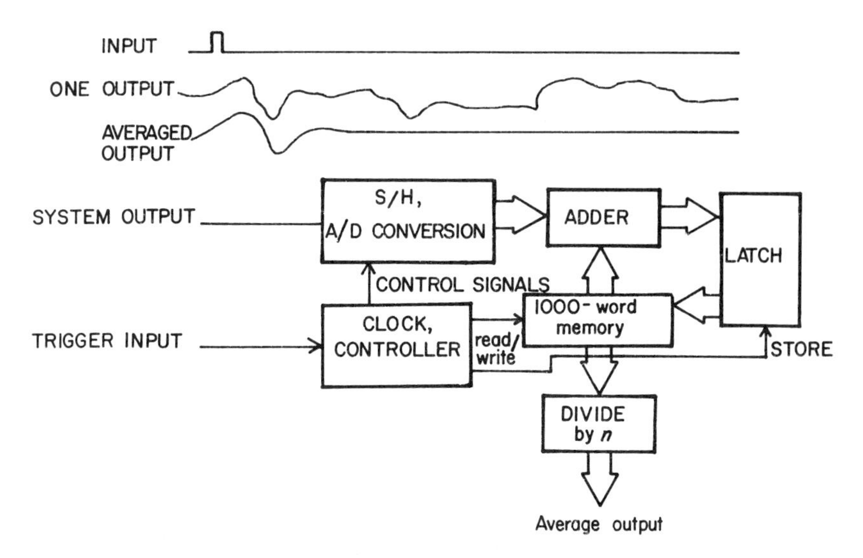

Figure 7.14 An averager.

Averagers are available commercially, but they cost several times the cost of parts to build one. Simple, stand alone averagers are not too difficult to design and build, although the time it takes may justify buying one. The most difficult portion to construct is the digital divider: if a sum rather than an average is acceptable, the design is trivial.

7.6.2 Digital Filtering

If the impulse response of the desired filter (or, in fact, any kind of linear system) is known, a digital simulator of this system can be constructed using the impulse response. One advantage to digital filtering is that the basic circuit is always the same: all that needs to be changed are certain resistors. Other advantages include the ease with which filters can be constructed which are difficult or impossible to build using analog techniques, the ease with which they can be modified (by changing resistors), and the possibility of incorporating the properties of many cascaded analog filters into a single digital filter.

A generalized block diagram for a digital filter is diagrammed in Fig. 7.15. The device consists of a series of n-bit digital words, arranged as a shift register. That is, the clock advances all words forward one address. At the leftmost (input) word, the output of the ADC is shifted in on each clock cycle. A data value shifted in at the left at time t appears in the second word from the left at $t+\Delta t$, where Δt is the width of a clock cycle. If words are numbered from the left, starting at t, then the signal appears in word n at time $t+(n-1)\Delta t$.

If each word is converted by a DAC into an analog voltage, as shown in Fig. 7.14, and if all DACs are summed, then the output of the adder is the sum of the current voltage plus all the previous values from times $t-\Delta t$, $t-2\Delta t,\ldots,t-(n-1)\Delta t$. Note, however, that not all DAC outputs need be weighted by the same amount.

Now consider the consequences of passing an impulse through this system: at time t, the input to the ADC goes high (say, 1 V) for one clock cycle. At times $t-\Delta t, t-2\Delta t,\ldots,t-(n-1)\Delta t$ and at times $t+\Delta t, t+2\Delta t,\ldots,t+(n-1)\Delta t$, the input is zero. That is, the input is equal to zero at all times before and after the one clock cycle during which the input is high.

The impulse response of the system is determined by the weights assigned to the adder inputs. These weights are controlled by the corresponding input resistances (see Chapter 4 for

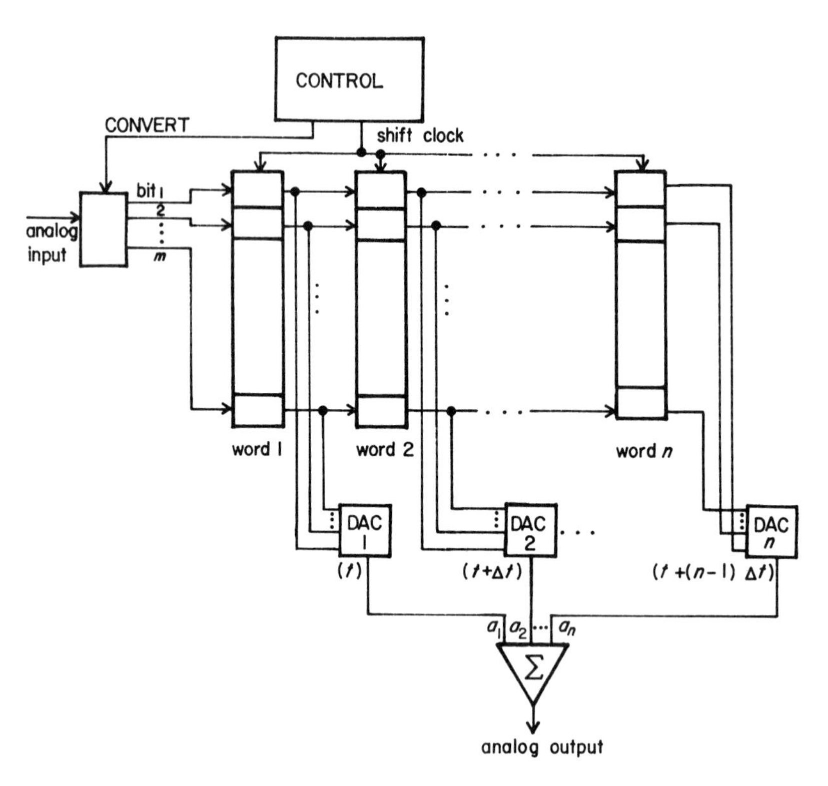

Figure 7.15 Generalized block diagram of a digital filter.

a description of analog adder circuits). By adjusting these values to determine the contributions of the impulse voltage at times $t + \Delta t, t + 2\,\Delta t, \ldots, t + (n-1)\,\Delta t$, the past history of the signal interacts additively with the current value of the signal, as is the case with linear systems containing energy storage elements (capacitors and inductors). The shape of the impulse response can be adjusted to approximate any desired impulse response by adjusting the relative values of the input resistors so that for each input i, the weight is proportional to the magnitude of the impulse response at $t + i\,\Delta t$. The number of words n is determined by the length of time over which the impulse response extends, within the resolution of the system (number of bits per word). The clock frequency is equal to double the desired Nyquist frequency, which should be two or three times the desired upper frequency cutoff in most systems.

Having designed the system for the impulse response of the linear system to be simulated, the digital system will now replicate the response of the linear system, regardless of the waveform at the input, as long as the ADC's operating limits are not exceeded, and within the time and voltage resolution limits set by word length and clock rate. Note that *any* linear system whose impulse response is known can be simulated, including very complex series of filters and gain elements. The impulse response can be determined empirically in a corresponding analog system, it can be determined using a computer simulation program, or it can be computed from the desired transfer function.

Obviously, all these DACs are expensive. It is cheaper to use the system of Fig. 7.16, where a weighted sum of the n words is obtained digitally, and the weighted digital sum is converted using one DAC. Even using this system (which is somewhat harder to modify if variable transfer characteristics are desired), the cost is prohibitive for most applications. However, the use of digital processing permits the implementation of nonlinear and complicated linear

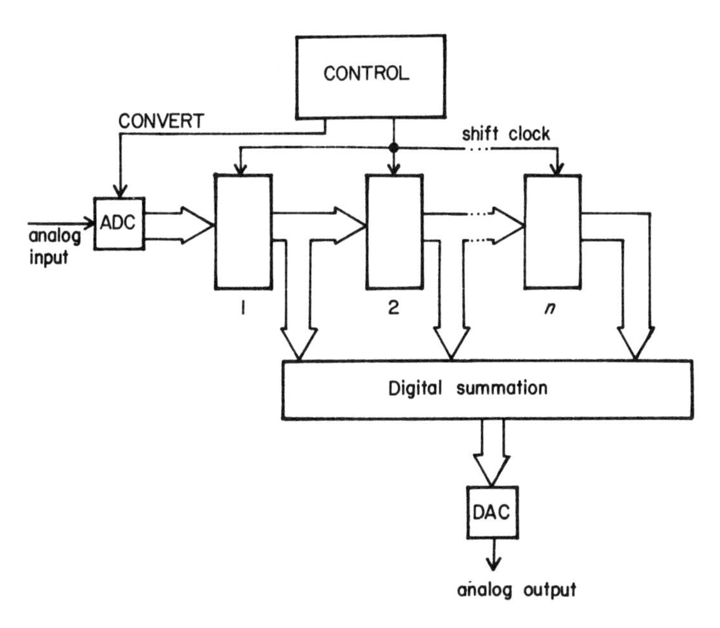

Figure 7.16 A digital filter using only one DAC.

systems which could be even more expensive using analog methods. Digital filter modules will probably appear on the market which will make this approach less expensive.

A less expensive approach which retains most of the advantages of digital filters is illustrated in Fig. 7.17. This device avoids the expense of A/D and D/A conversion by performing all operations in the analog domain, but still retaining the advantage of discrete time operation. It relies on the use of a technology which has not been described before in this volume: an *analog* shift register.

Analog shift registers consist of storage elements which can store analog voltages in a number of analog storage elements. These are interconnected in such a way that the input of each element comes from the output of the element to the left, and the output of each element

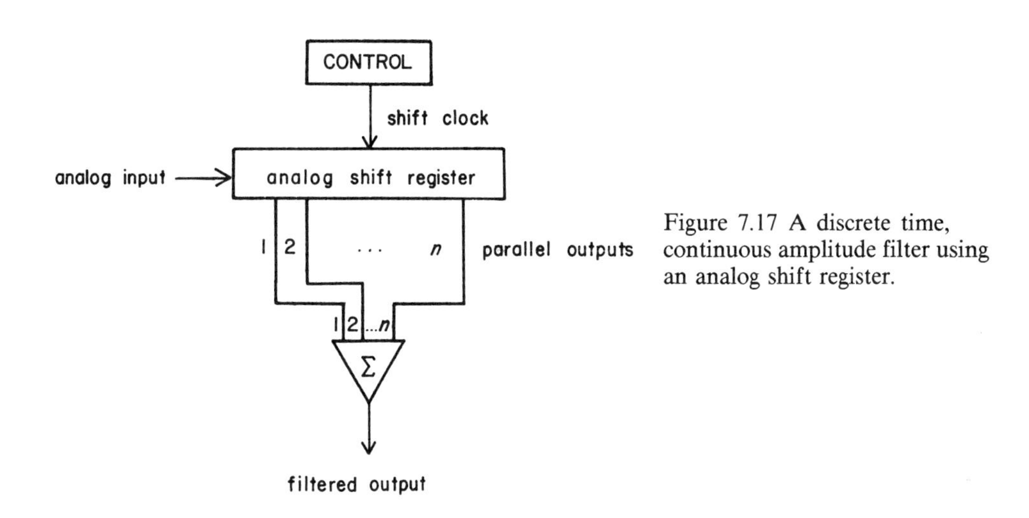

Figure 7.17 A discrete time, continuous amplitude filter using an analog shift register.

goes to the next element to the right, with the exception that the leftmost element has an external input (the serial input) and the rightmost element has an external output (the serial output). Thus, the analog shift register is entirely analogous to a digital shift register, except it stores analog rather than digital voltages.

The analog shift register of Fig. 7.16 has a serial input and parallel outputs. Each parallel output is equivalent to the corresponding DAC output of Fig. 7.14, except that it is continuous voltage rather than discrete voltage. All the parallel outputs constitute the inputs of the weighted summing amplifier. This method eliminates A/D conversion, digital processing, and D/A conversion, drastically reducing the cost of the device and retaining the advantages of discrete time devices (ability to simulate arbitrary linear devices by simulating their impulse responses). Analog shift registers constitute a relatively new technology, so it is difficult to predict the forms that such discrete time filters will take.

Problem

1 Design an averager, with 10-bit input, 1024 words, 20-bit sum capacity, and display of the sum. The display should be continually refreshed, at 100 refreshes/sec. Note that there must be two address counters: one used during an input sweep, and one during a display refresh. These must be multiplexed for memory address control; the input sweep must take precedence over the display refresh in accessing memory addresses.

References

Applications Engineering Department, Hybrid Systems Corporation. *Digital-to-Analog Converter Handbook*. Hybrid Systems Corp., Burlington, MA, 1970.

Brown, P. B., V. W. Maxfield, and H. Moraff. *Electronics for Neurobiologists*. MIT Press, Cambridge, MA, 1973.

Hoeschle, David F., Jr. *Analog-to-Digital-to-Analog Conversion Techniques*. Wiley-Interscience, New York, 1968.

Rabiner, Lawrence R., and Bernard Gold. *Theory and Application of Digital Signal Processing*. Prentice-Hall, Englewood Cliffs, NJ, 1975.

Rosenblith, A. (ed.). *Processing Neuroelectric Data*. MIT Press, Cambridge, MA, 1959.

Sheingold, Daniel H. (ed.). *Analog–Digital Conversion Handbook*. Analog Devices, Norwood, MA, 1972.

Stanley, William D. *Digital Signal Processing*. Reston Publ. Co., Reston, VA, 1975.

8

Principles of Radio Frequency Electronics and Electromagnetic Radiation

Howard Moraff

We know that an electrically charged particle exerts a force on other charged particles placed near it. The force represents a form of stored potential energy, which is distributed around the particle that produces it. This energy distribution, which is a function of position with respect to the source, is called a field. We observe that the stored energy, or field, has the properties of magnitude, or intensity, and direction at each point in space. The field set up by a stationary electric charge is called an electrostatic field. Similarly a magnet produces a field which can be observed by bringing another magnet nearby. If the first magnet is not moving, its field is called magnetostatic.

A static field is created by a charge that is fixed in position and intensity. If the charge (or magnetic source) is either moving or changing in intensity, a dynamic field will be produced. It is an interesting property of dynamic fields that a moving or changing electric field will produce a magnetic field. The result is a field which has both electric and magnetic components, and energy is continually transferred from one to the other form. Such a field is called an electromagnetic field.

An electric current consists of a stream of moving electrons. If we pass an electric current through a wire, an electromagnetic field is produced in the space around the wire. That field extends, with diminishing strength, over a distance which is theoretically infinite. If we place another wire, B, anywhere within the electromagnetic field produced by the current in wire A, then a current will be induced in wire B which is similar to (although smaller than) that in wire A. The effect has a wide variety of implications and applications in the design and use of electronic equipment.

On the one hand, it is often useful to be able to send signals from one place to another without using a direct connection between the origin and destination (in such applications as telemetry, telestimulation, and telecontrol). In radio telemetry,[1] for instance, a current, which is related to the signal of interest, is passed through a wire of suitable configuration at the location of the signal source, and the current induced in another wire at the receiving location is amplified and converted to a signal for measurement. The wires at the source (transmitting) and destination (receiving) sites are called *antennas*. These are designed for minimum loss of signal power in transmission between the sites, so that the maximum possible signal power reaches the receiving circuitry. The actual signal transmitted in radiotelemetry is normally a high-frequency (radio frequency or RF) sinusoidal signal called a *carrier*, upon which the signal of interest is impressed by a process called *modulation*. The receiver circuitry then can be tuned to amplify only signals in the frequency band of the modulated carrier signal, thereby rejecting signals such as those produced by radio stations, radiations from power lines, machinery, and other sources of interference and noise. The receiver demodulates the carrier signal, thereby extracting the modulation signal for measurement and analysis.

On the other hand, two wires which are reasonably close together in an electronic apparatus will to some extent transmit and receive signals related to the currents passing through those wires—an effect called crosstalk—and any wire in a circuit will receive signals (i.e., will have currents induced in it) due to the electromagnetic

[1]Telemetry, or the measurement of data over a distance, may use a variety of means to transmit signals, such as sonar, ultrasound, optical beams, etc., but radio transmission is currently the most common method.

fields surrounding it, from various interference and noise sources. The signal power received by an antenna is related to the length of the conductor with respect to the signal wavelength, and also to the orientation of the antenna in relation to the field of the transmitted signal. At higher frequencies, wavelengths are shorter and more signal power is captured in the receiving antenna. For example, the frequency range of the commercial FM radio broadcast band is 88–108 MHz. This corresponds to a wavelength range of about 2.8–3.4 m. Also, various electromechanical components (motors, relays, etc.) produce spikes with high-frequency components. A wire that is a meter or two in length may pick up a considerable signal from a local source; in many applications wires longer than a few millimeters must be shielded to minimize such pickup.

This chapter will present an introduction to the principles of radio frequency generation, transmission, reception, and processing, and the related areas of electromagnetic fields, antennas, and noise. An extensive reference list is provided for further study. The special considerations of the highest radio frequencies—the microwave frequency range—are beyond the scope of this book. Several references are given to volumes on microwave electronics, theory, and systems.

8.1 The Electromagnetic Spectrum

It is convenient to categorize electromagnetic fields according to the frequencies or range of frequencies they occupy in what is called the electromagnetic spectrum. The International Telecommunication Union has done this, in developing a set of Radio Regulations in 1959 in Geneva, Switzerland, which have been adopted by international treaty. Table 8.1 shows the frequency band designations and ranges. Note that the wavelength λ is related to the frequency f by the formula

$$\lambda = c/f,$$

where c is the speed of light, which is 3×10^8 m/s in a vacuum. Also note that the highest radio frequency, in band 12, overlaps slightly with the range of infrared waves which extend from 10^{12} to approximately 4×10^{14} Hz (3×10^{-4}–0.7×10^{-6} m). The visible and ultraviolet ranges lie just beyond that. In fact, in our ever expanding communications activities, we have already entered the visible range, and optical communications techniques are now widely used.

8.1.1 Usage

Electromagnetic waves perform various functions in the world of communications. Their transmission and propagation behaviors vary widely with frequency range, so that for each application there is normally a preferred range of frequencies which gives the best results. In addition, the electromagnetic spectrum is crowded with users, so that some systematic way of sharing it among those users is necessary. Hence, frequencies have been allocated, both by region of the world and by type of use. A short summary of frequency allocations within the United States is given in Table 8.2, which was derived from the more detailed listings given in Chapter 1 of the ITT Handbook (1975).

Authorities have by now designated radio frequency bands ranging from 30 Hz to beyond 3 THz (terahertz). Only a few years ago (in 1971) a Radio Amateur's

Table 8.1. Radio Frequency Bands in the Electromagnetic Spectrum

Wavelength (meters)	Frequency (hertz)	Frequency Band No.	Band Designation
10^{7}	3×10^{1}		
	}	2	ELF (Extremely Low Frequency)
10^{6}	3×10^{2}		
	}	3	VF (Voice Frequency)
10^{5}	3×10^{3}		
	}	4	VLF (Very Low Frequency)
10^{4}	3×10^{4}		
	}	5	LF (Low Frequency)
10^{3}	3×10^{5}		
	}	6	MF (Medium Frequency)
10^{2}	3×10^{6}		
	}	7	HF (High Frequency)
10^{1}	3×10^{7}		
	}	8	VHF (Very High Frequency)
$10^{0}\ (=1)$	3×10^{8}		
	}	9	UHF (Ultra High Frequency)
10^{-1}	3×10^{9}		
	}	10	SHF (Super High Frequency)
10^{-2}	3×10^{10}		
	}	11	EHF (Extremely High Frequency)
10^{-3}	3×10^{11}		
	}	12	
10^{-4}	3×10^{12}		

Handbook (published annually) regarded the domain of radio as extending from 15 kHz (being roughly the top of the audio range) to 30 GHz. A practical delineation would cover all frequencies for which useful transmission of electromagnetic waves has been accomplished, and that range will likely continue to expand.

8.2 RF Signals

In this section, we shall deal with the production of useful RF signals. Later, in Section 8.6, we shall discuss the problems of noise and interference caused by extraneous RF signals.

The general scheme for RF signal generation, transmission, and reception is depicted in Fig. 8.1. Several steps are involved in producing and using RF signals. The carrier signal is *generated* by circuitry which is basically some form of oscillator. The information-bearing signal is applied to the carrier in such a way as to vary, or *modulate*, some characteristic of the carrier such as its amplitude, frequency, or phase. The resulting signal, a modulated carrier, is amplified and supplied to some form of antenna. The antenna essentially transforms the electric signal to electromagnetic waves that can travel, or *propagate*, through air or space—a step called *transmission*. The transmitted waves may be *received* at the destination by another antenna which converts them back to an electric signal which is fed to the receiver ciruitry for amplification and *demodulation* or *detection*, the process by which the modulating signal is separated from the carrier. While these basic steps are covered

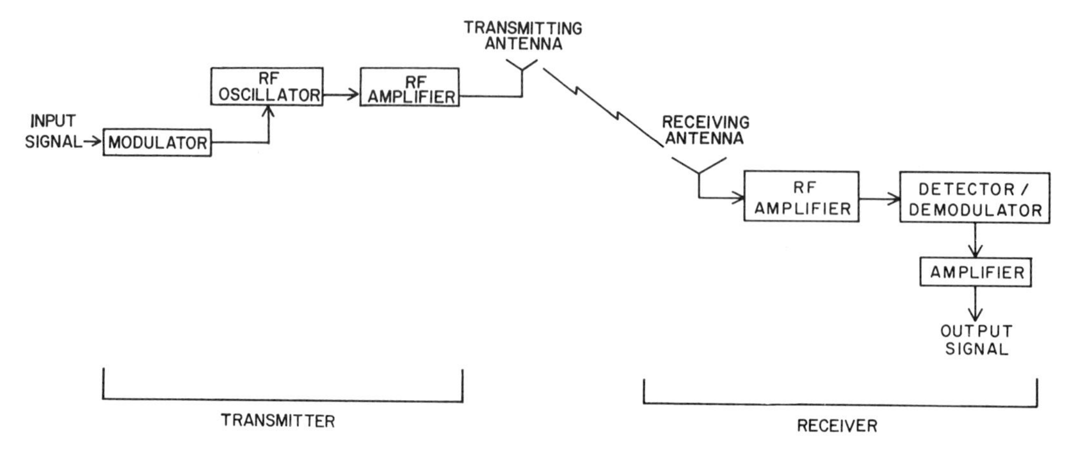

Figure 8.1. An RF system.

individually below, it should be noted that several may be combined in a single circuit module in a practical system.

8.2.1 Generation

There are many forms of circuits suitable for generation of RF signals. Each basically implements an oscillator which produces a carrier signal of the desired frequency. The basis of the oscillator is normally either a positive feedback circuit, a relaxation scheme, or a negative resistance element.

Within these three general categories, a variety of designs are appropriate for the RF range, including the Hartley, Colpitts, and Pierce feedback oscillators, astable multivibrators, and circuits which use a negative resistance device such as a unijunction transistor, IMPATT diode, Read diode, or tunnel diode. Basic schematics of some of these are shown in Fig. 8.2. At low frequencies, LC tuned oscillators tend to be unwieldy due to the size of the inductors. Other designs, such as the phase shift and Wien bridge oscillators are preferred over the tuned circuits for frequencies much below 1 MHz. Crystal oscillators, such as the Pierce, offer excellent frequency stability but cannot be directly frequency modulated. The prospect of integrated circuit implementation imposes additional constraints on the selection of a design. For instance, inductors for the lowest frequencies are difficult to fabricate as parts of integrated circuits.

Several texts provide descriptions and analyses of RF oscillator circuits, including Clynes and Milsum (1970), Zeines (1970), Miller (1966), Mackay (1970) and Caceres (1965). Zeines (1970) presents analyses of various FET and junction transistor oscillator designs, including Hartley, Colpitts, and other tuned circuits, phase shift, Wien bridge, and tunnel diode oscillators. Miller (1966) in Chapter 14 gives more detailed design considerations and techniques for some of these, and in Chapter 16 covers some specific designs for the VHF and UHF ranges.

In many telemetry applications, circuit size, weight, and power consumption must be kept as low as possible. The use of integrated circuits and the combining of functions such as oscillation, modulation, and amplification in the same circuit are two ways to satisfy these constraints. Clynes and Milsum (1970) present a survey of

Table 8.2. Summary of Frequency Allocations Within the United States

Use	Band										
	ELF 2	VF 3	VLF 4	LF 5	MF 6	HF 7	VHF 8	UHF 9	SHF 10	EHF 11	— 12
Aeronautical Mobile (around–air, air–air communication)				√	√	√	√	√			
Aeronautical Radio Navigation (radio beacons, ranges, landing systems, airborne radar)				√	√		√	√	√		
Amateur ("ham" radio)					√	√	√	√	√	√	
Broadcasting											
domestic AM					√						
international AM ("short wave")						√					
FM							√				
TV							√	√			
Citizens Band						√		√			
Fixed (nonmobile stations)			√	√	√	√	√	√	√		
Government (military, etc.)					√	√	√	√	√	√	
Land Mobile					√	√	√	√	√		
Maritime Mobile (ship–ship, ship–shore)				√	√	√	√				
Meteorological Aids (radiosondes, radar, satellite)							√	√	√		
Radio Astronomy						√	√	√	√	√	
Standard Frequencies			√	√	√	√					
Radio Location (coastal radar, tracking systems)				√	√			√	√	√	
Radio Navigation (radio beacons, shipboard radar, navigational systems, direction finding)			√	√	√			√	√	√	
Satellites							√	√	√	√	
Space Operation and Research							√	√	√	√	

oscillators for biotelemetry use; Mackay (1970), Caceres (1965), and Fryer (1970) show a variety of compact, low power consumption transmitter circuits for telemetry from inside living organisms and biotelemetry in space applications; and NASA (1972) has published a compilation of communications circuits in which they include (on p. 14) a schematic for a low-power compact 1–4 MHz voltage-controlled oscillator based on TTL NAND gates.

For many applications, packaged oscillators and other RF components are available from commercial vendors. These may be located via listings such as the annual Guide to Scientific Instruments (published by the American Association for the Advancement of Science), the Electronic Engineers Master catalog (United

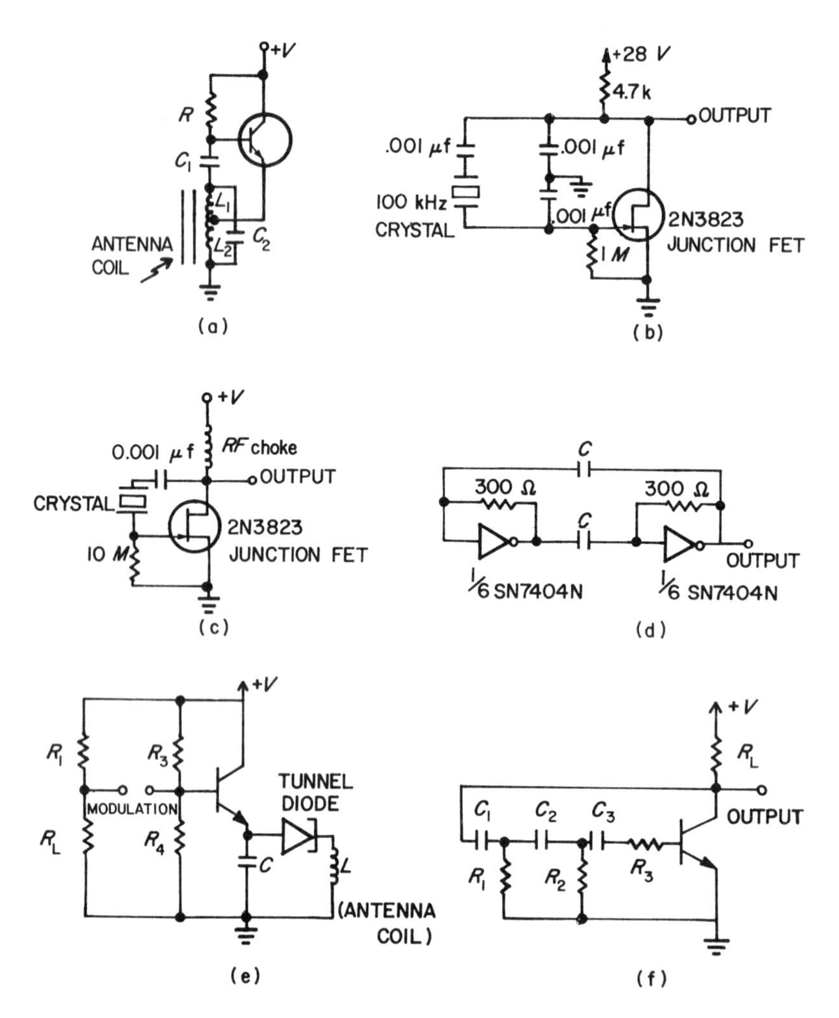

Figure 8.2. Some basic RF oscillator circuits. **(a)** Hartley oscillator. **(b)** Colpitts crystal oscillator (from Application Note AN32, February 1970, National Semiconductor Corp., by permission). **(c)** Pierce crystal oscillator (from Application Note AN32, February 1970, National Semiconductor Corp., by permission). **(d)** Oscillator using astable multivibrator with TTL integrated circuits (frequency is approximately 10 MHz for C = 150pf). **(e)** Transmitter using tunnel diode oscillator (from Clynes and Milsum, *Biomedical Engineering Systems*, Fig. 3.7a, p. 72, by permission). **(f)** Phase shift oscillator.

Technical Publications), the Electronic Design Gold Book (Hayden Publishing Co), and the Electronics Buyers Guide (McGraw-Hill). All are published annually and may be found in libraries and purchasing departments, or ordered from the publishers.

8.2.2 Modulation

Radio frequency transmission is essentially used to send information from one place to another. In some applications the signals are sent along a conductor (such as a microwave waveguide, a coaxial cable, or a pair of wires) called a transmission line.

In other applications, the signals are sent via a wireless route between two antennas. In all cases, it is the sending of information that is the objective of the process. There is very little information to be obtained from the reception of a steady carrier signal; all we know then is that the source is transmitting, and that neither the sender nor the transmission pathway is varying the signal.

For information to be transmitted, it must be impressed in some way upon the carrier signal. That process, in which the carrier signal is changed in some way by the information or input signal, is called *modulation*. It takes many forms. Any feature of the carrier signal may be modulated, but the features most commonly chosen are the amplitude, frequency, and phase.

For analysis, it is useful to regard the various techniques as being either *continuous* or *pulsed* modulation schemes, according to whether the modulated signal exists continuously or is switched on and off.

8.2.2.1 Continuous Modulation

For the purpose of analysis, it is useful to represent the sinusoidal carrier wave, $e_c(t)$, as the real part of a *vector* or *phasor* of length E_c (the peak amplitude of the sine wave) and angle $\theta_c(t)$ which is rotating as a function of time in the complex plane (Fig. 8.3). The trigonometry tells us that

$$e_c(t) = E_c \cos \theta_c(t) \tag{8.1}$$

The phasor rotates with an *angular frequency* (or angular velocity) ω_c, which is defined as

$$\omega_c = \frac{d\theta_c(t)}{dt} \tag{8.2}$$

Figure 8.3. Vector representation of RF signals: **(a)** sinusoidal carrier; **(b)** amplitude-modulated carrier; **(c)** frequency-modulated carrier.

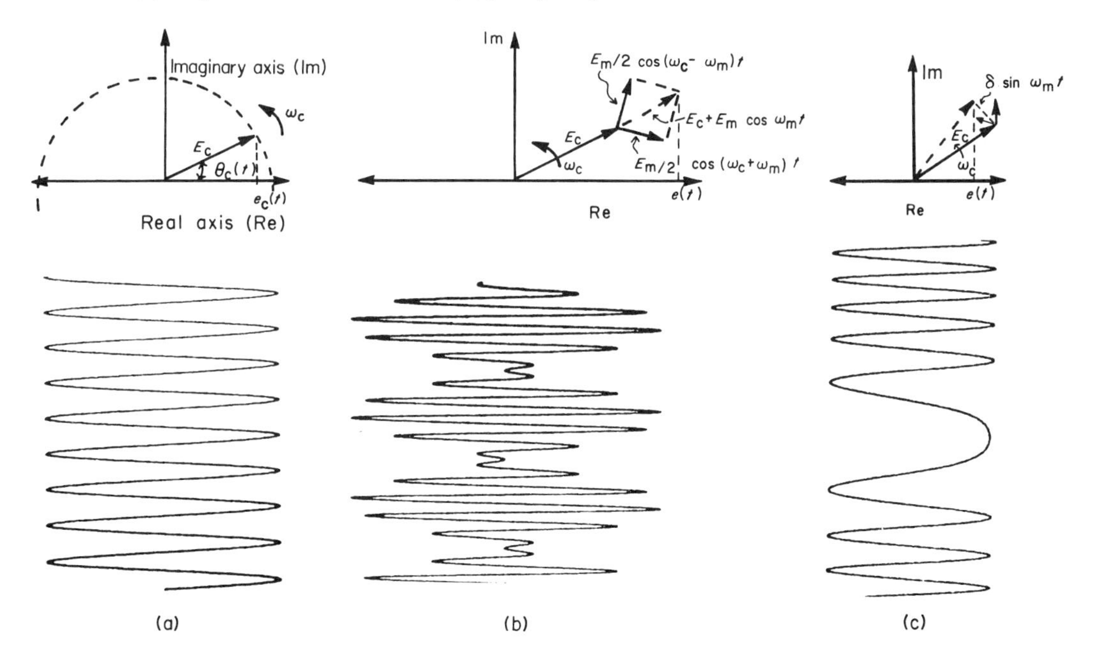

so that any given time t the instantaneous angular position of the vector is obtained by integrating (8.2). Thus

$$\int_0^t d\theta_c(t) = \int_0^t \omega_c \, dt$$

or

$$\theta_c(t) = \int_0^t \omega_c \, dt + \theta_0$$

where θ_0 is the initial angular position. Then

$$e_c(t) = E_c \cos\left[\int_0^t \omega_c \, dt + \theta_0\right] \tag{8.3}$$

Now, we observe that this signal has three parameters; its amplitude E_c, which is varied in amplitude modulation; its frequency ω_c, which is varied in frequency modulation; and its angular position or phase angle θ_c, which is varied in phase modulation.

8.2.2.1.1 Amplitude Modulation. If the frequency and phase variables in Eq. (8.3) are fixed, and the amplitude E_c is modulated, then we get a modulated carrier signal

$$e(t) = E_c(t) \cos \omega_c t \tag{8.4}$$

where the fixed θ_0 term does not matter and is therefore removed.

Let us consider a modulating signal

$$e_m(t) = E_m f(t)$$

where E_m is the peak amplitude and $f(t)$ is a time-varying function with amplitude $= 1$. Now let us impress this signal on the carrier in such a way that the amplitude of the carrier varies as a function of the modulation thus:

$$E_c(t) = E_c + E_m f(t)$$

or

$$E_c(t) = E_c[1 + mf(t)]$$

where $m = E_m/E_c$ is defined as the degree of modulation of the carrier.

Then Eq. (8.4) becomes

$$e(t) = E_c[1 + (E_m/E_c) f(t)] \cos \omega_c t \tag{8.5}$$

Now, consider the simple case of a modulating signal which is a pure sinusoid:

$$e_m(t) = E_m \cos \omega_m t$$

then

$$e(t) = E_c[1 + (E_m/E_c) \cos \omega_m t] \cos \omega_c t$$

or

$$e(t) = E_c \cos \omega_c t + E_m \cos \omega_m t \cos \omega_c t$$

We now use the identity

$$\cos x \cos y = \tfrac{1}{2}[\cos(x + y) + \cos(x - y)]$$

to obtain

$$e(t) = E_c \cos \omega_c t + \frac{E_m}{2} \cos(\omega_c + \omega_m)t + \frac{E_m}{2} \cos(\omega_c - \omega_m)t \tag{8.6}$$

Note that the resulting modulated signal is the sum of three terms: the original carrier, and two signals each of amplitude $E_m/2$ but with frequencies above and below the carrier frequency by an amount equal to the modulation frequency ω_m. These are called *sidebands*, for reasons that are apparent from the graph of the frequency spectrum of the modulated carrier (Fig. 8.4). In vector terms, the amplitude modulation process adds to the rotating carrier phasor of Fig. 8.3a the pair of counterrotating sideband phasors which comprise a cyclic increase and decrease in the amplitude of the carrier phasor. The modulated carrier is seen as the projection of this total phasor on the real axis. In practice, the modulation is not a simple sine wave, but may be complex. As a real signal, it may even be represented by an infinite series of sinusoidal components, so that the spectrum of the modulated wave will theoretically have an infinite array of upper and lower sidebands.

There are many ways to produce an amplitude modulated carrier signal. Perhaps the easiest is to allow the modulating signal to control the output amplitude of the carrier oscillator, typically by shifting the bias of some active element in the oscillator circuit. An example of this approach is shown in Fig. 8.2e. For higher output power requirements, an amplifier may follow the RF oscillator, and the modulation may be accomplished by causing the modulation signal to vary the gain of the amplifier. Various schemes are presented in the references; note especially Zeines (1970, Chapter 9).

Recall that the carrier does not itself convey information. This suggests that it need not be transmitted. Transmission of only the sidebands is called *suppressed carrier* transmission. This scheme may be implemented using a balanced modulator, whose inputs are the carrier and the modulation, and whose output is the sidebands of the AM waveform. If we can omit transmission of the carrier because it carries no information, why not also transmit only one set of sidebands? This may be done by filtering out the other sideband set. The difficulty here is in design of a filter sharp enough to cut off completely the bands on one side of the carrier while passing as fully as possible the bands on the other side. Methods and analysis are presented in Zeines (1970) and in Brown and Glazier (1964).

Double sideband, suppressed carrier transmission offers some advantage in signal to noise ratio over conventional AM by allowing the total transmitted power to be concentrated in the sidebands.

Figure 8.4. Frequency spectrum of amplitude-modulated carrier.

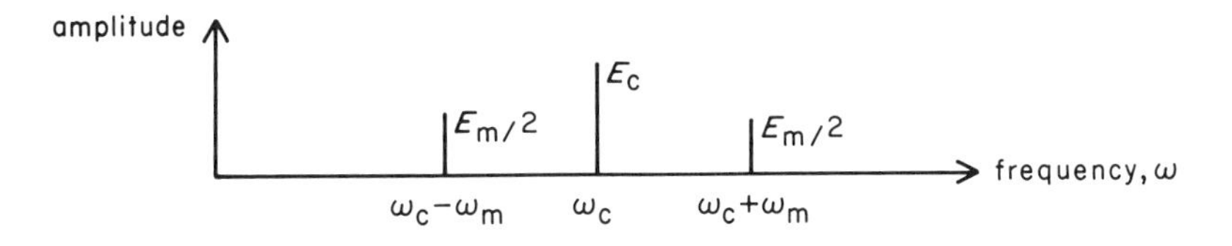

8.2.2.1.2 Frequency Modulation. Going back to Eq. (8.3), let us now hold the carrier amplitude E_c and the phase θ_0 fixed and modulate the angular frequency ω_c. Then the carrier may be described by

$$e_c(t) = E_c \cos\left[\int_0^t \omega_c(t)\,dt\right]$$

again discarding θ_0 as we did for the AM case. Now suppose the modulation signal is

$$e_m(t) = E_m \cos \omega_m t$$

and it is applied to the carrier signal so as to modulate its frequency thus:

$$\omega_c(t) = \omega_c + K_f E_m \cos \omega_m t$$

so that

$$e(t) = E_c \cos\left[\int_0^t (\omega_c + K_f E_m \cos \omega_m t)\,dt\right]$$

or

$$e(t) = E_c \cos(\omega_c t + K_f E_m / \omega_m \sin \omega_m t)$$

let $\delta = K_f E_m / \omega_m$; then

$$e(t) = E_c \cos(\omega_c t + \delta \sin \omega_m t) \tag{8.7}$$

Equation (8.7) may be expanded using $\cos(a+b) = \cos a \cos b - \sin a \sin b$. Then

$$e(t) = E_c[\cos \omega_c t \cos(\delta \sin \omega_m t) - \sin \omega_c t \sin(\delta \sin \omega_m t)] \tag{8.8}$$

Using the series expansions for sine and cosine,

$$\sin x = x - \frac{x^3}{3!} + \frac{x^5}{5!} - \cdots$$

and

$$\cos x = 1 - \frac{x^2}{2!} + \frac{x^4}{4!} - \frac{x^6}{6!} + \cdots$$

we can write

$$\cos(\delta \sin \omega_m t) = 1 - \frac{1}{2!}(\delta \sin \omega_m t)^2 + \frac{1}{4!}(\delta \sin \omega_m t)^4 - \cdots$$

and

$$\sin(\delta \sin \omega_m t) = \delta \sin \omega_m t - \frac{1}{3!}(\delta \sin \omega_m t)^3 + \frac{1}{5!}(\delta \sin \omega_m t)^5 - \cdots$$

Some algebra and collection of terms, aided by a bit of judicious precognition, lead us to the form

$$\begin{aligned}\cos(\delta \sin \omega_m t) = {} & [1 - (\delta/2)^2 + (\delta/2)^4/2!\cdot 2! - (\delta/2)^6/3!\cdot 3! + \cdots] \\ & + 2\cos(2\omega_m t)[(\delta/2)^2/2! - (\delta/2)^4/3! + (\delta/2)^6/3!\cdot 5! - \cdots] \\ & + 2\cos(4\omega_m t)[(\delta/2)^4/4! - (\delta/2)^6/5! + (\delta/2)^8/2!\cdot 6! - \cdots] \\ & + \cdots\end{aligned} \tag{8.9}$$

and similarly

$$\begin{aligned}\sin(\delta \sin \omega_{\mathrm{m}} t) = & 2 \sin \omega_{\mathrm{m}} t\left[(\delta/2) - (\delta/2)^3/2! + (\delta/2)^5/2!\cdot 3! - \cdots\right] \\ & + 2\sin 3\omega_{\mathrm{m}} t\left[(\delta/2)^3/3! - (\delta/2)^5/4! + (\delta/2)^7/2!\cdot 5! - \cdots\right] \\ & + 2\sin 5\omega_{\mathrm{m}} t\left[(\delta/2)^5/5! - (\delta/2)^7/6! + (\delta/2)^9/2!\cdot 7! - \cdots\right] \\ & + \cdots \end{aligned} \tag{8.10}$$

We have essentially derived, by relatively straightforward algebraic means, a Fourier series expansion for the modulated carrier wave. A more elegant approach exists which involves solving a differential equation which is related to the integral form of expression of the Fourier coefficients. This equation is called Bessel's equation, and has the form

$$x^2 \frac{d^2 J_n(x)}{dx^2} + x\frac{dJ_n(x)}{dx} + (x^2 - n^2) J_n(x) = 0 \tag{8.11}$$

A solution to Eq. (8.11) has the form

$$J_n(\delta) = \sum_{m=1}^{\infty} \frac{(-1)^m (\delta/2)^{2m+n}}{m!(m+n)!} \tag{8.12}$$

which depicts a family of functions called Bessel functions. Some of the details of the Bessel function treatment of this problem are given in Appendix B of Brown and Glazier (1964).

The series of terms in (8.9) and (8.10) convey a striking suggestion of relationship to the form obtained from (8.12). Indeed, matching the corresponding powers of $(\delta/2)$ yields the following substitutions:

$$\cos(\delta \sin \omega_{\mathrm{m}} t) = J_0(\delta) + 2\left[J_2(\delta)\cos 2\omega_{\mathrm{m}} t + J_4(\delta)\cos 4\omega_{\mathrm{m}} t + \cdots\right] \tag{8.13}$$

and

$$\sin(\delta \sin \omega_{\mathrm{m}} t) = 2\left[J_1(\delta)\sin \omega_{\mathrm{m}} t + J_3(\delta)\sin 3\omega_{\mathrm{m}} t + \cdots\right] \tag{8.14}$$

Inserting the expressions of (8.13) and (8.14) in (8.8) and collecting terms yields the final result:

$$\begin{aligned} e(t)/E_{\mathrm{c}} = & J_0(\delta)\cos \omega_{\mathrm{c}} t - J_1(\delta)\left[\cos(\omega_{\mathrm{c}} - \omega_{\mathrm{m}})t - \cos(\omega_{\mathrm{c}} + \omega_{\mathrm{m}})t\right] \\ & + J_2(\delta)\left[\cos(\omega_{\mathrm{c}} - 2\omega_{\mathrm{m}})t + \cos(\omega_{\mathrm{c}} + 2\omega_{\mathrm{m}})t\right] \\ & - J_3(\delta)\left[\cos(\omega_{\mathrm{c}} - 3\omega_{\mathrm{m}})t - \cos(\omega_{\mathrm{c}} + 3\omega_{\mathrm{m}})t\right] \\ & + J_4(\delta)\left[\cos(\omega_{\mathrm{c}} - 4\omega_{\mathrm{m}})t + \cos(\omega_{\mathrm{c}} + 4\omega_{\mathrm{m}})t\right] \\ & - \cdots \end{aligned}$$

or

$$e(t)/E_{\mathrm{c}} = \sum_{n=-\infty}^{\infty} J_n(\delta)\cos(\omega_{\mathrm{c}} + n\omega_{\mathrm{m}})t \tag{8.15}$$

The vector representation and waveform for the FM case is depicted in Fig. 8.3c. Note that the peak amplitude of the modulated carrier is constant; only the frequency varies with time. It is also clear, from Eq. (8.15) and Fig. 8.5, that an FM signal contains an infinite number of sidebands, whose spacing is equal to the

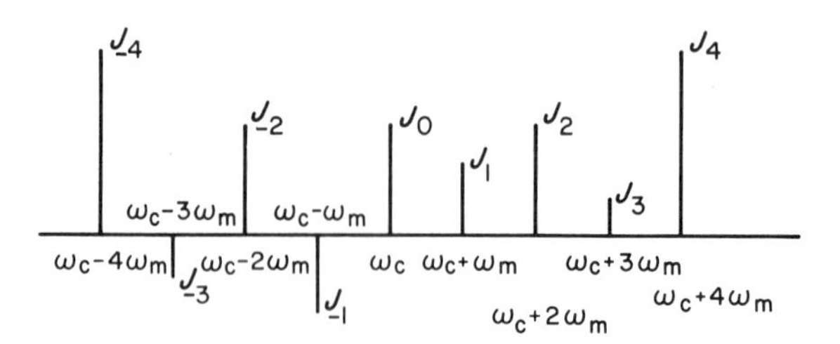

Figure 8.5. Frequency spectrum of frequency-modulated carrier.

modulation frequency. This suggests that the bandwidth of the FM signal is infinite, but as Brown and Glazier (1964) show, the contributions of the sidebands which are distant from the carrier frequency are small, so that relatively narrow band transmission is practical.

8.2.2.1.3 Phase Modulation. We saw, in Section 8.2.2.1.2, that we could modulate the frequency of an RF carrier signal while holding its amplitude and phase angle constant. This type of modulation may be referred to as *direct* frequency modulation. If, on the other hand, we were to hold the carrier amplitude and frequency constant, and modulate its phase, the resulting signal would still, for practical purposes, be regarded as an FM signal, but this phase modulation (PM) is regarded as *indirect*. FM and PM are two types of modulation of the angular position of the rotating phasor which represents the carrier signal. To see how FM and PM are related, we recall the relation between carrier frequency and phase given by Eq. (8.2):

$$\omega_c(t) = \frac{d\theta_c(t)}{dt}$$

This equation tells us that the instantaneous frequency ω_c of the rotating phasor is equal to the present rate of change of its angular position θ_c. In FM, the maximum excursion $\Delta\omega_{c(max)}$ of the instantaneous frequency is held constant over all modulating signal frequencies, and the maximum excursion $\theta\Delta_{c(max)}$ of the carrier's phase angle is inversely proportional to the modulation frequency ω_m and directly proportional to the modulating signal.

In PM, the maximum phase excursion $\Delta\theta_{c(max)}$ is held constant over all modulating frequencies, and the maximum frequency excursion is proportional to ω_m. For a fixed modulating frequency ω_m it turns out that

$$\Delta\omega_{c(max)} = \omega_m \, \Delta\theta_{c(max)}$$

and FM and PM are indistinguishable. There are differences in practical use and circuitry implementation, but those are outside the scope of this text. Brown and Glazier (1964) point out that a phase modulated carrier can be produced by a frequency modulation system, by using the derivative of the modulating signal as the input to the frequency modulator.

It should be noted that in both FM and PM, the transmitted power is constant, and both schemes offer immunity to the amplitude distortions that plague AM transmission systems.

8.2.2.2 Pulse Modulation

We have seen that information may be imposed on an RF carrier signal by modulating its amplitude, frequency, or phase. With an ever increasing need for expansion of communications capabilities, and particularly a need for increased range and reliability of transmission, the limitations of the traditional approach of continuous modulation of sinusoidal carriers have stimulated development of other methods, which may be categorized as discontinuous or pulse modulation techniques. In these schemes, the carrier is a periodic pulse train. The information-carrying signal is caused to modulate some feature of the pulses, such as their amplitude or duration, or their positions in time relative to their expected periodic occurrence times (Fig. 8.6). Alternatively, the transmitted signal may consist of a pulse code which carriers amplitude information about the modulating signal.

8.2.2.2.1 Pulse Amplitude Modulation (PAM). If the amplitudes of successive pulses in the RF carrier are made proportional to the modulation signal amplitude at successive sample times, we have produced pulse amplitude modulation (Fig. 8.6c). Since the information (i.e., the amplitude of the modulating signal) is transmitted in analog form, it is subject to degradation due to noise and distortion imposed by the transmission process. Also, in any sampling process, care must be taken to assure that the sampling rate exceeds twice the highest frequency of the

Figure 8.6. Some forms of pulse modulation.

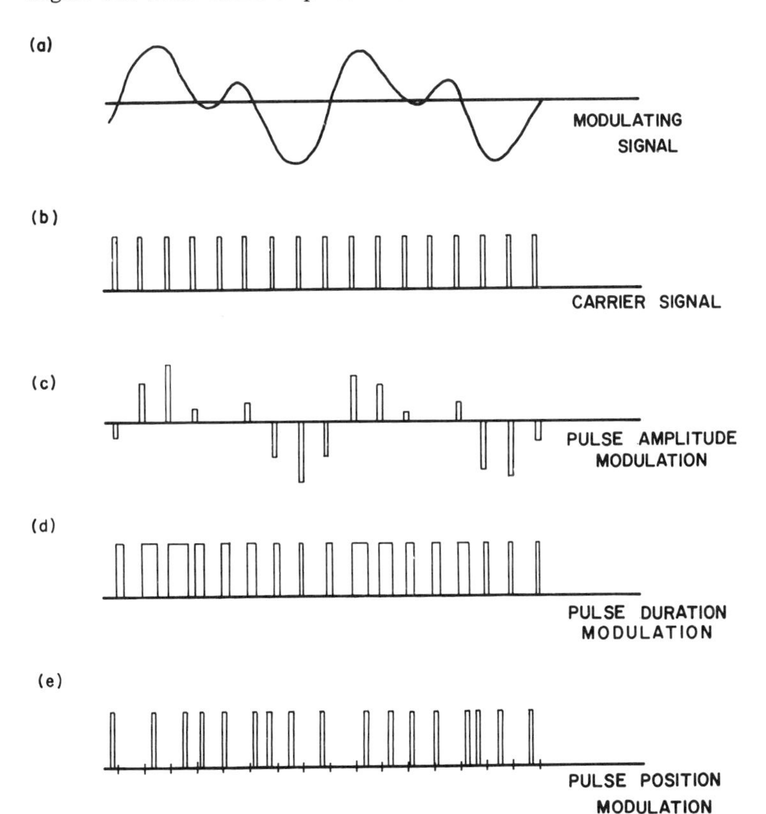

input (modulating) signal, in accordance with Shannon's sampling theorem, and in actual systems a factor of at least five is normally used. This represents an important consideration in determining the limit of bandwidth of information to be transmitted by pulse modulation schemes.

8.2.2.2.2 Pulse Duration Modulation (PDM). If the amplitudes of the pulses in the carrier train are held constant, and their durations are varied in proportion to the amplitudes of the corresponding time samples of the modulating signal (Fig. 8.6d), we have pulse duration modulation (also called pulse width modulation). An advantage of PDM over PAM is that since in PDM the transmitted pulses are all supposed to have the same amplitude, the effects of amplitude distortions due to transmission are eliminated; however, the bandwidth requirements of the transmitting and receiving systems are higher due to the need for precise timing information.

8.2.2.2.3 Pulse Position Modulation (PPM). If both amplitude and duration of the carrier pulses are held constant, and successive carrier pulses are advanced or delayed in time in proportion to the amplitudes of the modulating signal samples, we have pulse position modulation (Fig. 8.6e). In this scheme, the effects of all types of distortions of pulse shape are eliminated, but again, extra bandwidth is required to preserve the precise timing information.

8.2.2.2.4 Pulse Code Modulation (PCM). In this scheme, the modulating signal amplitude is again samples periodically, but instead of using a periodic pulse train as the carrier, we translate each amplitude sample into a digital code consisting of a series of pulses and spaces as shown in Fig. 8.7. Now, the receiving equipment simply decides whether a pulse is present or not, at each point in time during the code sequence. This scheme therefore achieves better immunity to noise and distor-

Figure 8.7. Pulse code modulation. Arrows indicate sample times.

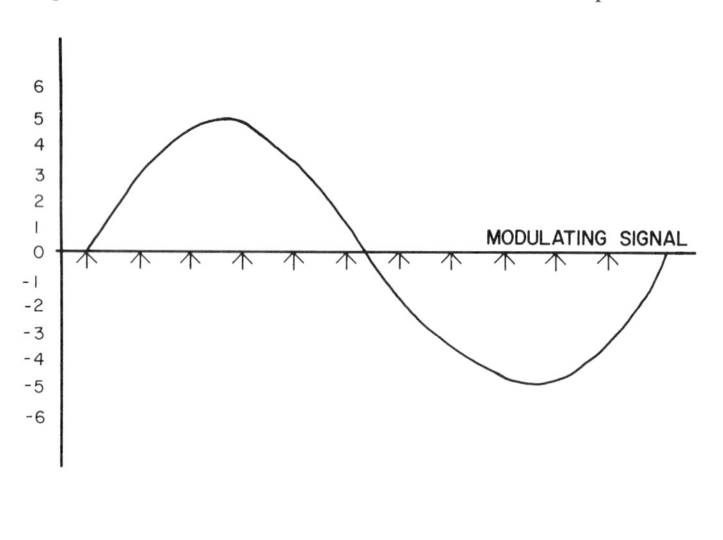

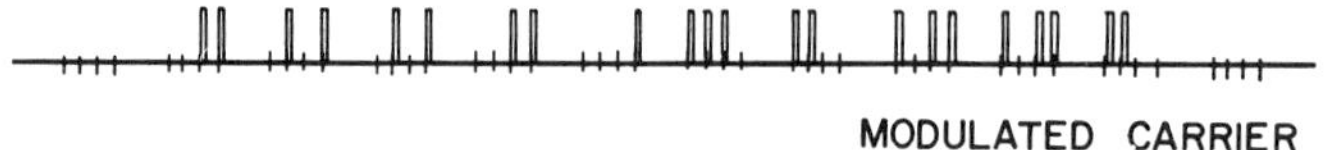

tion of pulse shapes, and even to propagation delays up to a limit, than do the preceding schemes. Again, due to the number of pulses transmitted per sample, a relatively high bandwidth is required compared with that of the PAM scheme. Also, in contrast with all of the preceding schemes, which transmitted more or less faithful analogs of the successive samples, PCM introduces a new error component due to quantization of the sample amplitude. The amount of such quantizing error that is acceptable dictates the length of the code required to represent the signal.

8.2.2.2.5 Other Pulse Modulation Schemes. For a variety of reasons, other pulse modulation schemes have been developed. For example, in pulse interval modulation (PIM), the length of the interval between successive carrier pulses is made proportional to the amplitude of the signal sample at that time. In a more complex scheme, called pulse interval ratio modulation (PIRM), the ratio of the lengths of two successive intervals between carrier pulses is made proportional to the signal sample. The PIRM scheme is more resistant to errors due to transmission distortions and voltage fluctuations in the transmitter power source (which in telemetry applications may be a battery).

A scheme of pulse modulation which reduces the quantizing noise inherent in PCM is differential PCM. In this scheme, the differences between amplitudes of successive samples are encoded, rather than the sample amplitudes themselves. It should be noted that a problem with differential PCM schemes is that errors tend to be cumulative.

A special case of differential PCM in which the coding is limited to one binary digit (or bit) is called delta modulation (Fig. 8.8). In operation, the integrators in

Figure 8.8. Delta modulation: **(a)** system; **(b)** signals.

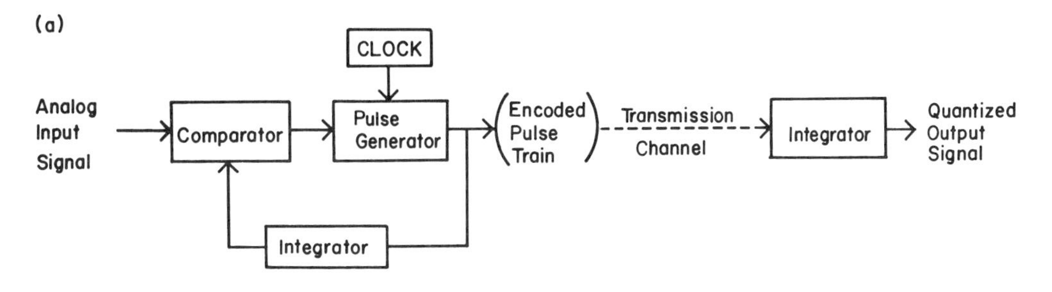

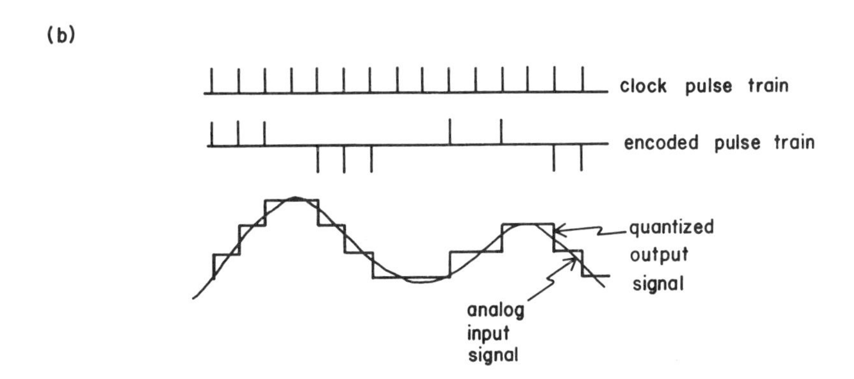

both the encoder and decoder reconstruct a stepped or quantized approximation of the analog input or modulating signal. The comparator causes the pulse generator (synchronized by the clock) to produce a positive pulse when the input exceeds the output of the integrator, and a negative pulse when the input is less than the integrator output. Delta modulation offers the advantages of PCM in general. In addition, where bandwidth is limited, its natural characteristic of precisely coding slow signal changes and imprecisely coding the fastest changes is well suited to signal transmission applications such as black and white television. Some additional material on delta modulation is given by Maschhoff (1964).

8.2.2.3 Multiplexing

We have looked so far only at a class of modulation schemes by which the information signal is impressed directly on the RF carrier signal, resulting in a modulated carrier that is referred to as AM, FM, PCM, etc.

In many data transmission applications, more than one information signal must be transmitted simultaneously. Because space among the usable radio frequency bands is quite limited, it may be impractical to transmit a separate modulated carrier for each information signal. Instead, schemes are often used in which the information signals are *multiplexed* (added together or combined) in a way such that they may be *demultiplexed* (separated from each other) at the receiver. One way of doing this is to sample each of the several information signals in turn, periodically, and impress each sample in sequence onto the carrier. This approach is termed *time division multiplexing*. At the receiver, the demodulated signal is dissected back into the samples corresponding to the original array of input signals, assuming that some means is used for synchronizing the receiver's sampling circuitry with that of the transmitter.

Figure 8.9. A subcarrier system: **(a)** transmitter; **(b)** receiver.

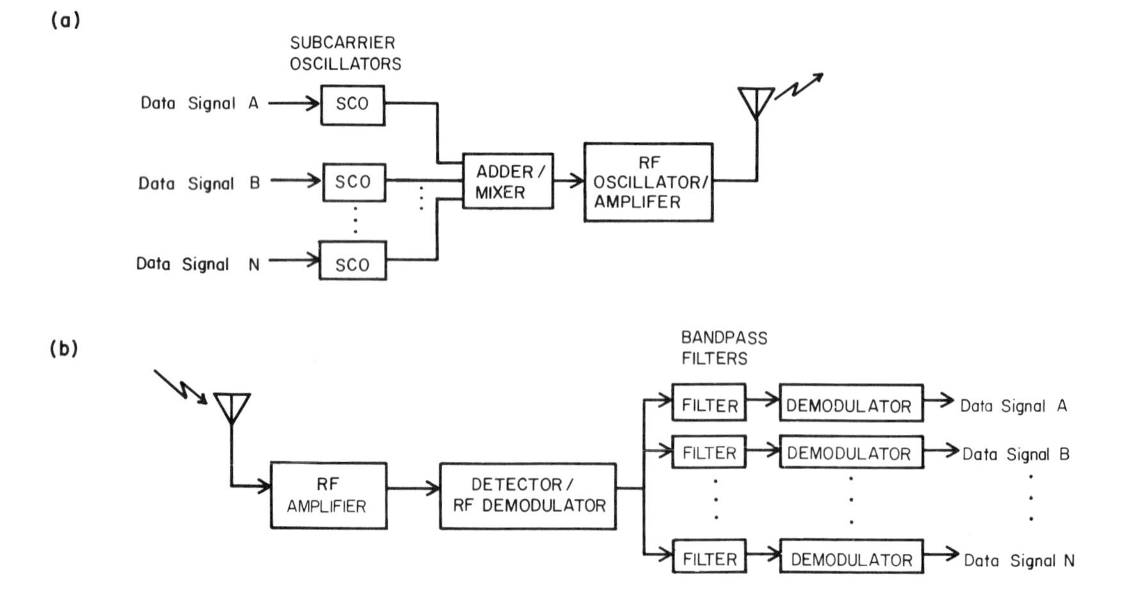

Because of the synchronization problem in time multiplexing, a preferable way of combining signals is by the use of *subcarriers* (Fig. 8.9). A subcarrier is a signal at a frequency intermediate between that of the information signal and that of the main RF carrier. Each information signal modulates its own subcarrier, and the modulated subcarriers are added together. The combined signal is then used to modulate the RF carrier. This approach is termed *frequency division multiplexing*. Since each subcarrier has a different frequency, separation of the signals at the receiver simply involves demodulation of the combined subcarrier signals from the RF carrier, followed by bandpass filtering for each information channel and then demodulation of each information signal from its subcarrier. Because of the reduction in transmitted bandwidth used per information signal (which is itself a benefit), subcarrier transmission reduces the exposure of the transmitted signal to interference and noise.

It is not necessary that the same type of scheme be used for subcarrier modulation as for modulation of the RF carrier. In fact, a wide variety of combinations is used in practice. For instance, the subcarriers may be frequency modulated and the RF carrier may be amplituded modulated. Such a scheme would be designated as FM/AM. The convention used in such designation is that the scheme for subcarrier modulation is given first, followed by the scheme for RF carrier modulation.

Signals which are combined by time division multiplexing may be transmitted by continuous modulation schemes. Thus, we can have a PCM/FM system, in which one or more signals are encoded via pulse code modulation, and the resulting composite signal frequency modulates an RF carrier. We may even encode individual signals using PCM, combine them by frequency modulating a collection of subcarriers, and use the composite of the modulated subcarriers to frequency modulate the RF carrier: such a scheme would carry the designation PCM/FM/FM.

8.2.3 Transmission

Recall the overall RF system diagram of Fig. 8.1. Having *generated* an RF carrier signal (Section 8.2.1) and *modulated* it (Section 8.2.2), we are now ready to *transmit* it.

Depending on the means of generation and modulation, and on the transmitted power required (more for long distance transmission, less for local), it may or may not be necessary to incorporate an RF power amplifier in the transmitter. One notable advantage of employing such an amplifier is that without it, the RF oscillator is directly coupled to the antenna or transmission line, which represents a load impedance that may fluctuate, causing unwanted variations in the transmitted frequency. Practical design information for RF amplifier circuits is outside the purpose of this text, but the subject is well covered in Miller (1966).

How do we get the signal from the transmitter to the receiver? Basically, there are two ways. For reasonably short distances, in some applications we may connect the transmitter and receiver together directly by means of a *transmission line* (Section 8.5). For longer distances and for applications in which direct wire connections are unacceptable, the signal will be sent directly through the ambient medium, e.g., air, space, and ocean, in which case the signal will be coupled to the medium by means of a specialized conductor called an *antenna* (Section 8.4).

8.2.4 Reception

When a signal arrives at its destination, having traveled some distance over transmission lines and antennas, and through air or space, it has typically become considerably attenuated. Moreover, with our crowded airwaves, it may be competing for its place in the electromagnetic spectrum with a host of other signals, some deliberately and some inadvertently transmitted, and with just plain noise in a variety of forms.

The receiving antenna can often help the situation by optimizing its directional sensitivity and its size relative to signal wavelength, as described in Section 8.4. But the real burden of signal recovery rests with the receiver electronics. Signals ranging from millivolts to tiny parts of a microvolt must be selectively detected and enlarged and converted to audible or otherwise usable amplitudes. Modulation and coding must be separated from the RF carrier signal, without introducing undue distortions. Fluctuations due to varying conditions of propagation may need to be compensated for.

In this section, we will depict the basic form of the RF receiver and discuss its performance characteristics, and in the following section we will describe various methods for separating the modulation from the carrier.

8.2.4.1 The Receiver

The basic receiver system is shown in Fig. 8.10. Not all elements shown here will be incorporated in every design: the original crystal radio employed an antenna, a semiconducting crystal as the detector or demodulator, and a set of headphones, which enabled operation without audio amplification.

The RF amplifier, when used, may consist of one or more stages of amplification. It is designed to increase the amplitude of the received signal over an appropriate range of frequencies, while introducing as little noise and signal distortion as possible, thereby presenting a substantial signal to the demodulation section so that it may extract the original modulating signal with maximum fidelity. In one simple design for an AM receiver, signal (station) selection is done by tuning the RF amplifier. The tunable RF section is followed by the detector and audio amplifier. Such a scheme is called a *tuned radio frequency*, or *TRF*, *receiver*. A problem with this design is the requirement for high gain in a tunable range: if multiple stages of amplification are needed, each must be individually tuned, and all tuned stages must track closely in frequency. A great deal of narrow band RF gain may be obtained by

Figure 8.10. Basic receiver.

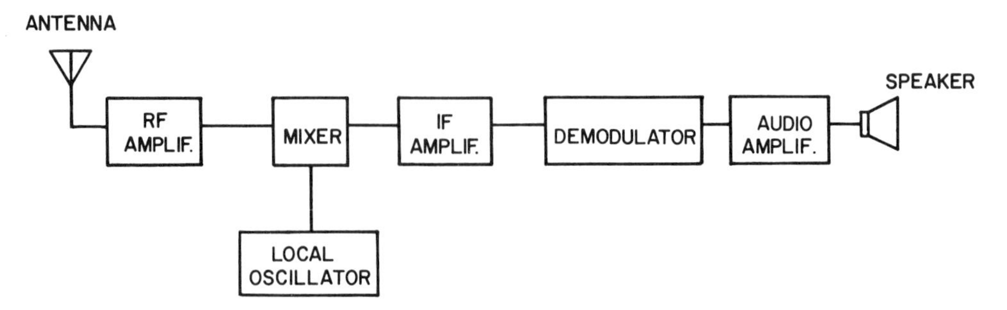

using a controlled amount of positive feedback in the RF amplifier. The active device in this amplifier may also provide detection; this scheme is called *regenerative detection*, and a receiver which utilizes positive feedback in this way is called a *regenerative receiver*. Since the objective is to maximize the sensitivity and selectivity, the positive feedback is made as great as possible, just short of inducing oscillations. A tendency for oscillations is inherent in this design; such oscillations will produce a local RF signal which radiates back through the receiving antenna and may interfere with local reception of other RF signals.

If the receiver is to cover a substantial band of frequencies, achievement of sufficiently high gain over the wide RF frequency range may be difficult. A popular alternative is to use only a modest amplification in the RF section, and then use a *mixer* circuit to beat, or *heterodyne*, the RF signal against a signal of a somewhat higher frequency, produced by a local oscillator in the receiver. Such a scheme is called a *superheterodyne receiver*, and is depicted by the full diagram of Fig. 8.10. The mixer circuit essentially converts the modulated carrier from its original high frequency to a considerably lower *intermediate frequency* (IF), which is the difference between the original carrier frequency and that of the local oscillator. For instance, in reception of the commercial AM broadcast band, the local oscillator is tuned to 455 kHz above the frequency of the RF carrier as broadcast by the desired station. Then the mixer output will have a center frequency of 455 kHz.

The output of the mixer is coupled to one or more stages of IF amplification. Here, we see the major advantage of the superheterodyne receiver: the amplification is mainly done at the single lower, intermediate frequency, and may be made very narrow band, achieving good sensitivity *and* selectivity among nearby signals. Tuning is done by adjusting the local oscillator, which may be designed to be quite stable.

Heterodyning is especially useful in FM receivers, where the high sensitivity normally required would involve more gain at RF than could readily be managed with stability.

In FM systems, there is an inherent insensitivity to noise, much of which appears as amplitude modulation of the FM carrier. However, most FM demodulation circuits will respond to such amplitude variations to some extent. To maximize the noise immunity of the receiver, an amplitude *limiter* section is usually incorporated after the last IF stage. The limiter is essentially a saturating amplifier whose output signal is of fixed amplitude as long as the input exceeds a certain threshold amplitude (Fig. 8.11). In the absence of such a signal, a rushing or hissing sound of approximately white noise may be heard at the receiver output. The receiver is tuned to a carrier signal of adequate strength, the limiting takes effect and the receiver *quiets*: that is, the noise at the output is substantially reduced or eliminated. The sensitivity of an FM receiver may be specified in terms of the amplitude of input signal required to achieve a given degree of quieting, as described in the next subsection on receiver performance.

The demodulator section of a receiver is where the information-carrying modulation signal is finally separated from the carrier. A variety of techniques may be employed, as described in Section 8.2.5. The demodulator may be a separate circuit, or its function may be incorporated within a stage of amplification.

In an FM receiver, demodulation consists of converting the modulation, in the form of fluctuations of the carrier frequency, to amplitude fluctuations that comprise

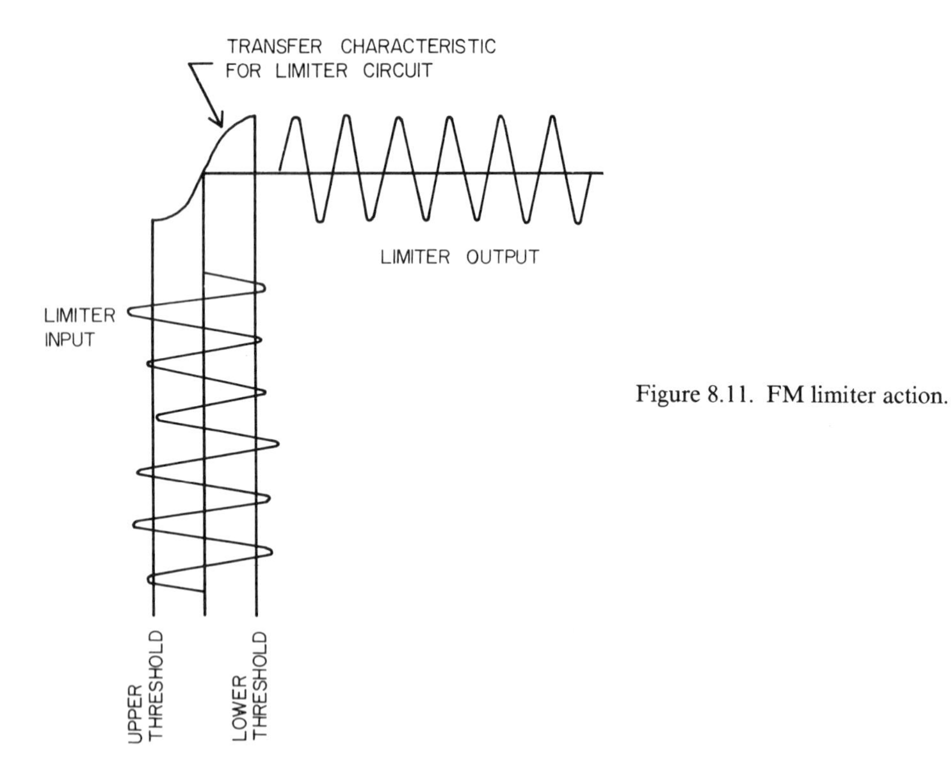

Figure 8.11. FM limiter action.

a signal in the audio range. Any drift in receiver tuning will produce a signal in the subaudio frequency range at the output of the demodulator. In many FM receivers, a section of circuitry called *automatic frequency control* (AFC) takes the demodulator output, extracts the slowly changing signal due to drift by the use of a low-pass filter, and generates a proportional correction signal which is fed back to adjust the receiver tuning.

In the final step of the reception process, the modulation signal, as recovered by the demodulation circuitry, is normally passed through one or more stages of low-frequency amplification. This may be a simple audio amplifier driving a speaker, or a special purpose amplifier driving a recorder, actuator, or other device as desired.

8.2.4.2 Receiver Performance

Since the function of an RF receiver is to recover an information-bearing signal from a complex signal transmitted over a distance, we may discuss the *performance* of a receiver in terms of how well it does that job, i.e., how small a signal it can usefully recover (its *sensitivity*); how well it can extract the signal of interest from the crowded spectrum of other transmitted signals and noise (its *selectivity*); and, how well it can maintain its hold on the narrow, precise frequency band which contains the signal of interest (its *stability*).

8.2.4.2.1 Sensitivity. The quality of the RF amplifier section in a receiver is a major factor in determining the sensitivity of the receiver, for it is here at the front end that the most delicate operation is accomplished—recovering the modulated carrier signal, which may be tiny, after its journey from the transmitter.

Since the gain–bandwidth product is essentially fixed for a given amplifier design and device selection, RF amplifiers are usually tuned to as narrow a frequency range as possible, compatible with signal distortion considerations, in order to maximize the gain, and therefore the receiver's sensitivity, at the frequencies of interest. A receiver which is to handle more than one transmitted frequency (e.g., a broadcast band receiver) will typically provide for tuning of its RF amplifier stage(s) by adjustable capacitors or inductors.

The *sensitivity* of a receiver is defined as the minimum signal strength in microvolts at the receiver input to produce a specified output. Thus, sensitivity is the input signal required to produce a given output ratio of signal to noise, or of signal to noise and distortion (SINAD); for example, a receiver might be said to have a sensitivity of 0.6 μV for 10-dB SINAD. This gives an indication of how well the output signal stands out against the background noise, but is not an ideal measure because it is affected by the bandwidth of the receiver. Just as any resistor or conductor produces a tiny noise potential due to thermal agitation of its electrons (known as *Johnson noise*), so any practical amplifier contributes to its output signal a noise component due to thermal and other random effects in its circuitry. For simplicity of handling this phenomenon, a measure called the *equivalent noise resistance* is often applied to amplifiers. The equivalent noise resistance is the resistance which, placed at the input to the amplifier, would produce (at room temperature) an output noise level equal to that produced by the amplifier itself. This measure of sensitivity is independent of frequency, but depends on bandwidth. In terms of evaluating noise performance of a receiver, we may note that the noise produced in the first RF amplification stage will normally be the overriding determinant of the noise performance of the receiver, due to the successive amplification factors. An ideal receiver would add no noise to the received signal. Then the smallest usable signal would be determined solely by the noise produced in the antenna and in the propagating medium. How closely an actual receiver approaches this noise-free ideal is indicated by the measure called *noise figure*. The noise figure of a receiver is defined as the ratio in decibels of the signal-to-noise power ratio of the ideal receiver with the same bandwidth to the signal-to-noise power ratio of the actual receiver. A figure of a few decibels is typical in quality receiver performance in the low megahertz range. This measure is independent of receiver bandwidth.

All these measures of sensitivity evaluate only the linear operations of the receiver (i.e., amplification). In FM systems, the mode of demodulation also affects the noise performance of the receiver, and so for FM receivers, a different type of measure is used. A characteristic of the typical FM receiver is the white noise which appears at its output in the absence of a carrier signal at the receiver input. The degree to which that output noise is reduced, or "quieted," by an input signal is an effective measure of the receiver's sensitivity. Thus, for example, a receiver may be rated at "2 μV for 20 dB of quieting", meaning that a 2-μV input signal will reduce the output noise by 20 dB.

8.2.4.2.2 Selectivity. Another key measure of the performance of a receiver is its *selectivity*, i.e., its ability to detect and enhance (amplify) signals in the desired frequency range, while rejecting all other inputs. This is typically accomplished by the use of one or more cascaded tuned bandpass filters operating within the amplifying section of the receiver. The sharpness of the frequency cutoff of the overall tuned circuitry in a receiver determines the selectivity of the receiver. It depends on the tuning sharpness of each stage and on the number of stages. Selectivity is usually specified as a graph of the signal attenuation over a frequency range which includes the band of interest, or may be given as the degree of attenuation at a given frequency deviation from the tuned center frequency, e.g., "20 dB down at 10 kHz," or, in the case of multichannel receivers, simply the decibel attenuation of channels adjacent to the one in tune. Ideally, the receiver would have a bandwidth just wide enough to receive the carrier frequency and the sidebands associated with the modulation, so that signal distortion is minimized, and it would also have a cutoff sharp enough to attenuate adjacent frequencies so that they are not noticeable in the received signal.

8.2.4.2.3 Stability. Finally, the *stability* of a receiver is an indication of its ability to remain tuned to a precise frequency under conditions of varying gain settings, line or supply voltage, and thermal and mechanical environment. This may be specified in terms of drift of tuning frequency, in hertz per hour. The overall receiver stability is affected by stability of power supplies, local oscillators, and tuned circuits.

8.2.5 Demodulation

The process of separating the modulating or information signal from a modulated carrier is called *demodulation* or *detection*. The objective of the demodulator is to recover a signal which approaches being an exact replica of the original modulating signal.

Various types of demodulators are used in receivers, depending on the type of modulation scheme, the intended end use of the receiver, and cost and other design considerations.

8.2.5.1 AM Demodulation

In amplitude modulation applications, the simplest form of demodulation circuit is the diode detector, shown in basic form in Fig. 8.12. This circuit detects the positive envelope of the modulated carrier, which, as long as the degree of modulation m (as defined in Section 8.2.2.1.1) does not exceed 100%, is a replica of the original modulating signal. The diode rectifies the carrier wave, and the capacitor–resistor combination serves as a low pass filter, or integrator, which removes the carrier frequency and produces the envelope waveform.

Figure 8.12. Diode detection for AM.

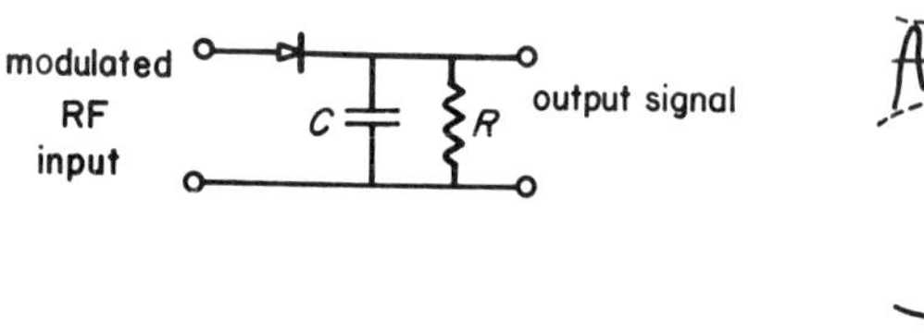

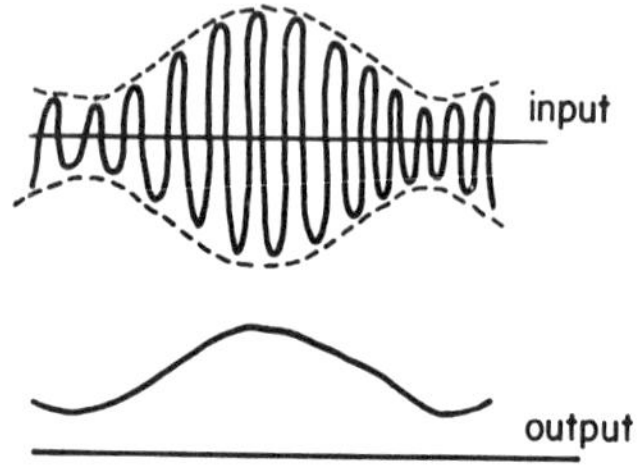

Note that if m reaches unity, the envelope minima will reach the zero point. If m should exceed unity, a distortion consisting of a folding or reflection of the envelope wave about the axis would occur. For a treatment of this situation and its consequences, see Brown and Glazier (1964, pp. 58–60). Normally, this overmodulation condition is avoided, and m is held below unity.

We may observe that the diode detector just described operates on one polarity of the modulated carrier in a linear fashion, if we assume that the diode is an ideal rectifier (i.e., has zero forward resistance and infinite reverse resistance). However, a real diode has a nonlinear forward characteristic: the forward resistance is significant at signal potentials up to several hundred millivolts. This threshold effect makes the diode, or envelope, detector useful only in cases where the carrier signal is much larger than the background noise. This condition is normally satisfied in conventional AM transmission. In single- and double-sideband systems, however, the carrier may not be transmitted, and other detection schemes are used. Also, even in conventional AM systems, it is often convenient to combine the detection process with RF amplification, for circuit economy. In the latter case, the amplification reduces the threshold effect. In any event, the detection is often accomplished by an active device which also provides amplification.

In the cases of single-sideband (SSB) and double-sideband (DSB) transmission, the detection process is essentially the reverse of the modulation process. A carrier signal is required in the receiver: it may be produced there by an oscillator, which is phase synchronized to the transmitted carrier signal if present. Such systems involve *synchronous* detection, whereby the incoming signal is multiplied by the locally generated signal, then filtered and detected. The scheme works whether or not the local signal is synchronized to a transmitted carrier. The synchronization improves the signal-to-noise ratio of the receiver by a factor of 3 dB.

8.2.5.2 FM Demodulation

Basically, the objective in demodulating an FM carrier signal is to convert the deviations from the center frequency of the carrier to proportional amplitude fluctuations. A circuit having the transfer characteristic depicted in Fig. 8.13a would accomplish this. Such a circuit is called a *frequency discriminator*. Practical circuits normally behave more like Fig. 8.13b; the degree of modulation must be limited to lie within the useful range of the discriminator. Circuits which implement this transfer characteristic include the Foster–Seely discriminator and the slope detector, both of which utilize the voltage differences produced by frequency deviations when the FM signal is transformer-coupled to a pair of tuned circuits. A circuit called the *ratio detector* is similar in appearance to the discriminator and slope detector, but operates on the ratio of the voltages produced in the coupling transformer. The ratio of these voltages is of course far less affected by amplitude fluctuations in the input signal than is their difference, so that the ratio detector has an inherent limiting function that, in many applications, reduces the requirement on the limiter stages. The conventional frequency discriminator requires a difficult and critical alignment procedure, which has been eliminated in an improved circuit called the crystal discriminator. In this version, the tuning of the transformer secondary circuit is accomplished precisely by a quartz crystal.

Another approach to FM demodulation involves the use of a phase-locked loop (PLL) device, which is now available as a single integrated circuit chip. The basic

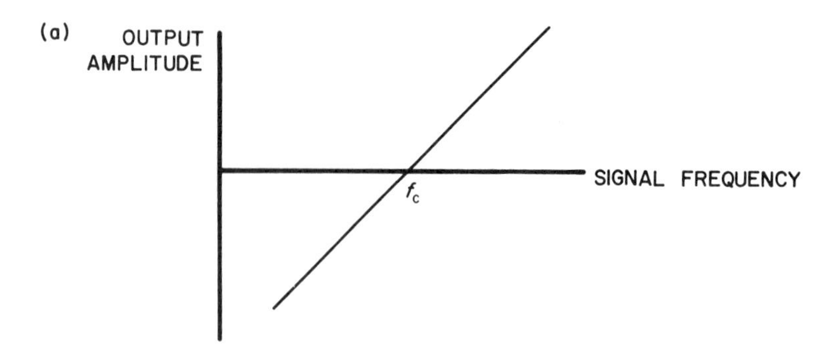

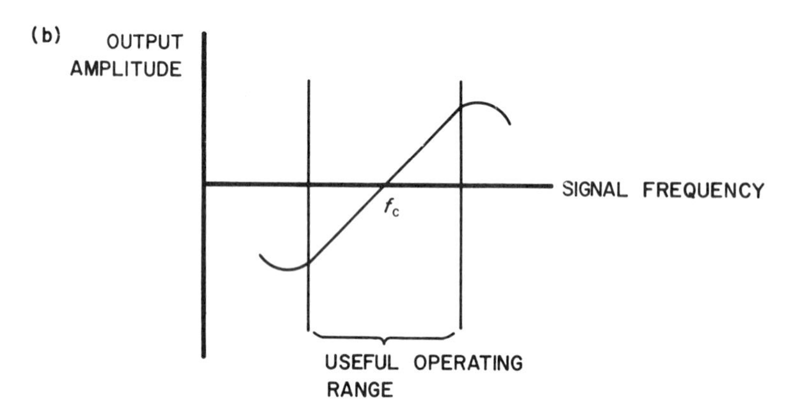

Figure 8.13. Frequency discriminator characteristics: **(a)** ideal transfer; **(b)** practical transfer. f_c = carrier frequency.

circuit is shown in Fig. 8.14. The voltage-controlled oscillator (VCO) produces a signal whose frequency is controlled to be very close to that of the FM input signal. Any frequency difference causes the phase detector to produce a voltage, which is smoothed, amplified, and fed back to the VCO to correct its output, thereby locking its frequency to that of the incoming signal. The voltage fed back to the VCO is proportional to the frequency deviation in the input signal and so is usable as the demodulator output.

A more complete presentation of theory and design of FM demodulators (as well as other aspects of FM systems) is given in Kiver (1960).

8.2.6 Standards

In search of compatibility of equipment among the various vendors, and compatibility of transmitted or recorded data, technologists have agreed upon a variety of standards. There are standards for transmission frequency utilization, standards for formats and characteristics of various transmitted and recorded signals, and standards for performance of RF equipment.

For instance, the International Telecommunication Union (ITU) in Geneva, Switzerland, an agency of the United Nations, allocates frequencies and designates

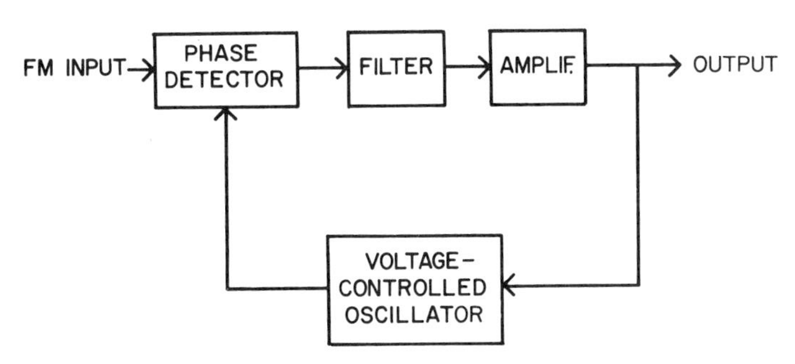

Figure 8.14. Phase-locked loop demodulator.

their uses (see Tables 8.1 and 8.2). The ITT Handbook (1975) devotes the whole of Chapter 30 to broadcast and recording standards, and Chapter 32 to standards in navigation systems. The Inter-Range Instrumentation Group (IRIG) has published a document (revised every 2 years) on telemetry standards, covering general frequency requirements, transmitter and receiver systems, pulse code modulation (PCM) and pulse amplitude modulation (PAM), and standards for magnetic tape recording and reproducing equipment and materials. The prospect of great increases in the use of satelite and space vehicle RF transmissions for public and commercial broadcasting as well as military and scientific applications suggests that the need for international standards will increase accordingly. An organization active in this regard is the International Radio Consultative Committee (CCIR) in Geneva.

Within the United States, the major responsibility for development of RF transmission standards and regulating their use is the Federal Communications Commission, in Washington, DC. Any RF transmission (above a specified low threshold of power, which depends on the frequency used) requires FCC licensing. The *Rules and Regulations* of the FCC are published periodically, and are available from the Superintendent of Documents, US Government Printing Office, Washington, DC.

Another aspect of standards is the transmission of standard signals and carrier frequencies. In the United States, the National Bureau of Standards operates a number of special broadcast stations, including WWV (at 2.5, 5, 10, 15, 20, and 25 MHz) in Colorado; WWVH (at 2.5, 5, 10, 15, and 20 MHz) in Hawaii; WWVB (at 60 kHz) in Colorado; and WWVL, an experimental multifrequency VLF station in Colorado. Details of the transmissions of these and numerous other standard frequency and time stations worldwide are given in Chapter 1 of the ITT Handbook (1975).

8.3 Electromagnetic Fields and Waves

In order to converse about antennas, transmission lines, and the propagation of RF signals, it will be useful to establish a basic vocabulary of symbols, terms, and concepts relating to electromagnetic fields and waves.

Recall that in the introduction to this chapter we introduced the notion that the electromagnetic field is produced by movement of an electric or a magnetic field: this movement in turn is produced by acceleration or deceleration of an electrically or magnetically charged particle or object. The electromagnetic field may be quantified in terms of four quantities: the *electric field strength* **E**, the electric flux density **D**, the magnetic field strength **H**, and the magnetic flux density **B**.

The (electric or magnetic) field strength (also known as *field intensity*) is defined as the force exerted by the field on a unit (electrical or magnetic) charge which is concentrated at a single point in the field. The units of **E** are volt/meter, and of **H**, newton/weber. The electric flux density **D** is related to the electric field strength **E** by

$$\mathbf{D} = \varepsilon \mathbf{E} \ (\text{coulomb/meter}^2)$$

where ε is the electric permittivity of the surrounding medium (in coulomb/volt-meter). The magnitude of the electric field strength due to an electric point charge Q at a distance r from that charge is $Q/4\pi\varepsilon r^2$. If another charge Q' were placed at distance r from charge Q, the force on Q' due to the electric field of Q would be

$$\mathbf{F} = Q'\mathbf{E}$$

The magnetic flux density **B** is related to the magnetic field strength **H** by

$$\mathbf{B} = \mu \mathbf{H} \ (\text{weber/meter}^2)$$

where μ is the magnetic permeability of the surrounding medium (in henry/meter). The magnitude of the magnetic field strength **H** due to a point magnetic charge (or pole) of strength M at a distance r from that charge is $M/4\pi\mu r^2$. If another magnetic charge or pole M' were placed at a distance r from charge M, the force on M' due to the magnetic field of M would be

$$\mathbf{F} = M'\mathbf{H}$$

It should be noted that the four basic quantities **E**, **D**, **H**, and **B**, which characterize the electromagnetic field, and the quantity **F**, are *vector* quantitities. That is, they represent spatial direction as well as quantity. Note that we gave expressions only for the *magnitudes* of electric and magnetic field strengths in the preceding discussion: the *direction* is the direction of the radius vector from one charge (or pole) to the other. Thus, for the above-mentioned cases, we may write

$$\mathbf{E} = \frac{Q\mathbf{r}}{4\pi\varepsilon r^3}$$

and

$$\mathbf{H} = \frac{M\mathbf{r}}{4\pi\mu r^3}$$

where **r** is the *vector* distance from Q to Q' or from M to M'. Note that since **r** has magnitude as well as direction, we must insert an additional distance factor in the denominator by raising the *scalar* r to the third power.

The four vector quantities **E**, **D**, **B**, and **H** are related by *Maxwell's field equations*, which form the mathematical basis for describing the behavior of electromagnetic fields.

8.3.1 Maxwell's Field Equations

Building on the work of Ampere and Faraday, Sir James Clerk Maxwell developed a set of four equations which set forth the basic relations among the electric and magnetic field variables.

Maxwell's first equation states that the work (or magnetomotive force) needed to move a unit magnetic charge around a closed path in a magnetic field of strength **H** is equal to the total current j_T linking that path, i.e., the total current passing through any surface bounded by that path. This current is the sum of two components; one being simply the current j_C *conducted* through the surface, and the other being a *displacement* current j_D due to the changing electric flux ϕ_D through the surface. The displacement current is equal to the partial time derivative of the electric flux, so that in equation form

$$\oint \mathbf{H} \cdot d\mathbf{l} = j_T = j_C + j_D = j_C + \frac{\partial \phi_D}{\partial t} \tag{8.16}$$

where $\oint$ denotes the line integral around a closed path and $d\mathbf{l}$ is a vector of length dl along the path.

The second of Maxwell's equations states that the work needed to carry a unit positive electric charge around a fixed closed path (i.e., the electromotive force induced within that closed path), in an electric field of intensity **E**, is equal to the negative of the time rate of change of the magnetic flux ϕ_B induced through that path. That is,

$$\oint \mathbf{E} \cdot d\mathbf{l} = -\frac{\partial \phi_B}{\partial t} \tag{8.17}$$

Maxwell's third equation states that the total electric flux due to a charge Q is equal to Q in magnitude, so that

$$\int_S \mathbf{D} \cdot d\mathbf{S} = Q \tag{8.18}$$

where $\int_S$ is the integral over the entire closed surface S, **D** is a vector expressing the electric flux density, S is any closed surface containing Q, and $d\mathbf{S}$ is a vector normal to an element of S (and representing that element).

Maxwell's fourth equation, also known as Maxwell's law of magnetic field flux, states in essence that all magnetic flux lines are closed paths, so that there are no free magnetic charges (i.e., sources or sinks of magnetic flux do not exist). In equation form,

$$\int_S \mathbf{B} \cdot d\mathbf{S} = 0 \tag{8.19}$$

where $\int_S$ is the integral over a closed surface S, **B** is the magnetic flux density, $d\mathbf{S}$ is a vector element of surface S as before. In other words, any magnetic flux line *leaving* through closed surface S must also *return* through it.

While Maxwell's equations are stated for individual electric charges and magnetic poles, they are valid for fields involving large numbers of charges and lines, as long as the fields are linear (so that the principle of superposition may be applied). These four equations completely specify the behavior of a time-varying electromagnetic field.

The four equations (8.16)–(8.19) as given above are the *integral form* of Maxwell's field equations. A derivative form is also useful, and is given below:

$$\text{curl } \mathbf{H} = j_C + \frac{\partial \mathbf{D}}{\partial t} \tag{8.20a}$$

$$\text{curl } \mathbf{E} = -\frac{\partial \mathbf{B}}{\partial t} \tag{8.20b}$$

$$\text{div } \mathbf{D} = \rho \tag{8.20c}$$

$$\text{div } \mathbf{B} = 0 \tag{8.20d}$$

Here, *curl* is a vector operator which is a directed partial derivative representing a rotational velocity per unit area, and *div* is a vector operator representing a flux or flow, per unit volume, out of a region.

Manipulation of Maxwell's field equations by means of vector mathematics, which is beyond the scope of this text, can lead to derivation of the wave equations for homogenous space:

$$\nabla^2\mathbf{E}=\mu\varepsilon\frac{\partial^2\mathbf{E}}{\partial t^2} \quad \text{and} \quad \nabla^2\mathbf{H}=\mu\varepsilon\frac{\partial^2\mathbf{H}}{\partial t^2} \tag{8.21}$$

For further explanations and details of the origins and derivations of Maxwell's field equations and the wave equations, the reader is referred to an excellent treatment by Johnk (1975).

Maxwell's field equations depict the relations among the electric and magnetic field variables. The wave equations represent solutions to Maxwell's field equations. The simplest wave solutions are *uniform plane-polarized waves*. If three-dimensional space is represented by orthogonal x, y, and z coordinate axes, we may consider electromagnetic waves as having components in the x, y, and z directions. In general, at any moment, the instantaneous electric (**E**) and magnetic (**H**) fields are perpendicular. For the special case of plane waves, **E** and **H** continually lie in some plane, such as the x, y plane. The wave would then travel, or propagate, in the z direction. In such a case, the *magnitudes* of **E** and **H** are related by

$$E/H=\eta \tag{8.22}$$

where

$$\eta=\sqrt{\mu/\varepsilon}$$

Such uniform plane waves are a good approximation for many real situations of practical application of RF transmission. The quantity η has the physical dimension of impedance (ohms), and is called the *intrinsic impedance* of the medium. Its value for vacuum (free space) is 377 Ω.

The transmission of a sinusoidal RF signal produces waves which are harmonic solutions to Maxwell's equations for the fields involved. The solutions involve harmonic factors of the form $\cos(\omega t+\theta)$, where θ is a phase angle and ω is an angular frequency. The *wavelength* λ of the traveling wave in free space is related to the angular frequency by

$$\lambda=2\pi/\left(\omega\sqrt{\mu_0\varepsilon_0}\right)=c/f$$

where c is the speed of light and f is the frequency of the wave in hertz. The phase velocity, or velocity of propagation v_P, of the wave in free space is given by

$$v_P=1/\sqrt{\mu_0\varepsilon_0}=c$$

8.4 Antennas and Propagation

We have seen that the transport of electric charges by a varying electric current produces electromagnetic waves. Whether this is the result of an intentional process which we have initiated for the purpose of communicating information at a distance without wires, an unintentional byproduct of our use of electronic apparatus, or a natural occurrence, it is worthwhile examining the conditions for producing and propagating these waves, as well as for inhibiting their propagation in unwanted directions. Means for inhibiting wave propagation by the use of electromagnetic shielding materials will be discussed in Section 8.6. In this section, we will discuss the principles of RF antennas and the propagation of waves through natural media. The patterns of radiation fields and waves may be predicted by the use of Maxwell's field equations together with suitable boundary conditions, as is done in Chapter 11 of Johnk (1975).

8.4.1 Basic Antennas

While any conductor through which a varying current is passed will produce a moving electromagnetic field, and while a current will be induced in any conductor which is placed in such a field, it is possible to enhance this action by prudent design of the configuration of the conductor.

A conductor which either radiates an electromagnetic field because of current flow, or develops a current due to a surrounding field, is acting as an *antenna*. If the objective is to make use of this antenna action (rather than to suppress it), then it is desirable to maximize the amount of radiation (for a transmitting antenna) or the amount of current induced (for a receiving antenna). An antenna has an impedance, which depends on its size and shape. The surrounding medium also has an impedance, which for free space is 377 Ω, essentially purely resistive. The antenna is connected to the transmitter or receiver by a conductor (transmission line) which also has an impedance associated with it. To transfer the maximum amount of signal power, these impedances should all be equal; otherwise some of the signal power will be reflected back at the connection point or interface, and lost. The impedance of the antenna itself is a function of its shape. The feed transmission line may be impedance—matched to the antenna at one end and to the transmitter (or receiver) at the other, by the use of special matching sections of line or by loading coils, using techniques described in the Radio Amateur's Handbook.

One of the simplest antennas (Fig. 8.15) is a straight wire, cut in the middle of its length and fed by a transmission line which consists of a pair of wires leading to the transmitter (or receiver, for a receiving antenna). This antenna configuration is called a *dipole* antenna. If a dc potential were applied continually to the dipole, the radiation pattern would be as shown. Only the **E** field is shown in the figure. The **H**-field of this static field at any point is everywhere perpendicular to it. The strength is inversely proportional to the cube of the distance of that point from the

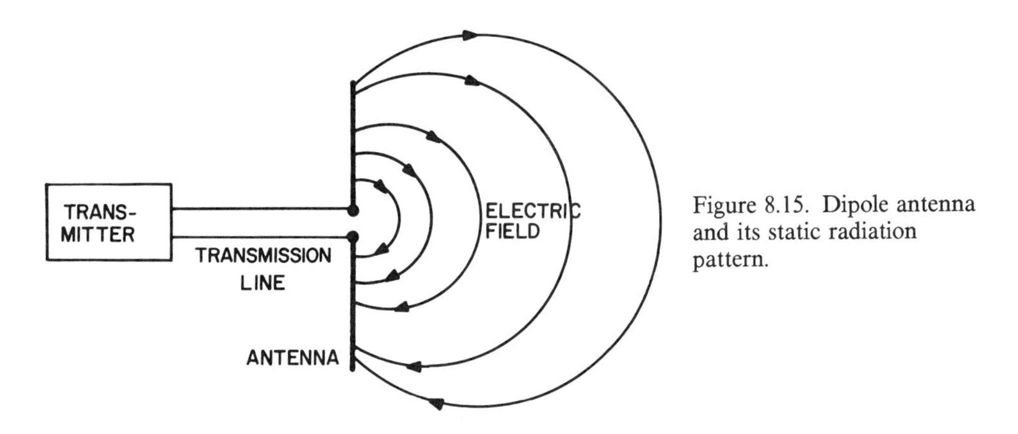

Figure 8.15. Dipole antenna and its static radiation pattern.

antenna. While the drawing does not show it, the field is really three-dimensional, with rotational symmetry about the antenna as an axis. The static field is too weak at any distance to provide useful transmission, and besides, the steady field does not carry any information.

For antenna current at radio frequencies, the field pattern is very different, as seen in Fig. 8.16. Because field propagation is not instantaneous, but occurs at a finite velocity (i.e., the speed of light), the actual field pattern varies dramatically during each cycle of the driving signal. The propagation velocity is fast enough that some of the energy does manage to return to the antenna during each cycle. This energy is effectively stored in the antenna and its nearby field. The rest of the energy emanating from the antenna gets far enough away that it effectively gets another "push" during the next cycle, so that it continues to radiate (propagate) away from the antenna, and never returns. Thus, it is the fact that energy is radiated at less than infinite velocity that enables any propagation at all.

As one might expect, observation of the field near an antenna (the *near field*) will show two components, the stored energy and that which will propagate, while observations at a distance from the antenna (*far field*) will show essentially only the propagated energy, since the stored energy falls off so rapidly with distance. The stored energy predominates at distances up to a small fraction of a wavelength, and is essentially gone when the distance reaches a few wavelengths.

The finite length of a dipole antenna also affects the field pattern. A simplified analysis procedure is based on the theory of the elementary dipole. This analysis makes the assumption that the current in the antenna has constant magnitude and phase everywhere along the length of the antenna. Hence, this analysis produces good approximations only if the antenna length is no more than a small fraction—of the order of one tenth—of the shortest wavelength of the signal. Expressions for field strength of small dipole antennas are given in Chapter 27 of the ITT Handbook (1975).

Let us consider a dipole antenna whose length is such that the current from one RF signal can traverse the length of the antenna in a round trip in exactly the duration of a single cycle of the signal. The round trip will cover a distance of exactly the wavelength of the signal. Thus the antenna length will have to be made exactly one-half wavelength. Since the next wave of signal will start exactly when the first round trip of current has finished, it is easy to see that the current is going to repeat this round trip during each succeeding cycle, and except for that portion of

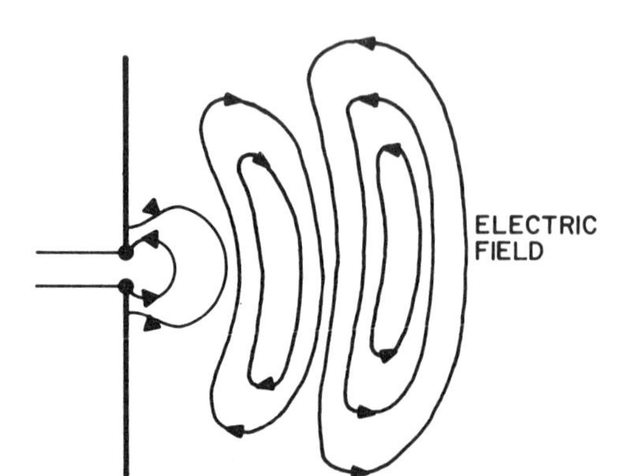

Figure 8.16. Dynamic radiation pattern of a dipole antenna.

the signal energy which is lost (i.e., leaves the antenna) by radiation, the current will in fact build up during successive cycles. We therefore have a situation of *resonance* in the antenna and its near field. Resonance enables the building up and storage of power in a circuit. A half-wave dipole antenna is in effect a resonant circuit. For such an antenna, if the impressed signal were a pure sinusoid, the current (and voltage) along the antenna would have fixed distributions in the form of *standing waves*, as shown in Fig. 8.17. Once the standing wave pattern for an antenna is known, one can endeavor to compute the far field patterns for that antenna. Johnk (1975, Chapter 11) does this for the straight dipole.

One thing the far field pattern tells us about a transmitting antenna is in what direction and to what extent the radiated power is concentrated—or, for a receiving antenna, in what direction the maximum sensitivity lies. For a dipole, the three-dimensional field pattern is essentially donut shaped, with the dipole forming the axis of the donut. Thus, the transmitted power (or receiving sensitivity) is concentrated in a circle whose plane is perpendicular to the dipole, and there is very little power (or sensitivity) in a direction along the dipole.

A commonly used modification of the simple dipole antenna is the single-wire or *whip*, antenna, such as used on automobiles, portable radios, and walkie-talkies. The whip as a transmitting antenna is driven by one polarity of the transmitter output: the other side of the transmitter output is connected to a *ground plane*, such as the earth. The ground plane acts like a mirror for electromagnetic waves, and thus effectively creates, by reflection, the other half of a dipole. In some cases, the unit is self-contained and not directly connected to earth: then the coupling to the ground plane is very loose. A typical AM broadcast antenna is a tower with a conductor which forms a quarter-wave whip, with buried wires radiating horizontally in the earth under the tower for improved contact with the earth as a ground plane, thus effectively making a half-wave dipole by reflection.

A vertically oriented whip or dipole transmits—or is receptive to—waves which are vertically polarized. (By convention, the direction of polarization of a field is taken as the orientation of the **E** field component.) Since commercial AM broadcast stations use vertical towers, their signals are vertically polarized, and a receiving antenna should normally be oriented vertically for maximum reception (although reflections from obstacles such as hills or buildings may shift the polarization to some other orientation). With respect to locations on the surface of the earth, a

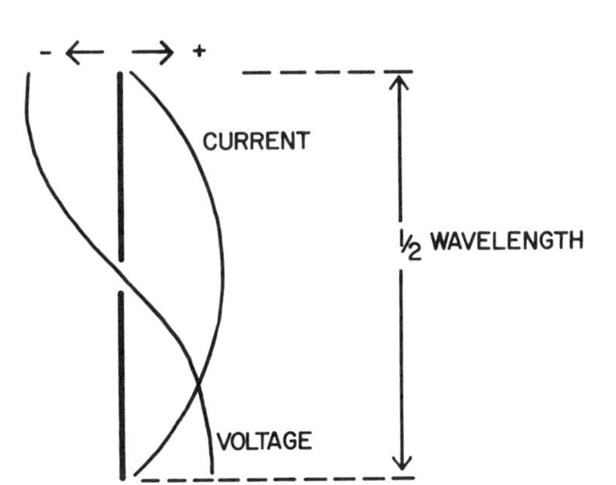

Figure 8.17. Standing waves on a half-wave dipole antenna.

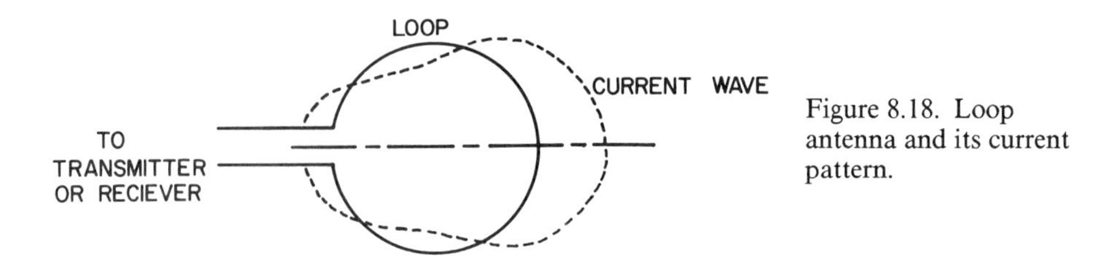

Figure 8.18. Loop antenna and its current pattern.

vertically oriented whip is omnidirectional; that is, the power transmitted (or receiving sensitivity) is equal in every horizontal direction.

If we take a dipole antenna, bend the two elements around, and connect them to form a circle (Fig. 8.18), we have a *loop* antenna. For the dipole antenna, the current must be zero at the far ends. For the loop, the current wave is merely symmetric about a diameter drawn through the feed point (dashed line in Fig. 8.18). The radiation pattern of such a loop antenna is concentrated in the two directions perpendicular to the plane of the loop. A loop antenna is quite directional, and in fact finds much use in radio direction-funding (RDF) equipment and in short-range telemetry.

Antennas can take many shapes, designed to enhance their directivity, achieve a particular radiation pattern, or achieve a particular pattern of polarization. Aside from simple horizontal and vertical polarization, some more complex antennas produce other forms of polarization as the resultant of a complex combination of horizontal, vertical, and skewed polarizations; for instance, a commonly used rhombic-shaped antenna produces elliptic polarization, whereby the tip of the electric field vector traces an elliptic path once per cycle.

The *directivity* of an antenna is often stated relative to that of a reference antenna. Sometimes the reference is a hypothetical point source which produces a uniform (isotropic) radiation in every direction. A more commonly used reference is the half-wave dipole. For VHF, UHF and microwave frequencies, special types of reference antennas are used. The directivity of an antenna is most usefully expressed in terms of the *gain* pattern of the antenna. The gain is defined as the ratio of the power density (i.e., power per unit solid angle) radiated in a given direction, compared with that of a reference antenna. Examples of gain patterns may be found in the ARRL Antenna Book (published periodically).

To achieve a desired radiation pattern, it is often necessary to use an array of antenna elements. The resulting field is a linear combination of the radiation patterns from each element. Complex circuitry, and sometimes computer control, may be used to adjust the phasing of each component, thereby achieving a *steerable* beam from a mechanically fixed antenna. The VHF Yagi antenna, discussed later in this section, is a common example of a highly directive pattern achieved by an array design.

At commercial AM broadcast frequencies, wavelengths are so long (typically several hundred meters) that for many purposes, even a half-wave dipole is impractical (the quarter-wave broadcast tower is pretty tall), and most AM receiving antennas are essentially elemental (short) dipoles. As we pass 50 MHz, and enter the VHF and UHF ranges, wavelengths become short enough to make more elaborate

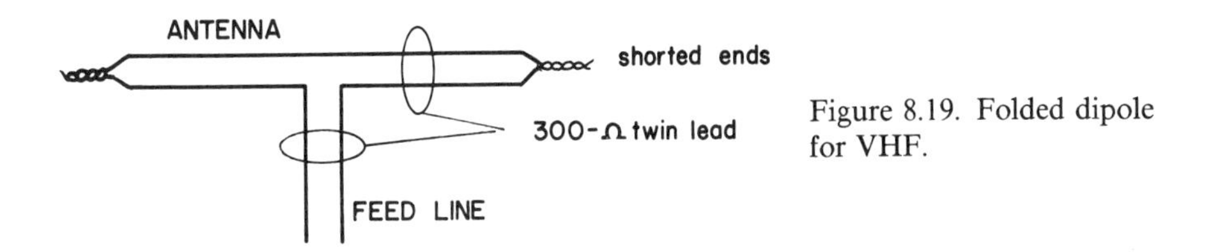

Figure 8.19. Folded dipole for VHF.

high-gain antenna designs practical. Also, at these high frequencies, working ranges are limited because as we shall see later in this section, we get very little propagation beyond line of sight, and a high-gain antenna can substantially extend the operating range of a system.

The most common antennas in these ranges are the dipole, in the form of "rabbit ears" as used for VHF television; the folded dipole constructed of TV twin lead (Fig. 8.19) and used for broadcast FM reception as well as for VHF television; and the loop antenna used for UHF television reception. In addition, a wide variety of multielement antennas finds use on rooftops for television and FM reception. Because propagation is essentially limited to line of sight, it is advantageous to mount a VHF or UHF antenna as high as possible above the ground. Potential interference with television and FM broadcast reception by other VHF communications activities is minimized by the fact that the television and FM broadcasts are horizontally polarized, while the other VHF communications are normally vertically polarized to accommodate the vertically oriented whip antennas, which are most practical for mobile stations such as Citizens' Band or public service bands.

A very popular multielement array for VHF is the Yagi array (Fig. 8.20). A Yagi may have three or more, and up to eight or nine dipole elements (five are shown in the figure), mounted on a horizontal boom. One element, the *driven element*, is connected to the transmitter (or receiver) and behaves as a dipole should. At a certain distance from the driven element a *reflector* element is mounted. It, too, acts as a dipole. The current induced in the reflector by the surrounding fields produces a field of its own. The separation of driven and reflector elements is such that their combined field *cancels* in the direction from driven element to reflector, and *adds* in

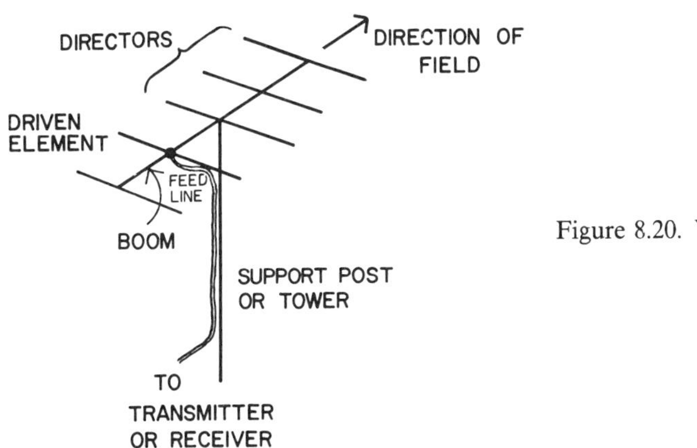

Figure 8.20. Yagi array for VHF.

the direction from reflector to driven element. One or more *director* elements also act as dipoles, and they are positioned so that their induced fields all add maximally to the total field of the antenna, which is thus highly concentrated in a single direction. Required spacings and design information for Yagis are given in the Radio Amateur's Handbook.

At higher frequencies (several hundred megahertz and up) it is practical to increase the gain and directivity of antennas by using plane reflecting screens or wire grids behind the antenna array. At microwave frequencies, where waves travel in pipes called waveguides, an antenna typically consists of a reflecting parabolic dish with a flange or dipole feed positioned at its focal point, achieving nearly perfect collimation of the radiation beam. The parabolic reflector must be large compared with signal wavelength, which accounts for its use being essentially limited to microwave frequencies.

Further information about antenna theory and design are presented in Raff (1977), Brown et al. (1973), and Weeks (1968).

8.4.2 Wave Propagation

In free space, the propagation patterns of electromagnetic waves are readily determined by application of the basic principles embodied in Maxwell's field equations and the wave equations. For earthbound transmission and reception, the presence of layers of atmosphere having different and varying field properties complicates the situation considerably, as does the proximity of the earth's surface, which behaves as a conductor.

Electromagnetic waves traveling over the earth can travel on or near the ground, in which case they are referred to as *ground waves*. They can also travel upward, hundreds of kilometers, where they may be bent or reflected by the ionosphere, thence to return to earth at a distance—these are called *sky waves*.

8.4.2.1 Ground Waves

Ground waves may travel in straight line-of-sight paths between two antennas over short distances (usually less than 100 km) or they may actually be bent by contact with the ground, which as a conductor slows the lower edges of the waves. A ground wave which travels in contact with the earth's surface, and thus follows it, is called a *surface wave*. Such waves must be vertically polarized, since it is the difference in propagation velocity of the upper and lower extremities of the waves which causes the bending, and because the earth would short circuit any horizontally polarized component in contact with it. It is the surface wave which is the predominant mode of propagation of AM broadcast waves in daytime, and of most low-frequency transmissions. Another form of close-to-the-ground transmission is line-of-sight transmission, as depicted in Fig. 8.21. It is mainly useful at high frequencies (VHF and UHF range), for the following reason. Along with the direct line-of-sight transmission, the waves are also reflected by the ground, at which point they undergo a 180° phase reversal. The result is that the direct and reflected waves tend to cancel at the receiving antenna. However, the reflected waves travel farther, and hence undergo an additional phase shift. At the relatively low frequencies of

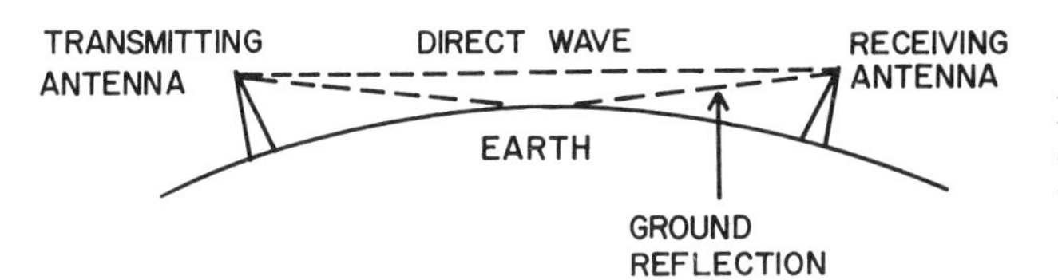

Figure 8.21. Combination of direct and ground-reflected waves.

broadcast AM, the effective path length difference is a small fraction of a wavelength, and so the cancellation is almost complete. In the VHF and UHF range, the path length difference can be large compared with the wavelength, and so can contribute a large amount of additional phase shift, reducing the cancellation effect of the reflection.

8.4.2.2 Sky Waves and the Ionosphere

The surface wave induces currents in the earth which constitute losses in transmitted power. These losses increase with frequency, so that surface wave propagation has a very limited range for frequencies above the AM broadcast band. Longer transmissions in this range (i.e., 2–30 MHz) are done mainly by means of the sky wave, which utilizes the ionosphere.

The ionosphere consists of a number of layers of air which has been ionized by solar radiation. In some of the lower layers, the air is sufficiently dense that recombination of ions and electrons is rapid, and the ionization lasts only through the daylight hours; hence these layers affect only daytime transmissions. On the other hand, the F layer, at about 325 km altitude, is sufficiently high that its ionization lasts through the night. This layer presents the main means of long-distance propagation. Seasonal increases in intensity of ionization occurring in the E layer (sporadic E) at about 130 km enable improved medium-range communications during those periods. The primary mechanism by which waves reaching ionospheric layers are returned to earth is a bending by refraction ; reflection also occurs, though to a lesser extent.

The amount of refraction accomplished in the ionosphere depends not only on the signal wavelength (longer waves are bent more), but also on the intensity of ionization, which varies with time of day, season, and solar activity. A further complication is due to the fact that the earth's magnetic field causes the ionosphere to be birefringent, that is, double refracting, so that multiple paths are traversed by waves propagating via the ionosphere. Waves which return to earth may be reflected back to the ionosphere for another hop.

The complexities of ionospheric action make sky wave propagation difficult to predict. On the other hand, ground waves are greatly affected by tropospheric disturbances, and climatology is not yet in a position to help us with our predictions of ground wave propagation. For more information on this subject, the reader is referred to the ARRL Antenna Book, the Radio Amateur's Handbook, the ITT Handbook (1975), and Brown et al. (1973). In addition, Kiver (1960, Chapter 4) presents a fine treatment of propagation.

8.5 Transmission Lines

We noted in Section 8.2.3 that RF signals may travel by wire lines, normally for relatively short distances, or, by the use of antennas, by propagation through the ambient medium. Even when an antenna is used, it is often necessary or convenient to locate it at a distance from the transmitting or receiving electronics (for instance to get around building shielding or to gain a vantage point for line-of-sight transmission). In these cases, some sort of transmission line is used to convey the signal, either directly from transmitter to receiver, or locally, from the transmitter or receiver to its antenna.

A transmission line is basically a system of conductors designed to convey radio frequency waves from one place to another, e.g., from transmitter to antenna, from antenna to receiver, or directly from transmitter to receiver. Transmission lines take various forms. One of the more common of these forms is a pair of wires which is separated and insulated by a more or less flat, insulated plastic strip. This is the familiar TV twin lead, which typically connects a TV receiver to its rooftop antenna. Another type is a coaxial cable, which consists of a center conductor wire, surrounded by insulation around which is wrapped a flat conductor constructed of braided wire. The whole cable is then insulated. At the high frequencies of the microwave bands, the transmission lines become solid conducting pipes called *waveguides*.

While at low frequencies (e.g., in the audio range) we normally consider wires over which our signals travel as being simply pure conductors, i.e., having essentially zero impedance, we cannot do so at the higher radio frequencies. Here, a conductor or transmission line has a fixed, finite, nonzero impedance, which is called the *characteristic impedance*, and which is designated by Z_c. For example, the typical flat TV twin lead has a characteristic impedance of 300 Ω, while for coaxial cables values of 50, 75, and 90 Ω are common (although others are also available).

A transmission line acts like a ladder network of low-resistance inductors and high-resistance (low-conductance) capacitors, as shown in Fig. 8.22. For this arrangement, the characteristic impedance may be shown to be

$$Z_c = \sqrt{\frac{R + j\omega L}{G + j\omega C}} \tag{8.23}$$

An ideal transmission line would be lossless—that is, the effective series resistance R of the wires would be zero, and the shunt conductance G between them would also be zero. Then Eq. (8.23) reduces to

$$Z_c = \sqrt{L/C}$$

Figure 8.22. Transmission line equivalent circuit.

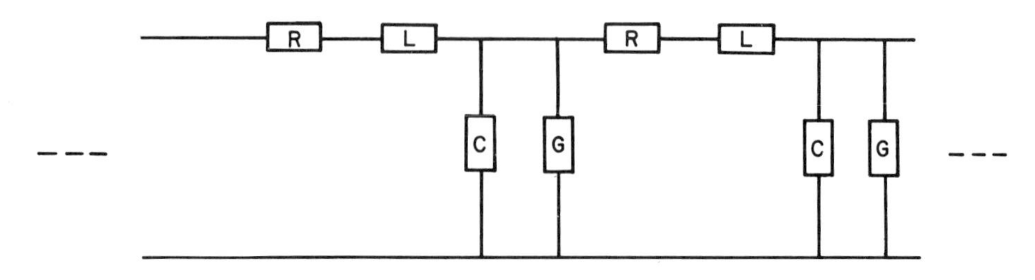

The inductance and capacitance per unit length of line depend on the configuration of the line and on the size and spacing of the conductors. Thus, for example, a transmission line having large-diameter wires with close spacing will have a low inductance and a high capacitance, and therefore a low Z_c.

If a transmission line of characteristic impedance Z_c is terminated in a purely resistive load whose impedance is equal to Z_c, then the impedance looking backward (toward the source) will be Z_c. Clearly, then, Z_c must be independent of the length of the line, since each added section of line may be regarded as a load of impedance Z_c.

The maximum transfer of signal power from source (transmitter) to load (receiver) via a transmission line occurs when the source and load impedance are *matched* to the line (i.e., when they are equal to the characteristic impedance of the line). In this condition, the load appears to be just an extension of the line, and all the power that reaches it is absorbed there. If, instead of connecting a matched load, we were to short circuit the transmission line at the load or receiving end, the voltage at that end would be zero, and the current high. All of the forward going signal power would be reflected at the short circuit, and would travel back toward the

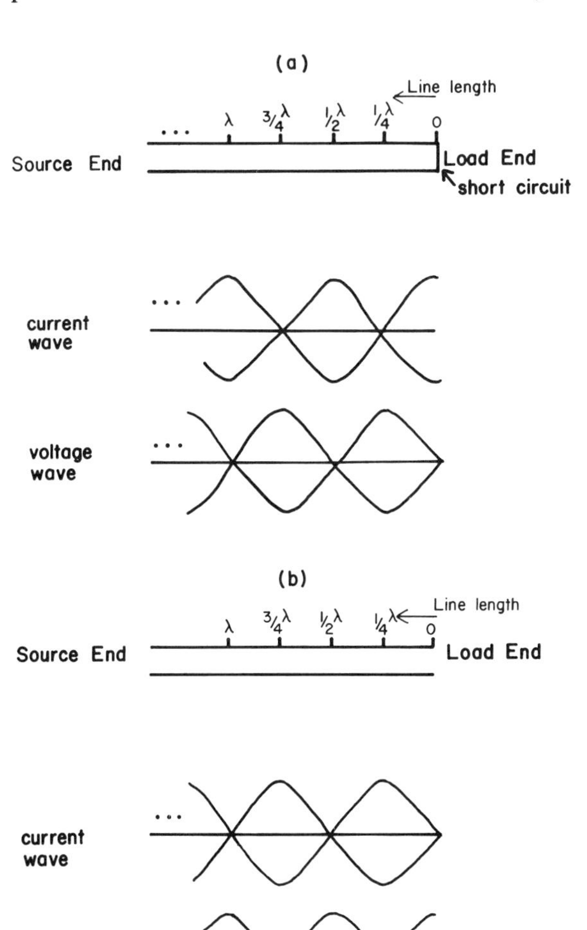

Figure 8.23. **(a)** Standing waves on a short-circuited transmission line. **(b)** Standing waves on an open-circuited transmission line.

source. The interference between this reflected wave and the originally transmitted forward wave would produce a pattern of *standing waves* along the line, as in Fig. 8.23a, which depicts the situation given that the transmitted signal is at a single frequency and has a wavelength λ.

If now we were to replace the short circuit with a variable load, and gradually increase the load impedance, the standing waves would decrease in amplitude until we reached a matched load (whose impedance equals Z_c) at which point the standing waves would disappear. If we then continued to increase the load impedance toward infinity, we would again observe standing waves, but now the peaks and nodes would have shifted a quarter-wavelength to the positions shown in Fig. 8.23b. In all cases, the standing waves are caused by the signal reflection from a mismatched load. The nodes are the points where the forward and reflected current (or voltage) cancel each other, while the peaks are the points where they add maximally. Their locations depend only on the signal wavelength λ and on the speed of propagation of the signal waves along the line, which is essentially the speed of light.

It should be noted that even though the signal waves propagate at the speed of light, which for most purposes is enormous, modern communication systems involve frequencies sufficiently high to make the travel time of the signals noticeable. A signal travels 300,000 km in 1 s, but only 300 m in 1 μs, and 30 cm in 1 ns. A light-nanosecond is equivalent to 30 cm (about 1 ft). Microwave systems can involve waves whose lengths are measured in millimeters or less. Thus, the design principles and devices used at microwave frequencies (about 1 GHz) are quite different from those which apply to lower radio frequencies. Raff (1977), Brown et al. (1973), Weeks (1968), Liao (1980), Soohoo (1971), and Adam (1969) provide an introduction to microwaves.

8.6 Radio Frequency Interference and Noise

Most of this chapter is concerned with radio frequency signals which are generated for some useful purpose—usually the transmission of information. But RF signals can—and often do—go where they are not wanted. In this section, we will examine the problems of *radio frequency interference*, or RFI (which may be created as side effects of useful transmissions) and *noise*, in the form of natural emissions such as solar and galactic radiations, atmospheric electric discharges, man-made noise such as that produced by current-switching devices, and noise produced in the receiver circuits themselves.

The term "RFI" refers to interference with the reception of wanted RF transmissions by unwanted ones. "Noise" refers to *any* unwanted signal, however produced; it not only limits the useful range of most radio communications, but also affects other forms of signal processing. In this regard, the "wanted" signal in one application (such as a radio broadcast) may simultaneously represent a troublesome "noise" source to an experimenter dealing with high-impedance amplification of low-level signals.

8.6.1 Sources and Effects of RFI

When an RF signal is transmitted across the surface of the earth, it may take several paths in reaching a given receiving antenna. Depending on distance, it may go straight from the transmitting antenna to the receiving antenna, or it may bounce off the ground or the ionosphere or both, perhaps several times, on its way. Hills, buildings, and other obstacles may also deflect the path of short radio waves by reflection or diffraction. This phenomenon is generally referred to as *multipath* propagation. Its effects on signal reception depend on the frequency and on the actual differences in path lengths: low-frequency signals traveling short distances by direct and reflected paths will arrive at the receiving antenna essentially 180° out of phase (since reflection produces a phase reversal) and effectively cancel, while in the VHF range signals traveling over the same paths may arrive with a more favorable phase relation and exhibit far less cancellation.

Frequency allocations for RF transmission are arranged to take into account the normal ranges of transmission and the potential for interference. Thus, for example, commercial radio stations in each locale are assigned frequency bands sufficiently far apart to enable effective separation of stations by a receiver, and a given frequency will be assigned to other stations ony at distances sufficiently far apart to have reasonable assurance of freedom from interference with each other. Because AM transmission has much greater range at night than in the daytime, certain frequencies have been set aside for nighttime use by so-called *clear-channel* stations, which typically operate at a maximum allowable power of 50,000 W with an evening range often exceeding 1000 km. Because of this, local stations which operate during the daytime on the clear channel frequencies are required to leave the air at night to avoid interference. The increased nighttime broadcast range of AM may still cause some interference, especially during periods of unusually good propagation.

Every oscillator is potentially a transmitter. The local oscillator of a high-frequency receiver may radiate enough power to cause interference with a nearby receiver at a lower frequency.

Medical RF appliances (such as diathermy units) and RF heaters are notorious RFI sources. The horizontal time base circuitry of a television receiver produces a 15.75 kHz pulse train, with harmonics extending well into the RF range.

8.6.2 Noise Sources and Effects

Noise comes from a variety of sources. Galactic noise includes solar flares, sunspots, and other celestial sources, mostly in the plane of our galaxy. These noise sources are significant mainly over the frequency range of 40–250 MHz. Atmospheric noise, due largely to lightning discharges, is of most significance below 20 MHz. Man-made noise, such as that from electric motors, automotive ignition systems, relays, and other current-switching devices, covers a wide range of frequencies. Receiver noise is mainly of two kinds. One, a thermal, or Johnson noise, is due to the random movements of free electrons in a conductor (resistor), and depends on the temperature, the resistance, and the bandwidth. The Johnson noise potential e_{n} is given by

$$e_{\text{n}} = (4kTR\,\Delta f)^{1/2}\ \text{V},$$

where k is Boltzmann's constant (1.38×10^{-23} J/°K), T is the temperature of the resistor in degrees Kelvin, R is its resistance in ohms, and Δf is the bandwidth of the circuit. This thermal noise has an extremely broad frequency spectrum, covering the entire RF range. A second source of noise internal to a receiver is called *shot noise*. Originally used to describe noise due to fluctuations in the random flow of electrons between elements in an electron tube, the term has been generalized to solid state devices. Shot noise is limited in frequency by the transit time between electrodes of the device. In addition to the thermal and shot noise sources, receivers also may suffer from microphonics, that is, noise produced by mechanical shocks or vibrations which impact on circuit components—especially transistors—and are converted to electric signals.

8.6.3 Control of RFI and Noise

Radio frequency interference is best controlled by avoiding it, that is, by selecting the operating frequency to be as clear as possible of potential interference sources, by using antennas which are as directional as the situation permits, and by using receiver circuitry having the greatest achievable selectivity. In some situations, the *source* of the interference can be minimized or eliminated, either by relocating it, aiming it, or shielding it. In strong AM signal areas, an experimenter designing an RF communication link might well choose frequency modulation or some form of pulse modulation for his system.

Noise control is best done at the source. A properly shielded motor, arc suppression devices for relays, or a carefully implemented noise suppression system for automobile ignition may save many nearby RF receivers from disruptive noise. Fluorescent lamps may radiate high-frequency signals due to sharp transients in their gas discharge waveforms. Lamp dimmers and motor speed controls using SCRs also produce sharp transients and may radiate considerable RF noise.

For many noise sources, control at the origin is not practical. Then the solution may involve filtering. For instance, often a troublesome dose of RF noise affecting operation of a high-gain, low-frequency amplifier may be removed by low-pass filtering.

Internal receiver noise has its greatest effect in the front end—that is, in the first stage of amplification. Prudent design using low-noise components and low resistance values in the front end can minimize the effect of internal noise.

In many cases, sensitive leads, devices, circuits, or even entire instrumentation setups must be shielded to control the effects of noise on an experiment. Electromagnetic waves have an intrinsic impedance of 377 Ω. The impedance of a metal enclosure or shield is far lower—normally a few ohms or less. The impedance mismatch prevents the waves from coupling a significant amount of energy into the metal, and thus prevents the wave from penetrating the shield: almost all of the incident wave energy is reflected. Perforated metal sheet or metal screening makes effective shielding material for an EM wave whose wavelength is large compared with aperture (or mesh) size. For wavelengths comparable to aperture size, the holes act somewhat like waveguides, and shielding effectiveness is reduced. Shielding effectiveness is also improved if the radiation source (antenna) is sufficiently far (i.e., a few wavelengths) away from the shielding enclosure, so that far-field conditions apply. When a shielding enclosure houses only a part of the instrument or circuit, it is often necessary to apply RFI filters to each line connecting the enclosed circuitry with the outside. The filters are connected to the signal lines at their points of entry into the enclosure. A good deal of valuable information on shielding principles, calculations, and materials is presented in an EMI/RFI Shielding Handbook and Catalog by the Metex Company (1976).

8.7 Television

Maxwell published his theory of electromagnetic radiation in 1873. The discovery of the light sensitivity of selenium and the invention of the telephone, both also events of the 1870s, set the stage for serious consideration of the possibility of recording moving visual images electronically and sending them over distances by radio waves. As early as the mid-1800s, still pictures were scanned and transmitted electronically by wire. In 1884, Paul Nipkow patented a spinning disk scanning system for television. The disk had a spiral array of holes. A selenium photocell behind the disk would convert the line-by-line tracings of light patterns into a signal which would be transmitted to a display unit. There it would modulate the intensity of a light source behind another spinning disk with the same pattern of holes. A viewer, looking through the disk at the light source, would see an element-by-element reconstruction of the original image. The disk rotated sufficiently rapidly to enable the viewer's eye to fuse the images of the individual elements into a steady visual scene. By the 1920s, the mechanical disk system of TV had reached its prime, and a system was used publicly, with the British Broadcasting Company (BBC) transmitting a disk-scanned video signal after regular broadcast hours. In 1927, the Bell Telephone Laboratories demonstrated their version of the spinning disk system, using a large neon glow lamp with 2500 individual elements as the "picture tube." In the next few years, regular television broadcasting was developed and the viewing public grew. One key to the ultimate success of television was the switch from mechanical to electronic scanning, using the cathode ray tube (invented by Ferdinand Braun in 1897) and the iconoscope camera tube (patented in 1923 by Vladimir Zworykin). Color television transmission was first achieved mechanically in the late 1920s. In 1940, Dr. Peter Goldmark of the Columbia Broadcasting System (CBS) demonstrated a system which used a disk containing color filters, spinning in front of a conventional black and white receiver. The image for each constituent color would be transmitted in turn, and the rotation of the disk synchronized so that the proper color filter was in position during the transmission interval for that color. At the same time, the Radio Corporation of America (RCA) was developing an all-electronic color system. The RCA system at first used three separate camera tubes and three separate picture tubes, each with its own color filter, and mirrors to merge the images. Aside from the unsatisfying mechanical nature of the CBS system, it also was incompatible with the accepted scanning standards for black and white television, thus requiring special receiving sets. Moreover, it required three times the transmission bandwidth of black and white television, and the RF spectrum was too crowded to accommodate that. After a hard fight, the RCA system finally prevailed. The three tubes were combined into one, which has a matrix of intermixed green, blue, and red phosphor dots, sequentially and selectively excited by an electron beam which is directed through a screen called a shadow mask, which contains a tiny hole aligned with each triplet of phosphor dots. More recently, large-screen (>1 m) video projection systems and a variety of video recording systems (using magnetic tapes or disks, or optical disks) have been developed, which have vastly increased the usefulness of video techniques in the laboratory.

The rest of the story of television is essentially that of a continuing search for improvements in hardware—for higher speed and bandwidth, improved camera sensitivity, better resolution, lower power consumption, and compact size. The story is well told in an article on television in a special issue of *Electronic Design* (1977) devoted to articles about communications.

8.7.1 The Video Signal

A television camera converts a visual image into an electronic signal by repeatedly scanning across the image from left to right in a series of horizontal sweeps, starting at the top of the image and ending at the bottom. During each sweep, the instantaneous light intensity is converted to a proportional voltage or current. The result is a signal which may be transmitted to a receiver together with suitable synchronizing pulses. At the receiver, the electron beam of a cathode ray picture tube is deflected horizontally by a sawtooth signal and vertically by the transmitted video signal. Thus, the original image is reproduced on the receiver screen as a raster of horizontal lines whose intensity patterns correspond with those of the original scan sweeps of the camera. The video image on the screen appears as 525 almost horizontal lines. The scan lines actually slope downward and the retrace (rapid return of the beam to the left side after each scan line) is horizontal. During the retrace, or flyback, the beam is blanked. In order to avoid flickering of the image, the picture is created in two passes of 262.5 scan lines each. The lines of the second pass are offset vertically from those of the first pass, thereby producing an *interlacing* of the alternating scan lines, as shown in simplified form in Fig. 8.24. Each pass is called a *field*, and the two interlaced fields make up a *frame*. Thirty frames

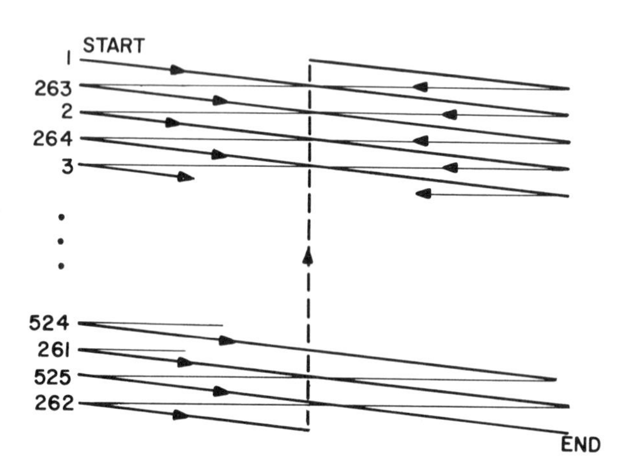

Figure 8.24. Interlaced scanning of video image.

Figure 8.25. Simplified composite video signal.

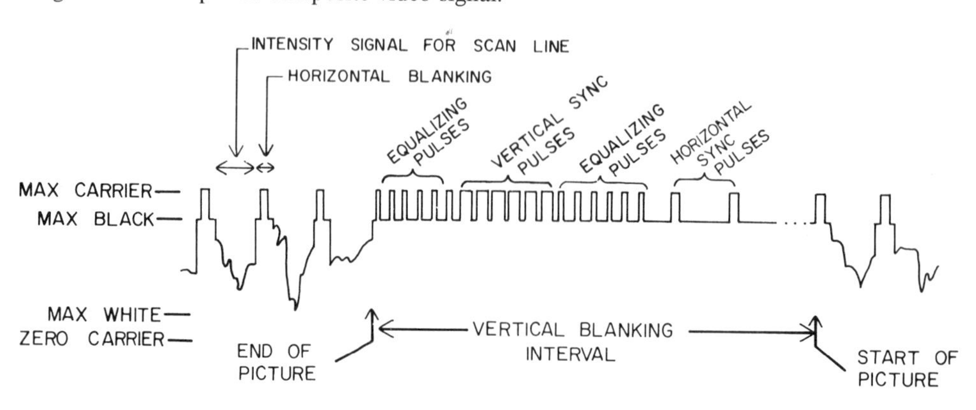

are produced per second. The signals for scan synchronization are combined with the intensity information, blanking and reference signals, producing a *composite video signal* which is used to amplitude modulate the RF carrier for transmission (Fig. 8.25). The accompanying sound is transmitted as an FM signal near the upper end of the frequency band allocated to that television channel.

Each television channel is allocated a 6-MHz frequency band. Commercial TV as a whole uses the spectrum from 470 to 806 MHz for VHF broadcast. A table of actual frequency allocations by TV channel is provided is Chapter 30 of the ITT Handbook (1975).

References

Adam, Stephen F., *Microwave Theory and Applications*, Prentice-Hall, Englewood Cliffs, NJ, 1969.

Brown, J., and Glazier, E. V. D., *Telecommunications*, Vol. 1, Chapman and Hall, London, 1964.

Brown, Robert G., et al., *Lines, Waves and Antennas*, 2nd ed., Wiley, New York, 1973.

Caceres, Cesar A. (ed.), *Biomedical Telemetry*, Academic Press, New York 1965.

Clynes, Manfred, and John H. Milsum, *Biomedical Engineering Systems*, Chapter 3, McGraw-Hill, New York, 1970.

Electronic Design 25, 18, September 1, 1977. Hayden Publishing Co., Rochelle Park, NJ.

EMI/RFI Shielding Handbook and Catalog, Bulletin EMI-300, Metex Corp., Edison, NJ, 1976.

Fryer, Thomas B., "Implantable Biotelemetry Systems," NASA Report No. SP-5094, 1970.

Johnk, Carl T. A., *Engineering Electromagnetic Fields and Waves*, Wiley, New York, 1975.

Kiver, Milton S., *F-M Simplified*, 3rd ed., D. van Nostrand, New York, 1960.

Liao, S., *Microwave Devices and Circuits*, Prentice-Hall, Englewood Cliffs, NJ, 1980.

Mackay, R. Stuart, *Bio-Medical Telemetry*, 2nd ed., Wiley, New York, 1970.

Maschhoff, Robert H., "Delta Modulation," *Electro-Technology*, Jan. 1964.

Miller, John R. (ed.), *Solid State Communications*, McGraw-Hill, New York, 1966.

NASA Technology Utilization Office, Electronic Circuits For Communications Systems, NASA Rept No. SP-5950(01), 1972.

Raff, Samuel J., *Microwave System Engineering Principles*, Pergamon, Elmsford, NY, 1977.

Radio Amateur's Handbook, Amer. Radio Relay League, Newington, CT 06111, published annually.

Reference Data for Radio Engineers, 6th Ed., IT&T, Howard W. Sams and Co., 1975.

Soohoo, Ronald F., *Microwave Electronics*, Addison-Wesley, Reading, MA, 1971.

"Telemetry Standards," IRIG Document 106-73, Inter-Range Instrumentation Group, Range Commanders' Council, White Sands Missile Range, NM 88002. Revised every 2 years.

The ARRL Antenna Book, published periodically by the American Radio Relay League, Inc.

Weeks, W. L., *Antenna Engineering*, McGraw-Hill, New York, 1968.

Zeines, Ben, *Electronic Communications Systems*, Prentice-Hall, Englewood Cliffs, NJ, 1970.

9

Transduction

Howard Moraff

The conduct of science can be viewed as a process of observing a system, developing hypotheses about the nature and functioning of the system, testing the hypotheses by observing the response of the system to external influences (including our own), and adjusting and modifying the hypotheses accordingly. Two basic types of operations are involved: the *communication* operations of delivering the inputs and control signals and making the observations, and the *cognitive* operation of developing the most appropriate model or explanation of the system's nature. Transducers play an enormous role in the communication processes. For while many observations may simply involve visual inspection of the system and the recording of what is seen, the scientist increasingly seeks to *instrument* the measurements in order to improve their accuracy, and to *automate* the observation process for greater speed, consistency, and reliability.

Communication involves energy. We look at images which are conveyed to the eye via light energy, or we feel or hear system actions which produce appropriate movements or vibrations. Some system variables are only observable when converted from one form of energy to another. Thus, if we wish to know the temperature of a solution during a chemical reaction, we need to use some kind of temperature transducer, such as a mercury thermometer, because our own thermal sensitivity is insufficiently precise. The mercury expands or contracts according to the temperature. By use of a calibrated scale, the length of the mercury column tells us the temperature of the solution. In this case, a small amount of heat energy has been converted into a mechanical form represented by the expansion of the mercury. The mercury serves as a *transducer*, which converts energy from one form to another for more convenient observation.

If we wish to monitor rapid temperature changes, the response of the mercury column may be too slow, and we may also have difficulty writing down the temperature and time often enough to get an accurate picture of its time course. If the experiment is a long one, we may wish to avoid the drudgery of recording the data by hand. In these situations, the mercury thermometer we have described may not be suitable. We could instead insert a bimetal strip in the solution. Such a device consists of strips of two metals bonded together side by side, having different coefficients of thermal expansion. As the temperature changes, one metal expands more than the other, causing the strip to bend. The amount of bend is related to the temperature. If one end of the strip is held fixed in the solution, the movement of the other end could be linked mechanically to the stylus of a chart recorder, which would then automatically record the temperature of the solution. In this case, the bimetal strip is the transducer. It converts heat energy to a mechanical deformation which may be conveniently recorded. This type of mechanism is actually used widely in recording thermometers, mainly for monitoring air temperature.

In the setup just described, it may not be very convenient to have the chart recorder close to the experimental solution, but extensive mechanical linkage to a more distant recorder introduces difficulties of loading due to the masses being driven and errors in measurement due to any looseness in connections or thermal deformation of the linkage itself. Most practical transducers used to measure physical variables convert the measured variable to electric form. This allows the data to be sent over distances with the conveniences of wire or even wireless transmission, electronic amplification, filtering and conditioning, fast and accurate display, high-fidelity reproduceable recording via magnetic tape, and computer processing. In the above setup, we might opt for a thermistor, which is a two-terminal semiconductor device whose electric resistance varies predictably with temperature. The thermistor could be connected as one leg of a voltage divider or bridge circuit controlling the input voltage to an amplifier, or it could be employed as a component in an oscillator to control the amplitude or frequency of oscillation. If we wish to send the temperature signal to the recording instrumentation by telemetry, the thermistor could be connected as a component of the transmitter.

Another major application of transducers is the conversion of electric signals to other physical forms. For example, we convert electric to auditory signals in audio equipment; the cathode ray tube is used to convert a video signal to an image; and in testing system responses, electric energy is converted into appropriate physical forms at the system input, and the system output is converted back to electric form for analysis.

In this chapter, we will discuss various types of transduction mechanisms and present sample techniques for transduction of various energy forms.

9.1 Principles of Transduction

The electric properties of electronic devices may be affected by variations in pressure, temperature, light intensity, and chemical composition of their environment, as well as by other factors such as local electric and magnetic fields, x-ray and ionizing radiation fields, and mechanical forces and vibrations acting on the body of the device. Good electronic design technique usually requires *minimization* of these effects, which are normally unpredictable and unwanted. In transducer technology, however, the situation is the reverse: the objective is to *utilize* the effect of a particular environmental factor on a selected electric property of the transducer, with the goal of obtaining, in the fluctuations of that electric property, a faithful representation of the behavior of the input factor. However, even a transducer should have the property of being relatively insensitive to forms of energy other than the one being measured.

The most common electric properties employed by transducers are resistance, capitance, inductance, and voltaic or amperic action. It is convenient to treat the transduction principles in terms of these properties. In addition, some specialized transducers employ nonelectric phenomena such as chemical or molecular changes, light production, etc.

9.1.1 Resistance

When an electric current flows in a substance, the amount of current passing through a 1-cm^2 cross section of the substance is defined as the *current density* j (A/cm^2). The current density is proportional to the potential gradient $\partial v/\partial x$,

$$j = -\sigma(\partial v/\partial x),$$

for current flow in the x direction. The proportionality constant σ is called the *conductivity* of the substance. The reciprocal of the conductivity is the *resistivity*, denoted by $\rho = 1/\sigma$. Now, if we use that substance to fabricate a conductor of length l and cross-sectional area A, then the resistance R, in ohms, of that particular conductor will be

$$R = \rho l/A.$$

Note that the resistance of a conductor depends on the resistivity of the substance and the shape of the conductor, i.e., its length and cross-sectional area. In order to use such a conductor as a transducer, we may employ changes in any of these factors. The length and/or cross-sectional area may be varied by exerting a force on the conductor to stretch or compress it. Stretching a conductor to increase its resistance is the basic operating principle of the resistance strain gauge. Using this principle, conductive material of moderate resistivity may be packaged in granular or liquid solution form in a more or less flexible container to transduce rather large deformations. Heat, light, and other forms of energy impinging on a conductor also affect its resistivity, and may thereby be transduced into electric form.

The resistivity of a substance varies with temperature. For most metals, for instance, that variation is fairly linear in a range around room temperature and is approximated by

$$\rho_t = \rho_0[1 + \alpha(T_t - T_0)],$$

where ρ_t is the resistivity of the substance at temperature T_t, ρ_0 is the resistivity at temperature T_0, and α is the temperature coefficient of resistivity for the substance. The temperature coefficient may be positive (i.e., ρ increases with temperature) or negative: it is negative for carbon, typically positive for metals. The resistive temperature coefficient is usually so small that metallic resistive temperature transducers require bridge circuits for operation. Thermistors, constructed by heating and compressing a mixture of various metals, can be designed to have greater temperature coefficients than those of pure metals, and also can have negative as well as positive coefficients, but the temperature coefficient of a thermistor is nonlinear and thus usually requires compensation in the circuitry.

Potentiometers may also be used to transduce forces or motion. The potentiometer may be set up as a voltage divider or as a variable resistor depending on the circuit. The shaft is connected to the device whose motion is to be transduced. This approach offers large signal output and linear or special taper functions, thus simplifying circuitry.

9.1.2 Capacitance

If two conductors are placed close together but not touching, and a potential difference is created between them, then the electric lines of force from one conductor will impinge on the other, and the conductors will become electrically charged with equal magnitude and opposite polarities. Such an arrangement is called a *capacitor*, and the ratio of the charge to the potential difference is the *capacitance*

$$C = q/v$$

The value of the capacitance depends on the physical configuration of the conductors and on the dielectric substance, if any, that lies between them. So, for two flat, parallel conductor plates, the capacitance is given by

$$C = \varepsilon_0 KA/d \tag{9.1}$$

where K is the dimensionless *dielectric constant* of the medium, or substance between the plates; A is the area of one plate in square meters; ε_0 is the permittivity of free space and has the constant value 8.85×10^{-12} C/V m (or 8.85×10^{-12} F/m); and d is the plate separation in meters (assumed to be small in comparison with plate size).

C is in units of farads. The dielectric constant K is actually the ratio of the permittivity ε of the dielectric to the free space permittivity ε_0:

$$K = \varepsilon / \varepsilon_0$$

Thus, Eq. (9.1) may be rewritten as

$$C = \varepsilon(A/d)$$

The common ways of using capacitance for transduction involve changing either the effective area of the plates A or their separation d, or even the permittivity of the dielectric medium ε.

Capacitive transducers have an advantage over other types in that the measurement may often be made without making physical contact to the subject, which may be arranged as one of the capacitor's conductive plates. This minimizes hindrances and errors due to friction, loading, and mechanical hysteresis. On the other hand, capacitive transducers have a relatively high impedance, which affects circuit design and cable-shielding considerations.

9.1.3 Inductance

If two wires, a and b, are placed in close proximity, and a current is passed through one wire, a magnetic field will be produced in the vicinity of that wire. If the current is changed, the magnetic field will change proportionally, and a voltage v will be induced in the other wire. The potential difference between two points is defined as the work done when an electric charge is moved from one point to the other. The magnitude of the voltage induced depends on the strength of the magnetic interaction between the two wires. That interaction is measured by the mutual inductance M, which is defined by

$$M_{ab} = \frac{v_b}{di_a/dt} \tag{9.2}$$

for a current i_a in wire a, inducing a voltage v_b in wire b. If the wire a is made into a long helical coil, or solenoid, of length l_a with N_a turns, and wire b is a coil wound coaxially around solenoid a with N_b turns, then the magnetic flux in the cave of solenoid b due to current i_a in solenoid a is given by

$$\phi_{ba} = \frac{\mu_0 i_a N_a A_a}{l_a} \tag{9.3}$$

where A_a is the cross-sectional area of the solenoid a and μ_0 is the permeability of free space (1.257×10^{-6} H/m). A change in i_a will induce in solenoid b the voltage

$$v_b = N_b \frac{d\phi_{ba}}{dt} = \frac{\mu_0 N_a N_b A_a}{l_a} \cdot \frac{di_a}{dt} \tag{9.4}$$

Then the mutual inductance (in henrys) between the solenoids is given by

$$M_{ab} = \mu_0 N_a N_b A_a / l_a \tag{9.5}$$

From Eq. (9.2), we can see that if the mutual inductance between two conductors were 1 H, then a current changing at the rate of 1 A/s in one of the conductors would induce a voltage of 1 V in the other.

We may also combine Eqs. (9.2) and (9.4) to obtain

$$M_{ab} = \frac{N_b d\phi_{ba}/dt}{di_a/dt} = N_b \frac{d\phi_{ba}}{di_a}$$

Differentiating Eq. (9.3) with respect to i_b gives

$$\frac{d\phi_{ba}}{di_b} = \frac{\mu_0 N_a A_a}{l_a} = \frac{\phi_{ba}}{i_b}$$

and

$$M_{ab} = \frac{N_b \phi_{ba}}{i_b}$$

By the same procedure we may obtain

$$M_{ba} = \frac{N_a \phi_{ab}}{i_a}$$

and while the derivation is beyond the scope of this book, it may be shown that it is always true that

$$M_{ba} = M_{ab}$$

therefore we may refer to the mutual inductance simply as

$$M = \frac{N_a \phi_{ab}}{i_a} = \frac{N_b \phi_{ba}}{i_b} \qquad (9.6)$$

Now consider that the two wires we started with could just as well have been two segments of a single wire: a changing current in a wire will also induce a voltage in the same wire—an effect which is called *self*-inductance, L. We could derive analogous expressions for the self-inductance of a solenoid:

$$L = v/(di/dt) \qquad (9.7)$$

$$L = \mu_0 N^2 A/l \qquad (9.8)$$

and

$$L = N\phi/i \qquad (9.9)$$

Now, let us see how the properties of mutual inductance and self-inductance may be put to use in transducers. First, we note that we have used the constant μ_0, the permeability of free space, in our formulations: we have assumed that our conductors are being used in a vacuum. If we introduce a ferromagnetic substance into the vicinity of the coil, the inductance may be changed considerably. So, for instance, the motion of a ferromagnetic core into and out of the center of the coil may be transduced into changes in inductance. From (9.5) and (9.8), we see that inductance also depends on the length and cross-sectional area of the coil; stretching and compression will affect these dimensions and effect changes in inductance. One popular transduction scheme is to use a movable core, for instance, one made of a high-permeability ferrite material, to vary the mutual inductance between two solenoids. If the two solenoids are connected in series (Fig. 9.1), the effective total inductance is

$$L_{\text{total}} = L_a + L_b + 2M$$

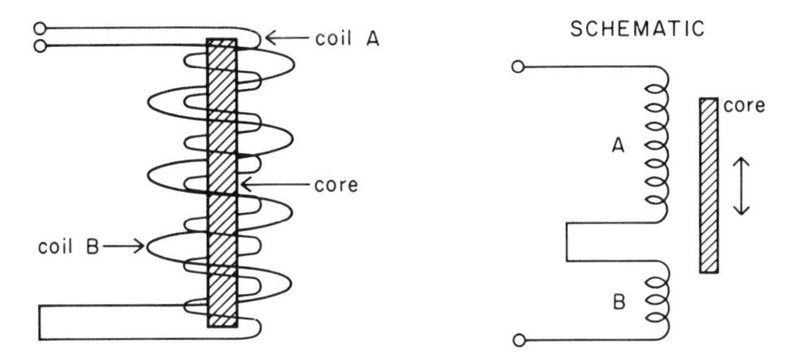

Figure 9.1 Solenoids wired in series.

and all three terms will vary with motion of the core. Alternatively, the two solenoids may be wired into separate circuits and used as a transformer (Fig. 9.2) with an alternating current excitation applied to one; the induced voltage in the other will then vary with the motion of the magnetic core which is in proximity to both solenoids.

An inductive transducer which makes use of three coils is the linear variable differential transformer, or LVDT (Fig. 9.3). An ac signal is applied to the primary coil, and opposing signals are induced in the secondary coils. If the core is exactly centered with respect to the secondary coils, the output should be null. Shifting the core away from the centered position toward secondary coil A will cause an increase in mutual inductance between the primary winding and secondary winding A, and a decrease in mutual inductance between the primary and secondary winding B, thereby producing an imbalance in secondary output signals, resulting in a net output signal. That signal will be linearly proportional to core displacement. If the core is displaced from center toward coil B, a proportional output signal is again produced, but now it is of opposite phase. In practical use, the LVDT may either be operated exclusively on one side of the center or null position, or a phase sensitive circuit may be driven by a centered LVDT. An excellent and comprehensive treatment of the theory and application of the LVDT is given in Herceg (1972).

Figure 9.2 Solenoids wired as a transformer.

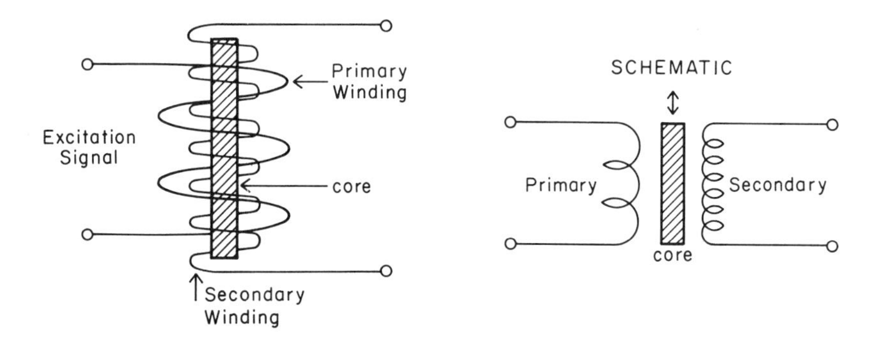

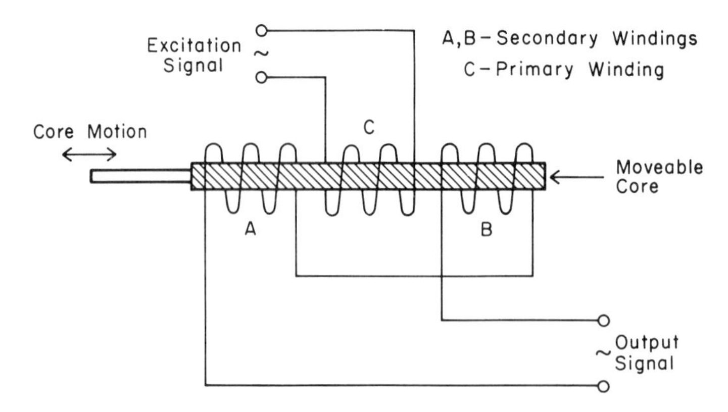

Figure 9.3 A linear variable differential transformer (LVDT).

9.1.4 Voltaic Transduction

Various transduction devices operate by the shifting or moving of electric charges within the material of the device. This may result in the production of a potential difference across the device. This effect may be regarded as voltaic transduction, and is manifested in such devices as piezoelectric crystals, photovoltaic cells, thermocouples, and Hall effect devices.

9.1.4.1 Piezoelectric Effect

When a crystal (such as Rochelle salt) is compressed, twisted, or otherwise deformed, an electric charge is produced between the crystal faces. This principle is called the piezoelectric effect. It is used, for example, in crystal pickup cartridges for phonographs and in crystal microphones. The effect also operates in reverse: applying an electric potential across a piezoelectric transducer will produce a mechanical deformation in the structure. Deflections of the order of a few microns are associated with potentials in the fractional volt range. The effect is transient, as the charge tends to dissipate by leakage through the crystal as well as in the connected circuits; it is thus primarily useful for ac and dynamic signal pickup and requires the use of high-input impedance charge amplifiers. Other crystalline materials, such as quartz and tourmaline and some synthetic ceramic materials, such as barium titanate, also exhibit the piezoelectric effect. A good description of principles and applications of piezoelectric transducers is given in Geddes and Baker (1968).

9.1.4.2 The Photovoltaic Cell

The selenium photovoltaic cell (Fig. 9.4) is a widely used device that typically consists of a substrate such as an iron or steel plate coated successively with a base layer of semiconducting material (usually selenium), very thin transparent layers of oxidized cadmium and clean cadmium, and a deposited silver alloy collecting bar or ring providing one electric contact (the other connection is made with the iron base). Light quanta reaching the interface between the cadmium oxide layer and the selenium are absorbed, resulting in the formation of electron–hole pairs. An electric

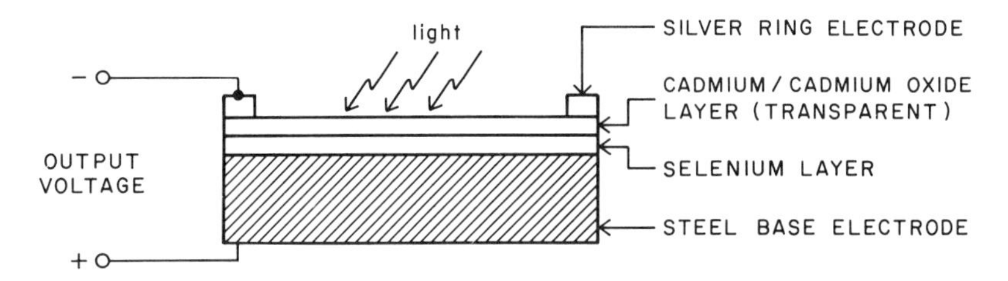

Figure 9.4 A cadmium–selenium photovoltaic cell.

field is built up in the oxide layer, causing separation of electrons and holes, and movement of electrons from the p-type selenium into the n-type cadmium oxide. Since it needs no external excitation power, this device is typically used in such applications as photographic exposure meters. The open circuit voltage of a photovoltaic cell increases roughly logarithmically with the intensity of illumination, while the short circuit current increases linearly with illumination and is also proportional to the active area of the cell.

9.1.4.3 The Thermocouple

When two different metals are physically joined, a contact potential is developed across the connecting interface. This potential, which was first demonstrated by Seebeck over a century and a half ago, is determined by the difference in work function of the two metals, and depends on the temperature of the junction. The basic arrangement is shown in Fig. 9.5. In practical applications, two thermocouples are used, as in Fig. 9.6. One thermocouple junction is typically held at a fixed reference temperature, such as 0 or 100°C, and the other is thermally coupled to the medium whose temperature is to be measured.

Connections to the thermocouple also involve different metals, and thus form additional thermocouple junctions. To avoid the measuring error which would result by the production of differing output potentials from these connections, both connections to a thermocouple should be kept at the same temperature. The material composition of the connecting wire may be chosen in relation to the thermocouple metals to minimize the potentials produced at these connections, thereby minimizing the error due to any temperature differences in those connections. The sensitivity of a thermocouple junction may range from a few microvolts to a few hundred microvolts per degree centigrade, depending on the materials. Thermocouples may be stacked up and wired in series for greater voltage output in a configuration called a thermopile.

Figure 9.5 Basic thermocouple.

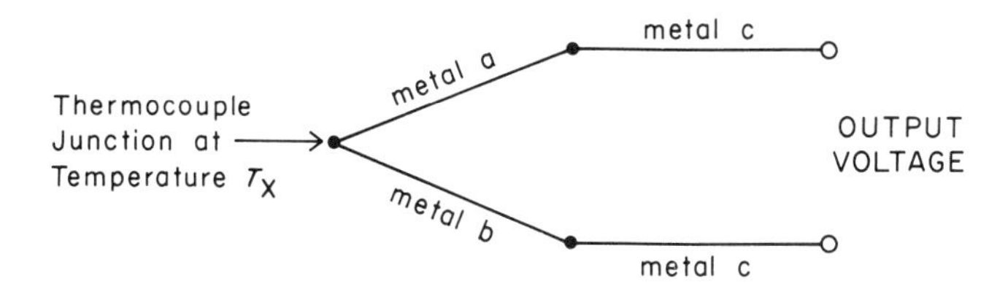

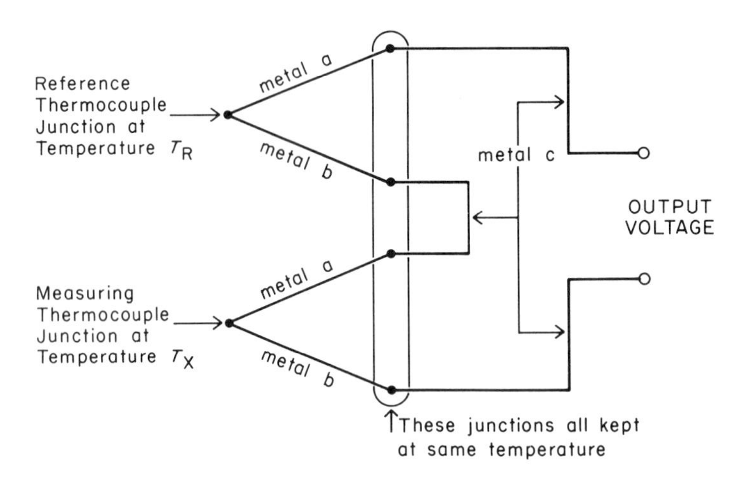

Figure 9.6 Practical thermocouple arrangement.

The voltage produced by a thermocouple is usually measured with the use of a resistance bridge or other high-impedance circuit, since current flow through the junction causes heating of the wires and junctions—the Joule effect—and also a transfer of heat between the junctions—the Peltier effect. While these effects are usually small and may be made negligible, the Peltier effect especially may be put to good use in pumping heat, in particular, to achieve well-controlled and localized heating or cooling. Devices which use the Peltier effect are called thermoelectric heaters and coolers—the direction of heat flow is determined by the direction of current flow.

9.1.4.4 The Hall Effect

If a current I is passed through a metal strip in a direction x (see Fig. 9.7) and a magnetic field B is applied in a direction z, perpendicular to the x, y plane of the sheet, then the moving electrons will be deflected by the force of the magnetic field in a direction y, perpendicular to x and z. This will create an electric field resulting in a potential difference E_H across the strip in the y direction. E_H is called the Hall

Figure 9.7 Hall effect device.

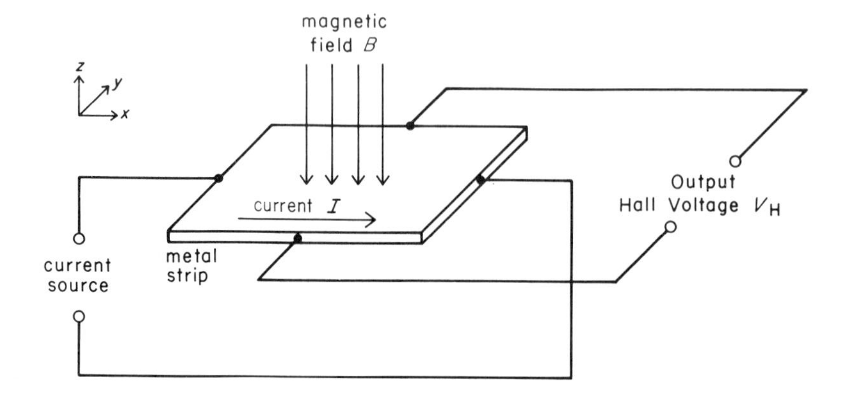

potential and is given by

$$E_H = K_H(IB/t)$$

where E_H is in volts, I is the excitation current in amperes, B is the magnetic flux density in gauss, and t is the thickness of the metal strip in centimeters. K_H is the Hall coefficient for that metal, in volt-centimeters per ampere-gauss. Germanium, bismuth, silicon, and tellurium have extremely high Hall coefficients compared with most other metals. K_H is positive for some metals and negative for others, and also depends on the temperature and on the purity of the metal. The Hall potential is normally in the millivolt or microvolt range, thus requiring a high-gain amplifier for practical use.

9.1.5 Amperic Transduction

We have seen in Section 9.1.4 that a device such as a photovoltaic cell may produce a movement of electrons. Whether we call such a device a voltaic or an amperic transducer really depends on the quantity of the electron (or current) flow, or, in other terms, it depends on whether it seems more appropriate to treat the device as a current or a voltage source in designing circuitry for it. The photovoltaic cell, for example, normally produces a relatively small current (microamperes) and thus requires relatively high impedance, voltage-measuring circuitry.

9.1.5.1 The Junction Photodiode

Another light sensitive device, the junction photodiode, behaves like a normal rectifier diode in the absence of light, but its back current (i.e., the current which flows when the junction is reverse biased) increases substantially, to many milliamperes per lumen, when the junction is exposed to light. This device is normally used as a current controller, and is properly regarded as an amperic transducer. Silicon is preferred for its wide range of operating temperatures, but germanium photodiodes find use when more sensitivity in the infrared region is needed. Photodiodes are fast: typical response times of silicon junction photodiodes are in the submicrosecond range.

9.1.5.2 The Thermionic Thermal Radiation Transducer

If a wire filament is heated in a vacuum, it will emit electrons. If an electrode located near the filament is given a sufficiently large positive electric charge with respect to the filament, the electrons will be made to flow to this *anode*, producing a current. The process is known as thermionic emission of electrons. The device is a vacuum diode, with the filament comprising the cathode, and the nearby electrode being the anode. The rate of flow of electrons from cathode to anode is essentially independent of the potential difference between the electrodes provided that the difference is sufficiently great. Electron flow does depend on the material of the filament and on its temperature. It is this temperature sensitivity, which can be large, that is exploited in the thermal radiation transducer, shown schematically in Fig. 9.8. Thermal radiation striking the cathode/filament raises its temperature, increasing electron emission and thus causing the plate current to increase. Lion (1959) discusses this device and indicates a source of references.

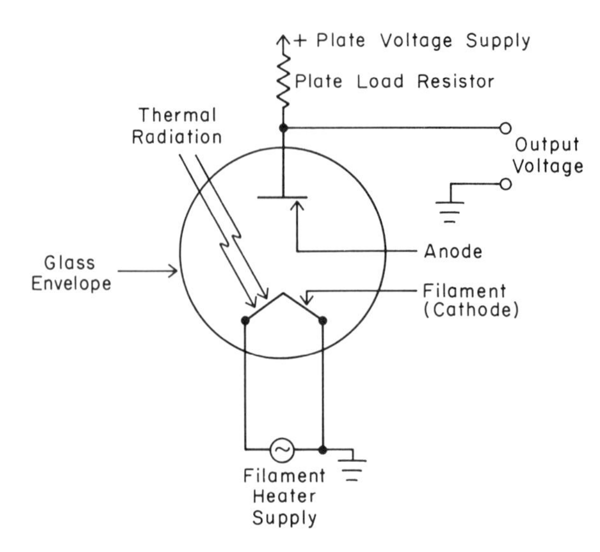

Figure 9.8 Thermionic radiation detector.

9.1.6 Transduction of Ionizing Radiation

Ionizing radiation includes all forms of radiation capable of producing ionization in the surrounding medium. Thus, it includes particles such as electrons (beta rays), positrons, protons, neutrons, and others; and electromagnetic waves such as x rays, gamma rays, etc. The transduction may occur via ionization of molecule of a gas in a chamber (e.g., ion chamber, Geiger–Muller tube, spark chamber) or via ionization of molecules in a crystal, scintillating phosphor, photographic emulsion, or semiconductor medium.

Basic descriptions of the ionization chamber, the Geiger counter, and the crystal counter are presented here. For more depth, and coverage of other devices, Section 5-2 of Lion (1959) and Part 3 of ISA (1969) are recommended.

9.1.6.1 The Ionization Chamber

The ionization chamber (described schematically in Fig. 9.9) consists basically of a chamber containing two electrodes, with an electric field between them produced by an external voltage source. The chamber is filled with a gas, and the radiation to be measured is beamed into the chamber and directed between the electrodes. The radiation impinging on gas molecules will remove electrons from some of them, producing positive ions which move toward the negative electrode. Some of the electrons and ions may recombine before they reach their respective oppositely charged electrodes, but at some high voltage level, the electrons and ions will be accelerated to sufficiently high velocities that such recombination becomes rare, and the resulting electrode current reaches a steady saturation value. The electric field strength required for current saturation increases with intensity of the incident radiation and with the pressure of the gas in the chamber. The value of the saturation current is essentially a function of the number of ionizations per second. Measurement of this current thus provides a means of measuring the radiation intensity. This current is normally very small, so that elaborate measuring methods may be needed. For low radiation intensity levels, an alternative means of measure-

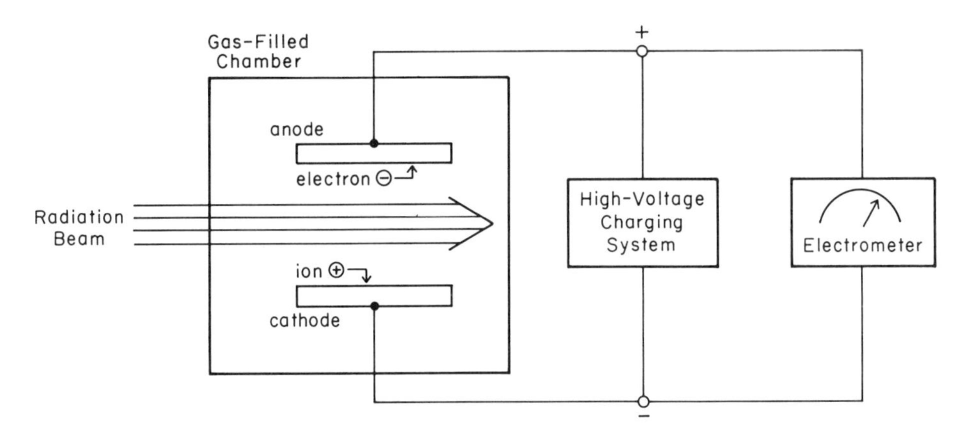

Figure 9.9 Ionization chamber.

ment may be used, in which the ionization chamber is charged to a high voltage E_1 and exposed to the radiation for a measured time interval t, during which it will discharge to a somewhat lower voltage E_2 (but still above the level required for current saturation). If the starting and ending voltage levels are measured without drawing much current, then the saturation current I_s is simply

$$I_s = C(E_1 - E_2)/t$$

where C is the capacitance of the chamber. This scheme is popular in radiation dosimetry applications, where the chambers, in miniature form, may be worn by personnel to monitor their exposure to ionizing radiation on the job.

9.1.6.2 The Geiger Counter

A form of ionization chamber, the Geiger counter is adapted to produce individual gaseous discharges for each molecular ionization resulting from the incident radiation. This is done by using a particular mixture of gases, such as argon plus a small amount of a polyatomic gas (e.g., alcohol), and raising the electrode voltage well above the range of ionization chamber operation (typically to 1000 V or higher). Then an ion produced by the incident radiation not only produces other secondary ionizations in the gas by colliding with molecules, but when an ion reaches the cathode and absorbs an electron, it emits ultraviolet photons of sufficiently high energy to cause photoionization of gas molecules, thus producing an avalanching gaseous discharge. The effect is terminated when the alcohol molecules, having a lower ionization potential than the counter gas, give up their electrons to the counter gas ions, so that the ions reaching the cathode are mostly those of the alcohol. Thus, the cathode becomes surrounded by alcohol molecules. These absorb strongly in the ultraviolet, but dissociate rather than ionize, and hence they serve to quench the ionization and terminate the impulse. The avalanche effect involves ions and electrons many more orders of magnitude in number, and hence a much larger current, than in the case of the ionization chamber of the previous section. Each molecule ionized by the incident radiation leads to a large current impulse, which may be used to drive a loudspeaker, headphone, or light bulb for qualitative indication of radiation intensity as the number of ionizations per second is converted to clicks or flashes per second. The impulse train may also be integrated to drive a meter for more quantitative indication.

9.1.6.3 The Crystal Counter

The crystal counter is a slab of a crystal such as sodium iodide, with thin metal film electrodes deposited on opposing surfaces. A high-voltage source is used to produce a field strength of several thousand volts per centimeter in the crystal. Ionizing radiation passing through the crystal will produce free electrons, in analogy to ionization in a gas. These free electrons will move through the crystal under the force of the electric field. Some will reach the anode, typically producing a pulse in the millivolt range. The pulse duration is normally in the submicrosecond range, therefore permitting high counting rates. These crystals are operated in darkness to avoid photoconductive response.

Semiconductor crystals are also used for radiation detection and measurement. In one version, a p-type silicon crystal has a thin layer of an n-type impurity diffused into it. The p–n junction is reverse biased, producing a depletion layer, from which ionizing radiation can liberate electrons, thus producing electron–hole pairs which move in the electric field, producing impulses. A brief description of semiconductor crystal detectors is given in ISA (1969, Part 3, pp. 122–123).

9.2 Transducer Circuitry Considerations

While some transducers can produce sufficiently large changes in electric properties as to permit their use with rather simple circuitry, many transduction mechanisms yield only small electric changes and so require large amounts of amplification; others require special conditions of excitation or bias; still others have very high impedance or are sensitive to extraneous environmental effects. In this section, some of the more common transducer circuitry design considerations are presented.

Generally, recent advances in IC technology permit the elimination of older schemes for processing transducer signals. For example, in transducing the forms of energy to electricity in so-called input transducers, it is no longer necessary to use Wheatstone bridges, transformers, or other devices to detect small changes in voltage or current or to provide the high-input impedances or low-bias currents required by transducers with low-current capacity or low-voltage outputs. One can now purchase operational amplifiers having input impedances exceeding 10^{12} Ω, and bias currents of less than 10^{-11} A. Such devices cost about a dollar. With appropriate compensation, dc offset voltages can be reduced to a few microvolts, with excellent stability. High-gain instrumentation amplifiers are available, with high-input impedance, low-bias current, low-offset voltage and drift, as well as fixed gains selected by external resistances. Since neither input of such a differential amplifier is used for feedback, both the inverting and noninverting inputs offer high impedance. Gains of such devices can be made as high as 10^6, with common mode rejection typically 100 dB or better.

When using transducers to convert electric waveforms to other forms of energy (e.g., displacement) in so-called output transducers, the transducers may require large voltages or currents to activate them. Current- and voltage-boosting techniques such as those in Chapter 4 can be used, or ICs can be purchased which are capable of handling large voltages or currents directly.

9.3 Effects of Transducer Properties on Systems

A transducer can be viewed as a device which responds to the application of input energy by producing an output in some other energy form. In many instances of the use of transducers, it is necessary to know the quantitative relation of output to input. This is accomplished by calibrating the transducer. Such calibration involves precise measurement of the output magnitude (and its time course, in some cases) in relation to a set of known inputs. Sometimes such measurement is readily done by simple mechanical means. In many instances, however, this measurement requires the use of reference (standard) transducers, which have in turn been calibrated by other reference devices. Ideally, such a "calibration chain" should be traceable back to some ultimate reliable standard, such as those maintained by the US National Bureau of Standards.

In calibration of a transducer, we endeavor to determine quantitatively whether the device is linear, and over what input range it is linear; and we need to know either its impulse or step response (time domain), its frequency response function (frequency domain), or its transfer function (s plane). Naturally, the transducer used to make the calibration must have a greater amplitude range and frequency range than the one being calibrated.

The calibration procedure may use any of the strategies outlined in Chapter 3 for analysis of a system's response: the forcing function may be an impulse, a step, a sine wave, or a white noise, to list the most common ones. The outcome of the calibration procedure is the specification of either (1) an impulse of step response or (2) a Bode plot or Nyquist diagram, which may then be expressed as a transfer function or differential equation.

The purpose of calibration is to quantify the response characteristics of the transducer. The simplest approach to transduction is to use a transducer which has already been calibrated reliably by the manufacturer and whose dynamic range (range of amplitudes over which distortion and noise are within acceptable limits) and frequency range are adequate to the application.

Many transducers exhibit nonlinear characteristics, due to the nature of the transduction mechanism, variations in loading, or other causes. In some cases, the output of a nonlinear transducer may be monitored by another transducer, producing a signal which may be used to modify the transducer input to achieve the desired output. In other cases, nonlinearities in a transducer's output may be compensated by external circuitry or waveform processing devices. Negative feedback from a monitor transducer may be used to make a servo system like those described in Chapter 3. This approach is particularly useful when the output transducer's characteristics are load dependent.

References

Cobbold, Richard S., *Transducers for Biomedical Measurements: Principles and Applications*, Wiley, New York, 1974.

Cooper, William D., *Instrumentation and Measurement Techniques*, 2nd ed., Prentice-Hall, Englewood Cliffs, NJ, 1978.

Geddes, L. A., and Baker, L. E., *Principles of Applied Biomedical Instrumentation*, Wiley, New York, 1968.

Herceg, Edward E., *Handbook of Measurement and Control*, Schaevitz Engineering, 1972.

Hougen, Joel O. (ed.), *Measurements and Control Applications*, 2nd ed., Instrument Society of America, 1978.

ISA Transducer Compendium, 2nd ed., Parts 1–3, Instrument Society of America, 1969.

Lion, Kurt S., *Instrumentation in Scientific Research*, McGraw-Hill, New York, 1959.

Lion, Kurt S., *Elements of Electrical and Electronic Instrumentation*, McGraw-Hill, New York, 1975.

Mansfield, P. H., *Electrical Transducers for Industrial Measurement*, Butterworth, London, 1973.

Neubert, Hermann K. P., *Instrument Transducers: An Introduction to their Performance and Design*, 2nd ed., Clarendon Press, Oxford, 1975.

Neuman, Michael R., et al. (eds.), *Physical Sensors for Biomedical Applications*, CRC Press, Boca Raton, FL, 1980.

Norton, Harry N., *Handbook of Transducers for Electronic Measuring Systems*, Prentice-Hall, Englewood Cliffs, NJ, 1969.

Index